GLACIOVOLCANISM ON EARTH AND MARS
Products, Processes and Palaeoenvironmental Significance

The study of volcano–ice interactions, or 'glaciovolcanism', is a field experiencing exponential growth. Explosive eruptions in Iceland at Gjálp (1996) and more recently at Eyjafjallajökull (2010), the latter of which caused major economic disruption across Europe, illustrate the importance of assessing possible future hazards associated with ice-clad volcanoes. Subglacial volcanoes are also thought to exist on other bodies in the Solar System, especially on Mars.

This comprehensive volume presents a discussion of the distinctive processes and characteristics of glaciovolcanic eruptions, and their products and landforms, with reference to both terrestrial and Mars occurrences. Supported by abundant diagrams and photos from the authors' extensive collections, this book outlines where eruptions have occurred and will occur in future on Earth, the resulting hazards that are unique to volcano–ice interactions, and how the deposits are used to unravel planetary palaeoclimatic histories. It has a practical focus on lithofacies, glaciovolcanic edifice morphometry and construction, and on applications to palaeoenvironmental studies. Also available online are a series of expertly illustrated and annotated lecture slides that can be incorporated into teaching materials.

Providing the first global summary of past and current work, this book also identifies those areas in need of further research, making this an ideal reference for academic researchers and postgraduate students in the fields of volcanology, glaciology, planetary science, and palaeoenvironmental studies.

JOHN L. SMELLIE is a professor in the Department of Geology at the University of Leicester, UK. He is the leading expert on Antarctic glaciovolcanism and has extensive experience working on Iceland's glaciovolcanoes. He is a prolific author, having produced over 190 publications, including 10 edited volumes. Professor Smellie is the co-founder and first Chair of the IAVCEI/IACS Commission on Volcano–Ice Interactions, and, in 2000, was co-convener of the first International Conference on Volcano–Ice Interactions on Earth and Mars. He has been awarded the Polar Medal, conferred by Her Majesty the Queen (UK), is co-Chair of the SCAR (Scientific Committee on Antarctic Research) Expert group on Antarctic Volcanism, and has three geographical features in Antarctica named after him.

BENJAMIN R. EDWARDS is an associate professor in the Department of Earth Sciences at Dickinson College, USA. He has over 24 years of field experience working in remote regions of British Columbia, Iceland, Alaska, Russia and South America on modern and ancient glaciovolcanoes, and is one of the leading researchers in the rapidly expanding field of large-scale experimental volcanology, specifically focused on lava–ice and lava–water interactions. Dr Edwards is also a co-founder of the IAVCEI/IACS Commission on Volcano–Ice Interactions, and has convened various special sessions on volcano–ice interactions at AGU, GSA, EGU and IUGG international conferences.

GLACIOVOLCANISM ON EARTH AND MARS

Products, Processes and Palaeoenvironmental Significance

JOHN L. SMELLIE

University of Leicester

BENJAMIN R. EDWARDS

Dickinson College, Pennsylvania

CAMBRIDGE
UNIVERSITY PRESS

CAMBRIDGE
UNIVERSITY PRESS

University Printing House, Cambridge CB2 8BS, United Kingdom

Cambridge University Press is part of the University of Cambridge.

It furthers the University's mission by disseminating knowledge in the pursuit of
education, learning and research at the highest international levels of excellence.

www.cambridge.org
Information on this title: www.cambridge.org/9781107037397

First published 2016

Printed in the United Kingdom by TJ International Ltd. Padstow Cornwall

A catalogue record for this publication is available from the British Library

Library of Congress Cataloguing in Publication data
Smellie, J. L. | Edwards, Benjamin R., 1967–
Glaciovolcanism on Earth and Mars : products, processes, and palaeoenvironmental
significance / John L. Smellie, University of Leicester, and Benjamin R. Edwards,
Dickinson College, Pennsylvania.
Glacio volcanism on Earth and Mars
Cambridge: Cambridge University Press, 2016. | Includes bibliographical references and index.
LCCN 2015048526 | ISBN 9781107037397
LCSH: Volcanism. | Volcanoes. | Magmas. | Ice. | Frozen ground. | Mars
(Planet) – Geology. | Mars (Planet) – Volcanoes.
LCC QE522 .S575 2016 | DDC 551.21–dc23
LC record available at http://lccn.loc.gov/2015048526

ISBN 978-1-107-03739-7 Hardback

Additional resources for this publication at www.cambridge.org/glaciovolcanism

Contents

Colour plates are to be found between pp. 212 and 213.

Preface

Nature and scope of the volume

This book is the first monograph on glaciovolcanism to be published. One of its principal intentions is to revivify the topic by celebrating 15 years of burgeoning research following the first international conference on volcano–ice interactions on Earth and Mars. That conference took place in 2000 in recognition that glaciovolcanism was a unique and new research area in its own right (Chapman et al., 2001). Also known as subglacial volcanism, glaciovolcanism has largely been omitted from current textbooks because of its youth. Although the study of interactions between volcanoes and glaciers dates back about a century, publications of scientific studies of glaciovolcanism before the 1990s were few. GeoREF searches (http://www.agiweb.org/georef/) show that prior to 2000, fewer than 30 literature citations were linked to studies of glaciovolcanism, subglacial volcanism, or volcano–ice interactions. However, the publication rate for articles on volcano–ice interactions has risen by almost an order of magnitude in the past decade, signalling the rapidly increasing interest in the topic. Major reasons for the rise in interest globally are threefold: (1) the occurrence, in 1996, of a spectacular and exceptionally well-documented glaciovolcanic eruption in Iceland (Gjálp) and associated huge glacial outburst flood (jökulhlaup) which wiped out a large part of the transport infrastructure in southern Iceland, followed by the Eyjafjallajökull eruption in 2010 which had a major economic impact across much of Europe; (2) the dawning realisation during the past decade that Mars is a water-rich planet hosting numerous glaciovolcanic edifices, a discovery that greatly enhances Mars' potential for future human colonisation and as an exobiological target; and (3) the development and rapidly growing importance of glaciovolcanic studies as a powerful new palaeoenvironmental proxy for reconstruction of planetary ice sheets. Although that importance is still largely unacknowledged by mainstream palaeoenviron-mental scientists, glaciovolcanic deposits are currently the most holistic repositories of information that can be used for determining *and quantifying* critical parameters of past ice sheets. In regions that have experienced glaciovolcanism, studies of the volcanic rocks should be at the core of any palaeoenviromental investigations. It is also now thought that glaciovolcanic centres may have played a pivotal role in enabling terrestrial species to survive and repopulate Earth's polar regions after multiple glacial periods. Glaciovolcanism is therefore not only of interest to multiple disciplines but it is also highly topical, and this book is the first to be published that describes and examines the eruptive processes and products involved, with reference to occurrences mainly on Earth but also on Mars.

Terrestrial glaciovolcanic deposits are globally widespread. However, the three most important geographical regions are Antarctica (the largest and longest-lived glaciovolcanic province, extending back at least 28 million years), Iceland (with the greatest concentration of glaciovolcanic edifices and largest number of observed historical eruptions) and British Columbia (where amongst the most insightful and earliest glaciovolcanic models and terminology were devised). Snow- and ice-capped volcanoes, found throughout much of the Pacific ring of fire including New Zealand, Japan, Russia (Kamchatka), Alaska, western Canada, western USA and South America, also potentially have a glaciovolcanic history and have a long history of hazards from glaciovolcanism. Indeed, every country with volcanoes and ice is potentially a glaciovolcanic province, even if glaciovolcanic products have not yet been recognised. One estimate puts the number of potentially active snow- or ice-covered volcanoes at more than 400. Mars is also now considered to have a large number of putative glaciovolcanic edifices, which have been used to document the water budget of the planet, and their discovery has led the way for postulating formerly extensive glacial surface ice and for identifying potential targets for exobiology investigations.

Glaciovolcanic eruptions can be devastating in fatalities and economic costs. For example, the 1985 eruption of Nevado del Ruiz volcano in Colombia produced snow/ice-generated lahars that killed c. 25 000 people. The 2010 eruption beneath the small ice cap on Eyjafjallajökull (Iceland) caused losses of several billion euros due to disruption of flights across the whole of Europe, whilst the meltwater flood associated with the 1996 eruption at Gjálp (also Iceland) was, for a period of a few hours, the second largest freshwater discharge on Earth, peaking at approximately four times the discharge of the Mississippi and exceeded only by that of the Amazon. The associated floodwaters caused some US$19 000 000 of economic damage principally to the Icelandic transport infrastructure which, for a country with a small population (c. 250 000 persons), is a traumatic and costly outcome.

This book is intended as a standalone monograph, and it is illustrated by line drawings and numerous photographs culled from the authors' extensive and unique collections, together with tables, graphs, maps and other figures as appropriate, many reproduced from the original papers. Its primary goal is to educate current and future earth, environmental, planetary and engineering scientists actively working in glaciovolcanic terrains about the distinctive characteristics of glaciovolcanic eruptions, their products and landforms. Throughout the writing of this book our preferred modus operandi was to use the best-described examples of glaciovolcanic sequences available and to present the published and some unpublished information in a simplified manner to illustrate how those sequences may have formed and the varied processes involved. In every case, we urge readers to go back to the original referenced descriptions to glean all of the available information, some of which we will have omitted for space considerations and which may ultimately prove to be more important than our current understanding suggests. We also include a glossary of the sometimes difficult-to-understand glaciovolcanic and related terminology in current use (often based on Native American and Icelandic terms unfamiliar to many scientists); abundant photographic illustration of the different constituent lithofacies; diagrammatic depiction of the eruptive and depositional processes; and evaluation of the distinctive sequence architectures and edifice morphometries. A special feature, available at www.cambridge.org/glaciovolcanism, is a series of expertly illustrated and annotated digital lectures that can be adapted or incorporated wholesale into existing university undergraduate and graduate teaching programmes. The inclusion of ready-made teaching materials is intended to substantially increase the outreach possibilities of the volume, to enhance the teaching of glaciovolcanism as a mainstream subject and considerably enlarge its profile in tertiary education worldwide. Currently, traditional curricula largely overlook glaciovolcanism. This is especially true in countries that lack the means to readily acquire and teach such knowledge, particularly in developing countries where many of the world's most dangerous glaciovolcanic systems occur.

Finally, the study of glaciovolcanism, like all scientific research, is like standing before a building facing a locked door. You unlock the first door and there is another, then another. The doors never end, *but you are further into the building*. This book is the key to the first few doors. What you do and where you go in the building after that is up to you. We wish you luck.

Acknowledgments

First and foremost, this book could never have been completed without the unstinting support of our families (Candy, Catriona and Kirsty Smellie; and Kim, Teagan and Kaelan Felknor-Edwards). We happily acknowledge that support with heartfelt thanks. In addition, we gratefully acknowledge the invaluable input and inspiration derived from countless conversations and collaborative projects with colleagues worldwide over the years. No list of colleagues can ever be complete or suitably appropriate and we apologise in advance for omissions, which are unavoidable, but among those we particularly wish to thank are the following (in no order): Magnus Gudmundsson, Sergio Rocchi, Kelly Russell, James White, John Stevenson, Mary Chapman, Joanne Johnson, Philip Kyle, Hugh Tuffen, Melanie Kelman, Lionel Wilson, Mike Hambrey, Ian Skilling, Bill McIntosh, Angela Walker, Cathie Hickson, Claire Cousins, Wes LeMasurier, Jim Head, Matt Patrick, Alan Haywood, Dave McGarvie, Sue Loughlin, Harry Pinkerton, Malcolm Hole, Tenley Banik, Warren Hamilton, Jennie Gilbert, Meagan Pollock, Steina Hauksdóttir, Tina Neal, Björn Oddsson, Kirstie Simpson, Jeff Karson, Lucy Porritt, Sasha Belousov, Jersy Marino Salazar, Barry Cameron, Chris Waythomas, Marina Belousova, Robert Wysocki, Hugo Delgado Granados, Bob Anderson, Marie Turnbull, Lucia Capra, Erica Massey, Nelida Manrique, Thor Thordarson, Carol Evenchick, Jim Reed, and too many Dickinson College undergraduates to name. We also wish to acknowledge the organisations that have financially supported much of our own research, in particular the British Antarctic Survey and University of Leicester, Programma Nazionale di Recherche in Antartide (Italy), Antarctica New Zealand, Transantarctic Association, US National Science Foundation, the National Geographic Committee for Research, and Dickinson College. Finally, we thank Emma Kiddle and, especially, Zoë Pruce, our editors at Cambridge University Press, for their patience and editorial support during the writing of this book.

1

Introduction

1.1 What is glaciovolcanism?

This book is about a class of volcanoes, active and formerly active, that erupt in association with ice in a glacierised setting. They are collectively known as glaciovolcanic centres and their distinctive eruption style is called glaciovolcanism. Glaciovolcanic eruptions are also often referred to as 'subglacial'. The terms glaciovolcanism and subglacial volcanism are commonly used interchangeably, but the inclusion of volcanic features such as subaerial lavas that simply banked and chilled against ice and were not truly erupted subglacially suggests that, of the two, glaciovolcanism is etymologically the more correct. However, we recognise that many scientists have favourite terms that they are loathe to give up, and geologists are no exception. Indeed, one of the respondents involved in the review of the proposal for this book criticised our original title that used glaciovolcanism in preference, arguing that subglacial was more obviously descriptive, more commonly used and therefore more acceptable. Both words will be used in this book interchangeably, but with a preference for glaciovolcanism.

Glaciovolcanism describes *the interactions between magma and ice in all its forms, including snow, firn, ice and any meltwater* (Smellie, 2006). The interaction may occur under ice (subglacial; Fig. 1.1a), above ice (supraglacial; Fig. 1.1b) and/or proximal to ice (ice-constrained, ice-impounded or ice-contact; Fig. 1.1c). The cryosphere (from Greek, *cryos*, 'cold') is the term that is used to describe glaciers, snow cover, floating ice and permafrost, although glaciers are the most visible component. It comprises those areas of the Earth's surface where water occurs in solid form. Today, glaciers (a catch-all term that includes ice sheets and ice caps; Neuendorf et al., 2011) cover about 15.9 million square kilometres of the Earth's land surface, equivalent to slightly less than the size of South America and about 1.5 times the size of Australia. Most of that ice is stored as ice sheets in Antarctica and in Greenland, which comprise c. 96% of all of the Earth's glacier areas and 99.4% of glacier volume, but glaciers are present on all of the continents today except Australia. Similarly, active volcanoes occur on all continents except Australia. For example, Europe has Mt Etna and Stromboli; Russia has Shiveluch and Klyuchevskoy; North and South America have their Pacific coast sections of the Ring of Fire with many active volcanoes (e.g. Mt Redoubt, Alaska; Villarrica, Chile); Africa has Mt Nyiragongo in the

Fig. 1.1 Examples of recent glaciovolcanic eruptions. (a) Aerial view of a crater formed during the 2004 Grimsvötn eruption within the Vatnajökull ice cap, Iceland. The ice cauldron is c. 0.7 km in diameter and is flanked by ash formed during explosive activity. (b) Aerial view of the southern flank of the intracaldera cone during the 2013 eruption of Mt Veniaminof, Alaska. This is a large stratovolcano with a subaerial tephra cone situated within its ice-filled summit caldera, whose margin is visible in the background. The tephra cone protrudes above the surrounding ice and it emitted several 'a'ā lavas that flowed downslope until they abutted against and/or flowed on top of the surrounding ice, but rates of melting and drainage were such that no lake formed, unlike in 1983–4. The cone is c. 300 m high. (c) Sheet lava flowing over snowpack during the 2013 Tolbachik eruption, Kamchatka. The foreground lava lobe is c. 5 m wide. Note how minimal snow melting has occurred below the moving lava and the lava (with a temperature of over 1000 °C) is flowing over the snow surface. (A black and white version of this figure will appear in some formats. For the colour version, please refer to the plate section.)

East African Rift; even Antarctica contains a prominently active volcano, Mt Erebus, within a major volcanic province largely contained in the West Antarctic Rift System. The likelihood, therefore, of future interactions between volcanism and ice on Earth is high (Fig. 1.2).

The term 'glaciovolcanism' was first used by Kelman et al. (2002), but was only formally defined in 2006 (Smellie, 2006). It is a young science topic and involves multiple scientific fields, including volcanology in its various forms, sedimentology, glaciology, geomorphology, geochemistry, biology (including exobiology), climatology and planetary science. It is thus truly multidisciplinary. Its history extends back only about a hundred years, to early descriptions of subglacially erupted volcanic edifices in Iceland and, later, in British

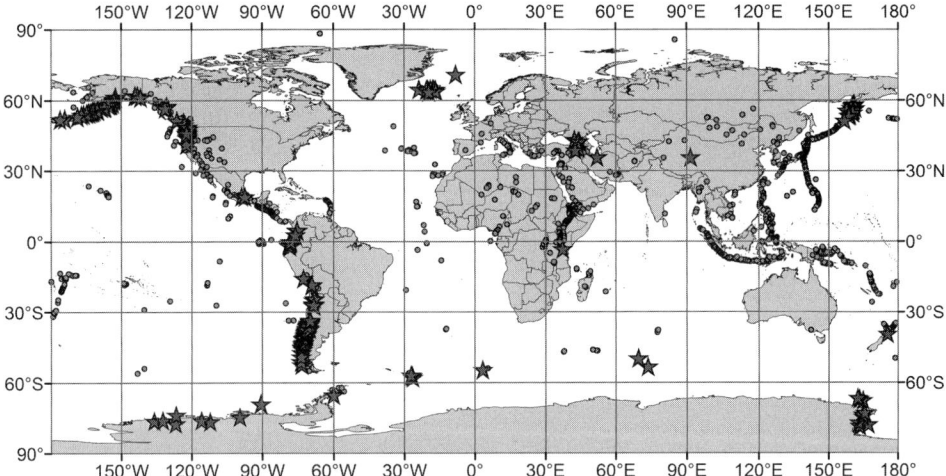

Fig. 1.2 Global map showing the distribution of ice-clad (stars) and ice-free (dots) volcanoes on Earth.

Columbia (see Section 1.3, below). Published accounts were few throughout much of the twentieth century and the early research was undertaken by only a handful of workers until the late 1990s. That was all set to change when the first meeting was convened in Reykjavik in 2000 to consider volcano–ice interaction as a topic in its own right. The meeting attracted 120 participants, indicative of the growing recognition of the increasing importance of major discoveries on planets such as Mars of evidence for past surficial water that could have hosted life. The meeting deliberately targeted terrestrial and planetary researchers and it was the first formal coming-together of the two communities. Following a meeting of interested parties at the IAVCEI (International Association of Volcanology and Chemistry of the Earth's Interior) General Assembly in Pucón in 2004, the establishment of a working group on volcano–ice interactions was proposed to IAVCEI in 2005 and accepted that year. It subsequently became a Joint Commission with the IACS (International Association of Cryospheric Sciences) and it hosts its own very well illustrated and informative website (http://volcanoes.dickinson.edu/iavcei_iacs_viic/index.html). A series of 'Volcano–Ice Interactions' meetings have followed: Reykjavik in 2006 (hosted by the International Glaciological Society); Vancouver in 2007; and Anchorage in 2012. In addition to abstract compilations, three thematic volumes from the meetings have been published thus far: Smellie and Chapman (2002); Clarke and Smellie (2007); and Edwards et al. (2009a).

Our understanding of subglacial eruption dynamics has several significant shortfalls. Because of generally insurmountable difficulties of access and minimal monitoring, subglacial eruptions are far less well understood than their subaerial and submarine counterparts. Many details of glaciovolcanic eruptions are seldom or never observed. This particularly applies to the outset, when the eruptions are often obscured by hundreds of metres of ice. Their products are impossible to observe and sample until the edifice builds

up above the surrounding ice surface. With rare exceptions such as well-monitored eruptions at Grimsvötn and Eyjafjallajökull in Iceland (e.g. Gudmundsson et al., 1997, 2012a;
Jude-Eton et al., 2012; Magnusson et al., 2012), our understanding of glaciovolcanic
processes relies heavily on interpretations from ancient landforms together with detailed
analysis of their eroded lithofacies exposed after their associated ice sheets have long since
melted away. Our starting point for this book, therefore, is that descriptions of the lithofacies and interpretations of the volcanic processes involved in their formation, in concert
with observations of modern and historical eruptions, are vital for a holistic understanding
of glaciovolcanism. The knowledge gained will improve our abilities to understand past
climate changes (Chapter 13), identify volcano–climate links (Chapter 14), and mitigate
associated glaciovolcanic hazards (Chapter 15). It will also significantly improve our ability
to interpret past environmental conditions in order to constrain planetary climate histories
and to accurately assess the water inventory not just on Earth but on other planetary bodies
(Chapter 16).

1.2 The importance of glaciovolcanism

Research on glaciovolcanism has grown extensively in recent years for several reasons. (1)
The 2010 eruption of Eyjafjallajökull highlighted the combination of at least two hazards
that are uniquely prominent in glaciovolcanism: local massive flooding by meltwater and
associated tephra production capable of disrupting regional travel (Chapter 15). Airborne
ash from the comparatively small eruption grounded more than 100 000 flights across wide
swathes of Europe for a period of a few weeks and caused substantial costs to the airline
industry estimated at c. 1.3 billion euros (£1.1 billion, US$1.7 billion). (2) The geological
history of many parts of the planet, particularly Iceland, British Columbia and Antarctica,
cannot be fully understood without taking into account the increasingly important role
glaciovolcanism has in shaping landscape and its pivotal importance in preserving evidence
of critical parameters of palaeo-ice sheets (Chapter 13). (3) The simple presence of an ice
cover and variations in its thickness over geological time, especially during glacial cycles,
can significantly modulate processes of mantle melting and rates of eruption. The resultant
sudden increase in atmospheric CO_2 during rapid deglaciation has been cited as a possible
positive feedback loop that strengthens global warming, causing further ice melting and
triggering more eruptions (Chapter 14). (4) Research on the geology of Mars using various
remote sensing techniques has called for improved understanding of terrestrial analogues,
including observed glaciovolcanic eruptions and associated deposits (Chapter 16).
Finally, (5) the combination of high heat flow and water together with nutrient-rich volcanic
glass that are characteristic of glaciovolcanic environments also means that glaciovolcanic
environments can likely support 'extreme-life', not only on Earth but also possibly on Mars
and other planetary bodies. Subglacially active volcanoes may also have played a pivotal
role in biological evolution by providing warm, wet ice-free areas sustained by very long-
lived geothermal systems. The volcanic refugia thus created may have re-established

terrestrial biodiversity in Earth's polar regions, permitting life to survive through multiple glaciations, and the same may be true of other planets (Fraser et al., 2014).

1.3 History of glaciovolcanic research

1.3.1 Iceland

Scientists in Iceland were the first to attempt to interpret the origins of widespread, distinctive volcanic deposits and their relationships to glaciation (see Jakobsson and Guðmundsson, 2008 for a review). Pjetursson (1900) recognised evidence that central Iceland had been glaciated multiple times, and that volcanic units were interstratified with glacial sedimentary deposits. Peacock (1926) discussed the origins of the Palagonite Formation, an extensive suite of volcanic deposits covering much of central Iceland, and he deduced that their distinctive properties were a direct result and indication of interactions between volcanism and glaciation. By the early 1940s, Noe-Nygaard (1940) had identified specific regions where the volcanic deposits had formed subglacially (e.g. Kirkjubæjarheiði) and he presented one of the first step-wise models for a subglacial eruption sequence. Kjartansson (1943) suggested that ridges of móberg, comprising a variety of lithified volcaniclastic deposits, should be referred to as 'hryggir', and flat-topped móberg mountains be called 'stapar'. While a large group of Icelandic and European geoscientists were investigating the subglacial origins of Icelandic volcanoes, similar research was being conducted simultaneously and independently by Mathews (1947) in Canada, deducing the origins of Quaternary volcanoes in northern British Columbia (Section 1.3.2). A terminology of glaciovolcanism emerged mainly based on the studies in Iceland (Table 1.1; also Chapter 8). The large, flat-topped subglacially erupted volcanoes in Iceland were referred to as stapar and table mountains – an approximate English translation of stapar. Likewise, elongate ridges with glaciovolcanic origins were referred to as hryggir (Kjartansson, 1943), tindars (Jones, 1969b, 1970), móberg ridges (Kjartansson, 1943), and hyaloclastite ridges (Chapman et al., 2001; see Glossary). While hyaloclastite (*sensu lato*) was used as a synonym for the Icelandic term móberg, the namesake Móberg Formation comprises a diverse array of volcaniclastic rocks (Peacock, 1926; Kjartansson, 1959; Jakobsson and Guðmundsson, 2008). Indeed, few of these volcaniclastic deposits would now be considered hyaloclastite based on modern usage (e.g. deposits formed during effusive volcanism and dominated by vitric fragments formed by quenching and mechanical fragmentation of the flowing lava; cf. Rittmann, 1952; Fisher and Schmincke, 1984; White and Houghton, 2006; see Glossary and Chapter 9).

Other important European contributions to Icelandic glaciovolcanism include the wide-ranging studies by van Bemmelen and Rutten (1955) and, particularly, Jones (1966, 1969a, b, 1970). Van Bemmelen and Rutten (1955) proposed that the table mountains formed as a result of eruptions from central vents or fissures beneath a relatively thick ice sheet (>450 m). They invoked ponding of lava against the enclosing ice to explain over-thickened

Table 1.1 *Varied nomenclature used to describe glaciovolcanic landforms (modified from Russell et al., 2014)*

Terms	Morphology	Sources[a]
stapi/stapar	equant and flat-topped	1, 4, 11, 15, 16
table mountain		4, 5
tuya		2, 3, 9, 21, 23, 26, 31
flow-dominated tuya		19, 21, 26
effusion-dominated tuya		20
flat-topped tuya		25
subglacial mound	equant and conical	17, 26
palagonitic cone		4
conical tuya		25
tephra-dominated tuya	linear and flat-topped	20, 21, 26
hyaloclastite ridge		4
tindar	linear but not flat-topped	8, 9, 21, 23, 24, 26, 31
hyaloclastite ridge		18
palagonitic ridge		4
hryggir		1
móberg ridge		1, 11
linear tuya		25
sheet-like sequence	thin sheet or sinuous ribbon	26, 27
sheet-flow sequence palagonite		28
breccia mass		29
hyaloclastite flow móberg sheet		30
		31
pillow mound/ridge pillow sheet	low oblate smooth mound	21, 26
		21, 26, 31, 32
subglacial dome/lobe	tall steep-sided dome or lobe	19, 21, 26

[a] 1 Kjartansson (1943); 2 Mathews (1947); 3 Mathews (1951); 4 van Bemmelen and Rutten (1955); 5 Kjartansson, (1959); 6 Jones (1966); 8 Jones (1969b); 9 Jones (1970); 10 Allen (1979); 11 Allen et al. (1982); 12 Smellie and Skilling (1994); 13 Hickson et al. (1995); 14 Moore et al. (1995); 15 Werner et al. (1996); 16 Werner and Schmincke (1999); 17 Hickson, (2000); 18 Gudmundsson et al. (2002); 19 Kelman et al. (2002); 20 Tuffen et al. (2002a); 21 Smellie (2007, 2013); 22 McGarvie et al. (2007); 23 Jakobsson and Gudmundsson (2008); 24 Edwards et al. (2009a); 25 Russell et al. (2014); 26 Smellie (2009); 27 Smellie (2008); 28 Smellie (2001); 29 Walker and Blake (1966); 30 Bergh and Sigvaldason (1991); 31 Jakobsson and Gudmundsson (2008); 32 Snorrason and Vilmundardóttir (2000)

masses of lava. Van Bemmelen and Rutten (1955) also established the importance of meteoric water–magma interaction for eruptive explosivity, and described the multiple sequences of lithofacies that result from the draining and refilling of associated englacial lakes during ongoing eruptions.

Jones (1969b, 1970), working in south-central Iceland, was largely responsible for erecting the 'standard model' for mafic tuya construction (Section 1.5). Jones' work remains one of the most influential of all glaciovolcanic studies and his model has been used extensively by practically all subsequent workers. Jones (1969b) applied the term tuya to several flat-topped Icelandic volcanoes and proposed the term tindar for ridges formed during glaciovolcanic eruptions. He deduced that the magma–water interactions resulted from water stored in englacial lakes derived from melting of the surrounding ice. Building on the Canadian-based work of Mathews (1947), Jones (1970) developed a model comprising an initial phase of subaqueous effusion producing basal pillow lavas and associated hyaloclastite created by quench fragmentation within an englacial lake (Section 1.5). He suggested that, as the volcanic pile approached the surface of the lake, the eruption style became explosive prior to transitioning to subaerial lava effusion. Critically, Jones (1969b) suggested that the mappable transition between the subaerial lavas and the subaqueous deposits, a boundary also observed but not named by Mathews (1947), served to demarcate the water level in the ancient englacial lake. Jones (1969b, 1970) called the surface a 'passage zone' (cf. Jones and Nelson, 1970; Skilling, 2002; Smellie, 2006). Passage zones are now one of the signature tools that allow glaciovolcanic deposits to be used as palaeoenvironmental proxies (e.g. Smellie, 2000, 2006; Edwards et al., 2009b, 2011; Skilling, 2009; Russell et al., 2013; Smellie et al., 2013a).

1.3.2 Canada

Independently of the early Icelandic studies, F.A. Kerr and W.H. Mathews published landmark studies describing the stratigraphy and morphology of a series of steep-sided and flat-topped volcanoes in the Iskut and in the Tuya–Kawdy regions of northwestern British Columbia (Kerr, 1948; Watson and Mathews, 1944; Mathews, 1947).

Kerr's observations from fieldwork in the 1920s along the Iskut River (published postumulously in 1948) are the earliest known descriptions and interpretations of volcano–ice interactions in Canada. Kerr gave a general description of the intermediate composition lavas that form much of Hoodoo Mountain volcano, along the central reach of the lower Iskut River in northwestern British Columbia. He recognised that the over-thickened sections of lava flows along the flanks of the volcano were evidence of confinement by surrounding glacial ice at a time when present-day glaciers in the area (Hoodoo and Twin glaciers) must have been much more extensive. Kerr's observations, while not widely distributed by publication in an international journal, presaged the more detailed and widely distributed work by Mathews in the following decades in northern and southern British Columbia.

In the Tuya–Kawdy area, Watson and Mathews (1944) documented numerous, small, apparently young, volcanoes hosting a variety of distinctive features including flat tops formed from subhorizontal lava flows, circular plan-views, and the predominance within the lower stratigraphy of prominently outward dipping fragmental units. Mathews (1947) suggested that the lavas capping the summits of these mountains did not correlate with each other nor could erosion explain their morphology. Instead, he suggested that they represented individual volcanoes. Mathews (1947) also recognised that these volcanic edifices shared common stratigraphical elements previously described at Icelandic volcanoes (e.g. Peacock, 1926; Nielsen, 1937; Noe-Nygaard, 1940), including: pillow lavas and breccias, massive to bedded deposits of outward dipping fragmented glassy basalt ('hyaloclastite'), and caps of horizontally bedded basaltic lava. He proposed the term tuya for these flat-topped, steep-sided volcanoes after a local aboriginal term used to name several local geographical features. He also interpreted the morphology and attendant volcanic lithofacies of these tuyas as indicative of volcanic eruptions from beneath late Pleistocene glacial ice sheets. Mathews (1947) also noted similar-aged, cone-shaped volcanoes lacking flat tops and comprising only pillow lava, dykes and fragmental deposits. He postulated that these non-flat-topped volcanic edifices were also glaciovolcanic in origin but that they had not breached the surface of the enclosing englacial lake. Later workers in British Columbia have referred to these edifices as subglacial mounds or conical tuyas (Hickson et al., 1995; Hickson, 2000; Russell et al., 2014; Table 1.1).

1.3.3 Antarctica

Because of its remoteness and its status as the last continent to be discovered, with a history of human exploration only dating back to 1819 (Campbell, 2000), progress on glaciovolcanism was later in Antarctica than elsewhere. Surprisingly, the earliest references to glaciovolcanic outcrops were published almost simultaneously from localities extending right across the continent (Hamilton, 1972 (northern Victoria Land); LeMasurier, 1972a, b (Marie Byrd Land); Bell, 1973 (Antarctic Peninsula)),. Thereafter, apart from numerous studies in Marie Byrd Land by LeMasurier (e.g. LeMasurier and Rex, 1982; LeMasurier, 1990) and a study of isolated outcrops in the Transantarctic Mountains (Stump et al., 1980), there is a significant gap until the late 1980s, when numerous papers mainly on Antarctic Peninsula glaciovolcanic outcrops were published (Wörner and Viereck, 1987; Smellie et al., 1988, 1993; Skilling, 1994; Smellie and Skilling, 1994; Smellie and Hole, 1997; Smellie, 1999) together with a major volume summarising the basic outcrop characteristics of all Cenozoic volcanism in Antarctica, irrespective of its eruptive environment (LeMasurier and Thomson, 1990).

1.3.4 Subsequent work (2000-on)

Publications on glaciovolcanic outcrops and glaciovolcanism have generally increased exponentially since 2000, the year of the first conference on volcano–ice interactions, and the subsequent 15 years can be viewed as a period of increasingly vigorous interest in the topic. Four main studies have reviewed and discussed the classification and nomenclature of glaciovolcanic edifices and their deposits (Hickson, 2000; Smellie, 2007, 2009,

2013; Jakobsson and Guðmundsson, 2008; Russell et al., 2014). Hickson (2000) presented a brief overview of terms and provided a list of examples mainly from western Canada based on morphological diversity (Table 1.1). Jakobsson and Guðmundsson (2008) gave a succinct review of glaciovolcanic terms and the Icelandic literature. In particular they advocated a specific geometrical criterion (>2:1 length to width ratio) to distinguish tindars from tuyas based on a survey of measurements from Iceland. Using a greatly expanded dataset, Smellie (2007, 2009, 2013) proposed a hierarchical classification scheme for subglacial landforms based on morphology and magma composition using 98 examples from Iceland, Antarctica, Canada and Russia. His classification scheme recognised seven types of glaciovolcanic landforms and he discussed how the different landforms reflected differences in lithofacies, magma properties and the intrinsic properties of the enclosing ice sheet. His morphometric analysis showed that mafic tuyas can have much larger volumes than felsic tuyas, but that felsic tuyas may have higher aspect ratios. He also postulated that lava-fed deltas developed within polar (cold-based) ice sheet regimes would likely be smaller than those emplaced in temperate (warm-based) ice, that tuyas would be taller if erupted through polar ice, and that lavas may flow over the surface of the surrounding ice rather than always melting a passage and will either be advected away or collapse as rubble after ice sheet decay (see Chapter 8). Finally, Smellie (2007, 2009, 2013) observed that mafic tuyas are *defined* by the presence of lava-fed deltas as they give the landform its flat top. By contrast, Russell et al. (2014) suggested a significant modification to traditional classification schemes, and showed how broad-scale ice characteristics (e.g. the ability to trap water in an englacial lake) could exert fundamental controls on edifice morphologies and lithofacies. All of these studies provided important morphological summaries using the long-established classification for glaciovolcanic landforms that evolved progressively over many decades. However, apart from Smellie (2000, 2001, 2007, 2009, 2013), Edwards and Russell (2002), and Russell et al. (2014), few workers have attempted to relate glaciovolcanoes and their deposits to the fundamental physical controls (principally ice thickness, glacier structure, basal regime and hydraulics) imposed upon the volcanic systems by the volumetrically much larger enclosing glacial systems (see Chapter 4).

1.4 Styles of glaciovolcanism and classification of the products

1.4.1 Types of glaciovolcanic eruptions

In addition to endogenous factors (e.g. magma composition, volatile content, rheology), exogenous (i.e. environmental) factors such as the presence or absence of external water or different ice thicknesses can have a major impact on eruptive styles and are very important in glaciovolcanic eruptions. The first attempt to correlate different glaciovolcanic sequence types with exogenous factors, i.e. 'thick ice' versus 'thin ice', was by Smellie et al. (1993; also Smellie and Skilling, 1994; Smellie, 2000) using Mio-Pliocene monogenetic edifices in the Antarctic Peninsula. It was shown that the different ice thicknesses led to very different sequences of lithofacies and eruptive and depositional processes. Subsequently

Edwards and Russell (2002) showed that deposits at a polygenetic edifice could preserve evidence for eruption under different ice thicknesses (see also Smellie et al., 2008; Skilling, 2009). By contrast, Edwards et al. (2015a) created a classification comprising three generic types of glaciovolcanic eruptions based on a broader range of environmental conditions: those that originate beneath ice (e.g. Gjálp 1996 is type for 'thick' ice, Eyjafjallajökull 2010 is type for 'thin' ice; cf. Smellie and Skilling, 1994); eruptions that deposit material on top of ice (e.g. Redoubt 2009); and eruptions where lava erupts subaerially but is then transported into contact with snow and ice (e.g. Fimmvörðuháls (Eyjafjallajökull) 2010; Tolbachik 2013; see Section 3.3). The examples highlighted by Edwards et al. (2015a) are from observations of modern events and they were selected to encompass eruptions that demonstrate all of the main processes that occur in glaciovolcanism.

1.4.2 Classification of glaciovolcanic products

The existing classification of glaciovolcanic products includes several discrete landform or sequence types for monogenetic eruptions, each with distinctive morphological and lithofacies characteristics that reflect the eruptive conditions (especially ice thickness and thermal regime), magma composition, viscosity, etc. (Smellie, 2007, 2009, 2013; Russell et al., 2014). They comprise tuyas, glaciovolcanic tuff cones, tindars, sheet-like sequences, pillow mounds, and domes (see Glossary for an explanation of these terms); although subglacially erupted polygenetic volcanoes also exist, only rarely do they create distinctive landforms (Chapter 8). Distinctive mafic and felsic variants are also recognised. However, a new and significantly different classification has been proposed by Russell et al. (2014). In it, many of the glaciovolcanic landforms are recast as different types of tuya, and it does not apply to isolated deposits distal to eruptive vents such as sheet-like sequences. The most substantial change in the proposed new classification is the redefinition of what a tuya is. Originally defined by Mathews (1947) as a subglacially erupted edifice uniquely distinguished by its flat top (basaltic in the original but since expanded to incorporate intermediate-composition and felsic examples), it is the most distinctive glaciovolcanic landform. By contrast, in the Russell et al. classification flat-topped edifices are just one variety (called flat-topped tuyas) in a continuum of tuyas that includes at least nine different types. They include conical (cf. tuff cone, subglacial mounds), linear (cf. tindar, which is used as a synonym for linear by Russell et al., 2014) and complex (which applies to edifices that show more than one distinctive morphological characteristic attributable to glaciovolcanism, e.g. Kima'Kho tuya (Canada), which has both a large cone and a flat-topped plateau). Larger polygenetic volcanoes such as Hoodoo Mountain (Canada; Edwards and Russell, 2002) and Mt Haddington (Antarctica; Smellie et al., 2008) are also regarded by Russell et al. (2014) as complex tuyas. It is too early to say whether the system proposed by Russell et al. (2014) will gain widespread acceptance. Accordingly, and without prejudice to the new Russell et al. (2014) classification, this book follows the currently well-established existing classification of glaciovolcanic landforms described by Smellie (2007, 2009, 2013).

1.5 The 'standard tuya model'

Since the 1950s, many workers have presented lithostratigraphical models to broadly summarise the sequence of events that form monogenetic glaciovolcanic edifices (e.g. van Bemmelen and Rutten, 1955; Jones, 1969b; Allen et al., 1982; Skilling, 1994; Smellie and Hole, 1997; Hickson, 2000; Jacobsson and Gudmundsson, 2008; Edwards et al., 2015a). However, the most commonly cited model is that established by Jones (1969b, 1970; Fig. 1.3). Despite its lack of lithofacies detail, it has come to be regarded as a 'standard model' for tuya eruptions and it has underpinned very many glaciovolcanic studies. Here we present a brief review of Jones' standard model as an introduction to the fundamental architecture of tuyas, the most distinctive glaciovolcanic landform (see also

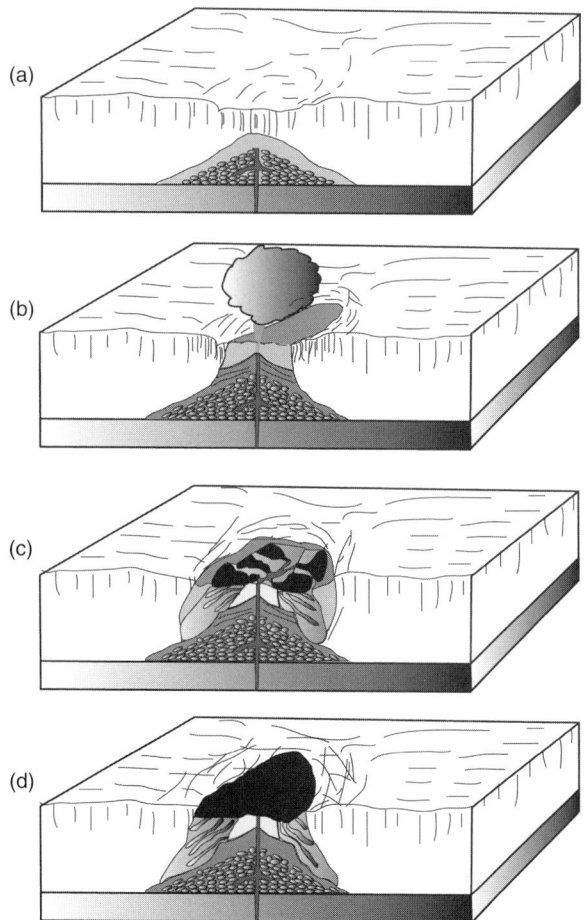

Fig. 1.3 Series of cartoons showing the sequential development of a tuya based on the 'classic' model described by Jones (1969b). See text for description of the constructional stages depicted. (A black and white version of this figure will appear in some formats. For the colour version, please refer to the plate section.)

Section 10.4.5). Tuyas and the many other glaciovolcanic sequence types now recognised are described in much greater detail in the following chapters of this book.

1.5.1 Eruption initiation: basal pillows?

Most historical models of glaciovolcanic eruptions envisaged initial eruptions dominated by pillow lava effusion. This is due to a common stratigraphical occurrence of pillow lavas in the cores and bases of many edifices (Jones, 1969a, b, 1970; Werner and Schmincke, 1999; Skilling, 2009) and is consistent with the inference from early studies of subaqueous eruptions that water depths greater than 500 m were sufficient to suppress most magmatic vesiculation and hence also suppress fragmentation. However, subsequent studies have shown that in the marine environment explosive fragmentation, driven by either magmatic volatiles or ingested external water, can occur at depths more than 1000 m below sea level (e.g. White et al., 2003 and references therein; Head and Wilson, 2003), and some glaciovolcanic edifices in Iceland (e.g. Helgafell: Schopka et al., 2006) and British Columbia (e.g. Ryane et al., 2011; Russell et al., 2013) show evidence for explosive initiation. Even pillow ridges can contain significant volumes of fragmental material potentially produced explosively (e.g. Pollock et al., 2014).

1.5.2 Subaqueous fragmentation

Virtually all historical glaciovolcanic models include a thick sequence of fragmental material, generally inferred or observed to have been emplaced on top of a basal pillow sequence. The wide range of lithofacies and processes involved is discussed later (Chapters 9–12). When a volcano erupts in an englacial meltwater lake at least three basic fragmentation processes are dominant: rapid chilling of lava in water ('quenching'), disruption of the lava due to rapid growth and expansion of bubbles driven by changes in pressure-controlled gas solubility, and explosions caused by rapid conversion of external water to steam as it is heated by lava. Amongst other complex processes (e.g. Wohletz, 2003; Zimanowski and Büttner, 2003; Wohletz et al., 2013), these three can occur simultaneously during a glaciovolcanic eruption and detailed stratigraphical studies are usually required to determine their relative contributions during the formation of glaciovolcanic edifices. Changes in englacial water levels, gravitational slumping and reworking by subaqueous traction currents also create complexities in the preserved record and will affect its interpretation. The processes involved in constructing glaciovolcanoes can be more complex than in many other volcanic environments.

1.5.3 Passage zones

Voluminous and long-lived eruptions will eventually build above the surface of the associated ice sheet, culminating in dominantly subaerial eruptive activity. The boundary demarcating the transition to subaerial conditions is known as a 'passage zone' (Jones 1969b). This transition will be discussed in more detail in Chapter 13 (Section 13.4.1), but the complexities of passage zones are critical, in particular, for extracting palaeoenvironmental information from ancient glaciovolcanic sequences and hence are worth emphasising (e.g. Jones and Nelson, 1970; Nelson, 1975; Smellie, 2006). Five types of subaqueous–subaerial transitions were described by Smellie (2006) associated with lava-fed deltas, and Russell et al. (2013) have described pyroclastic passage zones.

1.5.4 Effusion-dominated lava caps

The final phase of the standard tuya model is production of a broadly horizontal subaerial lava capping unit (actually a subaerial lava shield; e.g. Mathews, 1947; van Bemmelen and Rutten, 1955; Skilling, 2009; Chapter 10). In mafic tuyas, the lava flows are either 'a'ā or pāhoehoe, whilst block lava may form intermediate-composition lava-fed deltas. Studies specifically focused on the lava caprocks are few and are usually combined with descriptions of the underlying water-deposited cogenetic breccias (delta foreset beds; e.g. Skilling, 2002; Smellie, 2006; Smellie et al., 2013a). The presence of erosion-resistant subaerial lavas is probably critical for preserving many ancient tuyas.

1.6 Compositional classification used in this book

In order to manage the plethora of compositions represented by glaciovolcanic products, we have adopted an empirical division into mafic, intermediate and felsic groups (Fig. 1.4; see Chapters 10–12). Our division is based on the recommendation of the IUGS Subcommission on the Systematics of Igneous Rocks: that intermediate compositions are those with silica (SiO_2) content between 52 wt % and 63 wt % (LeMaitre et al., 2002). Although our usage is a convenience for description only, the divisions closely reflect variations in magma viscosity (Fig. 5.6). Viscosity is an important physical property that has a particular influence on transport mechanics in magmatic systems. It is affected by composition (particularly SiO_2), temperature, crystal and bubble content and dissolved gases (especially water), and it is an important variable for assessing volcanic hazards (Giordano et al., 2004, 2009; Chapter 5). It has many important consequences for glaciovolcanic eruptions, including how magma transport and degassing affect explosivity and melting rates of the overlying ice. Because the IUGS divisions are based solely on silica content, they cut across the compositional fields of mugearite, tephriphonolite and trachyte (Fig. 1.4). Thus, lavas in those fields with silica contents close to the IUGS boundaries may show viscosities of either of the adjacent compositional groups (mafic or intermediate; intermediate or felsic). In this volume all trachytes are grouped together with dacites and rhyolites as felsic rocks, and all mugearites and tephriphonolites are informally regarded as intermediate compositions, together with basaltic andesites, andesites and benmoreites. The IUGS division works least well for phonolites. Although they are clearly intermediate

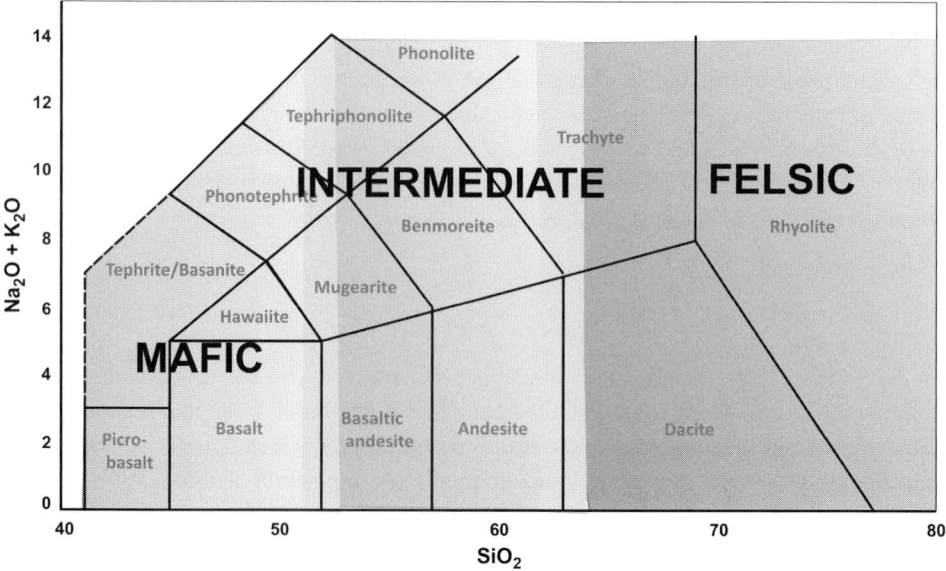

Fig. 1.4 Total alkalis versus silica diagram (after LeMaitre et al., 2002) showing division into mafic, intermediate and felsic compositional fields used in this book.

in composition based on silica content (Fig. 1.4), their viscosities span a particularly wide range, in part because of their wide range of eruptive temperatures (c. 775–830 °C). Measured phonolite viscosities overlap with those of trachytes and even rhyolites (Fig. 5.6). Although phonolites are included as intermediate-composition rocks in this volume, the more viscous varieties will show features more characteristic of felsic rocks (cf. Chapter 12).

2

Distribution of glaciovolcanism on Earth

2.1 Introduction

Glaciovolcanic provinces are conspicuously present in both polar regions on Earth, particularly Antarctica, Iceland and British Columbia (Canada). In addition, smaller and much less well studied provinces exist in other areas, such as Mexico, Patagonia, Africa, Indonesia and Siberia, and snow- and ice-capped volcanoes are prominent along the Pacific coasts of North and South America, in Japan, Kamchatka (Russia), and the South Sandwich Islands and other sub-Antarctic islands. However, even mid-latitudinal areas were glacierised during at least the Last Glacial Maximum, so that isolated volcanoes in Africa (e.g. Kilimanjaro), Hawaii (e.g. Mauna Kea) and even Indonesia (e.g. Mt Giluwe, Papua New Guinea) potentially have deposits formed by volcano–ice interactions. These examples are Neogene and Quaternary in age. However, with knowledge of Oligocene and Permo-Carboniferous glaciations, the possibility (disputed) of polar ice even in the Cretaceous and Eocene 'hothouse' periods (e.g. Miller et al., 2004; Tripati et al., 2005), and the advent of extreme-glacial hypotheses such as that of a 'snowball' or 'slushball' Earth (e.g. Hoffman et al., 1998; Cowen, 2005), it is likely that ancient (>100 Ma) glaciovolcanic deposits will also be identified in future.

2.2 Antarctica

Antarctica is the largest and longest-lived glaciovolcanic region on Earth, with a history of glaciovolcanic interactions extending back to 28 Ma, at least. It also contains the most continuous record of terrestrial glaciations in the Southern Hemisphere (LeMasurier, 1972a, b; LeMasurier and Rex, 1982; Smellie et al., 1993, 2009, 2011b; Smellie and Skilling, 1994; Smellie and Hole, 1997; Wilch and McIntosh, 2000, 2002, 2008). The volcanic centres are scattered over 5000 km from the sub-Antarctic South Sandwich Islands, via the Antarctic Peninsula, Ellsworth Land and Marie Byrd Land to northern Victoria Land in East Antarctica (Fig. 2.1). Most of the region became a within-plate tectonic province from mid-Cenozoic times, but subduction-related volcanism persists at the northern tip off the Antarctic Peninsula (Fretzdorff et al., 2004) and the remote snow- and ice-covered South Sandwich Islands are a classic intra-oceanic island arc (Leat et al.,

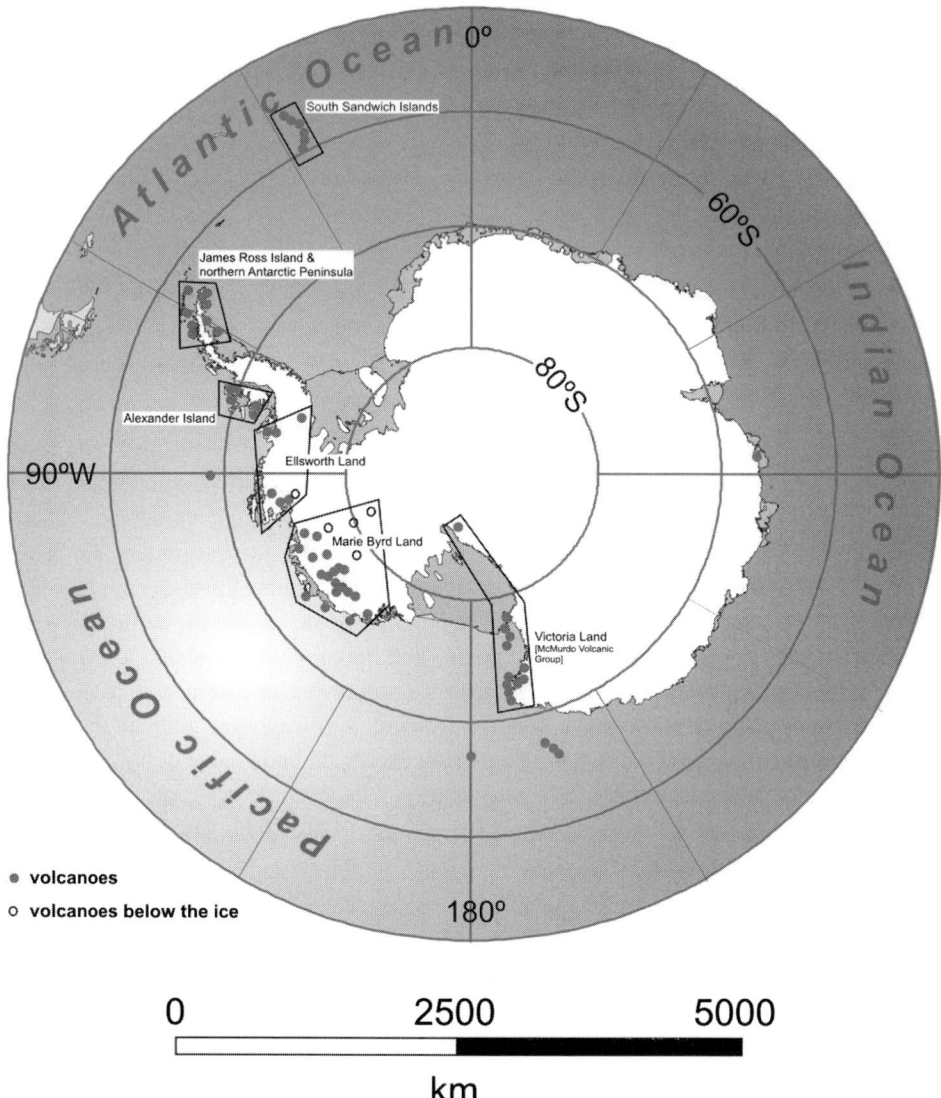

Fig. 2.1 Map showing the distribution of volcanic fields in Antarctica (modified after LeMasurier and Thomson, 1990).

2003). Mantle plume (thermal anomaly) activity has ensured that eruptions were frequent and occurred fortuitously during the period in which the Antarctic Ice Sheet was initiated and developed (LeMasurier and Rex, 1982; LeMasurier and Thomson, 1990; Haywood et al., 2009). Away from the Antarctic Peninsula, the volcanic centres lie within the West Antarctic Rift System (WARS), an extensional tectonomagmatic province with dimensions comparable to the East African Rift or the Basin and Range and characterised by a thin

(c. 27 km) continental crust and block faulting. The WARS may overlie one or more mantle plumes (Behrendt et al., 1992). The volcanoes are bimodal and many crop out along prominent linear chains orientated N–S and E–W. Alkaline volcanic centres in the Antarctic Peninsula are exclusively basaltic and mainly formed as a consequence of deep mantle flow into slab windows as a consequence of sequential ridge–trench collisions and subduction cessation along much of the Antarctic Peninsula trench (Hole et al., 1995). The region can be divided into several geographically separate glaciovolcanic provinces and fields, not all of which have been formally named, corresponding to the South Sandwich Islands, James Ross Island and northern Antarctic Peninsula, Alexander Island, Ellsworth Land, Marie Byrd Land and Victoria Land.

2.2.1 South Sandwich Islands

The South Sandwich Islands are a young and relatively small intra-oceanic consuming-margin volcanic arc situated at the east end of the Scotia Sea (Baker, 1990a; Leat et al., 2014). The arc contains 11 main islands all of which are either known to be active or are potentially active (Baker, 1990a; Patrick and Smellie, 2013). Most are extensively covered in snow and ice but the four smallest islands are too small and low to sustain a permanent ice cap. The islands may be classified as snow- and ice-capped volcanoes with the potential for glaciovolcanic eruptions. The most recent eruption was from Mt Belinda, Montagu Island (Patrick et al., 2005; Section 3.3.12), and Mt Michael, Saunders Island, contains a semi-permanent convecting lava lake, one of two known in Antarctica (Lachlan-Cope et al., 2001). The individual islands are small and dominated by lava flow eruptions although tephra-forming eruptions also occur and are principally represented in the marine record and in ice cores (Ninkovich et al., 1964; Hubberten et al., 1991; Basile et al., 2001). The largest, Montagu Island, is 12 km in diameter and the smallest, Leskov Island, measures just 930 by 550 m and is only 190 m high. However, the islands merely represent the exposed summits of several very substantial submarine volcanoes that rise up to 3 km from the ocean floor and have basal diameters of a few tens of kilometres (Baker, 1990a; Leat et al., 2014).

Evidence for volcano–ice interaction is minor, comprising coarse sediment gravity flow deposits with striated and facetted clasts and interbedded with lavas on Candlemas Island that are probably related to jökulhlaup event(s) during a subglacial eruption of Mt Perseus (unpublished information of JLS). Eruptions on ice-covered Bristol Island in 1935, 1956 and 1962 produced only scoria cones with no obvious glaciovolcanic imprint (Patrick and Smellie, 2013, Fig. 23). However, the eruption at Mt Belinda (Montagu Island) between 2001 and 2007 resulted in effusion of andesite lava flows that melted paths across the extensive summit icefield (Patrick et al., 2005). Lava entered the sea at the north coast, creating a small arcuate lava platform. Satellite images show how the adjacent valley glacier surged by a few hundred metres, probably caused by the release of abundant meltwater during the eruption. Where the lava meets the coast it is a block lava.

Although the eruption has undoubted glaciovolcanic characteristics, the vent responsible was a small subaerial pyroclastic cone (probably a scoria cone) and only the lavas are glaciovolcanic. It is unclear whether a meltwater lake was created at the lava snouts, although no lake is evident in the images. The lavas advanced as a thick coherent mass, without obvious development of the underlying hyaloclastic lithosome characteristic of lava-fed deltas.

2.2.2 James Ross Island and northern Antarctic Peninsula

The James Ross Island Volcanic Group (JRIVG) in northern Antarctic Peninsula is one of the great Cenozoic volcanic fields of Antarctica. Extending over 7000 km^2, it has a conservatively estimated erupted volume exceeding 4500 km^3 (Fig. 2.2; Smellie et al., 2013b). Eruptions were basaltic and commenced at least 12 m.y. ago (from dated clasts in tillite; Marenssi et al., 2010) but *in situ* outcrops only go back to 6.2 Ma. The region is

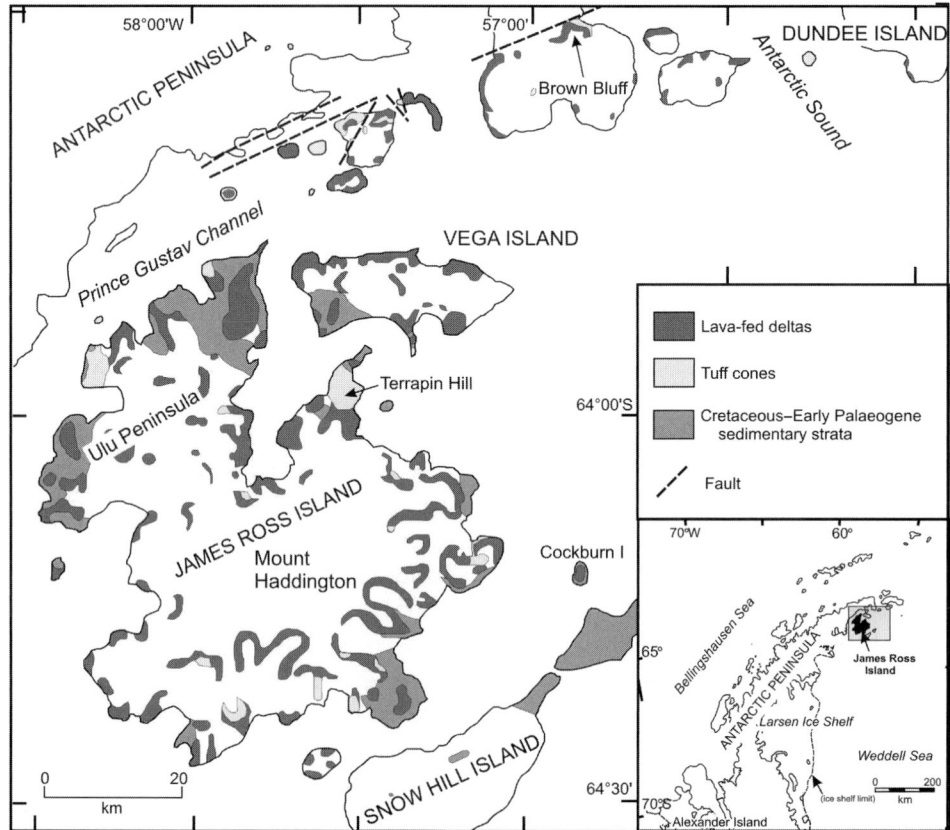

Fig. 2.2 Map of the James Ross Island Volcanic Group in northern Antarctic Peninsula (adapted from Smellie et al., 2013b).

probably still active. It is one of the most intensively dated volcanic areas in Antarctica (Smellie et al., 2008, 2013b). The volcanic field is dominated by spectacular large-volume pāhoehoe lava-fed deltas sourced mainly in the very large Mt Haddington central volcano, but tuff cones and tuya satellite centres are also present (Skilling, 1994; Smellie et al., 2013b). The deltas were originally considered to have flowed into the sea (Nelson, 1975), but more recent work has demonstrated that they are almost all glaciovolcanic (Skilling, 2002; Smellie et al., 2006b, 2008). By contrast, most of the tuff cones were marine-emplaced during ice-poor periods ('interglacials'; Smellie et al. 2006a; Williams et al., 2006). The lava-fed deltas show many types of subaqueous–subaerial transitions, such as rising and falling passage zones, some of which were described for the first time and are diagnostic of a glacial eruptive setting (Smellie, 2006). The sequences were interpreted to have been erupted in association with a relatively thin, wet-based ice sheet (mostly a James Ross Island ice dome, confluent with the Antarctic Peninsula Ice Sheet) that ranged in thickness between c. 200 and 800 m (Smellie et al., 2008, 2009). Mount Haddington is the largest glaciovolcanic polygenetic central volcano known on Earth. It has a low shield-like profile because of its construction from multiple lava-fed deltas that were emplaced in a draping ice sheet. Together with the evidence for marine conditions in the tuff cones and information derived from interbedded glacial sedimentary deposits (tills and debris flow deposits, some fossiliferous), the JRIVG has become the most important terrestrial geological unit for characterising the Antarctic Peninsula Ice Sheet and deducing palaeoclimatic characteristics, especially for the Pliocene (Hambrey et al., 2008; Nelson et al., 2009; Smellie et al., 2008, 2009; Clark et al., 2010; Williams et al., 2010; Nývlt et al., 2011).

Seal Nunataks are situated 125 km southwest of James Ross Island. They comprise a series of small basaltic nunataks that were formerly surrounded by the Larsen Ice Shelf (Hole, 1988, 1990). With the spectacular disintegration of the latter in 2002, they are now largely surrounded by the sea. The nunataks rise to 368 m a.s.l. (mostly <150 m) but their bases are c. 500 m b.s.l. They differ from other glaciovolcanic centres in Antarctica in that they are mainly composed of a central dyke encased in pillow lava and locally have a minor carapace of subaerial rocks (mainly scoria and agglutinate) (Smellie and Hole, 1997). Although the evidence for eruptive setting is ambiguous, they are interpreted as a series of small monogenetic mafic glaciovolcanic centres that were erupted beneath a substantial ice cover at least 600 m thick (and possibly much thicker). Their development mostly ceased at the pillow volcano stage with a few progressing briefly to the tindar stage; no tuyas are present.

Subglacially erupted volcanic sequences are also present on Brabant and Anvers Islands, on the west side of northern Antarctic Peninsula. Access to both islands is particularly difficult but the volcanic sequences are interpreted to represent the products of at least three large glaciovolcanic shield volcanoes, of which just one has been investigated (Smellie et al., 2006c). With the exception of a single lava-fed delta, the sequences are dominated by thin sheet-like sequences, which indicate eruption in association with a thin draping ice sheet less than 150 m thick in late Pleistocene time (<200 ka).

Deception Island is an active volcano also situated on the west flank of northern Antarctic Peninsula, constructed on a spreading centre in Bransfield Strait marginal basin (Keller et al., 2002; Smellie et al., 2002). About 60% of the island is covered in snow and ice (López-Martínez and Serrano, 2002), and it was presumably greater in the past, yet the volcanic record shows no clear glaciovolcanic influence. The volcano erupted in 1967, 1969 and 1970 (see Section 3.3.3). The 1969 event was a fissure eruption that transected the Mt Pond ice cap on the east side of the island and was glaciovolcanic. The eruption was unusually well documented at a time when glaciovolcanic eruptions were not recognised as distinctive events (Baker et al., 1975; Smellie, 2002). The vents were situated under ice just 70 m thick. The eruption was associated with a large jökulhlaup of meltwater highly charged with ice blocks, glacial debris and volcanic materials which overflowed the fissure and burst through ice portals lower down, swept downslope and extensively damaged a scientific station. Fall tephra buried and burned down another station on the island. A series of pyroclastic (scoria) cones was constructed. They are Strombolian and preserve no obvious indication of their subglacial origin, although the earliest-erupted basal parts of the cones are unexposed.

2.2.3 Alexander Island

Outcrops in southern Alexander Island (Beethoven Peninsula) consist of several basaltic tuyas scattered across a volcanic field over 2000 km^2 in area (possibly over 7000 km^2; Smellie, 1999). Some are entirely covered in snow and ice and their volcanic origin is only inferred (Hole, 1990). The nunataks are extensively glacially eroded. They comprise substantial thicknesses of stratified waterlain lapilli tuffs ('tindar stage') overlain by breccia foreset beds and overlying subhorizontal pāhoehoe(?) lavas of a lava-fed delta caprock. Together with Seal Nunataks, the outcrops were used as a basis for proposing a comprehensive genetic model for glaciovolcanic 'Surtseyan' (i.e. shoaling and emergent) volcano edifice construction, which substantially fleshed out the original cruder 'standard' model for tuyas by Jones (1969b) (Smellie and Hole, 1997).

By contrast, apart from two tiny basaltic occurrences on southern Rothschild Island comprising stratified waterlain lapilli tuffs that are possibly comparable in structural position with the basal sequences seen on Beethoven Peninsula (i.e. a 'tindar stage'), glaciovolcanic outcrops in northern Alexander Island are very different. They consist of four small isolated basaltic outcrops on the flanks of Mt Pinafore (Elgar Uplands) and Ravel Peak (Debussy Heights; Smellie et al., 1993). They are historically important for being the localities selected as sequence holotypes for glaciovolcanic sheet-like sequences of Mt Pinafore type, characteristic of eruptions under 'thin' ice (Smellie and Skilling, 1994; Smellie, 2000; see Section 10.6.1). Each of the sequences rests unconformably on pre-volcanic basement, either deformed Cretaceous metasedimentary rocks of an accretionary prism or calc-alkaline volcanic strata, and they appear to occupy small valleys cut in that bedrock. They are all broadly similar and comprise some combination of basal diamict (till), stratified waterlain (fluvial) sandstones and gravels largely derived from phreatomag-matic eruptions, and a thick section of blocky- or columnar-jointed lava encased in hyaloclastite breccia. The evidence for flowing meltwater was thought to be a particularly significant difference with models for tuya eruptions, which are dominated by lava-fed deltas flowing into ponded water in glacial vaults sealed hydraulically around their bases (Björnson, 1988). Thus, the conspicuous absence of lava-fed deltas was thought to signify that the glacial cover was permeable, i.e. composed of some combination of snow, firn and/or fractured (crevassed) ice, for which there are practical maximum thick-nesses of c. 150–200 m (Smellie, 2001). However, although the existence of a permeable upper layer is confirmed, it is now known that eruption-related meltwater vaults in wet-based ice are also leaky and undergo persistent basal drainage (Gudmundsson et al., 1997; Smellie, 2006; see Section 4.5.2).

2.2.4 Ellsworth Land

There are three important volcanic areas in Ellsworth Land: Snow Nunataks, Jones Mountains and Hudson Mountains. Snow Nunataks are a poorly known series of isolated volcanic outcrops that probably represent several small glaciovolcanic centres of uncertain age (Late Miocene or younger; Smellie, 1999). They are mostly pillow mounds overlain by waterlain lapilli tuffs that occasionally show spectacular slump-scar unconformities, but two nunataks have subaerial caprocks and may be tuyas. The Jones Mountains outcrops are 500–700 m thick and Late Miocene in age (7.63 Ma; Rutford and McIntosh, 2008). At least two basaltic erupted units have been identified, each comprising basal stratified (including cross stratified) lapilli tuffs overlain by a much greater thickness of chaotic pillow lava and breccia (Hole et al., 1994; Rutford and McIntosh, 2008). The volcanic rocks rest on discontinuous lenses of diamict interpreted as tillite which, in turn, rest on a well-exposed glacially striated surface showing chatter marks and grooves. The outcrops are historically important as they provided the first evidence for pre-Pleistocene glaciations in Antarctica (Craddock et al., 1964), although the discovery has been widely ignored. An origin for the volcanic sequences by glaciovolcanic eruptions has never been proven although it is very likely (Rutford and McIntosh, 2008). The association with basal tillite and a glacially eroded surface is important, as are the presence of striated and facetted boulders dispersed within the stratified lapilli tuffs and abundant palagonitised and fresh glass in the tillite matrix. Speculatively, the closest resemblance is to sheet-like sequences (Haywood et al., 2009).

The Hudson Mountains are a very remote and rarely visited series of small basaltic nunataks that crop out over an area of 8400 km^2 at elevations of 200–700 m a.s.l. Three main eruptive centres have been suggested, at Teeters Nunatak, Mt Moses and Mt Manthe, and small cones, some with craters, have also been postulated (Rowley et al., 1990). The nunataks have a generalised composite stratigraphy comprising basal pillow lava and minor hyaloclastite; a substantial section of stratified lapilli tuff, subordinate breccia and lava; and upper subhorizontal lavas (likely pāhoehoe). Although no features diagnostic of eruptive environment have been described, the nunataks crop out at significant elevations above present sea level and the stratigraphy resembles that of tuyas. The volcanism took place mainly in the Late Miocene and Pliocene (c. 8–4 Ma). A possible eruption at Webber Nunatak was postulated in 1985 based on satellite data but remains unconfirmed. However, an acoustic layer in ice was discovered during a radio-echo survey east of the nunataks whose distribution suggested it might be volcanic ash dated to 207 BC ± 240 years (Corr and Vaughan, 2008). A prominent basement high beneath the layer outcrop was identified as a possible subglacial source volcano ('Hudson Mountains Subglacial Volcano').

2.2.5 Marie Byrd Land

Marie Byrd Land contains the largest glaciovolcanic province in Antarctica, covering an area of more than 180 000 km^2. The potential volume of subglacial (but not necessarily all subglacially erupted) volcanic products exceeds 10^6 km^3 making it a Large Igneous Province (Behrendt et al., 1994). The region also contains the greatest number of large central volcanoes (Fig. 2.3), many with summit elevations over 3000 m a.s.l.; a wider compositional range than in any other volcanic province in Antarctica apart from Victoria Land, with an abundance of intermediate and rare felsic rocks (trachyte, phonolite, rhyolite) in addition to less evolved compositions; and a greater age range, extending back to 36 Ma (LeMasurier and Thomson, 1990; Wilch and McIntosh, 2000). At least 18 large shield and stratovolcanoes are present, together with a variety of small mafic flank and satellite centres. The volcanism is thought to be active at five centres, i.e. Mt Takahe, Mt Berlin, Mt Waesche and possibly Mt Kauffman and Mt Siple. In addition, a 700 m-high active

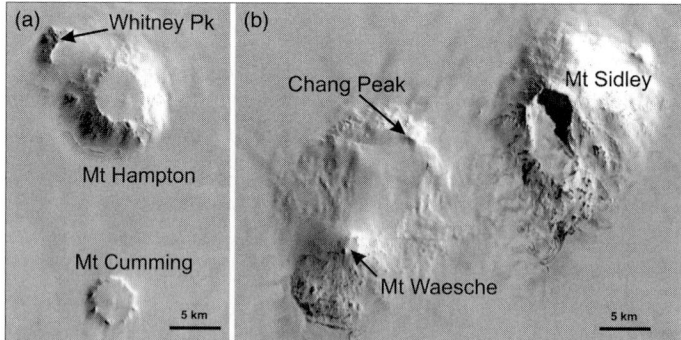

Fig. 2.3 GoogleEarth satellite images showing typical stratovolcano centres in Marie Byrd Land, Antarctica. (a) Whitney Peak (3003 m a.s.l.; 14 Ma), Mt Hampton (3323 m a.s.l.; 11–10 Ma) and Mt Cumming (2612 m a.s.l.; mainly 10 Ma). (b) Mt Sidley 4181 m a.s.l.; 6–4 Ma), Chang Peak (2920 m a.s.l.; c. 1.6 Ma) and Mt Waesche (3292 m a.s.l.; <1 Ma). Mt Sidley is the highest volcano in Antarctica. Mt Waesche is probably still active and is the source for several tephra layers found in the ice nearby. Ice-filled calderas are prominent on most of the volcanoes.

subglacial volcano (named Mt CASERTZ volcano) with a large caldera and elevated heat flow was discovered by remote sensing beneath 3 km of ice. Together with recent highly localised subglacial seismicity (Lough et al., 2013), these discoveries suggest that other active subglacial volcanoes may be present beneath the West Antarctic Ice Sheet. Three putative active subglacial centres have been provisionally identified and many others are possible (Blankenship et al., 1993; Behrendt et al., 1998, 2004; cf. Corr and Vaughan, 2008; Behrendt, 2013). A prolonged history of volcano–ice interactions in Marie Byrd Land has been inferred beginning in Oligocene time (LeMasurier, 1972a; LeMasurier and Rex, 1982; LeMasurier et al., 1994; Wilch and McIntosh, 2000, 2002, 2008).

The oldest exposed evidence for glaciovolcanism is from 29–27 Ma and occurs at Mt Petras near the centre of the Marie Byrd Land volcanic province. It consists of four different scoria and tuff cones at elevations varying between 2530 and 2862 m a.s.l. They are interpreted to imply only limited, thin syn-eruptive ice draping a mountain rather than a thick regional ice sheet at the time (Wilch and McIntosh, 2000). Thereafter, the glacio-volcanic record is not as well represented as might be expected, mainly because most of the volcanoes are shrouded in ice that typically exceeds 2 km in thickness. The volcanoes are generally poorly dissected and only the uppermost and latest-erupted products are available for examination, usually much obscured by a drape of thin ice and snow (Rocchi et al., 2006). Mount Murphy and its satellite centres are exceptions. The Mt Murphy volcano (9–8 Ma) is a complex polygenetic edifice constructed from at least two large central volcanoes, of which the westernmost is exposed in spectacular cliffs c. 2000 m high. The basal few hundred metres comprise alternating sequences of stratified waterlain lapilli tuffs overlain by much thicker chaotic blocky- and prismatic-jointed lava and variably stratified to massive breccia (LeMasurier et al., 1994). Each sequence is c. 100 m thick and is bounded by sharp glacially eroded surfaces on which diamict (interpreted as tillite) is often present. The lava–breccia units in some sequences are overlain depositionally by subaerial 'a'ā lavas. Although some of the basal sequences are strikingly similar to sheet-like sequences of Mt Pinafore type (Smellie, 2000, 2008), LeMasurier (2002) interpreted them as lava-fed deltas, without describing the type of feeding lavas. A more recent reappraisal (JLS, unpublished) broadly supports LeMasurier's views. The sequences are probably lava-fed deltas fed by 'a'ā lava, and the confusion with sheet-like sequences arose because some have had their subaerial lava caprocks removed by glacial erosion, yielding truncated sequences. The upper part of the Mt Murphy section comprises interbedded subaerial basalt and trachyte lavas.

Small satellite centres are also present on the west side of Mt Murphy. Much younger in age than the main volcanic massif (i.e. 6.8–4.7 Ma; Wilch and McIntosh, 2002), they are all flat-topped and tuya-like, in varying stages of glacial erosion. However, they are all polygenetic and probably represent the highly eroded fragments of a single larger edifice (unpublished information of JLS). Only the outcrop at Icefall Nunatak is well described and the study was the first to propose a model for glaciovolcanic (tuya) eruptions in which meltwater overflowing from a glacial vault through a permeable snow and firn layer was a primary control on water levels in the vault (Smellie, 2001; see also Smellie, 2006). Icefall Nunatak was constructed in three main stages between 6.8 and 6.5 Ma. A lowest mainly effusive phase of subaqueous sheet lava and cogenetic breccias was probably constructed within a subglacial vault. A second stage comprising lapilli tuffs deposited by a variety of eruption-fed density flows constructed a subaqueous tuff cone ('tindar stage') showing spectacular evidence for syn-eruptive slope failure. It was succeeded by lava-fed delta progradation. Finally, subaerial sheet lava effusion occurred and was overlain by a small scoria cone. Of particular interest is the presence of current-deposited volcaniclastic beds at the base of stage 2. Their presence was the first published evidence for subglacial meltwater drainage, albeit envisaged as temporary (corrected by Smellie, 2006), published at a time when meltwater vaults in subglacial eruptions were thought to be sealed (Björnsson, 1988). The two other glaciovolcanic outcrops in the area are highly eroded and dominated by lava-fed deltas: at Turtle Peak and Hedin Nunatak, (Fig. 13.18).

Sequences at Mt Rees and Mt Steer, Crary Mountains, are similar in age to Mt Murphy (i.e. 9–8 Ma; Wilch and McIntosh, 2002) but situated about 250 km inland and mostly above 2000 m a.s.l. They show many of the features of the Mt Murphy basal section, comprising alternating 'wet' and 'dry' lithofacies. They are reinterpreted here as the products of 'a'ā lava-fed deltas (cf. Smellie et al., 2013a). The apparent absence of unconformities and glacial tills led Wilch and McIntosh (2002) to suggest that cold ice conditions may have prevailed in the Crary Mountains compared with wet-based ice at coeval Mt Murphy (cf. Smellie et al., 2011a).

Mount Sidley is a complex composite basalt–phonolite–trachyte stratovolcano which rises over 2000 m above the surrounding ice sheet (Panter et al., 1997). It is the best-exposed volcano in the interior of Marie Byrd Land and was constructed in three major episodes of overlapping edifice construction between 5.7 and 4.2 Ma (Panter et al., 1994). Each episode was characterised by caldera formation. The youngest caldera exposes about 1000 m of section cut deep into the heart of the volcano. The sequences are dominated by subaerial lavas, tephra fall and local ignimbrite deposits. Despite the amount of rock exposed, evidence for volcano–ice interaction is minimal and comprises thin hyaloclastite breccia bases to some lavas, suggesting minor interaction with snow and a much-reduced West Antarctic Ice Sheet, if present at all (Haywood et al., 2009).

Finally, seven eroded small monogenetic centres exposed along the Hobbs Coast gave ages between 11 and 2.5 Ma (Wilch and McIntosh, 2008). The outcrops vary from tuya-like to tuff cone deposits, subaerial lava and tephra fall that rest on glacially eroded pre-volcanic surfaces and local diamict (likely tillite). They were interpreted as evidence for locally variable syn-eruptive ice levels and relatively thin ice in Mio-Pliocene time. The outcrops suggest former ice thicknesses similar to present, but the picture is complicated by their occurrence on high interfluves situated between lower-lying local ice streams.

2.2.6 Victoria Land

With the exception of two poorly known geographically isolated Middle Miocene glaciovolcanic centres (tuyas?) situated only a few hundred kilometres from the South Pole

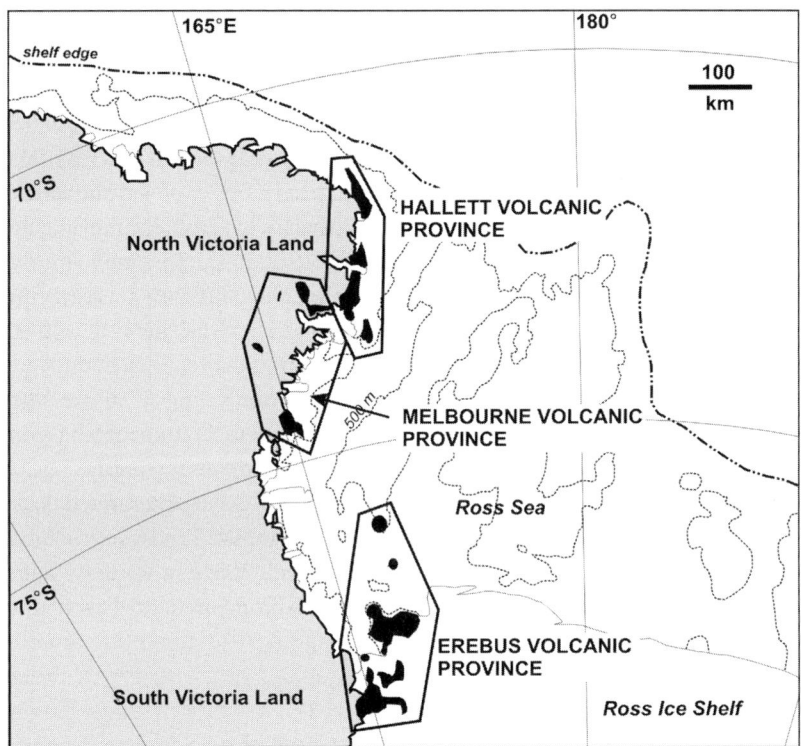

Fig. 2.4 Map showing the principal outcrops of the McMurdo Volcanic Group, Victoria Land, Antarctica (after LeMasurier and Thomson, 1990).

(Stump et al., 1980), the volcanic rocks in Victoria Land are all included in the McMurdo Volcanic Group (MVG). Those in northern and central Victoria Land are the most recently investigated in Antarctica. The MVG is subdivided into the Hallett, Melbourne and Erebus volcanic provinces (Fig. 2.4). Glaciovolcanic sequences are widespread and dominant in the Hallett Volcanic Province, but they also occur in the Melbourne Volcanic Field (i.e. that part of the Melbourne Volcanic Province close to Mt Melbourne; Wörner and Viereck, 1987). In the Erebus Volcanic Province, only sequences at Minna Bluff are definitely known to be glaciovolcanic.

Glaciovolcanic sequences form several large shield-like central volcanoes on the Ross Sea flank of the Transantarctic Mountains at Minna Bluff (near Mt Discovery), Mt Melbourne and between Coulman Island and Cape Adare. The volcanism ranges in age between 11.5 and 1.7 Ma (Hamilton, 1972; Smellie et al., 2011b, 2014). The outcrops are dominated by lava-fed deltas characterised by gently dipping subaerial 'a'ā lavas overlying similar or greater thicknesses of subaqueous lithofacies (chaotic lava lobes, hyaloclastite and lithic breccia, collectively called lobe-hyaloclastite; Smellie et al., 2011a, 2013a; see also Hamilton, 1972). They are the first examples of 'a'ā lava-fed deltas to be recognised and described. Fewer examples of sheet-like sequences are also present and include the first felsic examples to be described. The latter occur at a single locality on Daniell Peninsula (Smellie et al., 2011a) and a few possible examples at Minna Bluff (unpublished information of JLS). They are essentially identical in lithofacies to the mafic examples but are dominated by thick (few to several tens of metres) felsic lava. In many of the felsic examples, the lava upper surfaces were exposed subaerially. Syn-eruptive environmental conditions in the region were overwhelmingly glacial and indicate a relatively thin ice sheet, typically less than 300 m thick, which simply draped the landscape (forming large ice domes over each volcano) but with sparse evidence for rare overriding events by thicker, landscape-drowning ice (Smellie et al., 2011b, 2014 and unpublished information of JLS). Characteristics of the volcanism deduced from the glaciovolcanic sequences indicate that the basal thermal regime of the ice varied between wet-based (probably sub-polar rather than temperate) and cold-based through the period and that the ice was polythermal overall (Smellie et al., 2014). The outcrops are the first glaciovolcanic sequences for which eruptions under cold-based ice are likely (cf. Wilch and McIntosh, 2002). The deduced temporal and geographical varia-tions in thermal regime characteristics conflict with traditional views that Antarctic ice underwent a profound single step-change from wet-based to cold-based during the period, although the timing of that purported change is ill-defined (cf. Smellie et al., 2014).

2.3 Iceland

Iceland is almost entirely volcanic (Saemundsson, 1979; Thordarson and Höskuldsson, 2002). The volcanic rocks are generated in a plexus of rift zones that are the subaerial surface expression of the northern Mid Atlantic ridge, an oceanic spreading centre formed by the Cenozoic separation of Greenland from Europe. A major mantle plume or thermal anomaly is also present under Iceland and interacts with spreading processes and magma generation at the ridge. It currently resides beneath western Vatnajökull and may be responsible for the bifurcations and jumps seen in the several rift zones present. Rift jumps and capturing of older oceanic crust are probably responsible for Iceland being 1.5 times wider than would be expected from calculated ridge spreading rates (Martin et al., 2011). The oldest dated rocks are 16–15 Ma in the northwest and 15–12 Ma in the east. At present, active volcanism and extension are focused principally in six zones, the three most active of which are the Western, Eastern and Northern volcanic zones, and are linked by a region of more diffuse seismicity and volcanism (South Iceland Seismic Zone) that corresponds to a transform fault (Fig. 2.5). The three less active zones include the Snæfellsnes volcanic zone in the southwest, the Mid-Iceland volcanic zone (or Mid-Iceland Belt), and the Öraefajökull volcanic zone in the east (Hardarson et al., 2008; Thordarson and Höskuldsson, 2008). The volcanism is tholeiitic mid-oceanic ridge basalt (MORB) in character with off-axis mildly alkaline magmas erupted in the Snæfellsnes and Öraefi volcanic belts and at the southern end of the South Iceland Seismic Zone.

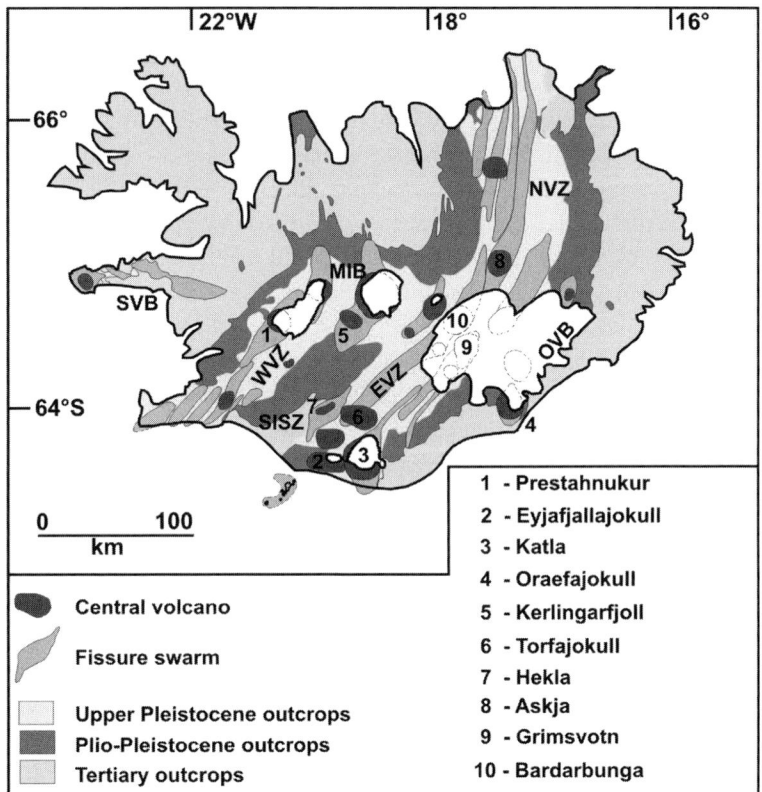

Fig. 2.5 Map of Iceland showing the volcanic zones and selected large volcanic centres (modified after Thordarson and Höskuldsson, 2008). Abbreviations: WVZ – Western Volcanic Zone; EVZ – Eastern Volcanic Zone; NVZ – Northern Volcanic Zone; MIB – Mid-Iceland Belt; SISZ – South Iceland Seismic Zone; OVB – Öraefajökull Volcanic Belt; SVB – Snæfellsnes Volcanic Belt. (A black and white version of this figure will appear in some formats. For the colour version, please refer to the plate section.)

The magmatism is overwhelmingly basaltic but subordinate evolved rocks (andesites, dacites, rhyolites) are also present and have played an important role in yielding good-quality isotopic ages for the volcanism (McGarvie et al., 2006; Flude et al., 2008, 2010; McGarvie, 2009; Martin et al., 2011).

The rocks occur in volcanic systems with a common lifetime of about 1–1.5 m.y. (Saemundsson, 1979). They erupt in two main ways: from fissure swarms and in central volcanoes. The fissure swarms form elongate strips 5–20 km wide and 50–100 km long parallel to the axis of the host rift zone. They are the surface expression of elongated magma chambers at depths greater than 10 km (base of the crust; Furman et al., 1992; Thordarson and Höskuldsson, 2002). Eruptions are fed by dykes and may be high volume (tens of km^3; Óskarsson and Riishuus, 2013; e.g. Laki 1873). The central volcanoes form basaltic shields commonly 20–25 km in diameter. They represent a major centralised focus of volcanic activity sourced in evolving magma chambers that typically lie about 3 km below the surface, although possibly much deeper (Sturkell et al., 2006). The central volcanoes generally have a core of evolved rocks (Furman et al., 1992), and a few erupted some high-volume ignimbrites (e.g. Deildargil ignimbrite: 20–30 km^3; Thorsmörk ignimbrite: Saemundsson and Noll, 1974; Jørgensen, 1980).

The abundance and variety of mafic and felsic glaciovolcanic edifices and sequence types in Iceland is unmatched in any other volcanic province (cf. Schopka et al., 2006; McGarvie, 2009; Skilling, 2009; Smellie, 2013). With initiation of Northern Hemisphere glaciations extending back to between 10 and 5 Ma, at least, and large-scale glaciations prevalent from c. 2.75 Ma (Geirsdóttir et al., 2007), the potential for volcano–ice interactions in Iceland is high. Indeed, during the past 1100 years, eruptions from ice-covered volcanoes have dominated the eruptive activity in Iceland, comprising at least 120 historical eruptions that represent c. 60% of all recorded volcanic activity (Larsen, 2002). This figure is almost certainly a minimum since an unknown number presumably did not reach the overlying ice and thus left no exposed record. Because of the high eruption rates, Iceland also contains the most continuous and detailed terrestrial glacial record in the Northern Hemisphere (Geirsdóttir and Eiríksson, 1994; Helgason and Duncan, 2001). The oldest glacial sedimentary deposits and glacial surfaces may be 4.7–4.6 Ma in age but, across Iceland as a whole, the glacial deposits are younger than c. 3.8 Ma (Geirsdóttir and Eiríksson, 1994; Geirsdóttir et al., 2007). Glaciovolcanic strata are prominent in Pliocene–lower Pleistocene sections dated or inferred to be younger than c. 3.5 Ma (Helgason and Duncan, 2001; Smellie, 2008). However, a basaltic lapilli tuff in eastern Iceland dated at 10.72 Ma was inferred to be derived from an explosive subglacial eruption (Duncan and Helgason, 1998). The subglacial origin of the lapilli tuff deposit is ambiguous but, if it is confirmed, it would represent evidence for the oldest glaciation in Iceland. The oldest glaciovolcanic sequences are best seen south and southwest of Vatnajökull where, at eruption, they would have been relatively elevated because of their proximity to the Iceland plume close to the centre of the coeval ice sheet (Duncan and Helgason, 1998). Other sequences of similar age elsewhere in Iceland are characterised by subaerial lavas and contain far fewer glacial sedimentary interbeds (Geirsdóttir, 1991). Glaciovolcanic sequences in those regions are mostly restricted to local highs formed by central volcanoes (Saemundsson and Noll, 1974). By contrast, glaciovolcanic products are widespread and characteristic of the upper Pleistocene outcrops throughout Iceland, signifying the regional expansion of a pan-Icelandic ice sheet (e.g. van Bemmelen and Rutten, 1958; Jones, 1969b, 1970; Geirsdóttir et al., 2007). The geographical distribution of the Pleistocene tuyas and their summit elevations have been used to reconstruct the form and thickness of the late Pleistocene ice cover (Walker, 1965; Bourgeois et al., 1998; Licciardi et al., 2007).

The description of Icelandic outcrops does not lend itself to a geographical breakdown. Instead, they are divided here into polygenetic (large) and monogenetic (small) edifices.

2.3.1 Polygenetic glaciovolcanic edifices

The oldest and most areally extensive glaciovolcanic sequences are grouped within the Plio-Pleistocene (c. 3.3–0.7 Ma) Sida Formation within the Sida–Fljotshverfi district of south Iceland, between Myrdalsjökull and Vatnajökull (Fig. 2.5; see Section 10.6.2). They are very distinctive and, although initially interpreted as products of multiple subglacial basaltic fissure eruptions that were emplaced on a marine shelf (Bergh and Sigvaldason, 1991), they have been reinterpreted either as entirely subglacial or subglacial with a subaerial component (Smellie, 2008; Banik et al., 2014). The outcrops were named glaciovolcanic sheet-like sequences of Dalsheidi type after a well-exposed and well-described Icelandic locality whose description has historical precedence (Walker and Blake, 1966). Each eruptive unit consists of a regular pattern of vertically stacked lithofacies including, from base up, laterally continuous basal lava with very well developed colonnade and entablature prismatic jointing; massive and stratified lapilli tuff usually described as hyaloclastite; and a capping unit of tuff. The sequences are truncated by glacial surfaces that are locally overlain by diamict (likely till). The basal lava was envisaged emplaced as a sill at the ice–bedrock interface and they have been called interface sills (Smellie, 2008). They may provide verification for a theoretical mode of emplacement of mafic magma as sills at the base of ice sheets proposed by Wilson and Head (2002; see also Wilson et al., 2013). The overlying vitroclastic units were deposited mainly from hyperconcentrated density flows during major jökulhlaups, with the tuff capping beds laid down during the waning stages of each flood event. The calculated erupted volumes are large (typically 13 km^3, possibly up to 31 km^3; Bergh and Sigvaldason, 1991).

Eyjafjallajökull is a large polygenetic mafic volcano situated on the south coast of Iceland. It rises to 1651 m a.s.l., extends over 30 km east–west and 15 km north–south and has a small ice-filled summit caldera and an ice cap that extends down to c. 1000 m a.s.l. It has been active for over 0.78 m.y. Sheet-like sequences are very common on Eyjafjallajökull (Loughlin, 2002; see also Carsewell, 1983). They show numerous variations on the standard sequence represented by sequence holotypes defined in Antarctica. They are the proximal deposits of subglacial eruptions beneath thin ice and those examined were laid down c. 5–10 km from source. The individual sequences are typically less than 20 m thick and are separated by glacially modified surfaces locally overlain by glacial diamict. The lavas are commonly sheet-like and show a variety of features indicative of water cooling and interaction (e.g. blocky or entablature jointing, glassy and lobate or pillowed margins, intimate association with cogenetic hyaloclastite breccia). The associated clastic deposits are dominated by sideromelane mainly generated explosively in phreatomagmatic eruptions and deposited by concentrated and hyperconcentrated flows and traction currents. Differences between sequence types comprise variations in the order of stacking of the constituent lithofacies and variations in lava type (pillowed, sheet, etc.). Subaerial lavas are present but are not described and details of their relationships to the subaqueous lithofacies are scant.

Öraefajökull is Iceland's largest active volcano, and is constructed from the products of mafic and felsic eruptions (e.g. Thorarinsson, 1958; Prestvik, 1985). It is situated on the southeast coast of Iceland and contains the highest peak in Iceland (Hvannadalshnúkur; 2110 m a.s.l.) and a summit ice cap with a 5 km-wide caldera. Subglacially emplaced rhyolite crops out at Godafjall and Hrútsfjall, southwestern Öraefajökull, and may be linked to a ridge-confined rhyolite lava forming the crest of nearby Hrútsfjall (Walker, 2011). Rhyolite lava locally overlies fine rhyolitic ash erupted phreatomagmatically in an ice cavity but it directly overlies bedrock elsewhere. In places large dome-like lobes crop out downslope of the main ridge-forming rhyolite mass and were probably emplaced in cavities melted between the mafic pre-volcanic bedrock and valley-occupying ice. The rhyolite lava is characterised, in particular, by spectacular columnar jointing and curved cooling master joint surfaces that give several outcrops an 'onion-like' appearance (Fig. 9.5). However, rhyolite outcrops are subordinate to basaltic (Prestvik, 1985). Although much less well studied, the latter include sheet-like sequences as well as evidence for glacial and inter-glacial periods (Helgason and Duncan, 2001; Stevenson et al., 2006; Walker, 2011; unpublished information of JLS).

2.3.2 Monogenetic glaciovolcanic edifices

Glaciovolcanism in Iceland has been recognised or inferred since the early twentieth century (e.g. Peacock, 1926), and today a greater number of detailed modern studies of glaciovolcanic centres have been published for Iceland than anywhere else (e.g. Werner et al., 1996; Werner and Schmincke, 1999; Tuffen et al., 2001, 2002a, b, 2008; McGarvie et al., 2006; Schopka et al., 2006; Stevenson et al., 2006, 2009, 2011; Bennett et al., 2009; Skilling, 2009; Owen et al., 2012; Pollock et al., 2014). Many are described in detail in this volume (Chapters 10–12). Basaltic tuyas and tindars are the commonest glaciovolcanic sequence types in Iceland. Prominent well-known examples include Kalfstindar, Hlödufell, Helgafell and Herdubreid (Jones, 1969b; Werner and Schmincke, 1999; Schopka et al., 2006; Skilling, 2009) but there are many others (e.g. van Bemmelen and Rutten, 1955). The tindars represent products of explosive tephra-forming eruptions from subglacial fissures whereas the tuyas are either individual monogenetic edifices or else were constructed at locations of focused magma supply on the tindar ridges. The 'standard model' for mafic tuya eruptions described by Jones (1969b, 1970) was based on outcrops in the Laugarvatn area (Section 1.5).

The most extensive study of a pillow-dominated basal sequence is by Pollock et al. (2014), who examined well-exposed outcrops at Undirhlíðar quarry on the Reykjanes Peninsula. They showed that even within relatively thin sequences (<100 m), multiple, distinct pillow lava emplacement units could be identified. Their stratigraphical analysis also showed that these sequences can have interlayered tuff breccia, lapilli tuff and tuffaceous units, consistent with a more complex eruptive environment than continuous passive effusion of pillow lava. Their detailed geochemical analysis also confirmed that it is possible for multiple geochemically distinct magma batches to be present in what otherwise appears to be a 'simple' (i.e. monogenetic) eruptive sequence. Helgafell is a well-preserved mafic tindar 2 km long and 300 m high situated in southwest Iceland immediately north of Undirhlíðar (Fig. 10.6). It lacks a well-developed basal pillow mound. It is formed largely or entirely of lapilli tuffs, suggesting that eruption was essentially explosive from the outset (Schopka et al., 2006; see Section 10.4.3). Such an early onset of explosivity was attributed to lower-than-glaciostatic vault pressures caused by basal meltwater drainage and sagging of the ice roof. However, most Icelandic tindars and tuyas are inferred to commence with a basal pillow mound/ridge (Jones, 1969b, 1970; Werner and Schmincke, 1999; Skilling, 2009). Basal pillow lava is prominent on the flanks and in the core of the well-exposed mafic Hlödufell volcano, southwest Iceland (Skilling, 2009; see Section 10.4.2). The pillow pile comprises mounds and ridges that were erupted from several short fissures. They locally preserve evidence for cooling by direct contact against enclosing ice and contain numerous voids (called *ice-block meltout cavities*) formed when lava engulfed blocks of ice. The presence of the cavities distinguishes glaciovolcanic pillow volcanoes/mounds from those erupted under the sea or in pluvial lakes. The pillow lavas are overlain by explosively generated eruption-fed lapilli tuffs deposited by subaqueous sediment gravity flows in an ice-bound meltwater lake (tindar stage). Traction current deposits are locally present and are attributed to currents related to coeval basal meltwater drainage (cf. Smellie, 2001). The eruptions culminated in a pāhoehoe lava-fed delta followed by a rise in the lake water level and construction of a second, higher pāhoehoe lava-fed delta. Both phases of lava-fed delta progradation are associated with prominent subhorizontal surfaces that are characteristic of the summit region (Fig. 8.4b).

Evolved glaciovolcanic sequences are commoner in Iceland than anywhere else and have been studied in considerable detail (McGarvie, 2009). They are distinctively different from their mafic counterparts and include andesite, dacite and (mainly) rhyolite compositions. Two main rhyolitic types are present: (1) lava-dominated and (2) tephra-dominated tuyas (*sensu* Smellie, 2007, 2013). The only Icelandic glaciovolcanic edifice that may be a lava-dominated felsic tuya is that at Prestahnúkur (McGarvie et al., 2007; see Section 12.3). Prestahnúkur is just 2 km in diameter, rising to 570 m. It comprises a pedestal of perlitised non-explosively erupted glassy rhyolite breccias and lava lobes ('lobe-hyaloclastite') overlain and dominated by thick lava sheets. The lavas are extensively columnar jointed and form sheet-like bodies rather than steep-sided domes or coulées. There is no evidence for subaerial conditions, the edifice lacks a flat top and effusion was entirely subglacial, comparable with subglacial felsic domes.

Tephra-dominated felsic tuyas are commoner and better known. Examples have been described in detail at Raudufossafjöll and Kerlingarfjöll (Tuffen et al., 2002a; Stevenson et al., 2011). That at Raudufossafjöll is a flat-topped rhyolite ridge 1.5 km long in the Torfajökull volcanic centre that rises c. 450 m above the surrounding landscape (1200 m a.s.l.; see Section 12.4.1). Its width may have been much wider but has been reduced by post-eruptive sector collapse(s) that substantially reduced the width of the summit lavas. The subaerial capping lavas are well exposed and have shapes and joint orientations consistent with ice-impounding. By contrast, the underlying tephra pile, which may comprise about half of the edifice volume, is very poorly exposed but consists of rhyolite ash with characteristics of phreatomagmatic eruptions. Its fineness despite being present in a vent-proximal position may be due to ice-confinement. Polymict volcaniclastic sediments occupy a small channel cut into one of the edifice flanks and may have been deposited from meltwater draining subglacially during the eruptive period. By comparison, the tephra pedestal is much better exposed at Kerlingarfjöll, where a series of rhyolite tuyas and tindars are present (Stevenson et al., 2011; see Section 12.4.2). Initial subglacial eruptions in these edifices were violent and phreatomagmatic but the influence of water rapidly declined and the later tephra was magmatically fragmented, illustrating ice confinement but in a 'dry' ice cavity. Most of the tephra was deposited from pyroclastic density currents. Multiple irregularly shaped intrusions with peperitic and hyaloclastic margins were also emplaced in the tephra pile and activity in the tuyas culminated in the effusion of comparatively thin (relative to the thickness of tephra) ice-confined subaerial rhyolite lava caprocks.

Some small-volume felsic glaciovolcanic edifices defy a simple classification as particular sequence types (Smellie, 2013). Bláhnúkur (Torfajökull) is a small pyramidal edifice with an erupted volume of less than 0.1 km^3 (dense rock equivalent; Tuffen et al., 2001; see Section 12.7). It drapes a previously formed volcanic ridge as a veneer 50 m thick and consists of numerous irregular to conical rhyolite lava lobes dispersed in poorly sorted glassy breccias, perlitised in part. The outcrop was explained as the product of numerous lava lobes sourced from a subsurface plexus of anastomosing conduits and emplaced on the ridge slopes in cavities melted in the overlying ice. That was followed by collapse of the lobes while still hot to create the glassy breccias followed by further lava emplacement as lobes and in subglacial tunnels trending down the bedrock slopes. Broadly similar lithofacies also occur in a rhyolite centre at Dalakvísl (Torfajökull; see Section 12.5) but are accompanied by unusual obsidian sheets and pumiceous breccia representing intrusive vent-proximal pyroclastic deposits, with syn-emplacement collapse of the more pumiceous domains to form the obsidian sheets (Tuffen et al., 2008). A conceptual model was presented in which the space available in the ice cavity determined whether the eruption was explosive or intrusive.

Eruptions of felsic and intermediate-composition magma at Vatnafjall, Öraefajökull, resulted in the emplacement of rhyolite and trachydacite lava flows (see Section 12.2.2). Although subaerial, they were laterally buttressed by valley-filling ice and became ridge-confined (Stevenson et al., 2006; cf. Lescinsky and Sisson, 1998). Meltwater drained down the flanks of the adjacent valley walls and some trachydacite magma was emplaced down tunnels eroded by that meltwater. Spectacular ridge-confined rhyolite lavas are also exposed nearby to the northwest, at Rotarfjall.

The only described products of subglacial eruptions in Iceland involving intermediate magma compositions (andesite and dacite) occur at Kerlingarfjöll (Stevenson et al., 2009). In their lithofacies and architecture, they more closely resemble products of mafic glaciovolcanism (tindars) rather than felsic (see Section 11.5). The sequences are overlain by rhyolitic tuyas. A wide variety of andesite lithofacies were erupted in two separate edifices.

They include products of subaqueous fire-fountaining, subaqueous effusion and collapses of unstable pillow lava pile(s). Conversely, volumetrically dominant lapilli tuffs probably represent products of vigorous magma–water interaction, with clasts formed by rapid quench-granulation and/or steam explosivity. Dacite lithofacies mostly consist of a pillow lava pile with intrusions, but they are draped by crudely bedded lapilli tuff probably formed similarly to the andesite equivalents. Overall, the intermediate-composition sequences resemble pillow volcano/mound lithofacies overlain by Surtseyan lapilli tuffs erupted in a water-filled englacial vault, with the upward transition to vitroclastic lithofacies interpreted as a response to decreasing hydrostatic pressures. This interpretation was unexpected since other intermediate-composition glaciovolcanic eruptions produce flow-dominated tuyas and subglacial domes lacking evidence for ponded water consistent with continuous basal meltwater drainage during eruption and the thermodynamic limitations of evolved magmas compared with (hotter) mafic magmas (e.g. Kelman et al., 2002). At Kerlingarfjöll, the combination of poor meltwater drainage caused by a low pre-eruption relief together with a poor hydraulic connection with the glacier snout may have enhanced meltwater accumulation in the vault (Stevenson et al., 2009).

Modern glaciovolcanic eruptions in Iceland have occurred mainly at four volcanic systems: Bárdarbunga, Grímsvötn, Katla and Eyjafjallajökull. These eruptions are discussed in more detail elsewhere (Chapter 3), but we note here that these active systems are major sources of observational information on the eruptive and depositional processes involved in glaciovolcanic centres and they are as important as outcrop studies for correctly identifying and interpreting ancient deposits, and vice versa.

2.4 North America

Together with Iceland, studies of glaciovolcanism in North America are some of the oldest and most influential (e.g. Mathews, 1947, 1951). All occurrences of Plio-Pleistocene glaciovolcanic deposits occur in the western part of North America, with the most extensive published work from Canada (e.g. Allen et al., 1982; Jackson et al., 1996; Souther, 1992; Hickson et al., 1995; Moore et al., 1995; Russell and Hauksdóttir, 2001; Bye et al., 2000; Hickson, 2000; Edwards et al., 2002, 2006, 2009b, 2011; Kelman et al., 2002; Harder and Russell, 2007; Russell et al., 2013; Ryane et al., 2013; Hungerford et al., 2014; Fig. 2.6). These studies have focused mainly on British Columbia, which hosts the bulk of Plio-Pleistocene volcanism in Canada, although a few studies have also been published on deposits in the Yukon Territory (e.g. Jackson, 1989; Jackson et al., 2012). Western North America has a range of tuya morphologies and compositions to rival Antarctica as it covers two different volcanic arcs (Cascades and Aleutians) and a broad area of extension-generated magmatism (Northern Cordilleran Volcanic Province (NCVP): Wells Gray Volcanic Field). Thus, rock compositions range from basanite to peralkaline rhyolite within alkaline suites, and tholeiitic basalt to dacite within the calc-alkaline series. Studies of the transtension-related NCVP glaciovolcanism have focused predominantly on basaltic edifices with the exceptions of Hoodoo Mountain

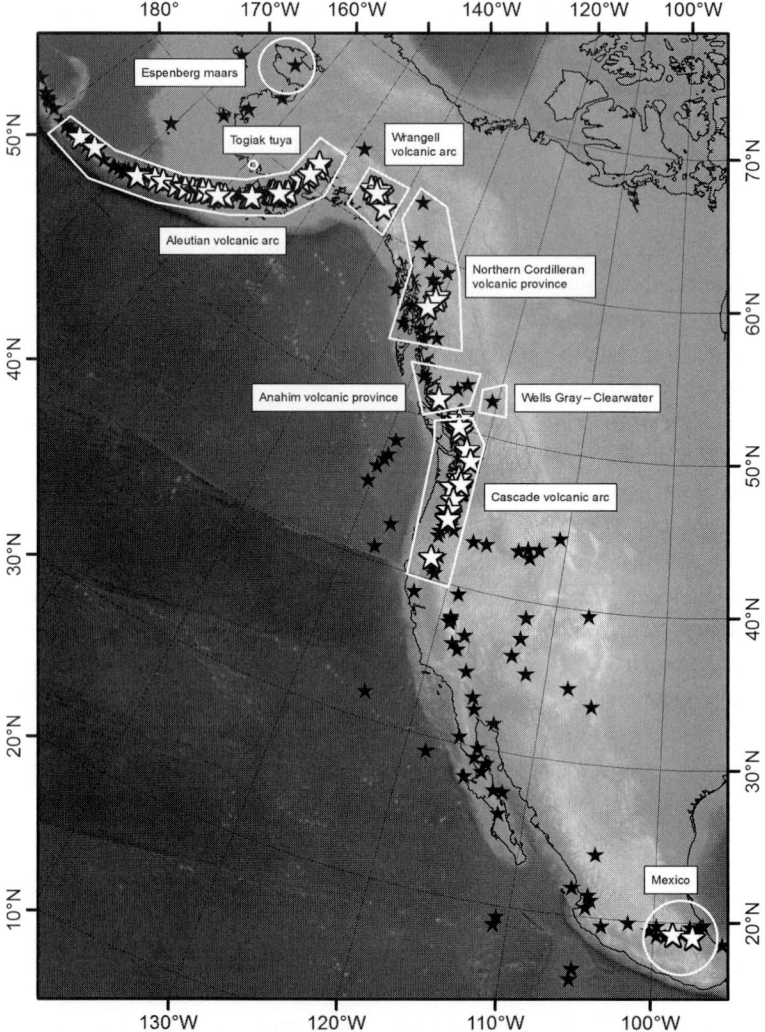

Fig. 2.6 Map of western North America showing the major volcanic provinces, volcanoes classified as active (black stars), and volcanoes that are presently ice clad (white stars).

(Edwards et al., 2000, 2002; Edwards and Russell, 2002) and Mt Edziza (Souther, 1992), both of which have peralkaline trachyte and phonolite. Mount Edziza also has peralkaline rhyolite that is thought to be glaciovolcanic (Souther, 1992). At least one example of basanitic glaciovolcanism has also been reported from the NCVP (Harder and Russell, 2007). Glaciovolcanism in the northern part of the Cascade arc has also been well documented (Mathews, 1951, 1952a, b, 1958; Bye et al., 2000; Kelman et al., 2002), and ranges in composition from subalkaline basalt to andesite and dacite. Much less detailed work has been carried out further south in the Cascades (Hammond, 1987; Lescinsky and Fink, 2000;

Bacon, 2008; Bacon and Lanphere, 2006), or on edifices in the Aleutian arc (e.g. Larsen et al., 2007). However, what little work has been undertaken demonstrates that they have significant potential for further glaciovolcanic studies (e.g. Bacon, 2008).

Volcanism in western North America has been ongoing through the Cenozoic, driven largely by subduction of the Farallon and Pacific plates beneath the North American plate. Remnants of this process still drive volcanism in the Aleutian and Cascade arcs, although that in the region between the two arcs (>3000 km long) is driven mainly by extensional processes (Souther and Yorath, 1991; Edwards and Russell, 1999; Thorkelson et al., 2011). Alpine glaciations began in western North America during the Plio-Pleistocene, although the details of the early glaciations are poorly known (Ruddiman and Wright, 1987). However, it has long been recognised based on ice-rafted debris that local ice was present in the Northern Hemisphere by 6 Ma (e.g. Jansen and Skoholm, 1991). By 2.8 Ma, extensive ice sheets were developing in Iceland (Geirsdóttir et al., 2007) and probably in western North America as well (Froese et al., 2000; unpublished data of BRE). During the maximum extent of ice cover, virtually all of the volcanism in western North America as far south as Mt Baker in the Cascades was probably affected by regional ice sheets. The coincidence of volcanism and glaciation from Alaska to northern California during the Quaternary led to the production of a wide array of glaciovolcanic landforms and deposits. Isolated glaciovolcanic occurrences are present as far north as the Arctic Circle in Alaska (Espenberg maars). They extend throughout the west–east length of the onland portion of the Aleutian arc (including some of the Aleutian islands) and the Wrangell arc, through the 2000 km-long Northern Cordilleran Province (Yukon to central British Columbia) and into the Cascade arc (which extends to northern California).

2.4.1 Espenberg maars

The Espenberg maars are a group of four maar-type volcanoes that are thought to have formed by interaction of magma with permafrost on the Seward Peninsula of Alaska (Begét et al., 1996); they are some of the only known examples of 'permafrost' glaciovolcanism. The volcanoes are all basaltic in composition, and represent the most northerly glaciovolcanic deposits in North America (c. 66.5° N). Estimated ages for the maars range from c. 17 ka to >200 ka. The four maars are, in order of increasing crater diameter: North Killeak (4 km), Whitefish (4.3 km), South Killeak (5 km) and Devil Mountain (8 by 6 km). The craters are up to 250 m deep, and were locally excavated into underlying basaltic lava flows. The maar cones comprise mainly lapilli tuff and breccia, interpreted to have formed by air fall and dilute pyroclastic density currents (e.g. surges; Begét et al., 1996).

2.4.2 Togiak tuya

Togiak tuya (59° N/160° W) is an isolated edifice described as a basaltic tuya situated in the middle of Togiak Valley, just north of the northern shoreline of Bristol Bay, Alaska (Hoare and Coonrad, 1978). The tuya appears to be moderately dissected, but still retains its lava cap, underlying palagonitised tuff breccia and minor pillow lava. The top of the tuya is 353 m above the floor of a much larger, glaciated valley. Its age is thought to be less than 750 ka, which is the radiometric age of flat-lying basalts upon which the tuya was built. Hoare and Coonrad (1978) inferred that the tuya pre-dated the Last Glacial Maximum as it has been glaciated. They mapped pillow basalt and pillow breccia along the western side of the tuya, immediately underlying the gently dipping subaerial lava cap, and a steeply dipping (30°) sequence of palagonitised tuff/tuff-breccia forming its northern end. They also identified two different vents (north and south), and inferred a minimum ice thickness during eruption of 300–500 m.

2.4.3 Aleutian volcanic arc

The Aleutian arc is one of the longest and, in many respects (e.g. geochemically), one of the best studied volcanic arcs on Earth (Fournelle et al., 1994). It stretches more than 2500 km east–west, with 37 mainland edifices (Miller and Richter, 1994) and a long island archipelago (see http://www.avo.alaska.edu for extensive references; Fournelle et al., 1994; Siebert et al., 2010). However, given its northerly latitude (entirely north of 51° N), and the large number of volcanoes (more than 70, 41 of which have been historically active), it is also one of the least well-studied glaciovolcanic provinces because of the remote locations and present-day glacial cover. Many of the onland volcanoes in the easternmost one-third of the arc are presently glacierised, and reconstructions of Quaternary glaciations for Alaska suggest that at the Last Glacial Maximum all of the Aleutian volcanoes east of 170° W longitude were ice-covered (Kaufman and Manley, 2004). However, much of the history of volcano–ice interactions over the past 4 Ma in the arc awaits detailed investigations. Because the western part of the arc comprises islands that are at least partly accessible by boat, they dominate the published literature. However, the islands also have much lower elevations, and so while all receive snow in the winter, most do not host glaciers. From Veniaminof to the west, the arc is dominated by mafic products (basalt to basaltic andesite), while to the east the large stratocones are more silicic (andesite to dacite; Fournelle et al., 1994). Because many of the eastern volcanoes are glacierised, and at least one eruption occurs per year (Brantley, 1999), the likelihood of new glaciovolcanic activity is always high.

Individual Aleutian volcanoes range in volume up to 300–400 km^3 (Miller and Richter, 1994) and include calderas more than 10 km in diameter. Recent eruptions from remote Pavlof, Westdahl and Veniaminof have produced contrasting styles of volcano–ice/snow interactions that are discussed in more detail in Chapters 3 and 15. Mount Redoubt and Mt Spurr volcanoes are of greater prominence due to their proximity to Anchorage. Both of these composite volcanoes are glacierised, and both have had numerous eruptions in the twentieth century. The eruptions are well studied (see Chapter 3 and 15 for more information on Mt Redoubt), and both tend to produce andesitic to dacitic lava domes in their summit craters, which can collapse onto surrounding glaciers and generate large floods (e.g. Waythomas et al., 2013).

Although studies of pre-historic glaciovolcanic eruptions are sparse in the Aleutians, two studies have documented possible involvement of ice during the evolution of at least one island (Umnak Island; Byers, 1959; Larsen et al., 2007). The Late Pleistocene caldera formed on top of older Pleistocene deposits that are variably palagonitised, and exposures of pillow lavas within the caldera walls show that parts of the older edifice may have also had glaciovolcanic eruptions (Larsen et al., 2007). Byers (1959) produced the first detailed geological description of Umnak Island and its three volcanoes (Okmok, Recheschnoi, Vsevidof). He surmised that all of these volcanoes had Pleistocene eruption histories, and he recognised that several units on Okmok were variably palagonitised. Byers (1959) proposed that some of the palagonitisation in tuff breccia units might have resulted from melting of snow and ice that intermixed with pyroclastic density currents during emplacement of the units. Larsen et al. (2007) give a summary of the two large caldera-forming eruptions from Okmok volcano. They interpreted the interbedding of basaltic andesite to andesite surge and lahar deposits as indicating the presence of snow/ice cover during the first and largest of two caldera-forming eruptions (designated as Okmok I; 30 km^3 DRE (dense rock equivalent)), and suggested that units of chaotic breccias could be analogous to ice- and snow-rich debris flow deposits described by Waitt et al. (1994) from the 1989 eruption of ice-clad Mt Redoubt volcano.

2.4.4 Wrangell volcanic arc

The Wrangell Volcanic Field is located 400 km to the east of the eastern Aleutian arc, and appears to be petrogenetically distinct (Miller and Richter, 1994). It comprises at least seven principal volcanic centres (Wrangell, Drum, Sanford, Capital, Tanada, Jervis, Gordon) with elevations ranging from c. 2300 m a.s.l. to almost 5000 m a.s.l. and spanning a compositional range from basalt to rhyolite. The oldest deposits extend back to the Miocene, but activity at most of the edifices continued into the Quaternary. Given their northerly latitudes (<62° N), they have probably been continuously glacierised for much of their history. At least local glaciovolcanic deposits have been recognised within the volcano stratigraphies (Miller and Richter, 1994). Given their elevation and location, it is possible that these volcanoes locally hosted pre-Pleistocene glaciers as well (Hamilton, 1994). Their remote location, extensive glacial cover and rugged terrain have impeded extensive detailed studies.

2.4.5 Northern Cordilleran Volcanic Province

The largest province is the Northern Cordilleran Volcanic Province (NCVP), which stretches from central British Columbia/coastal Alaska through the western Yukon Territory and into eastern Alaska (Edwards and Russell, 1999; 2000). The NCVP stretches for more than 2000 km north to south, and includes one of the largest areas of glaciovolcanism outside of Iceland and the Antarctic. The province results from transtensional forces between the Pacific and North American tectonic plates, and petrologically is very similar to the West Antarctic Rift System (Section 2.2): both are dominated not only by alkaline mafic magmas but they also host peralkaline intermediate to felsic volcanism. The NCVP includes a number of separate volcanic fields and volcanic complexes, many of which have not been studied in detail. However, two of the large examples, the Mt Edziza and Hoodoo Mountain volcanic complexes, have been the targets of detailed work (Souther, 1992; Edwards et al., 2002, 2009b; Edwards and Russell, 2002; Hungerford et al., 2014). Several other smaller volcanic fields have been investigated in varying detail, but the Tuya–Kawdy area in particular has the longest history of research on volcano–ice interactions of any place outside of Iceland (Mathews, 1947; Allen et al., 1982; Moore et al., 1995; Edwards et al., 2011; Ryane et al., 2013; Russell et al., 2013).

Fort Selkirk

In the Yukon segment of the NCVP, glaciovolcanism has been most thoroughly documented at the Fort Selkirk volcanic complex (Jackson, 1989; Jackson et al., 1996, 2012), although isolated occurrences downstream and to the north along the Yukon River may also be glaciovolcanic (unpublished data of BRE). The Fort Selkirk volcanic complex (Francis and Ludden, 1990) is a long-lived complex of mafic alkaline volcanism ranging in composition from olivine nephelinite to alkali olivine basalt. It is located at the junction of the Pelly and Yukon rivers in central Yukon Territory, Canada. The eruption history of the complex began at c. 4 Ma, and Jackson et al. (2012) suggested that the eruptions dammed the Yukon River numerous times, producing subaqueous tuffs and pillow lava. However, an eruption at the Ne Ch'e Ddhawa centre at c. 2.14 Ma produced a volcanic pile that sits 300 m above the floodplain of the river and has been interpreted as subglacial. The deposits include pillow breccia, palagonitised tuff and pillow lava. Breccias near the summit also contain 'exotic' pebbles and glaciogenic diamicton. This is one of only two known glaciovolcanic centres built of olivine nephelinite and basanite (see also Llangorse, below).

Atlin

The Atlin volcanic district (Edwards et al., 2003) comprises multiple mafic volcanic fields that are Plio-Pleistocene in age in northwestern British Columbia. The most recent volcanism occurs within the Surprise Lake Volcanic Field, where three different tephra cones were erupted in large U-shaped glacial valleys. The state of preservation of the cones suggests that they are either pre-Last Glacial Maximum or postglacial in age (Edwards and Bye, 2003). However, three separate deposits in the Llangorse Volcanic Field have been interpreted as representing basanitic glaciovolcanic deposits (Harder and Russell, 2007; see Section 10.2). The largest deposit is an over-thickened (100 m) basanite lava flow at Llangorse Mountain that was emplaced on top of a polymict diamictite. The diamictite is interpreted as having formed as a syn-eruption mass flow, and it contains rounded bedrock boulders as well as peridotite fragments similar in composition to those found in the overlying lava. The lava is postulated to have overridden and heated the diamictite, thus facilitating rapid palagonitisation of vitric clasts in the underlying deposit. The dimensions and jointing of the basanite lava are consistent with ponding adjacent to a wall of ice. The presence of irregular, fine-scale (10–30 cm diameter) cooling joints in lava masses at Lone Point and Hidden Point are also indicative of ice-contact.

Tuya–Kawdy

The Tuya–Kawdy volcanic fields are located in north-central British Columbia and comprise more than 55 separate volcanic outcrops, most of which are thought to have formed beneath various stages of the Cordilleran Ice Sheet (Mathews, 1947; Allen et al., 1982; Moore et al., 1995; Dixon et al., 2002; Edwards et al., 2011). The deposits are formally considered part of the Tuya Formation, which is a stratigraphical unit mainly confined to the region depicted on the Dease Lake and Jennings River 1:250 000 map sheets (Gabrielse, 1969, 1998). The area no longer hosts any glaciers, but many of the tuyas hosted small cirque glaciers until very recently. Watson and Mathews (1944) gave the first accounts and suggested the designation 'Tuya Formation' for all volcanic deposits thought to be of Pleistocene age based on stratigraphy and degree of glacial modification. The edifices of the Tuya Formation are scattered over a broad area, with relatively concentrated clusters around Tuya Lake (six individual centres) and on the Kawdy Plateau (six individual centres). Those in the Tuya Lake area have been studied in the most detail, although work on the Kawdy Plateau tuyas is presently (2007–15) ongoing (Russell et al., 2013; Ryane et al., 2013). Mathews (1947) drew on his impressions of several of these edifices (Tuya Butte, Ash Mountain, 'Kawdy' Mountain) when he formulated one of the first coherent lines of reasoning for the formation of volcanoes during eruptions beneath ice sheets. Several of the volcanoes with 'flat tops or benches' were interpreted as having erupted beneath ancient ice sheets that no longer exist. While his stratigraphical and lithofacies descriptions were brief, he noted pillow lava, pillow lava breccia and foreset bedding in the fragmental material underlying some of the flat lava caps. He also noted that the contact between the overlying lavas and the underlying fragmental material comprised broad, 'nearly horizontal treads' and shorter 'risers' that were parallel to the underlying, inclined bedding. This is the first published description of what is now termed a 'passage zone' (e.g. Smellie, 2006), and Mathews (1947) noted that it implied eruption into a lake with a relatively constant lake level. He also realised that the subaerial tops to tuyas in the region were at different elevations. He also connected his studies to those of workers in Iceland who had reported the 1934 eruption of Grímsvötn and described older glaciovolcanic deposits (e.g. Nielsen, 1937; Noe-Nygaard, 1940), and to work done in the Columbia River Plateau describing the formation of basaltic lava-fed deltas where basaltic lava flowed into palaeo-rivers or lakes (Fuller, 1931, 1934). This ground-breaking work established a terminology for use of the word 'tuya'. Tuya is a local geographical word used in northern British Columbia for specific mountains (Tuya Butte), nearby lakes (Tuya Lake, High Tuya Lake), and the region in general (Tuya Region). Mathews (1947) applied it to describe volcanoes with unusual geometries resulting from eruptions beneath ice sheets. While studies of glaciovolcanism were sparse from the 1940s to the 1970s, isolated workers from around the world cited Mathews' (1947) work for making interpretations in Iceland (Jones, 1969b, 1970), Indonesia (Blake and Löffler, 1971), and Africa (Downie, 1964). All of the tuyas in the Tuya–Kawdy area are basanites and alkali olivine basalts.

Since Mathews' (1947) study, six of the tuyas in the Tuya Lake area have been examined in varying levels of detail, including lithofacies, architecture and geochemistry: Tuya Butte, Ash Mountain, Caribou Tuya, South Tuya, Mathews Tuya and Tanzilla Butte (Allen et al., 1982; Moore et al., 1995; Simpson, 1996; Dixon et al., 2002; Edwards et al., 2011; and unpublished data of BRE). Tuya Butte was first described in detail by Allen et al. (1982), who conducted the first direct comparison between tuyas in British Columbia and Iceland. They suggested that, based on the dips of lava delta bedding and a general thinning of the capping lava flows from northeast to southwest, the eruption that formed the 390 m-high Tuya Butte had been from a dyke-fed fissure, partly exposed at the northern end of the edifice. The lava-fed delta prograded to the south and west and was capped by subaerial lavas 5–40 m thick. They did not observe any pillow lavas or 'massive tuff' below the deltaic 'palagonitised hyaloclastite agglomerate' at Tuya Butte, although Moore et al. (1995) observed pillow lava on the northeastern flank. Allen et al. (1982) also noted 'filled lava tubes' interpreted as locations where lava flows entered into a surrounding englacial lake. The ash-size fractions at Tuya Butte contain both vesiculated and dense fragments, interpreted by Allen et al. (1982) as evidence for water quenching in addition to magmatic fragmentation. They also identified calcic zeolites, smectite and calcite in areas where sideromelane was more highly palagonitised. Moore et al. (1995) re-examined the Tuya Butte stratigraphy and estimated its volume to be 2.6 km^3. They also noted tube-like masses of jointed lava near the base of the lava cap that they interpreted as feeder-tubes for lavas entering water at the palaeo-lake shoreline. Locally they traced a down-dip progression from subaerial lavas to tube-like lava masses, foreset-bedded pillow breccia and fine-grained hyaloclastite. The primary layering was found to be locally interrupted by steep planar surfaces interpreted as slump features. Moore et al. (1995) also reported numerous whole rock and glass chemical analyses from Tuya Butte samples, and used analyses of sulphur (S; c. 0.1 wt %) from pillow lava samples to estimate that the water depth for the lake into which the pillow erupted was c. 300 m. Edwards et al. (2011) showed that on a regional scale the passage zone at Tuya Butte is at the lowest elevation compared with other Tuya–Kawdy centres, suggesting that it either erupted through a thinner ice sheet, or formed in a shallower englacial lake. The age of Tuya Butte is 140 ka (unpublished information of BRE). Granitic boulders presumed to be glacial erratics have been found on top of the lava cap, implying that the edifice was covered by ice after 140 ka; Moore et al. (1995) speculated that a large amount of the eastern side of the edifice may have been removed by subsequent glacial erosion.

Mathews Tuya (referred to originally as 'Mathews #2 tuya' by Mathews, 1947; also Allen et al., 1982) has also been investigated (Moore et al., 1995; Edwards et al., 2011). The petrology and lithostratigraphy of this much smaller edifice (1.5 km by 3 km) are described in detail by Simpson (1996) and Edwards et al. (2011). Five lithofacies were mapped. The most unusual features are massive sloping lava bodies with pervasive, radiating cooling joints on the northwestern flank, and the unique position of the edifice within a perched U-shaped valley adjacent to a much larger valley. The passage zone is only 400 m above the floor of one valley (north of the tuya), whereas it is over 700 m above the floor of another valley (to the south). Edwards et al. (2011) also reported an eruption age of 730 ka. The tuya is the only evidence for relatively 'thick' ice at that time in the northern Cordillera. It corresponds in timing to eruption of tuyas in central Russia, which were also erupted within relatively thick ice (Komatsu et al., 2007a). Mathews Tuya also has boulders interpreted as glacial erratics on its flanks and summit and has been significantly eroded on its northern side, which is consistent with having experienced several subsequent glaciations. The petrological variations within the magmas are consistent with closed system, low-pressure fractionation.

Ash Mountain and South Tuya (also called Southern Tuya) lack flat-lying, capping subaerial lava flows. Ash Mountain is one of the larger tuyas in the region, rising 700 m above the surrounding terrain and with an estimated volume of 3.2 km^3 (Mathews, 1947; Moore et al., 1995; Fig. 8.7). The basal part comprises a thick sequence (>100 m) of pillow lavas that form a prominent plateau (e.g. Russell et al., 2014), upon which a partially palagonitised tephra pile has been built (Allen et al., 1982; Moore et al., 1995). South Tuya is constructed similarly and has an estimated volume of 1.6 km^3. The summits of both Ash Mountain and South Tuya are cut by dykes, which are locally pillowed (Allen et al., 1982; Moore et al., 1995; unpublished data of BRE). Both edifices formed by Surtseyan eruptions that transformed with time to Strombolian. The presence of glacial erratics on the summits also indicates subsequent overriding by ice. Measured concentrations of S and H$_2$O (c. 0.1 wt % S and 0.5 wt % H$_2$O in basal pillows, 0.06–0.07 wt % S and 0.2 wt % H$_2$O in overlying tephra) suggest eruptions in englacial lakes up to 300 m deep. After the eruptions breached the lake surface, degassing triggered a slight shift in the composition of the erupted products (Moore et al., 1995).

Tanzilla Butte is a subglacially erupted edifice formed mostly of pillow lava and tuff breccia located at the southeastern end of the Tuya Volcanic Field. Analyses of H$_2$O, S and Cl contents in pillow rim glass imply that coeval ice thicknesses were up to 1 km (Dixon et al., 2002).

The Kawdy Plateau contains a cluster of six basaltic tuyas erupted onto older basaltic lavas (Mathews, 1947). The edifices have been heavily modified by numerous glaciations (Gabrielse, 1998; unpublished data of BRE), and only one has been studied in detail (Russell et al., 2013; Ryane et al., 2013). Kima'Kho Tuya (originally called 'Kawdy Mountain' by Mathews, 1947) comprises a large tephra cone at the south end and a large lava-fed delta. It is the subject of ongoing stratigraphical and petrological studies (Turnbull, Russell, Edwards, unpublished data, 2015). A putative pyroclastic passage zone has been tentatively identified within the upper 100 m of the cone (Russell et al., 2014). Locating the passage zone position is based on a dramatic upward increase in the frequency of beds of armoured lapilli and laminated deposits interpreted to have formed from subaerial dilute pyroclastic density currents (e.g. surges). The northern part of the edifice has been extensively glaciated and is well exposed. The tuyas on Kawdy Plateau record some of the oldest evidence for large-scale ice sheet development in North America, with several glaciovolcanic eruptions occurring between 2.8 Ma and 1.1 Ma (unpublished information of BRE).

Level Mountain and Mount Edziza

The two largest volcanic complexes in the NCVP are located south of the Tuya–Kawdy area: Level Mountain to the southwest and Edziza to the south. They both represent enormous erupted volumes within a regional context, each probably exceeding 600 km^3. Level Mountain studies are limited (Hamilton 1981) due to its remote location and large size (>750 km^3). Gabrielse (1998) included all of the complex's volcanic lithofacies as part of the Miocene–Recent Tuya Formation. Hamilton (1981) suggested that the complex had formed over the past 4.5 Ma, first building a large basaltic plateau (locally 1000 m thick), and subsequently erupting more evolved lava compositions (trachyte to comendite). The complex has been heavily glaciated, and reconnaissance work suggests that it may have glaciovolcanic deposits locally (unpublished information of BRE).

The Edziza volcanic complex is a large plateau (20 × 60 km) with an estimated volume of c. 600 km^3 situated to the southeast of Level Mountain (Souther, 1992). It has a well-documented history of volcano–ice interactions (Souther et al., 1984; Souther and Hickson, 1984; Souther, 1992; Lloyd, 2007; Edwards et al., 2009b; LaMoreaux, 2008; Hungerford et al., 2014). The complex has probably been active since the Late Miocene (c. 10 Ma), and has had numerous Holocene eruptions. It still has a significant ice cap, and so future glaciovolcanic eruptions are possible. Eruptions were largely bimodal, comprising c. 60 vol % mafic lavas (mainly alkali olivine basalt and hawaiite) and 40 vol % more evolved lavas (trachyte and subordinate comendite and pantellerite). Souther (1992) defined thirteen stratigraphical formations (Raspberry, Little Iskut, Armadillo, Nido, Spectrum, Pyramid, Ice Peak, Pillow Ridge, Edziza, Klastline, Arctic Lake, Kakiddi, Big Raven), seven of which show some evidence of an association with ice (but not all are glaciovolcanic). Most of the description and interpretations presented below are based on his work. Recent studies have focused on the evolution of selected glaciovolcanic features: a large glaciovolcanic ridge (Pillow Ridge; Edwards et al., 2009b) and a smaller, subglacially emplaced cone and lava complex (Tennena Cone; Hungerford et al., 2014). Souther (1992) recognised five broad-scale eruptive cycles, each initiated by voluminous basaltic lava to form broad shield volcanoes, followed by more localised eruptions of trachyte and peralkaline rhyolite. The complex also contains glaciogenic sediments at several different stratigraphical levels that have been characterised locally (Spooner et al., 1995; Endress, 2007), and one deposit may record a glaciation at c. 1.1 Ma.

Starting in the middle of the second magmatic cycle at Edziza, Souther (1992) suggested that the eruption environment began to shift to one periodically ice dominated. He interpreted deposits of pillow lavas and tuff breccia within the Pyramid Formation (c. 1.1 Ma; c. 11.4 km^3) as glaciovolcanic. The Pyramid Formation, which erupted from vents at the northeastern edge of the complex, comprises an initial small field of basaltic subaerial lava flows and later silicic domes and minor lava flows that constructed Pyramid Dome. The younger, rhyolitic Sphinx and Pharaoh domes were erupted beneath ice, with the latter eventually emerging above the ice surface. Most of Souther's interpretations are based on the presence of interbedded glaciogenic sediments, as opposed to analysis of the volcanic facies. However, some of the interbedded sediments have clasts considered to be non-local and related to a regional ice sheet. The Pyramid Formation underwent extensive erosion by regional ice prior to emplacement of the overlying Ice Peak Formation.

The Ice Peak Formation (<1 Ma), which marks the start of the third magmatic cycle, was erupted from Ice volcano and comprises about 75 km^3 of basaltic and silicic lava flows, pyroclastic deposits and lahar deposits. It is divided into a dominantly basalt lower assemblage, and a basaltic to trachytic upper assemblage. Although the Ice Peak Formation overlaps spatially and temporally with the Pyramid Formation, it must be younger than the Pyramid Formation based on field stratigraphic relationships. The Ice Peak eruptions represent the last large (>20 km^3) outpouring of lava at the complex. Spooner et al. (1995) suggested that diamictons interbedded with lower assemblage basaltic lavas were derived from a regional ice sheet that originated in the Coast Mountains to the west of the complex. Two isolated basaltic vents to the southwest of Ice Peak, Cache Hill and Camp Hill, have basal pyroclastic deposits interpreted to be subaqueous. At Camp Hill, the deposits directly overlie glaciogenic sediments interpreted as regional tills. The upper assemblage contains proximal and distal deposits related to Ice volcano, with at least three lobes of trachytic lava, Ornostay, Koosick and a third unnamed nunatak, interpreted as having erupted through ice based on lava flow morphologies and accompanying volcano-genic and glaciogenic sedimentary deposits.

The Pillow Ridge Formation (c. 0.9 Ma) is confined to two elongate ridges, Pillow Ridge and Tsekone Ridge, located to the north of the most abundant Ice Peak exposures (Souther, 1992; Lloyd, 2007; Edwards et al., 2009b). Both ridges have abundant pillow lavas, with the better-exposed Pillow Ridge also exhibiting outward dipping beds of partially palagonitised 'sideromelane tuff breccia'. This formation may be slightly younger than Ice Peak. Despite its small volume (c. 3 km^3) the formation of Pillow Ridge and its two tindars is important because they undoubtedly erupted in a glacial environment in which the coeval ice surface (which the tindars breached) probably lay at least 450 m above the surrounding 2250 m-high plateau. It was probably a regional ice sheet. Edwards et al. (2009b) described the distribution and detailed character of the lithofacies and lithofacies associations, and geochemical studies are also in progress on the petrological evolution (Edwards and Pollock, unpublished data). Pillow Ridge is one of the few pillow-dominated tindars (*sensu* Russell et al., 2014; called pillow ridges or mounds in this book (Chapter 8)) described outside of Iceland and Antarctica (cf. Smellie and Hole, 1997; Pollock et al., 2014). Edwards et al. (2009b) recognised more than 13 lithofacies, many of which may represent secondary, gravity-driven transport of frag-mental materials while the ridge was still encased in ice. Analyses of H$_2$O in pillow rim glass (up to 0.8 wt %) were used to infer that the elevation of the enclosing englacial lake fluctuated with time between episodes of pillow lava effusion. Five distinct stratigraphical units were also identified.

The end of the third magmatic cycle is marked by the eruption of c. 18 km^3 of compositionally uniform trachytic lava flows, domes and breccia that together form the Edziza Formation (c. 0.9 Ma). Souther (1992) surmised that most of this lava and pyroclastic debris appears to have erupted from a central volcano at the northern end of the summit region, which today has a 3 km circular summit crater filled with ice. He speculated that the Edziza eruptions formed on an extensively glaciated landscape some time after the recession of the regional ice sheet through which Pillow Ridge erupted. The lower third of the central Edziza volcano comprises chaotic volcanic breccias enclosing irregularly shaped and 'crudely' jointed masses of lava. Similar deposits near the summit of Hoodoo Mountain have been interpreted as glaciovolcanic (Edwards and Russell, 2002). The upper sections of the volcano are dominated by massive lava with well-developed slender columnar joints. The lavas vary from relatively thin (1–5 m) to very thick (150 m), probably related to changes in viscosity due to degassing.

The fourth magmatic cycle comprises the Arctic Lake (0.7 Ma), Klastline (0.6 Ma) and Kakiddi (0.3 Ma) formations, which together represent a volume of only c. 15 km^3. The older two formations are predominantly alkaline basalt, erupted perhaps over an extended time interval spanning several glacial cycles. The Arctic Lake Formation erupted from at least seven different vents along the southwestern flanks of the Spectrum Range, south of 57.5° N. The Klastline Formation erupted from vents north of 57.75° N, at the northern end of the main plateau. Both units probably erupted over a long period, possibly over the entire course of a local(?) glacial advance. They include pre- and syn- to late-glacial lavas, tephra cones and glaciovolcanic pyroclastic deposits. Although the trachytic Kakiddi Formation is thought to be relatively younger than the Artic Lake and Klastline formations, marking the end of the fourth magmatic cycle, it could be coeval with the youngest parts of the Edziza Formation. Kakiddi trachytic lava flows and pumiceous pyroclastic deposits are indistinguishable from deposits of the Edziza Formation but are located south of the Edziza central cone and give younger radiometric ages (Souther, 1992).

The fifth and final magmatic cycle, which may still be ongoing, comprises the Big Raven Formation (c. 1.7 km^3) and Sheep Track member (<1 km^3), both of which probably erupted less than 20 k.y. ago (Souther, 1992; Hungerford et al., 2014). The Big Raven Formation erupted from more than 30 vents and forms at least two post-glacial basaltic lava fields with tephra cones: Desolation field in the north and Snowshoe field to the southwest. The vents are all considered to have formed either post-Wisconsin or during the waning stages of glaciation. The Sheep Track member represents a trachytic fall tephra that covers much of the youngest lava surfaces but whose source is unknown. An additional small but important glaciovolcanic sequence was described by Souther (1992) and Hungerford et al. (2014) at Tennena Cone. The deposit comprises a moderately dissected tephra cone that protrudes from the western edge of the Edziza ice cap and a series of lavas extending to the west that are interpreted as emplaced beneath a much thicker-than-present ice mass. The cone mainly comprises lapilli tuff and tuff breccia, but has basal pillow lava lenses. It is cut by dykes along the summit. The lavas west of the cone have a number of unique aspects, including 'vertical pillow lavas' whose origins are enigmatic but thought to be a distinctive ice-contact feature (see Section 13.4.2). The Tennena outcrop is important because it contains (1) an unusually wide range of products characteristic of a small-volume eruption beneath 'thick' ice, and (2) associated subglacially emplaced lavas that can be traced further than 3 km away from the vent area. The lavas appear to have flowed in tunnels created by eruption-generated meltwater escaping at the base of the enclosing ice.

Bell–Irving

To the south and east of Mt Edziza, a small volcanic field in the Bell–Irving area has been described by Edwards et al. (2006). It comprises nine isolated outcrops interpreted as emplaced beneath ice sheets. Aerial investigations show that several of the deposits now occupy arête-shaped ridgelines and comprise predominantly palagonitised tuff breccia, pillow lava and dykes. The vent area for one of the centres, at Craven Lake, is marked by isolated mounds of pillow lava on the upper flank of a ridge, with crudely bedded tuff breccia and lapilli tuff located in a stream valley several hundred metres below, associated with pillow lava at the upper end of the stream valley. Most of the medial and distal deposits are fragmental. The Craven Lake centre has an age of 430 ka (Edwards et al., 2006).

Iskut, Hoodoo Mountain

The Iskut Volcanic Field comprises nine mafic centres of alkali olivine basalt to hawaiite composition (Hauksdóttir et al., 1994; Russell and Hauksdóttir, 2001) scattered along the Iskut River and its main tributaries. It also includes Hoodoo Mountain, which is a unique, large polygenetic phonolite–trachyte volcanic centre in the Stikine sub-province of the Northern Cordilleran Volcanic Province of British Columbia, Canada (Edwards et al., 2000, 2002; Edwards and Russell, 2002). The Iskut mafic centres include both subaerial (Iskut River, Second Canyon, Canyon Creek, Snippaker Creek, Lava Fork) and glaciovolcanic deposits (Tom MacKay Creek, Cinder Mountain, King Creek, Cone Glacier, Little Bear Mountain) that range from 70 ka to 150 y BP, including the youngest eruption in Canada (Lava Fork). Most of the evidence for glaciovolcanism at these centres is based on the presence of pillow lavas, although Cone Glacier (Hauksdóttir et al., 1994) and Little Bear Mountain (Edwards and Russell, 1994; Edwards et al. 1995) also have palagonitised tuff breccia and hyaloclastite. Little Bear Mountain comprises predominantly fragmental litho-facies including bedded hyaloclastite, and coarse breccia that fills erosional channels. It crops out along a ridgeline that now separates two valley glaciers (Hoodoo and Twin glaciers), and its overall morphology is drumlinoid, with mega-grooves carved into the resistant, palagonitised deposits. It has an eruption age of 146 ka (by U-Th/He dating on apatite grains in a syenitic xenolith; Blondes et al., 2007), which is consistent with younger ages for overlapping units from the adjacent Hoodoo volcanic complex situated close by to the south.

The evolution of Hoodoo Mountain volcano was strongly controlled by interaction between the volcanism and local and regional ice sheets up to 2 km thick over the last

85 ka (Edwards et al., 2000, 2002; Edwards and Russell, 2002). The volcano has a flat ice-covered summit, a 6 km basal diameter and rises to 1850 m a.s.l. It is mainly composed of phonolite and trachyte and has a volume of c. 17 km^3. Its volcanic history can be divided into at least six eruptive cycles. The basal units are aphyric, strongly jointed lava and altered tuff breccia emplaced beneath thick ice at c. 85 ka. The basal deposits are capped by overthickened lavas (up to 200 m) thought to have been ice confined (Kerr, 1948; Edwards et al., 2002; Edwards and Russell, 2002). The glaciovolcanic episodes were followed by apparently subaerial explosive eruptions that produced welded lapilli tuff up to 100 m thick between 80 and 54 ka, partly capped by more subaerial lavas at c. 54 ka. Eruptions resumed as local ice levels rose again to include two more periods of glaciovolcanic activity, the first between 54 and 40 ka that resulted in an upper sequence of lava flows and domes enclosed by dense, blocky tuff breccia, presumably erupted through regional ice more than 2 km thick. The last glaciovolcanic deposits are thought to have formed via fissure eruptions through thinner ice on the summit and northern flank of Hoodoo Mountain, based on the elongate morphology of the deposits, the pervasive jointing of the lava, and a lava-capping carapace of pumiceous lapilli in a highly altered matrix (essentially trachytic palagonite; Edwards and Russell, 2002). Hoodoo Mountain has also had at least two Holocene eruptions. Edwards et al. (2002) showed that the effects of assimilation–fractionation processes on magma densities during the production of the phonolite–trachyte magmas could either inhibit or enhance subglacial and supraglacial eruptions.

Nass

Numerous smaller basaltic glaciovolcanoes have been mapped and studied in the southern part of the NCVP (Edwards et al., 2006), including deposits in west-central British Columbia and in southeastern Alaska in the Nass Valley region (unpublished data of BRE; personal communication, S. Karl). Isolated deposits of pillow lava have been mapped on the floor of glaciated valleys (Evenchick et al., 2008), and hillocks of palagonitised hyaloclastite occur along the coastal areas of southeastern Alaska (unpublished data of BRE). The Alaskan deposits are particularly interesting because they may have erupted through tide-water glaciers or ice shelves, based on present-day elevations and locations of deposits.

2.4.6 Anahim Volcanic Province

The Anahim Volcanic Province comprises three large volcanic complexes (Rainbow Range, Itcha Range, Ilgachuz Range; Bevier et al., 1979; Charland et al., 1993) and several smaller volcanic fields (e.g. Kuehn et al., 2015) whose volcanic stratigraphy has been investigated in limited detail. However, existing ages indicate that all of these centres were probably active during the Plio-Pleistocene, and they are all located in areas thought to have been periodically covered by incarnations of the Cordilleran Ice Sheet. While significantly more work is needed, reconnaissance studies demonstrate that much glaciovolcanism is probably present.

2.4.7 Wells Gray–Clearwater

Basaltic glaciovolcanism in southeastern British Columbia mainly occurred in an area that is now largely within the Wells Gray–Clearwater Provincial Park (Hickson and Vigouroux, 2014). The basaltic Wells Gray–Clearwater Province is in east-central British Columbia, in an anomalous tectonic setting that has produced a small region of volcanism active during the Pleistocene. It is unclear whether the volcanism is generated by back-arc processes, a mantle plume or transtension. However, the lavas are alkali olivine basalts, and several contain peridotite xenoliths. The area contains nine individual centres as well as several valley-filling lavas (Hickson, 1987; Metcalfe, 1987; Hickson et al., 1995; Hickson and Vigouroux, 2014). All are thought to be younger than 3.5 Ma and some are likely to be Holocene. Eight of the centres have been interpreted as glaciovolcanic, and K-Ar ages show that they span the Plio-Pleistocene. Five of the centres (McLeod Hill, Fiftytwo Ridge, Gage Hill, Hyalo Hill, Mosquito Mound) are flat-topped tuyas, which record minimum ice thicknesses during eruption of 250–550 m. Another six centres lack lava caps but have either morphological or lithofacies evidence for glaciovolcanic origins (Mount Ray, Jack's Jump, Pyramid Mountain, Spanish Mump, Spanish Bonk, Sheep Track Bench). Neuffer et al. (2006) suggested that scalloped surfaces on the southeastern flank of Pyramid Mountain resulted from slope failures related to rapid evacuation of a surrounding englacial lake. However, unlike Iceland and areas in northern British Columbia, the tuyas in the Wells Gray area are mostly below the treeline and exposures are limited.

2.4.8 Cascade volcanic arc

The Cascade volcanic arc is approximately 1250 km long and according to recent work hosts over 2300 individual edifices, although only 22 form major volcanic structures (Hildreth, 2007). The arc results from the relatively slow subduction (c. 3–4 cm per year) of the Juan de Fuca and Gorda plates beneath the North American plate. Each of the 22 major edifices are or have until recently been glacierised and are at least snow-clad during winter months. Even though some of the earliest studies of volcano–ice interactions by Mathews (1951, 1952a, b) were conducted in the northernmost segment of the arc, studies of glaciovolcanism in the rest of the arc are still relatively few considering the relative accessibility of the volcanoes and their extensive Quaternary eruption histories. Several workers have identified likely glaciovolcanic features in specific areas of Washington and Oregon (Hammond, 1987; Lescinsky and Fink, 2000; Lescinsky and Sisson, 1998; Lodge and Lescinsky 2009a, b; Bacon and Lanphere, 2006), including at least two purported tuyas in Oregon (Hayrick Butte, Hogg Rock). Most of the large stratovolcanoes in the Cascades presently host glaciers, which probably means that they all host ancient glaciovolcanic deposits awaiting discovery.

The northernmost segment of the Cascades volcanic arc, located in southwestern British Columbia, is locally referred to as the 'Garibaldi Belt'. It has been the subject of several classic studies of glaciovolcanism by Mathews (1951, 1952a, b), as well as more recent work by Green (1990), Bye et al. (2000), and Kelman et al. (2002). Mount Garibaldi is a highly dissected stratovolcano largely formed by glaciovolcanism. Mathews' description is one of the first for a large composite volcano in an ice-dominated environment. He suggested that the largely dacitic volcano periodically had Peléan-style eruptions, where dome collapses shed debris onto the surrounding coeval ice surface. He also used elevations of erratic boulders to infer that most of the volcano formed after the Last Glacial Maximum, and that some of the ice may have recovered after eruptions ended to locally deposit erratics on top of dacitic tuff breccia. He also documented ice-dammed lavas ('The Barrier'), an effusion-dominated tuya ('The Table'), and lavas occurring at lower elevations that exploited meltwater drainage cavities to form lava 'eskers' (Mathews, 1987; see Sections 11.4 and 12.2). Green (1990) looked more broadly at the volcanology and petrology of the area, as well as providing age information. Watts Point is a glaciovolcanically emplaced dacite lava dome (Bye et al., 2000). Fifteen andesitic and dacitic glaciovolcanic edifices are present in the Mt Cayley area, to the west of Mt Garibaldi. Most of these deposits are jointed lava masses (Kelman et al., 2002). That they lack fragmental material may be a consequence of vents located at high elevations, where efficient basal meltwater drainage is able to rapidly transport fragmental material away from the vent site.

While all of the Cascade composite volcanoes are glacierised, only limited work has been done to identify glaciovolcanic deposits (Lescinsky, 1999; Lescinsky and Sisson, 1998; Lescinsky and Fink, 2000; Bacon and Lanphere, 2006; Lodge and Lescinsky, 2009a, b). Evidence for lava–ice interactions occurs at Mt Baker and Mt Rainier (Washington State), and at South Sister and Mt Jefferson (Oregon; Lescinsky and Fink, 2000). Most of the described examples at those localities are lavas with distinctive morphologies and cooling fractures. However, all of the main volcanic edifices have been mapped in detail, and studies are increasingly identifying likely glaciovolcanic deposits. For example, Bacon and Lanphere (2006) identified at least nine glaciovolcanic deposits in the Mt Mazama/Crater Lake region that either had evidence for ice-contact (Redcloud Cliff, Grotto Cave, Pumice Point, Dutton Cliff, Munson Ridge), were ice-bounded (Grouse Hill, Roundtop), or formed tuyas (Bear Bluff, Arant Point). Equivalent work elsewhere in the Cascades is sparse despite their extended Pleistocene eruption histories. By contrast, the 1980 eruption at Mt St Helens is one of the best-studied modern explosive glaciovolcanic eruptions, and interaction with lava and glaciers continued to produce lahars at Mt St Helens until 1982 (Waitt et al., 1983; see Section 3.3.4).

Finally, the importance of volcano–ice/snowpack interactions has long been recognised in the Cascades, where an intensive programme of hazard mapping by the United States Geological Survey has identified likely lahar paths at all of the major volcanoes (e.g. hazard maps of Rainier, Shasta; Miller, 1980).

2.5 Other locations

Studies of glaciovolcanism are generally less extensive outside of the regions discussed above. However, more detailed studies of Quaternary deposits around the Earth increasingly show that high-elevation glaciations have occurred at most latitudes. For many of the areas discussed below, studies of glaciovolcanism are nascent or non-existent, yet all contain volcanoes that were active during the Quaternary as well as glaciers. For many, the glaciovolcanic evidence is circumstantial (e.g. lavas interbedded with till or presence of deposits that formed in the presence of water, e.g. lahars). However, given the coincidence of ice and volcanism, all of the locations warrant further investigations.

2.5.1 Mexico

Mexico hosts numerous Quaternary volcanoes, and at least 13 of those in the Trans-Mexican volcanic belt were active and hosted glaciers during the Pleistocene (Vázquez-Selem and Heine, 2004), including Nevado de Colima (4180 m a.s.l.), Tancítaro (3842 m a.s.l.), Nevado de Toluca (4690 m a.s.l.), Ajusco (3952 m a.s.l.), Popocatépetl (5452 m a.s.l.), Iztaccíhuatl (5230 m a.s.l.), Telepón (4090 m a.s.l.), Tláloc (4120 m a.s.l.), La Malinche (4461 m a.s.l.), Sierra Negra (4600 m a.s.l.), Pico de Orizaba (Citlaltépetl; 5675 m a.s.l.), Las Cumbres (3950 m a.s.l.; Höskuldsson, 1992) and Cofre de Perote (4282 m a.s.l.). Most of these edifices comprise andesitic to dacitic lavas, domes and pyroclastic deposits. Although studies of volcano–ice interactions are sparse (Capra et al., 2004, 2015), Late Pleistocene glacial chronologies for eight of the volcanoes summarised by Vázquez-Selem and Heine (2004) provide a useful framework for future glaciovolcanism research. For example, Iztaccíhuatl has evidence for five different periods of glacial advance, including one that is estimated to have covered more than 200 km^2 of the volcano with ice that extended down to 3000 m a.s.l., well below its present-day summit of 5230 m, between c. 205 and 175 ka. Nevado de Toluca volcano was extensively glaciated during the Last Glacial Maximum and is partly mantled with extensive volcaniclastic deposits including many interpreted as originating from lahars. La Malinche, Ajusco and Pico de Orizaba all also show evidence for extensive Last Glacial Maximum glacial activity. On the other hand, Nevado de Colima, Cofre de Perote and Tancítaro are older edifices and have potential for records of volcano–ice interactions as early as 800 ka (Vázquez-Selem and Heine, 2004). Three of Mexico's highest Holocene stratovolcanoes are still glacierised: Volcán Pico de Orizaba (also known as Volcán Citlaltepetl) hosts nine glaciers, Popocatépetl hosts three glaciers, and Iztaccíhuatl hosts twelve named glaciers (White, 2010). While there are few glaciovolcanism studies for volcanoes in Mexico, studies of hazards from lahars (Popocatepetl: Capra et al., 2004; Julio Miranda et al., 2005; Iztaccihuatal: Delgado Granados et al., 2005; Scheinder et al., 2008), enhanced explosivity due to excess hydrothermal fluids during or immediately following glaciations (Colima; Capra and Macias, 2002), and avalanche generation (Carrasco-Núñez et al., 1993) all involve possible glacier–volcano interactions. Capra et al. (2015) have also documented the glaciovolcanic emplacement and disintegration of a large lava dome at Nevado de Toluca.

2.5.2 South America

The Andes volcanic arc hosts at least 200 Quaternary volcanoes (Stern, 2004), many of which were snow or ice clad during their evolution (Fig. 2.7). The arc is subdivided into four segments that are spatially separated by gaps in volcanism; the northernmost three zones result from subduction of the Nazca plate beneath the South America plate, while the southernmost zone results from subduction of the Antarctic plate beneath South America. According to Stern (2004), the arc can be broken into four segments including the Northern volcanic zone (5° N–2° S; 19 active volcanoes in Colombia and 55 in Ecuador), the Central

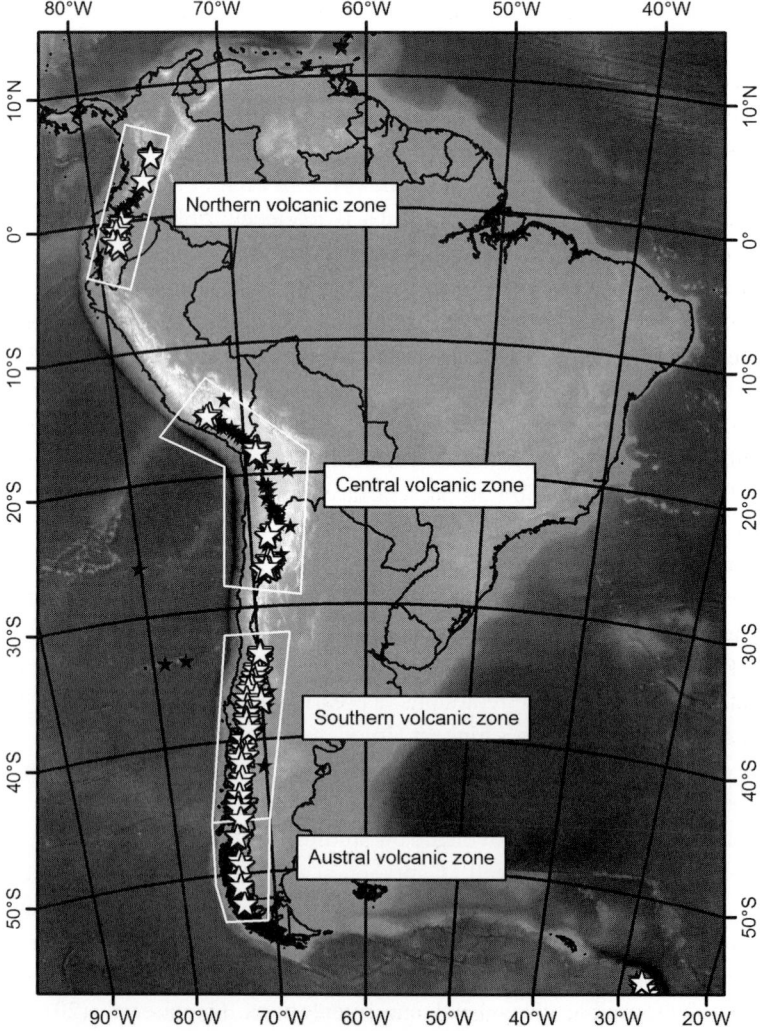

Fig. 2.7 Map of South America showing the major volcanic provinces, volcanoes classified as active (black stars) and volcanoes that are presently ice clad (white stars).

volcanic zone (14–27° S; 44 active volcanoes in total spanning Peru, Bolivia, Chile, Argentina), the Southern volcanic zone (33–46° S; at least 60 active volcanoes in Chile and Argentina), and the Austral volcanic zone (49–55° S; 6 active volcanoes in Chile). The volcanoes are predominantly composed of calc-alkaline andesite and dacite, with lesser volumes of rhyolite and basalt or basaltic andesite. The Quaternary history of volcanism in the arc is not well known, but many studies have been published on the active volcanoes and many of the better-known historical eruptions (e.g. 1985 Nevado del Ruiz, Colombia; see Section 3.3.6). Even on the equator in Ecuador, several volcanoes are seasonally covered with snow (e.g. Cotopaxi, Chimborazo) and thus they probably hosted glaciers during the Pleistocene. Most of the northern Andes arc segment only hosts snow-covered volcanoes today. However the southern part of the arc still has volcanoes with glaciers and ice caps.

There are few studies of Pleistocene glaciovolcanism in the Andes, partly because many of the volcanoes are still active and they are not extensively eroded (Larsen, 1940). While many Pleistocene volcanic complexes have been subjected to detailed geochronological and geochemical studies (e.g. Singer et al., 2008), less work has been done to identify ice/snow-contact lithofacies (e.g. Mee et al., 2006, 2009). However, a major feature is the importance of lahar deposits (e.g. Cotopaxi volcano; Mothes et al. 2004; Hall and Mothes, 2008). Nevado del Ruiz has a long history of lahar-generating eruptions, and its ancestral base ('Old Ruiz') formed at c. 2 Ma (Thouret, 1990). Several general stratigraphical studies of Pleistocene Chilean volcanoes have been published, including Hudson and Lautaro (Orihashi et al., 2004; Motoki et al., 2006; Weller et al., 2014), Lanín (Lara, 2004; Lara et al., 2004), Nevados de Chillan (Mee et al., 2006, 2009), Parinacota (Clavero et al., 2004) and Sollipulli (Gilbert et al., 1996). González-Ferrán (1995) provided summaries (in Spanish) of virtually every volcano in Chile, which extends to the South Sandwich Islands and Antarctic Peninsula.

Many studies of historical eruptions from Andean volcanoes have focused on understanding lahars and their deposits, as these pose major hazards at many active ice-clad volcanoes (e.g. Calbuco, Cotopaxi, Nevado del Ruiz, Villarrica; see Section 3.3). Most commonly the lahars have formed as a result of pyroclastic density currents travelling over and interacting with snow and ice, although lava flows have also produced catastrophic melting on steep slopes (e.g. Villarrica 1984; see Section 3.3.7).

2.5.3 Russia

Kamchatka–Kurile volcanic arc

The Kamchatka–Kurile volcanic arc is one of the most productive volcanic arcs on Earth (Fedotov and Masurenkov, 1991; Ponomoreva et al., 2007). It comprises three spatially separated areas of Holocene volcanism, organised from west to east, including the Sredinny Range, the Central Kamchatka Depression (CKD), and the Eastern Volcanic Zone (EVZ). The arc contains 68 active volcanoes, with 29 active centres on the Kamchatka Peninsula that range from rhyolitic calderas (e.g. Uzon, Academy Nauk; Bindeman et al., 2010) to large andesitic lava domes (e.g. Shiveluch and Bezymianny; Gorbach et al., 2013; Girina, 2013) and fissure systems that erupted high-Mg and high-Al basalt (e.g. Tolbachik; Churikova et al., 2015). Most of the Kurile Islands volcanoes are relatively low in elevation and do not presently host glaciers, so it is not clear whether they are likely to host older Pleistocene glaciovolcanic deposits. However, Kamchatka Peninsula volcanism has been pervasive throughout the Plio-Pleistocene (Ponomoreva et al., 2007) and Kamchatka almost certainly has an extensive glaciovolcanic history. At present the annual snowline is estimated at 1000 m a.s.l. (Bindeman et al., 2010). Most of the Kamchatka–Kurile stratovolcanoes are at least partly snow-clad in the winter and many host glaciers. The active ice-clad volcanoes are mainly located in the EVZ and the CKD, although a few of the Sredinny Range centres have also had Holocene eruptions (Fedotov and Masurenkov, 1991; Ponomoreva et al., 2007).

Few studies of the physical volcanology of pre-Holocene volcanoes have been published. However, it has been estimated that the Sredinny Range includes more than 120 polygenetic volcanoes, which, combined with numerous small volcanic edifices, are distributed across more than 19 000 km^2 in central and western Kamchatka (Fedotov and Masurenkov, 1991). Most of these volcanoes are thought to have been active in the Quaternary and are thus likely to include glaciovolcanic activity. Despite the limited evidence for pre-Holocene glaciation in Kamchatka, extensive ice cover probably occurred during Marine Isotope Stage (MIS) 6 (c. 130 ka) when virtually the entire peninsula may have been glacierised (Barr and Solomina, 2013). There is less evidence for widespread ice during MIS 4 and MIS 2, and that may be more consistent with several separate ice sheets centred on the Sredinny and Vostocny mountain ranges (Barr and Solomina, 2013). On the other hand, many of the active volcanoes in the EVZ and the CKD are presently glacierised, and volcano–ice/snow interactions are well documented at Klyuchevskoy (Ozerov et al., 1997; Belousov et al., 2011) and Tolbachik (Edwards et al., 2014a, 2015b). Much of the volcanism in these areas started by c. 250 ka, and extensive glaciations are known for at least two different phases (MIS 2 and 4). Overall, the diverse volcanism across Kamchatka represents one of the larger areas (along with Alaska and South America) that is likely to yield extensive new insights into glaciovolcanic processes.

Azas Plateau Volcanic Field

Unlike the extensive volcanism in Kamchatka, the Azas Plateau Volcanic Field in south-central Siberia, Russia (52° 25′ N; 98° 30′ E), near Lake Baikal, comprises only nine moderately sized tuyas that have been dated as Pleistocene in age (Yarmolyuk et al., 1999, 2001; Komatsu et al., 2007a, b). They comprise the Late Pliocene–Quaternary Azas Plateau Volcanic Field (2000 km^2, c. 600 km^3), the largest part of the Tuva Volcanic Province. The glaciovolcanic examples are middle to late Pleistocene in age (< c. 750 ka) and have trachybasalt and basanite compositions. They were constructed under former ice a few to several hundred metres thick. Most are tuyas but tindar (ridge-like) and glaciovolcanic tuff cone edifices are also present. All are relatively large, with basal dimensions mainly ranging between 2 and 10 km and heights of 150–600 m. Studies in the region have focused on geomorphological aspects and they emphasised the occurrence of slope failures of over-steepened flanks, gully formation due to streams and debris flows, and presence of cirques, U-shaped valleys and rock glaciers that together have significantly modified the primary landform characteristics. Six of the edifices are 'classic' flat-topped tuyas, and three have conical or ridge-like ('tindar') morphologies. Some of the tuyas have been modified by cirque glaciers, which is consistent with ages that are older than the Last Glacial Maximum. However, the range of ice thicknesses only varies from 300 to 600 m over time. Komatsu et al. (2007a, b) suggested that ages range from c. 750 ka to c. 50 ka, and potentially document up to seven different glaciations in the region.

2.5.4 Japan

Many of Japan's larger volcanoes (e.g. Mt Fuji, Akan (1499 m a.s.l.)) are seasonally snow covered, although none are presently glacierised. Mount Tokachi volcano in central Hokkaido has a summit elevation of 2077 m, which is well above the 1500 m equilibrium line altitude estimated for the Hidaka Range immediately southeast of Tokachi (Ono et al., 2005). Interestingly, it appears that the Last Glacial Maximum in Japan coincided with MIS 3/4 and not MIS 2 (Ono et al., 2005). The main part of the northern 'volcanic front' comprises Quaternary volcanoes that in total erupted over 800 km^3 of magma during the Quaternary (Umeda and Ban, 2012). The generation of lahars by melting of snowpack during historical eruptions has been common at least at Tokachi; as recently as December 1988 and January 1989 it had a winter eruption that was characterised by phreatomagmatic eruptions and lahars, although only limited snowmelt was noted (Katsui et al., 1990). The source of water for the eruptions was not determined but was probably melted snowpack.

2.5.5 Hawaii

Despite being situated in the centre of the Pacific Ocean, the Big Island of Hawaii has supported a succession of small ice caps during the last several glacial cycles of the Pleistocene (Stearns, 1945, 1966). In a remarkably perceptive series of publications, Porter (1979a, b, c, 1987) described detailed evidence for Hawaii's glacial past, including widespread glacial landforms, striated surfaces and till deposits interbedded with tholeiitic, alkali basaltic and hawaiite lavas on the upper slopes of Mauna Kea volcano (Wolfe et al., 1997). Subglacial eruptions have also been postulated that occurred within at least two discrete episodes. The evidence includes glacially oversteepened pyroclastic cones composed of 'hyaloclastite tuff' within the Laupāhoehoe Group (later reassigned to the Hamakua Group; Wolfe et al., 1997). By analogy with the formation of littoral cones on the island and in contrast with other cones on the volcano formed of magmatic (scoria) products, eruption of the Mauna Kea tuff cones was inferred to be within meltwater lakes

100–170 m deep (implying similar ice thicknesses). The possible presence of (unexposed) pillow lava cores to the tuff cones was also postulated. In addition, some lavas with ice-scratched surfaces differed from other lavas on the volcano in having abrupt cliff-like margins up to 25 m high, deeply embayed and showing glassy chilled surfaces and local palagonite alteration. The presence of unusual closely spaced block-like jointing in the dense aphanitic lavas was compared with similar features seen in ice-cooled lavas in Iceland (cf. blocky jointing described in Section 9.3.7). Emplacement of the lavas by flowing (melting) through ice was inferred, down to at least 3200 m a.s.l. Both lavas and tuff cones were assigned to the local Waihu glaciation (c. 175–170 ka). Water-deposited gravels within the Waihu Formation were also attributed to outburst floods (jökulhlaups). Similar features occur in younger lavas near the summit and they also include lava pillows, local glassy breccia and spiracle structures at lava bases. Like the older lavas, the abrupt steep lava margins were assigned an ice-contact origin, inferred to be during an expansion of ice caps during the local Makanaka glacial episode (c. 69–9 ka). Ice thicknesses during the lava flow eruptions were at least 25 m.

2.5.6 Indonesia

Papua New Guinea, situated immediately south of the equator in the southwestern Pacific Ocean, is part of a volcanic archipelago resulting from subduction of the Australian plate beneath the Eurasian plate. It hosts numerous low-elevation, active volcanoes (e.g. Rabaul) along with several older volcanoes that have summit elevations above 3000 m a.s.l. (Fig. 2.8). Several of these older volcanoes were glacierised during the Pleistocene, and limited work has been done on at least one that appears to have glaciovolcanic deposits (Blake and Löffler, 1971). Mount Giluwe has a present-day elevation of 4863 m a.s.l. and recent work by Barrows et al. (2011) shows that it experienced at least four periods of glaciation that affected the volcano down to elevations of 3200 m a.s.l. Using a combination of older K-Ar geochronology and ^{36}Cl exposure ages, they postulated episodes of ice cover at 306–293 ka (Gogon glaciation), 158–136 ka (Megane glaciation), 62 ka (Komia glaciation), and c. 20–12 ka (Tongo glaciation). Several other Pleistocene volcanoes in Papua New Guinea may also have been glacierised, including Mt Hagen (3778 m a.s.l.), Mt Ialibu (3408 m a.s.l.), Mt Kerewa (3491 m a.s.l.), Yelia (3340 m a.s.l.), and Doma Peaks (3514 m a.s.l.). At Giluwe, Blake and Löffler (1971) found multiple evidence for glaciovolcanism, including extensive palagonitised, mafic volcanic breccias interlayered with lava flows on the north side of the mountain (Gogon River drainage), less extensive deposits of palagonitised breccia on the east peak, and a dome to the north of the main peak that has glaciated sides and base but a no evidence for ice erosion on its top. Based on the thickness of the Gogon River palagonite mass, they suggested the ice was at least 100 m thick. Very little information is available about glaciovolcanic deposits at the other volcanoes (Mackenzie and Johnson, 1984), but they should be targets for future studies of equatorial volcano–ice interactions.

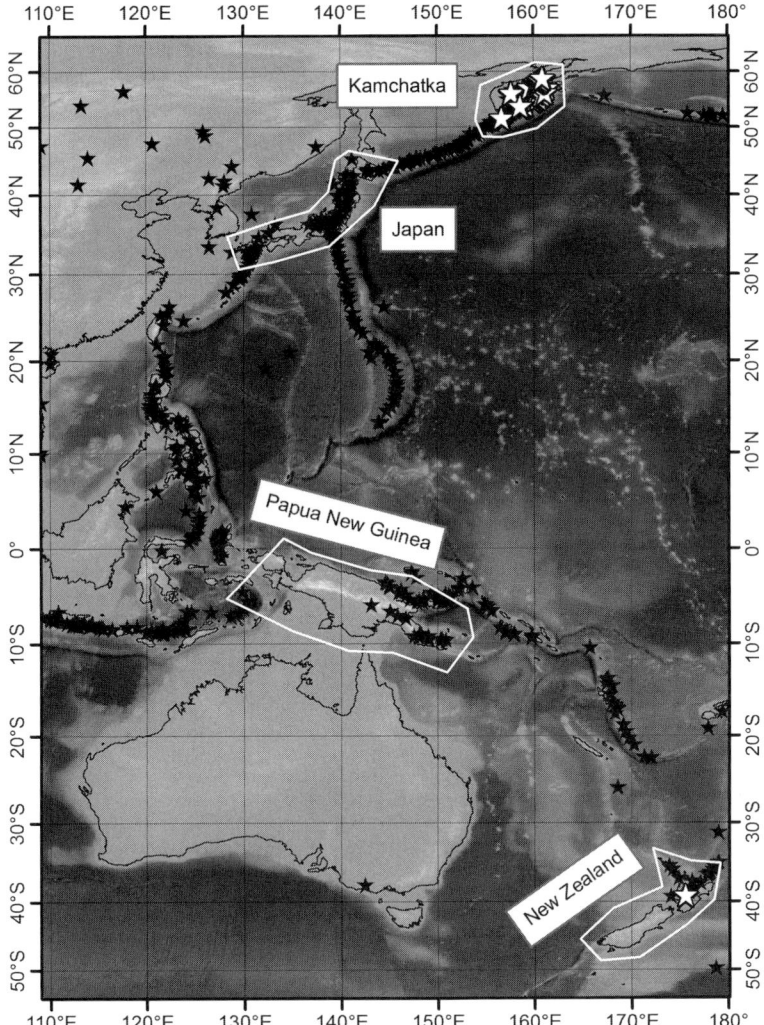

Fig. 2.8 Map of the western Pacific Ocean basin showing major volcanic provinces, volcanoes classified as active (black stars), and volcanoes that are presently ice clad (white stars).

2.5.7 New Zealand

While most of the glaciers in New Zealand are on South Island (Chinn et al., 2014), at least two of the North Island volcanoes presently host glaciers (Mt Ruapehu) or perennial snowfields (Mt Taranaki; Fig. 2.8). Significant recent work has been undertaken at Ruapehu to document glaciovolcanism during the Pleistocene, as well as documentation of historical lahars that have formed during eruptions that interacted with snow/ice on its summit (Cronin et al., 1996; Spörli and Rowland, 2006; Kilgour et al., 2010; Conway et al., 2015). New ^{40}Ar/^{39}Ar geochronology has been used to constrain ice-marginal lava flows erupted between 51 and 15 ka into valleys filled with glaciers. Lavas are overthickened and intercalated with glacial deposits. They also display pseudopillow fractures, kubbaberg joints and other features that are characteristic of ice-contact deposits.

2.5.8 Africa

Although Africa hosts many hundreds of volcanoes, only Mounts Elgon (4321 m a.s.l.), Kilimanjaro (5895 m a.s.l.), Kenya (5199 m a.s.l.) and Meru (4565 m a.s.l.) in the East African Rift are thought to have been glaciated during the Pleistocene (Mark and Osmaston, 2008). The volcanism comprises alkaline lavas that range from olivine nephelinite to peralkaline rhyolites, and it originates from active rifting, which has now produced the separate Nubian and Somalia plates. There is evidence for at least four different episodes of glaciation at Mt Kilimanjaro: at c. 500 ka, 360–240 ka, 150–120 ka and 20–17 ka (Mark and Osmaston, 2008). Palagonitised breccia, pillow lava and laharic deposits are present (Downie, 1964); the lahar deposits occur at lower elevations (c. 820 m a.s.l., below the Main Rhomb porphyries in the Garanga Tal valley) and the pillow lavas occur at c. 4500 m a.s.l. (beneath the Penck Rhomb porphyries). The presence of palagonitised tephra directly below the Nepheline Rhomb Porphyry series at 4000 m a.s.l. may be evidence for glaciovolcanism at Sud Ost Tal. The geochronology of glaciations at Mt Kenya are more uncertain, with one event possibly at 1.9 Ma; other glacial periods may include >780 ka, 420–355 ka, 300–150 ka and c. 50 ka (Mark and Osmaston, 2008, their Table 5).

2.6 Summary

While our most detailed knowledge of glaciovolcanism is derived from studies in three areas (Antarctica, Canada, Iceland), Pleistocene volcano–ice interactions are a global phenomenon. Many areas are subject to significant glaciovolcanic hazards, such as South America and western USA, but the volcanoes have not yet been closely studied. Moreover, Alaska and Kamchatka likely have dozens if not hundreds of deposits formed by volcano–ice interactions, and represent a treasure trove of potential information to aid in reconstructing the timing and distribution of global ice masses over at least the past 2.5 Ma. A large number of volcanoes have not yet been subjected to modern volcanological studies and their histories of interactions with ice or snow are essentially unknown. Documenting and understanding Earth's glaciovolcanic record during the last glacial period will be a critical guide in the search for similar deposits in ancient glaciations (e.g. Permo-Carboniferous, Ordovician, Neoproterozoic) and for continuing to unravel the climate history of other water-bearing planets such as Mars (see Chapters 13 and 16).

3

Observations of historical and recent glaciovolcanic eruptions

3.1 Introduction

Although an overwhelming majority of the glaciovolcanic deposits studied today formed more than 5000 years ago, increasingly important observations of volcanic edifice construction, the eruptive and depositional processes involved, and impacts on the associated glaciers have been made during modern and historical glaciovolcanic eruptions. These studies are critically important for at least two major reasons. Firstly, they demonstrate the associated hazards that are particularly important for eruptions in glacial settings. Those hazards can occur at volcanoes in proximity to large population centres (e.g. the heavily glacierised Mt Rainier is close to Seattle, USA, with a metropolitan population of 2–3 million people). Secondly, they provide insights that are vital for interpreting many of the processes involved in the formation of ancient deposits. Major and Newhall (1989) reviewed several dozen historical eruptions of snow- and ice-capped volcanoes and determined that explosive eruptions produced the most severe hazards (e.g. Eyjafjallajökull, Katla, Mt St Helens, Redoubt). However, they also included many reports of effusive eruptions that produced damaging jökulhlaups and lahars (e.g. Hekla, Klyuchevskoy, Llaima, Villarrica). In this chapter we present summaries of several important historical eruptions that display eruptive behaviours that have significantly influenced our knowledge of glaciovolcanism.

Although glaciovolcanic eruptions are an important source of information on volcanic processes (e.g. Nielsen, 1937), with rare exception (Baker et al., 1975; see also Smellie, 2002), opportunities for making detailed observations and measurements have only occurred since the 1980s. Since that time, well over a dozen eruptions have occurred in which lava interacted with snow or ice (Table 3.1). Those eruptions took many forms, including (1) truly subglacial eruptions under ice of varying thicknesses (e.g. summit phase of Eyjafjallajökull 2010; Gjálp 1996; Grímsvötn 1998, 2004, 2011; Mt Hudson 1991), (2) eruptions where ice was present but its impact was subordinate (e.g. Llaima 1994; Mt Hudson 1991; Klyuchevskoy, 1994; Mt St Helens 1980–6, 2004–6; Redoubt 1989–90, 2009; Villarrica 1971/1984–5), and (3) eruptions where lava erupted subaerially interacted with ice or snow (e.g. Etna 2006–7; flank phase of Eyjafjallajökull 2010; Hekla 1947; Tolbachik 2012; Veniaminof 1983–4, 1991, 2013; Fig. 3.1).

Table 3.1 *Summary of observations from selected historical glaciovolcanic eruptions (information from various sources but particularly Major and Newhall (1989), Lescinsky (1999), Siebert et al. (2010), and Edwards et al. (2015a))*

Eruption	Katla 1918	Hekla 1947	Villarrica 1971	MSH 1980	Veniaminof 1983	Nevado del Ruiz 1985	Redoubt 1989	Hudson 1991	Llaima 1994	Gjálp 1996	Grímsvötn 2004	Redoubt 2009	Eyjafjallajökull 2010
Magma composition	Basalt	Basaltic andesite	Basaltic andesite	Dacite	Basaltic andesite	Mixed (dacite/andesite)	Dacite	Basalt/trachybasalt	Basalt	Basaltic andesite	Basaltic	Dacite	Trachyandesite
Volcano type	Caldera/composite	Composite	Composite	Composite	Caldera/composite	Caldera/composite	Composite	Composite	Composite	Composite	Caldera/composite	Composite	Caldera/composite
Eruptions style(s)	Subglacial/explosive	Explosive/effusive	Explosive/effusive	Explosive/effusive	Effusive	Explosive	Explosive/effusive	Explosive/effusive	Explosive/effusive	Subglacial/explosive	Explosive	Dome/explosive	Explosive/effusive
Volcanic explosivity index	>4	4	2	5	3	3	3	>5	2	3(?)	3	3	<3
Subglacial phase (hours)	4	0	0	52 days*	0	0	0	na	na	31	0.5	0	4
Volume erupted (km^3 DRE)	na	na	na	~1	na	0.2	~0.15	4–7	na	0.45	0.02	0.10	0.17
Vol. fraction erupted subglacially	na	0	0	100/0	0(?)	0(?)	0(?)	na	na	~95%	~10%	Minor	~5%
Duration (days)	24	388	~1 hour	9 hours	318	<1 hour	184	80	104	13	6	13	39
Vol. airborne tephra (km^3 DRE)	na	0.1	na	~1	na	0.02	0.1	2.7	na	~0.02	0.02	0.02	0.14
Vol. of lava produced (km^3)	0.5	0.1	0.01	0	0.05	0	0.01	<1(?)	na	0	0	0.08	0.03
Initial ice thickness (m)	400	<20(?)	75–195	60–100	>100	<100 m	30–70**	50–100(?)	na	600–750	150–200	100	200
Max. cauldron width (km)	na	fissure	400 m fissure	~1	Intracaldera cone	750 m	Dome	400/800 m	Fissure	7	0.5–1.0	Dome	0.6
Vol. ice melted (km^3)	8	na	Open vent	0.12	na	0.05	~0.1	<0.1(?)	0.004	4.2	0.07	0.10–0.25	0.22
Peak water discharge (m^3 s^{-1})	300 000	880	20 000	68 000	0(?)	48 000	70 000	na	c. 1000	50 000	3000	~100 000	~5000

* estimated time during which cryptodome (100% of magma volume) was growing in upper edifice and deforming/possibly melting ice; it is difficult to estimate how much of the total volume was ultimately erupted subglacially

** estimated thickness of Drift Glacier downslope from the dome MSH – Mt St Helens

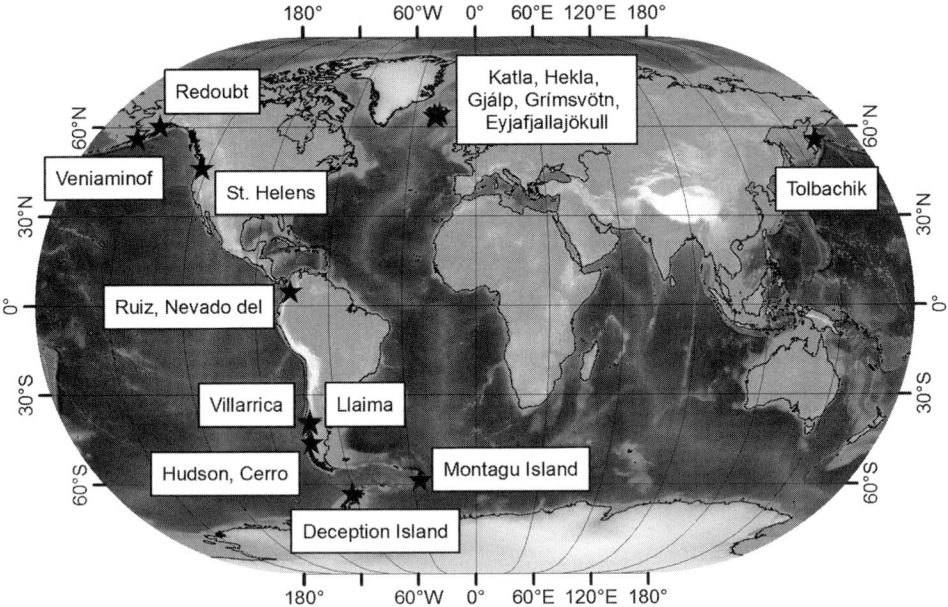

Fig. 3.1 Global map showing the locations of selected historical glaciovolcanic eruptions discussed in this chapter.

3.2 Classification of glaciovolcanic eruptions

Modern glaciovolcanic eruptions may be crudely divided into three broad classes based on eruptive setting (Edwards et al., 2015a). The first class includes eruptions that occur under glaciers and in which the presence of ice and meltwater exerts a major influence on the eruptive style (*ice-dominant interactions*). Modern examples are not common globally, but geological evidence indicates that they occurred frequently in volcanic regions during glaciations. The majority of Pleistocene tuyas are of this type. For this class two end-members exist: eruptions through thick ice (>500 m) and eruptions through thin ice (<200 m); eruptions through ice with intermediate thicknesses can have behaviour of both end-members depending on ice melting rates and pre-eruptive concentrations of magmatic volatiles. The first class of eruption is common in Iceland. It includes those at Katla (1918), Gjálp (1996), Grímsvötn (1998, 2004, 2011), and the second phase of the Eyjafjallajökull (2010) eruption, which originated from the summit crater. Eruptions from Hudson volcano in southern Chile also fall in to this category (Naranjo et al., 1993). Because of the continental scale of the Antarctic Ice Sheet and coincident presence of active volcanism, it is likely that subglacial eruptions occurred not infrequently in Antarctica in the recent past (i.e. last few k.y.), but due to the size, remoteness, and short observation periods for Antarctica, the scale and frequency of any eruptions is poorly known (e.g. Blankenship et al., 1993; Corr and Vaughan, 2008; Lough et al., 2013; Patrick and Smellie, 2013).

The second class includes glaciovolcanic eruptions in which the glacial cover at the vent is thin and extensive melting does not occur before the eruption becomes subaerial, but nevertheless the presence of snow and ice in the surrounding terrain exerts considerable influence on the characteristics and distribution of the eruption products (*volcano-dominant interactions*). These eruptions are common at ice- and snow-clad composite volcanoes and can pose significant hazards to populations, especially in South America. This style of eruption includes collapses of lava domes onto surrounding glaciers (e.g. Redoubt 1989–90, 2009), incorporation of large volumes of ice into volcanic avalanches/debris flows (e.g. Mt St Helens 1980), pyroclastic density currents emplaced onto ice/snowpack (e.g. Cotopaxi 1871, Nevado del Ruiz 1985), and effusion of lava flows onto ice that are significant enough to generate life-threatening lahars (e.g. many eruptions at Llaima, Villarrica, Hudson). This second class of glaciovolcanic activity includes, in addition to the examples given above, recent eruptions at Llaima (1994), Villarrica (1971, 1984) and other volcanoes in the Andes, Klyuchevskoy (1994), Etna (2006–7), Montagu (2001–7), and the second effusive phase of the Eyjafjallajökull eruption (2010).

The third class involves minor volcano–ice interactions that still produce identifiable glaciovolcanic deposits but where the interactions are confined to a local scale (*incidental interaction*). While this class does not construct a large edifice dominated by glaciovolcanic deposits and only generates local hazards (e.g. Edwards et al., 2015a), it is probably by far the most widespread type of glaciovolcanism, and has the potential to generate local deposits that can record the presence of snow/ice across a wide latitudinal range (e.g. many low-latitude stratovolcanoes are snow covered for at least part of the year now, such as Mt Fuji in Japan, Mt Taranaki in New Zealand, Mauna Kea in Hawaii, Mt Giluwe in Papua New Guinea, and many others presumably would have had significant snow/ice cover during glacial epochs).

3.3 Descriptions of observed glaciovolcanic eruptions

3.3.1 Katla 1918

Katla is a large composite volcano with a summit caldera located on the south central coast of Iceland (Fig. 3.2). It is at the southern end of the Eastern volcanic zone (see Chapter 2), and its summit is covered by Mýrdalsjökull, a large ice cap up to 600 m thick (Björnsson et al., 2000; Sturkell et al., 2010). It is the most productive volcano in Iceland (Thordarson and Larsen, 2007). In 1918 volcanic activity at Katla commenced with a large earthquake at 1300 hours on 12 October, followed by 30 minutes of continuous tremors and a major jökulhlaup (Table 3.1). The eruption itself was underway by 1500 hours, when a prominent tall eruption column was observed towering many kilometres above Mýrdalsjökull. The eruption broke through or melted ice c. 400 m thick over the vent. Although the eruption site was unobserved, it was suggested that pillow lava may have formed at the vent (Tómasson, 1996). The eruption produced at least 0.5 km^3 (dense rock equivalent) of basaltic magma and the flood released about 8 km^3 of meltwater and ice.

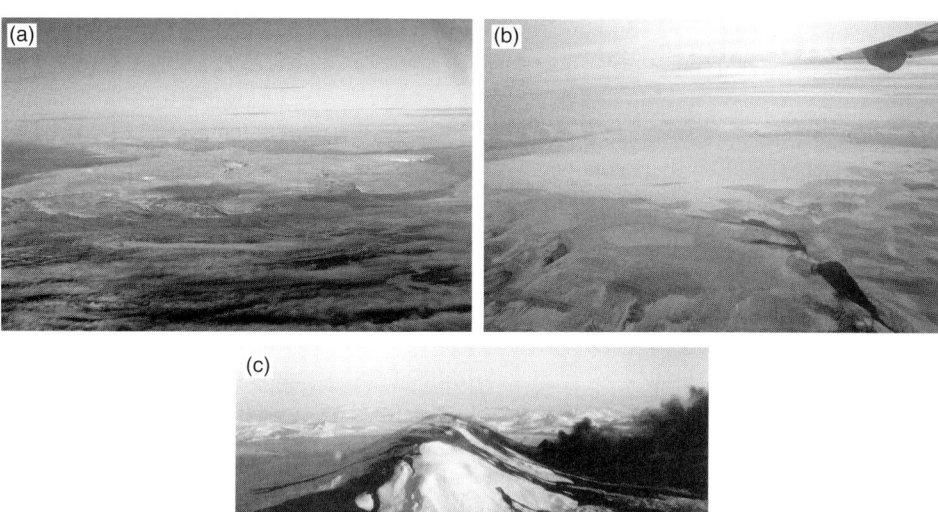

Fig. 3.2 Aerial images of Katla and Hekla volcanoes in southern Iceland. (a) Katla volcano viewed looking from the east showing the entire Myrdalsjökull ice cap. (b) Close-up aerial view of the ice-filled caldera in Katla. (c) View looking to the east-northeast showing lava flows and snow cover during the 1991 eruption of Hekla volcano (Fig. 3 from Gudmundsson et al., 1992).

It is the effects of the associated jökulhlaup for which the eruption is best known and, indeed, for which Katla eruptions generally are notorious. Eruptions of Katla volcano have had an average frequency of two per century during the last 11 centuries and many were associated with major jökulhlaups (Jónsson, 1982; Tómasson, 1996; Larsen, 2000). Although the jökulhlaup of 1918 was undoubtedly a major event, the jökulhlaups associated with eruptions in 1660, 1721 and 1755 may have been more voluminous and violent. Jónsson (1982), citing observations of icebergs that floated on the 1918 flood rather than submerged within it (suggesting that the flow had significant yield strength), believed that the 1918 jökulhlaup was so densely charged with detritus that it was effectively a debris flow. He referred to the jökulhlaups from Katla as *Kötluhlaups* (i.e. Katla floods). Conversely, Tómasson (1996) suggested that the sediment loads were much lower. Moreover, photographs of the resultant deposits on Mýrdalssandur (e.g. Jónsson, 1982, Figs. 3 and 4) show prominent planar stratification consistent with transport within concentrated or possibly hyperconcentrated density flows, categories of sediment flow unrecognised at the time when Jónsson (1982) undertook his study. At its peak, the jökulhlaup may have transported 25 kt s^{-1} of ice and a similar amount of sediment (Tómasson, 1996).

The effects of the 1918 eruption were unusually well documented. Contemporary observations indicated that water in the jökulhlaup burst out high on the glacier surface and flowed supraglacially in at least two prominent rivers but subglacial meltwater also escaped supraglacially from conduits lower down the glacier, and exited from several major portals at the glacier snout. Similar observations of meltwater escaping supraglacially, from supraglacial conduits, and probably subglacially were made during the 1969 eruption of Deception Island 1969 (Section 3.3.3). The force of the meltwater discharge ripped off substantial icebergs 40–60 m in diameter from the glacier terminus and swept them down to the sea, leaving many stranded on the sandur when the jökulhlaup diminished. Together, the flood inundated an area of c. 370–400 km^2 on Mýrdalssandur and travelled at speeds in excess of c. 10 m s^{-1}. Maizels (1993) estimated a flood velocity of 10–15 m s^{-1}, with the higher figure associated with deeper channels. Peak discharge across the entire sandur was either 2 × 10^6 m^3 s^{-1} (Jónsson, 1982) or 3 × 10^5 m^3 s^{-1} (Tómasson, 1996). It occurred in two pronounced stages, of which the second was the most voluminous. The flow was essentially steady for the first 2 hours. At 1700 hours, close to the glacier, it rapidly increased and was associated with a sudden influx of iceberg-sized masses of glacier ice. According to Jónsson (1982), the first phase may have been principally related to supraglacial discharge whilst the second phase may have been associated with supraglacial, intraglacial (through shear planes and crevasses) and subglacial discharge. However, Tómasson (1996) suggested that the change corresponded to the release of a logjam of large blocks of ice that blocked the gap between two small hills called Hafursey and Selfjall, situated just 4 km from the glacier front. The deluge also transported huge blocks of rock measuring over 6 m in diameter. Jónsson (1982) quoted flood depths of 60–70 m based on contemporary observations but Tómasson (1996) used a mean depth of 20–30 m for his calculations of discharge. Depths of c. 80 m probably occurred in constricted gorges close to the glacier front.

The flood continued until dark. It is unclear when it finished but it was over by dawn next day (c. 0700). Tómasson (1996) estimated that the main flood lasted about 5 hours before it began to abate. The jökulhlaup extended the local coastline at Kötlutangi by about 4 km where it ended in water depths of c. 100 m. The new land was initially about 14 km^2 in area but it retreated very rapidly. However, a strip of new land about 2 km wide was permanently added to the coastline between Kötlutangi and Vik. About 1 km^3 of volcanic material was deposited above sea level but an unknown volume ended up in the sea. About 0.3 m^3 of tephra was erupted into the atmosphere but most (c. 0.4 km^3?) was carried away by the jökulhlaup. The glacier front was temporarily floated and may have moved forward but a gorge 1.5–1.8 km long with walls 145 m high was also blasted out of the terminus by the force of the discharge. The eruption ended after 24 days (Larsen, 2000). The course of events described above is probably typical for the Katla volcano.

3.3.2 Hekla 1947 (1980, 1991, 2000)

Hekla volcano is located in south central Iceland within the South Iceland Seismic Zone, which links the Eastern and Western volcanic zones (Thordarson and Larsen, 2007; Fig. 3.2; see Chapter 2). The volcano has a complex morphology that when viewed from the south or north gives it a 'classic' composite volcano profile, but when viewed from the east or west shows it to be ridge-like. It has a long historical eruption record (>18 eruptions since AD 980; see Thorarinsson, 1967 for a detailed historical summary), and has had four well-documented eruptions since 1946 (in 1947, 1980–1, 1991, 2000). While it can produce dacitic to rhyolitic pumice, the largest volumes of effusive material are 'icelandite' in composition, which is similar to basaltic andesite but with somewhat higher concentrations of Al_2O_3. Because of its eruption frequency and its potential for large Plinian eruptions, the volcano has been studied in detail (cf. Grönvold et al., 1983; Gudmundsson et al., 1992; Höskuldsson et al. 2007). While Hekla has little if any ice-cover today, during the 1947 and 1980 eruptions at least small remnants of glaciers still existed near the summit, and the mountain is snow-covered during winter months. Written records of Hekla eruptions date from AD 1104, and the volcano's ice and snow cover are mentioned sporadically in descriptions of subsequent eruptions (Thorarinsson, 1967). While little direct evidence of flooding related to snow/ice-melting exists for eruptions prior to 1766, Thorarinsson noted that one observer of the April 1766 eruption, Hannes Finnsson, suggested that the Rangá River floods each time Hekla erupts, and Thorarinsson speculated that this was probably in part due to melting of summit snow and ice by eruptions (Thorarinsson, 1967, p. 111). The next major eruption, which started on 3 September 1845, seems to validate this speculation as large floods in the Rangá and the Markarfljót rivers accompanied the onset of the 1845 eruption, which observers attributed to melting of the extensive snow and ice cover on Hekla at that time (Thorarinsson, 1967). Observers also noted that the water in the Rangá was abnormally warm at the start of the flood.

The 1947 eruption of Hekla volcano marked one of the first modern eruptions where scientists were able to make detailed observations of almost all aspects of the eruption including measurements of lava temperatures, viscosities, compilation of eyewitness accounts, photographs of the growth of the initial eruption column, and geochemistry and petrology of the eruption products; reconnaissance aerial observations were also made. The eruption began on the evening of 29 March 1947, and was witnessed by many people who lived in the area. Two important sets of observations were especially relevant for studies of volcano–ice interactions: firstly, numerous lava flows were seen travelling over snow without causing rapid melting (Einarsson, 1949; Kjartansson, 1951). Several different accounts showed that lava flows moving at c. 1 m min^{-1} travelled across snow fields without causing significant immediate melting, although the surface of the snow was wet within 50 m of the radiant lava flow front (e.g. Einarsson, 1949). Likewise, where incandescent blocks rolled off the front of the advancing flows directly onto surrounding snow they generated small volumes of steam. Kjartansson (1951) dug into the edges of lavas up to 86 days after cessation of the flows and was still able to find unmelted snowpack more than 1 m thick, although in two of the four locations a layer of airfall tephra had been deposited on top of the snow prior to the arrival of the lava. The second set of significant observations pertain to the flood or hlaup that occurred at the very start of the eruption. While floods had been noted during earlier eruptions, the 1947 flood was documented in detail by Kjartansson (1951), who spent many visits in the field during and after the eruption examining deposits from the flood as well as compiling eyewitness accounts of the course of the flood. His reconstructions suggest that the flood had a very approximate total discharge of 3 000 000 m^3 and a maximum instantaneous discharge of up to c. 880 m^3 s^{-1}. The floodwaters are estimated to have locally reached temperatures of 40–50 °C and to have been the colour of rivers charged with glacier meltwater. Kjartansson (1951) also noted that the flood deposits had unusually high proportions of juvenile volcanic materials, especially volcanic bombs. A few observers noted at the beginning of the eruption that dark streaks appeared rapidly moving down the slopes of the snow-covered volcano, especially on the northwestern flank. When Kjartansson later visited that location on Hekla, he estimated that the flood had an aggregate width of over 3 km and had caused significant erosion to underlying remnants of glaciers. While he found evidence for significant melting of snow-pack and ice, Kjartansson (1951) also speculated that some of the water volume for the flooding might have originated from the magma. He was unsure how a mixture of lava and pyroclasts could cause such rapid melting. However, given the reports of steam clouds flowing down the flanks of the volcano, it may be that something akin to a base surge at the start of the eruption could have facilitated more rapid melting.

Subsequent eruptions in 1970, 1980–1, 1991 and 2000 have all been smaller in volume and lacking large hlaups. The 1980–1 eruption had two short phases. The first began on 17 August at 1327 hours with a moderate explosion (Grönvold et al., 1983). Photographs of the eruption column appear to show a surge-like cloud moving downslope accompanied by clouds of steam being produced by snow and ice melt. Floodwaters moved down the northern slopes of Hekla, and then were dammed when they encountered one of the 1970 lava flows. The overall meltwater volume was low, and the water dissipated relatively quickly into the underlying lava flows (Grönvold et al., 1983). The first phase of the eruption, which produced tephra and six different lava flows, ended around 20 August after just 4 days of activity. The second phase began 8 months later on 9 April 1981 with another set of explosions that produced a new crater row and one main lava flow. Although it is not described in detail, apparently a small lava flow travelled beneath a remnant of a summit glacier and appeared at the base of the ice at 740 m a.s.l. (Grönvold et al., 1983). The second phase ended by 16 April, and the two phases together produced c. 0.15 km^3 of lava mainly on the northwestern slopes of Hekla. The next eruption began on 17 January 1991 at 1700 hours after a repose time of just under 10 years, and, although it produced a similar overall volume compared to the 1980–1 eruption, most of the lava flowed down the eastern flanks of the volcano (Gudmundsson et al., 1992). This eruption occurred in a single phase that ended on 11 March after less than 2 months of activity. Initial lava discharge over the first 2 days was estimated to be 800 m^3 s^{-1}, from which it declined to an average of 1–2 m^3 s^{-1}. However, even though this eruption occurred during the winter when Hekla was visibly covered by snow (see Fig. 3 in Gudmundsson et al., 1992), no flooding was reported. The only indication of possible snowmelt took the form of a brief spike in F concentrations in the Rangá River on the fifth day after the eruption began, and this was attributed to snowmelt caused by rain (Gudmundsson et al., 1992). The most recent eruption of Hekla also began in the winter on 26 February 2000. The eruption was somewhat different from the previous five in that its initial explosive phase was minor and short-lived (Höskuldsson et al., 2007). The initial discharge was estimated to be briefly as high as 2600 m^3 s^{-1}, but it steadily declined over the first 12 hours to less than 300 m^3 s^{-1} (Höskuldsson et al., 2007). While this eruption was also short-lived, ending after just 12 days, it produced c. 0.19 km^3 of predominantly lava mainly on the southern flanks of the edifice. As with the 1991 eruption, even though the area was demonstrably snow-covered during the eruption (see Figs. 6 and 8 in Höskuldsson et al., 2007), no meltwater production was reported nor any evidence of lava–snow interactions.

3.3.3 Deception Island 1969

Deception Island is a volcano situated in Bransfield Strait, west of northern Antarctic Peninsula (see Chapter 2). A subglacial eruption took place on the island in February 1969. The total volume of ejected magma (mainly basaltic andesite in composition) was small, just 0.03 km^3. Because the island hosted three scientific stations and because of another eruption in 1967, there were closely spaced precursor and follow-up geological investigations. Thus, the 1969 eruption and its volcanic products were exceptionally well documented, including a seismic record and an extensive photographic database of the aftermath (Baker et al., 1975). Indeed, it took until the eruption at Gjálp, Iceland, in 1996, for a glaciovolcanic eruption to be better documented, although the importance of the Deception 1969 eruption was largely missed until recently (Smellie, 2002; Fig. 3.3). Because it occurred from a series of arcuate caldera-margin fractures that crossed a small ice cap formed of ice about 70 m thick (range: c. 50–100 m), the Deception eruption was glaciovolcanic but the preserved products (scoria characteristic of magmatic eruptions) do not reflect any glaciovolcanic influence. The eruption produced three en echelon gaping fissures carved in the ice cap, individually measuring 500–1000 m long and 100–150 m wide, one of which showed no sign of eruptive activity other than local passive gas emission. Several small pyroclastic cones were constructed within the two other fissures and spread onto the immediately surrounding ice surface.

The fissures and several associated vertical ice chimneys that were melted in the ice above vents, plus ice portals in the slopes below the fissures, were the sources for a major jökulhlaup that overflowed onto the ice cap (Fig. 3.3; see also Katla 1918; Section 3.3.1) and helped to destroy one of the scientific stations; a second station situated in a proximal position to one of the pyroclastic cones was largely buried and burned down by the thickness of accumulated hot tephra. Tephra and reworked glacial till (often comprising large blocks of frozen till a few metres in diameter) were carried by the flood and entered the sea in the island's flooded caldera, extending the coastline by 200 m in a series of short-lived coalesced sediment fans. The ice downslope of the fissures also suffered a short surge-like advance, presumably related to some subglacial meltwater escape simultaneous with the supraglacial discharge (as during Katla 1918; Section 3.3.1). The initial eruption column was white, signifying that it was composed of steam, and it did not transform to a grey tephra-bearing column until a few minutes later. About 76×10^6 m^3 of ice was melted at the fissures and the jökulhlaup probably had a mean discharge rate of 6.6×10^3 m^3 s^{-1}.

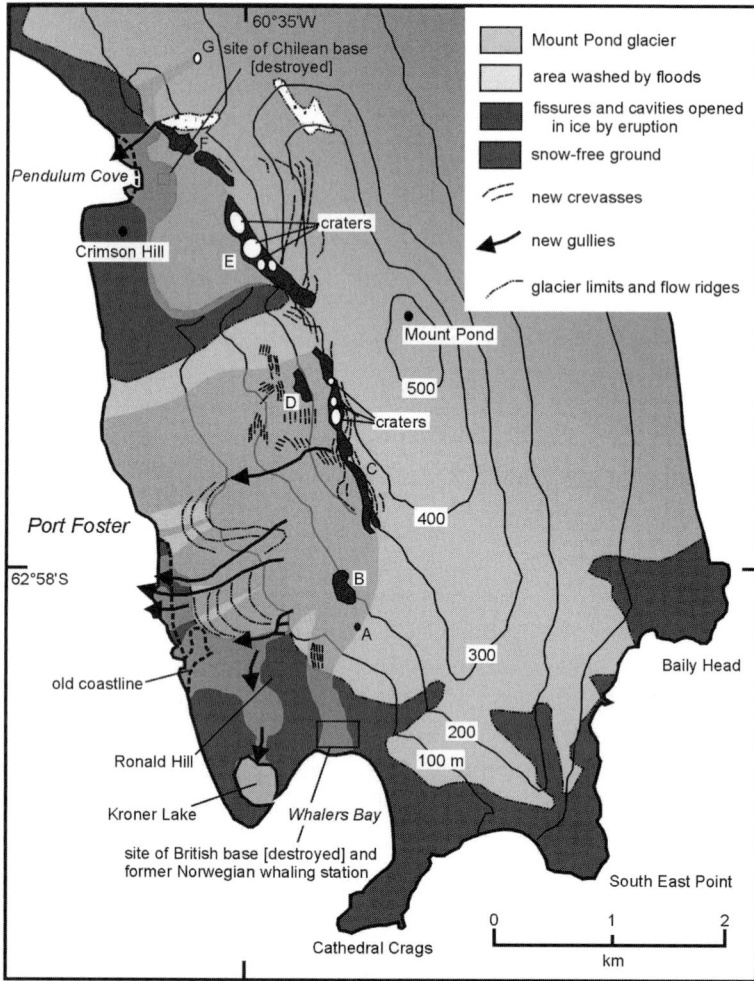

Fig. 3.3 Sketch map of eastern Deception Island showing the major features and effects of the 1969 subglacial eruption, including ice fissures and cavities opened up by the eruption, pyroclastic cones, the area washed by supraglacial meltwater flood, and the extended coastline in the floodplain below (from Smellie, 2002). (A black and white version of this figure will appear in some formats. For the colour version, please refer to the plate section.)

The eruption was characterised as one from a thin glacier (*sensu* Smellie and Skilling, 1994) but the supraglacial flooding was anomalous for such eruptions and more akin to eruptions within much thicker ice. This was ascribed to the eruptive fissures being situated within the ablation zone of the ice cap and the glacial cover being composed of ice (impermeable away from crevasses) rather than some combination of (permeable) snow, firn and ice. However, apart from several prominent ice portals on the glacier slopes, which were probably opened up by basal meltwater cutting up through the ice under pressure, no evidence was described for similar portals at the glacier front. Thus the extent of the subglacial meltwater flow is unclear and it is plausible that on such relatively steep slopes the escaping (mainly over-flowing) meltwater was mainly channelled to the surface via the abundant deep fractures (crevasses). Smellie (2002) postulated that the rising magma was preceded by juvenile and hydrothermal volatiles, as superheated steam, which initiated subglacial melting ahead of the magma. The gas-driven melting may have been responsible for the very rapid rates of ice melting observed (estimated as 0.02 m s^{-1}; i.e. about three times faster than at Gjálp (c. 0.005 m s^{-1}); cf. Gudmundsson et al., 1997). Speculatively, overpressures of at least several MPa associated with the gases may have lifted the overlying ice briefly, allowing early-formed water to escape subglacially, thus helping to create a largely dry cavity and promoting the eruption of 'dry' scoria.

3.3.4 Mount St Helens 1980–6, 2004–6

Mount St Helens (MSH) is a composite volcano in southwestern Washington State and is part of the Cascade volcanic arc (Fig. 3.4; see Chapter 2). It is one of the youngest large edifices in the Cascades, with a history that mainly started c. 40 k.y. ago (Mullineaux and Crandall, 1981). For most of its history, MSH has produced explosive eruptions of dacitic magma, with only recent excursions to less silicic andesite and basalt. A potential indication of the significance of glaciovolcanism in the history of the volcano is that lahar deposits are a regular part of the volcano's 40 k.y. stratigraphy (Mullineaux and Crandall, 1981); it has also been suggested that some of the eruptions prior to 1980 actually produced larger lahars than those witnessed after the 18 May eruption (Janda et al., 1981). Before the eruption began, the volcano hosted 11 named glaciers and numerous areas of perennial ice and snow; ice cover is estimated to have been 5 km^2, and the eruption is thought to have removed c. 70% of that total (Brugmand and Meier, 1981). From the beginning of its reawakening in 1980, volcano–ice interactions had a pivotal role for not only influencing the styles of eruption and transport/depositional processes, but also for monitoring of activity. Studies of the eruptive activities from 1980 to 1986 (e.g. Brantley and Waitt, 1988; Scott, 1988, 1989; Pierson, 1985; Pierson and Waitt, 1999; Waitt et al., 1983; Waitt, 1989), and again from 2004 to 2006 (Schilling et al., 2004; Walder et al., 2007, 2008) have shed many significant insights on glaciovolcanism at ice-capped composite volcanoes, which make up the significant majority of all glaciovolcanoes on Earth.

Research on the MSH eruptions and their deposits has produced one of the largest bodies of literature on any modern volcano (>500 publications). The first written descriptions based on the 1980 MSH eruptions were compiled as a Professional Paper by the United States Geological Survey (USGS Professional Paper 1250) and published in 1981, only a year after the eruption onset and while the volcano was still active (eruptions continued sporadically until 1986). That volume contained 62 papers, which form the basis for the general description of events given below. The 18 May 1980 eruption of MSH was one of the most destructive glaciovolcanic eruptions of the twentieth century (although see

Fig. 3.4 Views of volcano–ice interactions during eruptions from Mt St Helens. (a) Aerial view of the ice cap with a small crater melted through the ice during April 1980 (prior to the cataclysmic eruption). (b) View of the 2004–6 lava dome emplaced into the 1980 crater at Mt St Helens showing small phreatic eruption through the recently formed glacier within the 1980 crater. Note how the glacier ice has been displaced to the left by the growing dome. (All images courtesy of United States Geological Survey.)

Nevado del Ruiz; Section 3.3.6). The eruption killed 56 people and produced large debris flows that travelled more than 120 km downstream in the Toutle and Columbia rivers before entering the Pacific Ocean (Janda et al., 1981); it is estimated to have caused several billion US dollars in damages (Christiansen and Peterson, 1981).

Christiansen and Peterson (1981) gave one of the first detailed chronologies of the eruption, whose initiation was signalled by a series of earthquakes on 20 March 1980;

the following is a summary taken largely from their work. The seismicity within the edifice produced fractures in the summit glaciers around MSH as well as small snow/ice avalanches. On 27 March intermittent steam explosions began from the ice-filled summit crater and produced an opening estimated to be 60–75 m in diameter. From that date rapid deformation was noted on the north flank of the volcano, and a phreatic/phreatomagmatic phase of small explosions, which lasted from a few minutes to a few hours each, commenced and continued until 22 April. Evidence for an external water component to the blasts includes the combination of 'cocks-tail' jets and steam plumes (e.g. White and Houghton, 2000). Fumaroles were visible in the crater area from 22 April onwards, as well as emanating from fractures in the glaciers on the northern side of the volcano. The phreatic explosions began again on 7 May, and continued through to 14 May. The sequence of events that resulted in the largest historic volcanic avalanche and directed blast began at 0832 on 18 May. They were initiated by a magnitude 5 earthquake that caused the oversteepened north slope of the volcano to collapse, depressurising the dacitic cryptodome within the mountain as well as the overlying hydrothermal system. This synchronous drop in pressure generated a directed blast to the north of the volcano, probably fuelled by expansion of a combination of magmatic gases, externally derived hydrothermal fluids, and possibly some contribution from rapid melting of overlying ice and snowpack. After the initial blast and avalanche removed the volcano's summit, a vertically directed Plinian eruption column developed and was sustained for 9 hours. The complex set of events triggered over the span of a few minutes produced a variety of volcano–ice interactions, including (1) melting of snow and ice by the directed blast to add volume to the horizontally directed gas–ash mixture, in addition to an unquantifiable volume of melted snow and ice that was part of the hydrothermal system; (2) blocks of snow and ice incorporated into the avalanche deposit transported into the upper drainage of the North Fork of the Toutle River, which subsequently melted and contributed to the conversion of avalanche deposits to lahars; (3) melting of snow and ice by pyroclastic density currents that swept down drainages on the east (Muddy River and Pine Creek), northwest (South Fork of the Toutle River) and southwest (Kalama River) flanks of the volcano. Even though up to 70% of the 0.18 km^3 of ice cover was removed during the eruption (most presumably over the span of 9 hours), post-eruption studies of the glaciers by Brugman and Meier (1981) showed that while some were totally eliminated (Loowit and Leschi), others were hardly affected (Swift, Dryer). In fact, the ice cover on the south flank glaciers had unusually high positive mass balances for 1980 owing to the insulating effects of the ash (Brugman and Meier, 1981). Estimated discharge rates for lahars generated during the eruption vary up to 6800 m^3 s^{-1}. The total estimated volume of juvenile volcanic material erupted on 18 May was c. 1 km^3 (Christiansen and Peterson, 1981). While MSH produced five more explosive eruptions in 1980 (25 May, 12 June, 22 June, 7 August, 16–18 October), and continued growth of lava domes through 1986 produced sporadic explosions (e.g. Waitt et al., 1983), the most significant period of volcano–ice interactions occurred on 18 May. See Chapter 15 for more detailed discussions of hazards associated with MSH eruptions.

The end to the 1980s sequence occurred in October 1986 with cessation of dome growth after more than 17 episodes of activity, with the exception of minor phreatic events in 1989–90 and 1990–1 (Siebert et al., 2010). The renewed dome growth in 2004–6 was not nearly as dramatic as the 1980s activity, but it was also well documented and almost as instructive for glaciovolcanic studies (e.g. Schilling et al. 2004; Walder et al., 2007). A new dome was extruded into the pre-existing Crater glacier that began to form after the end of activity in 1986; the glacier is of note in that is it one of the only documented new glaciers to have formed in the late twentieth century. The emplacement of the new dome apparently physically displaced the ice with almost no melting (Walder et al., 2007). This is an important reminder that lava emplacement directly into ice will not necessary produce strong heating effects in all glaciovolcanic environments.

One of the long-term hazards related at least in part to eruptions of MSH is the amount of easily erodable material that was deposited in surrounding drainages. Scott (1989) estimated that the maximum discharges associated with the 18 May 1980 eruptions were 6800 $m^3 s^{-1}$ in the North Fork of the Toutle River, and 3600 $m^3 s^{-1}$ in the South Fork (earlier hydrograph analysis by Fairchild (1986) showed that the peak in the South Fork was more than an order of magnitude higher at c. 68 000 $m^3 s^{-1}$). However, Scott (1989) also examined ancient deposits generated by catastrophic collapses of the natural dam at Spirit Lake, immediately north of MSH, and found evidence for floods with maximum discharges of 200 000–300 000 $m^3 s^{-1}$ at a distance of c. 50 km downstream from the lake. These estimated discharge rates are as large as those from Katla, but are situated much further from the source; this magnitude of discharge is approximately equivalent to that of the Amazon River. Scott (1989) suggested that remedial engineering work on the Spirit Lake dam was critical to prevent future catastrophic collapses, which could be instigated by renewed glaciovolcanism at MSH.

3.3.5 Veniaminof 1983–4 (1993–4, 2013)

Veniaminof is a large, Pleistocene composite volcano with a summit caldera (Fig. 3.5). It is situated in the eastern part of the Aleutian volcanic arc (Yount et al., 1985; Welch et al., 2007). Its prehistoric history is poorly known, but it has had many historical eruptions from tephra cones that protrude up above the ice-filled summit caldera. The eruption geometry of Veniaminof is unique among most ice-capped volcanoes because eruptions through the cones emit lava flows that, emanating above the caldera ice, flow down onto the ice surface (Fig. 15.4). At least three such eruptions have been documented (1983–4 (Yount et al., 1985), 1993–4 and 2013), but only at the 1983 event was meltwater observed. During the 1983 eruption, a lake was formed at the base of the cone within the ice. It eventually drained, but its small volume meant that it caused little increase above the baseline water discharges coming from the large caldera-filling glacier (Yount et al., 1985). However, even though apparent drainage tunnels were visible, no significant increase in discharge from any of the summit drainages was noted. Ice radar studies of the caldera show that the ice is up to 400 m thick adjacent to the

Fig. 3.5 Veniaminof volcano, Alaska, viewed from the northeast showing the small englacial lake formed during the 1983–4 eruption caused by melted by 'a'ā lavas where they abutted the ice in the surrounding caldera (image courtesy of United States Geological Survey).

tephra cone (Welch et al., 2007), so it is possible that the total volume of ice melted during the eruption produced only a modest increase in the overall fluvial discharge from the caldera. The 1983–4 eruption created the only glaciovolcanic lava-fed delta ever observed. Meltwater escaped from the lake both subglacially and supraglacially (Smellie, 2006, Figs. 13, 14). Subsequent eruptions in 1993 and 2013 also produced visible melting of the ice around the base of the tephra cone, but rates of melting were low enough (or drainage fast enough) that no lakes were observed. A brief observation flight during August 2013 showed that lava flows were travelling down the eastern flanks of the cone into pits melted into the ice surface. An abundance of steam suggests that meltwater was draining onto the upper surface of the lava flows as they melted into ice tunnels. Comparison of aerial images shows that the 1983 eruption produced a large volume of lava, all of which came down the south side of the cone. This focusing of lava onto one area may have been responsible for the more extensive melting seen in 1983. During the 2013 eruption, lava flows were emplaced on the west and east sides of the cone, diffusing the overall volume of lava emplaced onto the surrounding caldera ice.

3.3.6 Nevado del Ruiz 1985

Nevado del Ruiz is a large composite volcano in Colombia situated within the northern Ecuador section of the Andes volcanic arc (see Chapter 2). The summit of the volcano is 5321 m a.s.l. and it is covered by nine named glaciers (Lagunillas, Azufrado, Guali, Molinos, Farallon, Nereidas, Alfombrales, Recio, El Oso) whose total estimated, pre-eruptive surface area and volume were c. 25 km^2 and 0.7 km^3, respectively (Thouret, 1990). The 1985 eruption of Nevado del Ruiz in Colombia produced one of the largest volcanic disasters of the twentieth century when a small eruption produced lahars during a relatively brief period of explosive activity that killed more than 24 000 people. The lahars travelled tens of kilometres down multiple river valleys and inundated several towns, including the town of Armero, where more than 23 000 people were killed. In part due to the significance of the disaster, the Nevado del Ruiz eruption has been well studied in spite of its relatively small size (overall <0.02 km³ DRE; Table 3.1) (e.g. Naranjo et al., 1986; Calavache, 1990; Gourgard and Thouret, 1990; Melson et al., 1990; Pierson et al., 1990; Sigurdsson et al., 1990; Thouret, 1990). Because they were so destructive, the formation of the lahars initiated by the 13 November 1985 eruption has been studied in detail (e.g. Pierson et al., 1990; see Chapter 15). Nevado del Ruiz had been essentially dormant since its last major eruptive phase on 12 March 1595, which is also thought to have produced large, destructive lahars (Pierson et al., 1990). Besides a small, poorly understood explosion in 1845 that produced another damaging lahar, the volcano had been quiet until renewed seismicity began in November 1984, followed by a small explosion on 11 September 1985, which produced a small lahar (Naranjo et al., 1986). The eruption sequence and events that followed on 13 November are described in detail by Pierson et al. (1990), but only briefly summarised here. A phreatic explosion at 1505 hows occurred in one of the several summit craters (Arenas), producing tephra fall and a small avalanche. Seismicity decreased over the next 6 hours until at 2108 hours a second explosive phase began. While this phase only lasted a few minutes, it produced both dilute (e.g. surge) and concentrated pyroclastic density currents as well as depositing large ballistic blocks and bombs on the summit glaciers. At least four surge and four pyroclastic flow deposits from this phase were later identified at distances up to 5.5 km from the main crater. The penultimate phase of this relatively brief eruption episode began at 2131 hours and was short-lived, but produced a Plinian eruption column with a volcanic explosivity index (VEI) of 3. A second series of surge deposits were formed by this final phase of the eruption, along with more tephra (ash to block-size). The entire eruption was over by 2200 hours local time.

Explosive glaciovolcanic interactions were studied in detail during this eruption. The amount of meltwater generated from different sources (e.g. glaciers and snowpack) and the specific effects of different volcanic processes (e.g. surge, flow, fall) were studied in detail, and sometimes produced conflicting results (cf. Pierson et al., 1990; Thouret 1990). However, the following observations summarised from Pierson et al. (1990) added critical knowledge to our understanding of how pyroclasts interact with the cryosphere during eruptions. Firstly, the surge deposits suggest that the dilute, fast moving pyroclastic density currents did not have enough thermal mass to produce significant melting. Although they did produce local scour, Pierson et al. (1990) speculated that it mainly occurred in the snowpack and did not extend into the underlying ice. Local surge deposits contained unmelted blocks of ice and even snow crystals, attesting to their low thermal mass. Secondly, the higher density pyroclastic flows not only produced more local erosion into the underlying glaciers (channels up to 4 m deep in the Nereidas, Azufrado and Lagunillas glaciers), but also incorporated ice and snow. While unmelted blocks were still found within flow margin levees, unmelted blocks were not reported from within flow channels.

Crevasses that were overrun by the denser pyroclastic flows also showed evidence of physical abrasion, and small craters produced by later steam explosions were also found. Locally large bombs produced melt pockets in surface snow up to 0.5 m deep (Pierson et al., 1990). Later discharge from englacial drainages at the termini of several glaciers also demonstrated that some of the meltwater generated by the eruption was temporarily stored subglacially (Pierson et al., 1990; Thouret, 1990). The eruption also generated a variety of secondary mass movements, including rock avalanches, 'slush' avalanches, and 'mixed' avalanches; the last of these was probably a significant contributor to the generation of lahars (Pierson et al., 1990).

Unlike the primary fragmental deposits, which were only locally preserved on top of the surrounding ice, floods and lahars generated by the eruption travelled down every drainage from the southwestern to the southeastern flanks of the volcano's summit (210° of the summit circumference; Pierson et al., 1990). Of the four major drainages affected, the channelised flows travelled 27 to 102 km from the crater at velocities generally between 5 and 15 m s^{-1}. Estimated peak discharges varied from as low as 710 m^3 s^{-1} (Lagunillas drainage) to 48 000 m^3 s^{-1} (Azufrado drainage). Pierson et al. (1990) estimated overall lahar volumes, which ranged up to c. 41 × 10^6 m^3, as well as 'bulking rates' along flow paths, which were as high as 680 m^3 m^{-1}. These estimates are particularly important for enabling better estimates for how rapidly volume can be added to flows along their transport paths. The only lahar that changed into a non-channelised flow was the one that buried the town of Armero; that flow resulted from the confluence of the Azufrado and Lagunillas flows, which arrived in multiple pulses. The first pulse inundated most of the town with debris up to 1.5 m thick travelling at c. 12 m s^{-1} and the final deposits covered more than 34 km^2 (data from Mojica et al., 1986 cited by Pierson et al., 1990).

The 1985 eruption was important for several reasons. First and most importantly, it demonstrated that even a minor eruption of an ice-capped volcano can produce a human disaster. It also showed that based on direct analysis of deposits a variety of different interactions could generate melting of snowpack and ice, and it demonstrated the importance of 'bulking' along flow paths to increase overall mass fluxes of lahars/debris flows. The eruption was one of the first documented glaciovolcanic eruptions of a mixed magma (andesite–dacite; Sigurdsson et al., 1990). Finally, the eruption established the critical importance of hazard mapping (Armero was known to have been built on ancient lahar deposits from the much larger 1595 and 1845 events) and of communication for preventing volcanic disasters; in the 7 hours between the eruptions that produced the initial meltwater and the arrival of the lahars at major towns, many of the people who were killed could probably have been easily moved to safety.

3.3.7 Villarrica (1971, 1984) and Llaima (1994)

Villarrica (39.42° S/71.95° W) and Llaima (38.70° S/71.73° W) are basaltic to basaltic andesite composite volcanoes in central Chile (González-Ferrán, 1995; Fig. 3.6); they are

Fig. 3.6 Views of Villarrica volcano in southern Chile. (a) View looking to the south from Pucon of the southern flank of Villarrica showing extensive spring snow cover. (b) View of the snow/ice-clad summit of Villarrica (people on crater rim for scale). (c) Satellite image of Villarrica volcano in March 2015 showing ash and lahar deposits on the north and east flanks, on top of snow/ice-clad summit (image from EOS/NASA).

part of the Southern volcanic zone of the Andes volcanic arc (see Chapter 2). While both have long records of historical eruptions that have produced damaging lahars from Strombolian activity, they also are two of the few volcanoes for which lava flows have been observed melting tunnels into ice (Moreno and Fuentealba, 1994; Naranjo and Moreno, 2004). Villarrica is a symmetrical composite volcano whose 2847 m a.s.l. summit frequently hosts an active lava lake within a crater 200 m in diameter. The seasonal snow depth is up to 20 m and the summit has ice above 1900 m a.s.l. that covers c. 30.2 km^2, with an estimated volume of c. 2.3 km^3 and local thicknesses up to 150 m (Castruccio et al., 2010). Lahars generated by a combination of effusive and explosive eruptions accounted for more than 100 fatalities during eruptions in 1908, 1948–9, 1963, 1964 and 1971. Earlier eruptions are also thought to have produced significant fatalities (e.g. 350 in AD 1575; González-Ferrán, 1995). Because Villarrica has an open summit crater that is surrounded by ice, any magma or hot country rock that leaves the crater can potentially produce melting of the surrounding ice and snowpack. The eruptions of 1971 and 1984–5 illustrate the main type of behaviour at Villarrica and its distinctive consequences. According to Naranjo and Moreno (2004), the 1971 eruption, which began in late October, significantly increased in intensity during the night of 29 December. An earlier phase of activity on 29 November had produced small lava flows on the southwestern flank of the volcano that melted channels into the surrounding ice up to 40 m deep at a relatively slow rate, and lacked associated floods or lahars. However, the 29 December fissure opened across the volcano's summit crater and lava fountains up to 400 m high were observed; the eruption rate was estimated to be 500 m^3 s^{-1}, and the activity produced enough melting of the summit glaciers to generate lahars in five different river drainages (Turbio, Zanjon Seco, Correntoso, Voipir, Chailupén) with maximum velocities up to $36 \, \mathrm{m \, s^{-1}}$ and maximum laharic discharges up to $20\,000 \, \mathrm{m^3 \, s^{-1}}$. By contrast, the 1984–5 Strombolian/Hawaiian eruption at Villarrica began on 30 October. It was episodically active until 20 December and produced multiple lava flows that melted channels in ice and snow in the north and northeastern sides of the summit but only had effusion rates up to $20 \, \mathrm{m^3 \, s^{-1}}$ (González-Ferrán, 1995; unpublished observations of Moreno et al. cited by Lescinksy, 1999; Naranjo and Moreno, 2004). The eruption produced three separate channels through the ice, one estimated to be 1 km long, 50 m wide and 30–40 m deep, the second 200 m long and 50 m wide, and the third 1 km long, 80 m wide and 40 m deep. Activity after 6–9 November produced a lava flow that travelled subglacially for c. 150 m before emerging onto the glacier surface. An increase in activity from 1 to 6 December apparently increased the rate of snow and ice melting sufficiently to produce a flood down the Río Pedregoso on the volcano's north lower flank. The lower effusion rate is thought to have prevented melting rapid enough to generate lahars. Other impacts of lava–ice interactions reported during these eruptions included generation of voluminous steam at the lava flow fronts, formation of crevasses in the surrounding ice, and mixed snow–rock avalanches.

Llaima volcano, whose summit is 3125 m a.s.l., is very similar in behaviour to Villarrica. It produces Strombolian and Hawaiian eruption on a frequent basis, and has had more than 30 eruptions since AD 1640 (González-Ferrán, 1995). Its summit is glacierised, and it is one of the largest volcanoes in the Andes, with a total estimated volume of more than 400 $\mathrm{km^3}$ (Naranjo and Moreno, 1991). Moreno and Fuentalba (1994) summarised the unique activity from the 1994 eruption that lasted 2 days (17–19 March). The eruption began on the morning of 17 March, at some time before 0500 hours. The source of the eruption, a 400 m fissure, was seen from the air to be producing lava fountaining and minor ash; the fissure fed a lava flow that exited the fissure by flowing beneath one of the western summit glaciers. The lava travelled subglacially for approximately 2 km before exiting from the lower end of the ice. It was estimated that c. $3–4 \times 10^6 \mathrm{m^3}$ of ice was melted, which generated somewhat smaller debris flows that travelled downslope into the Calbuco River. In the lower stretches of the Rio Quepe the lahar is estimated to have had a velocity of 14 km $\mathrm{h^{-1}}$, and it damaged several bridges and roads. The ice showed several features resulting from the sub-ice lava, including formation of new crevasses, explosions and 'strong' steam production when viewed from the air (Moreno and Fuentalba, 1994).

3.3.8 Redoubt 1989–90, 2009

Mount Redoubt is a 3110 m-high steep stratovolcano in Alaska (Fig. 15.5). It has an ice-filled summit crater and hosts ten glaciers on its flanks. Redoubt has had two large recent eruptions, in 1989–90 and 2009, both of which produced significant melting of snow and ice on the volcano as well as melting of the Drift Glacier (Brantley, 1990; Gardner et al.,

1994; Trabant et al., 1994; Waitt et al., 1994; Bleick et al., 2013; Schaefer, 2008; Waythomas et al., 2013). The 1989 eruption produced 23 different explosive episodes that ultimately resulted in more than US$160 million in damages to local infrastructure (including US$80 million to one commercial aircraft), making it the second most expensive volcanic crisis in US history (NOAA, 2010). Both eruptions had some distinctive types of glaciovolcanic activity, with the 1989 eruption producing more explosive phases, and the 2009 eruption being dominated by dome extrusion and collapse. However, much of the volcano–ice interaction during both eruptions occurred via interactions with Drift Glacier, which exits the crater of the volcano on its north side and extends for 8 km towards Cook Inlet. The upper part of the glacier is steep and valley-confined, but the lower reaches are much flatter and form a 'piedmont' surface (Brantley, 1990). Even though Redoubt is located in an area with few permanent habitations, it is close enough to Anchorage to significantly disrupt air traffic, and by unfortunate coincidence a major oil storage terminal is located 35 km downstream on the Drift River, which is the major drainage for Drift Glacier. Formation of large debris flows has apparently been one of the characteristics of Redoubt eruptions since at least the Holocene, and the volcanic history of the present edifice extends back at least to 200 ka (Brantley, 1990)

The onset of the 1989 eruption was abrupt, with only 24 hours of increased seismicity before the first explosive activity occurred (Brantley, 1990), although retrospectively small steam plumes observed by passing airline pilots between 20 November and 8 December probably signalled increased heat fluxes from rising magma. Subsequent changes in seismicity were effective for making short-term eruption forecasts preceding four different major explosions (on 14 December, 2 January, 23 March and 6 April), although some explosions were not forecast (Brantley, 1990). Activity produced especially large laharic flows on 2 January and 15 February down the Drift River, which caused local damage to the Drift River oil terminal (Brantley, 1990). During this eruption glaciovolcanic processes included interactions between snow/ice and (1) gas-driven pyroclastic density currents (surges and flows), (2) heavy ash fall, and (3) gravity-driven dome collapses (Gardner et al., 1994). Brantley (1990) summarised seven different periods where activity was relatively consistent over the 9 month episode. The eruption broke the surface of the summit crater at 0947 hours local time on 14 December. Four especially strong explosions were registered during the first 25 hours of activity, which initially deposited mainly lithic fragments but by the end of the sequence comprised mainly juvenile material. The second largest explosion of the entire eruption occurred at 1015 hours on 15 December and lasted for 40 minutes. The explosion was the main pumice-producing episode for the entire eruption (Scott and McGimsey, 1994), and it produced pyroclastic density currents and rock/ice/snow avalanches on Drift Glacier. These mass flows produced significant ice and snow melting that transported materials more than 35 km down the Drift River, including blocks of ice up to 10 m in diameter. Starting on 22 December extrusion of a lava dome in the summit crater was confirmed, with an estimated volume of 25×10^6 m^3. Changes in seismicity beginning on 28 December were interpreted as indicating that the dome was becoming pressurised, and on 2 January 1990 the dome was destroyed by two explosions. The explosions produced massive debris flows and extensive melting of Drift Glacier; the flows again extended more than 35 km down river and caused the first damage to the Drift River Oil Terminal along the lower reaches of the Drift River. The debris flows had two discrete pulses and produced a peak discharge of 30 000–50 000 $m^3 s^{-1}$ at a distance of 22 km downstream; this event also led to flight cancellations and delays in at Anchorage International Airport. The rest of the dome was destroyed by smaller explosions on 8 January, which was followed almost immediately by extrusion of a second dome. At this point in the eruption, approximately 0.1 km^3 of ice from Drift Glacier had been removed by erosion, melting and transport downstream (Brantley, 1990). Debris flow deposits also significantly affected channels in the Drift River. At 0410 hours on 15 February the second dome collapsed and generated an 'ash cloud' that travelled down the glacier to the north, hit the mountain buttress across the valley and continued downstream to the east for at least 4 km; estimates of flow velocities were up to 60 km h^{-1} at the base of the mountain and 30 km h^{-1} further down valley. This event caused new melting and avalanches on the flanks of Redoubt to the east and west of the main area that had been previously affected. Ten hours after the event the temperature of deposits on top of Drift Glacier was still 230 °C. From this point for the next 3 months small domes continued to form and be destroyed, generating relatively smaller avalanches and debris flows. The main lower channel of the Drift River sporadically shifted back and forth between other minor drainages (Rust Creek and Cannery Creek). The last major dome collapse occurred on 21 April, after which a replacement dome remained relatively stable in the summit crater until the end of the eruption in August 1990.

This eruption is one of the best-monitored and characterised glaciovolcanic eruptions (e.g. Brantley, 1990; Gardner et al., 1994; Scott and McGimsey, 1994; Waitt et al., 1994). Gardner et al. (1994) and Waitt et al. (1994) described in detail the glaciovolcanic pyroclastic density currents and their deposits, while Trabant et al. (1994) focused specifically on the origin of meltwater floods.

Mount Redoubt remained essentially inactive until another eruption in 2009, which was preceded by 8 months of unrest and escalating thermal activity causing some melting before the eruption commenced (Table 3.1; Bull and Buurman, 2013; Waythomas et al., 2013; Edwards et al., 2015a). About 0.1 km^3 (dense rock equivalent) of andesite was erupted, partly as lava domes. Although subsidence and crevassing occurred on the glacier surface, melting was not confined to ice cauldrons. The lava domes repeatedly collapsed producing hot debris avalanches that descended the Drift Glacier and melted large volumes of ice, just as had happened in 1989–90. These events resulted in short-duration lahars with peak discharges near the source of c. 100 000 $m^3 s^{-1}$ (more than 10 times the peak discharge of the Eyjafjallajökull floods; see Section 3.3.13, below) that flowed down the Drift Glacier valley towards the ocean. Some lahars were essentially slurries of meltwater and snow and ice derived from the surface of the glacier. Others were hyperconcentrated flows of meltwater carrying abundant juvenile detritus mixed with ice and snow. By contrast, meltwater in the Eyjafjallajökull eruption was mainly generated by hot pyroclasts flowing down the slopes and interacting with the surface of the glacier, including seasonal snow and firn.

3.3.9 Volcán Hudson 1991

Volcán Hudson (45.9° S/72.0° W; 1905 m a.s.l.), which is a very large and remote composite volcano with a 10 km summit caldera in the Austral volcanic zone of the Andes volcanic arc (see Chapter 2), was only recognised as being an active volcano in 1970 (González-Ferrán, 1995). A poorly documented eruption in 1971, which produced lahars that caused local damage and killed five people, was hardly preparatory for the massive eruption that occurred 20 years later on 8 August 1991 (Naranjo et al., 1993; González-Ferrán, 1995). Although several studies of the eruption products have been published (Naranjo and Stern, 1998; Scasso and Carey, 2005; Kratzmann et al., 2009, 2010; Amigo, 2013), the most detailed report of the eruption and its unique glaciovolcanic activities is by Naranjo et al. (1993). The 1991 eruption is significant for many reasons; its estimated erupted volume of 2.7 km^3 DRE makes it one of the largest eruptions of any kind in the twentieth century, and probably the largest glaciovolcanic eruption in several centuries. However, studies of ash in cores from lakes to the east suggest that a much larger eruption at c. 17 ka may have released more than 20 km^3 and resulted in the formation of the present-day caldera (Weller et al., 2014). The summit caldera is estimated to contain 2.5 km^3 of ice (Naranjo and Stern, 1998).

The 1991 eruption began on 8 August at 1802 hours, and was well-documented by Naranjo et al. (1993). It comprised activity from a 4 km fissure along the western side of the caldera that produced Strombolian explosions from a newly formed tephra cone and basaltic lava flows, which was followed on 12–15 August by a massive explosion of dacitic magma from the southwestern part of the ice-filled caldera that had a VEI of 4 (González-Ferrán, 1995; Naranjo and Stern, 1998). The initial stage of the eruption is thought to have produced lahars down the Huemules valley that drains the northwestern side of the caldera. The lahars travelled all the way to the Pacific Ocean (c. 35 km to the west). An ash column rose to 12 km, and gases released during the eruption had high fluorine concentrations that later had significant impacts on livestock within the area of ash deposition (Naranjo et al., 1993). A lava flowed beneath the Huemules glacier and produced voluminous steam clouds. The eruption is thought to have begun via phreatomagmatic explosions that produced a crater c. 400 m in diameter. The explosive phase of the initial eruption produced up to 0.2 km^3 of tephra, much of which was deposited within the caldera. However, ash-size particles were transported up to 150 km to the northeast. Based on the high degree of vesiculation, Naranjo et al. (1993) suggested that the basaltic magma was already highly vesiculated before it reached the base of the glacier and initiated phreatomagmatic activity. On 23 August a 3.5 km lava flow presumably from the first stage of the eruption was observed on top of the Huemules glacier (Amigo, 2013). Unfortunately, due to dangerous field conditions the lava was never investigated in detail.

The second phase of the eruption lasted from 12 to 15 August from a 1 km diameter crater formed in the southwestern part of the caldera. This eruption sent an ash plume to an elevation of 18 km and is estimated to have released c. 4 km^3 of tephra (not DRE; Amigo,

2013). Ash fall from this second phase is estimated to have affected more than 150 000 km^2 of land to the south and east in Chile and Argentina. Ash up to 10 cm thick fell on large parts of the Falkland Islands and fine ash was also observed to coat snow on Bird Island (South Georgia; unpublished information of JLS). Small, intermittent activity continued to occur until 27 October, after which only fumaroles remained active. A much larger lahar was emplaced on the Huemules Glacier on 11 October but was not clearly connected to new eruption activity; it may have been triggered by meltwater generated during the cooling of the tephra pile and lava flows, which accumulated within the caldera until October (similar to meltwater from the Gjálp eruption (Iceland) that collected in Grímsvötn and subsequently generated jökulhlaups).

While some details of distal ash deposition and compositions are known, essentially no information was collected on the proximal deposits. However, the lahar deposits were examined in more detail, and Branney and Gilbert (1995) documented features (called ice-melt collapse pits) that formed due to melting of large ice blocks transported within the lahar. Their observations form a critical basis for identifying evidence of former ice in ancient lahar deposits, and similar structures are likely to form hazards in future glaciovolcanic lahars. Four main criteria were inferred to indicate the former presence of buried and surface ice blocks: (1) circular collapse pits, (2) kettle holes, (3) ice rafted debris, and (4) crescent-shaped structures formed by erosion on the upstream side and deposition on the downstream side of ice blocks. Branney and Gilbert, (1995) also noted that similar features were still preserved 20 years after the 1971 Hudson eruption, suggesting that they might have relatively long-term preservation potential.

3.3.10 Gjálp 1996

The Gjálp eruption in October 1996 is the only recorded example of an eruption under a thick glacier (Table 3.1; Fig. 3.7). It remains one of the best observed, photographed, described and interpreted glaciovolcanic eruptions on Earth (e.g. Gudmundsson et al., 1997, 2002, 2004). As a result it has significantly influenced our understanding of glaciovolcanic processes, in particular by clearly demonstrating evidence for both subglacial and supraglacial meltwater drainage as well as quantifying, for the first time, rates of ice melting and heat transfer, magma discharge and timescales for tephra lithification. The eruption lasted 13 days and started as a 4 km-long fissure eruption under 600 m-thick ice in the northwestern part of the 8100 km^2 Vatnajökull ice cap, Iceland (Fig. 3.7). No preceding thermal activity was observed but the eruption quickly formed two ice cauldrons above the main vents, each about 2 km wide and 100 m deep. The cauldrons had crevassed margins but in cross section the depressions were V-shaped, indicating focused melting above the vents. On average 6000 m^3 s^{-1} of meltwater was generated during the first 4 days (melting rate of c. 0.5 km^3 day^{-1}). The meltwater drained away from the eruption site continuously. It travelled 15 km to the south and emptied into the Grímsvötn

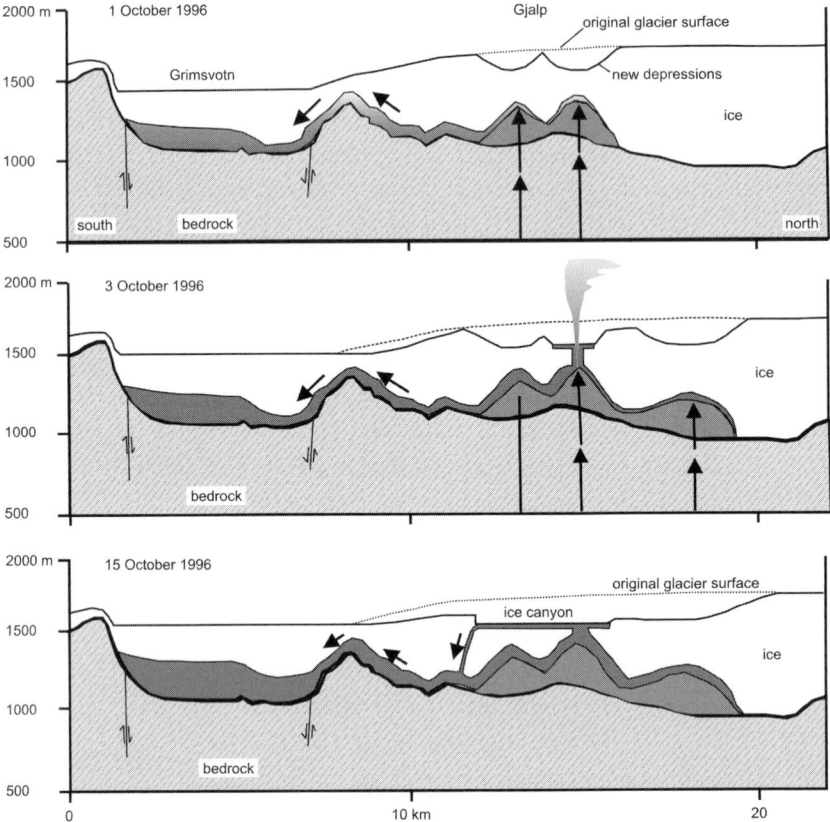

Fig. 3.7 Schematic cross sections illustrating processes that occurred during the 1996 subglacial eruption at Gjálp, Iceland (from Gudmundsson et al., 1997). Important features include the evidence for continuous subglacial meltwater drainage from the earliest stages (including flow up bedrock slopes); development of an overlying canyon and associated supraglacial meltwater escape; progressive subsidence of the surrounding ice sheet surface; and continuous accumulation of meltwater in the Grímsvötn caldera from where it subsequently escaped in a voluminous jökulhlaup.

caldera as a subglacial lake, where it created a fan formed of tephra derived from the eruption. The Grímsvötn lake is a permanent feature sustained by geothermal activity within the caldera and with a varying lake level determined by frequent jökulhlaups (Thorarinsson, 1953). The eruption melted through the ice overburden in 31 hours and two vents were active throughout. The volcanic fissure grew towards the north, reaching a length of 6 km on the second day, and forming an ice cauldron where the pre-eruption ice thickness was 750 m. After day four the rate of ice melting diminished substantially, but remained at $500–1000 \text{ m}^3 \text{ s}^{-1}$ until the end of the eruption, at which stage about 3 km^3 of ice had melted. A further 1 km^3 was melted in the 3 months that followed and another c. 0.5 km^3 in the next 5 years (Jarosch et al., 2008). About 80% of the melting happened at the eruption site, with the remaining 20% occurring above the subglacial path of meltwater flow into Grímsvötn, within Grímsvötn itself, and on the first 6 km of the path of the jökulhlaup out of Grímsvötn. The Gjálp eruption produced a 6 km-long, 0.5–2 km-wide and up to 450 m-high hyaloclastite ridge or tindar (bulk volume 0.7 km^3) of basaltic andesite. An additional 0.1 km^3 of pyroclasts was transported with the meltwater into the Grímsvötn caldera.

Although meltwater drainage was predominantly subglacial, meltwater generated later in the eruption also escaped supraglacially, within a 3.5 km-long ice canyon. Meltwater in the canyon plunged down a moulin to the base of the ice before entering Grímsvötn (Fig. 3.7). Most of the meltwater was released three weeks after the eruption, when the Grímsvötn lake drained in a large jökulhlaup onto Skeidararsandur situated to the south of the glacier. The total volume of floodwater was 3.4 km^3 (Stefánsdóttir and Gíslason, 2005). Despite being associated with a rather small-scale eruption, the jökulhlaup flood peaked at 50 000 m^3 s^{-1}, making it temporarily the second largest freshwater discharge on Earth, exceeded only by the Amazon and approximately equivalent to four times that of the Mississippi. It delivered at least 180 million tonnes of suspended solids to the sea in only 42 hours (Stefánsdóttir and Gíslason, 2005). This amounts to c. 1% of the total annual suspended flux delivered by rivers globally to the oceans. Only a small proportion of the detritus carried by the flood was tephra derived from the eruption and most was eroded along the subglacial path followed by the meltwater after exiting Grímsvötn lake. Measurements indicate that permeability of the edifice was reduced significantly in the first year after the eruption, possibly because of rapid palagonitisation and consolidation of the edifice (Jarosch et al., 2008).

It is worth noting that, because of the unique geographical circumstances of the Gjálp site (i.e. the bedrock configuration and proximity to Grímsvötn), the jökulhlaup associated specifically with the eruption at Gjálp was a *continuous* event dominated by subglacial meltwater discharge. The *sudden* release of meltwater onto Skeidararsandur described above, which is generally but erroneously regarded as the Gjálp jökulhlaup, was a secondary effect related to the temporary storage of meltwater from Gjálp accumulating in the Grímsvötn subglacial lake. Thus, whilst triggered by the influx of meltwater generated at Gjálp, the sudden flood at Skeidararsandur was a Grímsvötn jökulhlaup.

3.3.11 Grímsvötn (1934, 1938, 1983, 1998, 2004, 2011)

Grímsvötn is presently the most active volcanic system in Iceland (Gudmundsson and Björnsson, 1991; Thordarson and Larsen, 2007; Fig. 3.8), and also one of the most active glaciovolcanoes on Earth (along with Villarrica and Klyuchevskoy). It has had more than 64 historical eruptions (Thordarson and Larsen, 2007) and at least six significant eruptions between 1934 and 2011, some of which were closely observed by Icelandic scientists (1934: Nielsen, 1937; 1983: Gudmundsson and Björnsson, 1991; 1998: Sturkell et al., 2003a; 2004: Vogfjörd et al., 2005; Jude-Eton et al., 2012; Gudmundsson et al., 2012b). The volcano has a summit caldera that is filled by c. 62 km^2 of ice from the Vatnajökull ice cap, and basaltic eruptions frequently occur within the caldera. The Grímsvötn volcanic

Fig. 3.8 Aerial views of Grimsvötn caldera immediately after the 2004 eruption with Vatnajökull, eastern Iceland. (a) Aerial view of a crater 750 m in diameter formed through ice in the Grímsvötn caldera in 2004 showing ice extensively covered by ash; the small steam plume is rising from a meltwater lake in the crater. (b) View of a small tephra cone built during the 2004 eruption of Grimsvötn, Iceland, within an englacial lake formed by the eruption.

system comprises two composite volcanoes and an associated fissure system that extends a total of 90 km in a northeast–southwest direction. Grímsvötn eruptions have produced significant jökulhlaups in the past, although it is important to note that the high geothermal flux through the caldera floor also causes enough melting to produce periodic floods even in the absence of eruptions; in fact, several of the eruptions may have been triggered by pressure release due to draining of the caldera lake (Nielsen, 1937; Thorarinsson, 1953;

Albino et al., 2010). This is another unique type of hazard from glaciovolcanoes – they can produce hazards even when in a state of relative quiescence (e.g. Mt Chiginagak, Alaska: Schaefer et al., 2008; Ruapehu, New Zealand: Lube et al., 2009).

One of the earliest scientific descriptions of a Grímsvötn eruption was by Neilsen (1937), which is one of the earliest descriptions of any glaciovolcanic eruption (the description by Wolff (1878) of the 1877 eruption at Cotopaxi, including damage to the overlying ice and extensive lahar deposits, might be the earliest). The eruption described by Neilsen (1937) began on 30 March 1934 and produced a large jökulhlaup with a maximum discharge estimated at 80 000 $m^3 s^{-1}$. Observations from visits to the eruption site on 2 May recorded that tephra up to 25 m thick covered the ice and had temperatures of 50 °C. While the possibility that a catastrophic drainage of the lake within Grímsvötn could have triggered the eruption, Neilsen (1937) favoured another explanation: that the eruption triggered melting of the ice. While subsequent workers have shown that lake drainage can trigger eruptions (Thorarinsson, 1953; Albino et al., 2010), Nielsen published one of the first empirical thermal balance calculations to show that basaltic lava could potentially melt about 10 times its own volume of ice at maximum heat exchange efficiency (see Chapter 6). Smaller eruptions also occurred in 1983–4 (Gudmundsson and Björnsson, 1991) and 1998 (Gudmundsson, 2005; Sturkell et al., 2003a) but produced very limited ice melting (<0.1 km^3) and no major flooding.

The best-documented Grímsvötn eruptions have been more recent and the caldera has produced three eruptions within a 15 year period (1998–2011). In particular, the eruption in 2004 was exceptionally well monitored (Vogfjörd et al., 2005) and the fragmental deposits formed during the eruption were studied in detail by Jude-Eton et al. (2012). Both of these sources give detailed chronologies of the short-lived eruption, which started on 1 November and was over by 6 November. An increase is seismic activity in the caldera preceded the eruption by a few months, as did the start of a jökulhlaup on 28 October, whose discharge peaked on 2 November at 3000 $m^3 s^{-1}$. As the eruption had just started late on 1 November, and the water draining from the caldera has to travel over 50 km under the ice to its exit beneath the southwestern part of Vatnajökull (Skeidarárjökull), the flood must have preceded the eruption. Magma migration towards the surface began at 2020 hours, signalled by an intense earthquake swarm that lasted until 2150 hours. At that point, continuous volcanic tremor was registered, signalling the start of the eruption. Within 30 minutes the eruption had melted through less than 200 m of ice to begin an explosive phase. By 2250 hours the radar station in Reykjavik had measured an eruption plume that reached 6 km in height, which slowly built to 10 km over the next 4 hours. Two different vents were seen to be active during the eruption, which ultimately produced a 200 m diameter tephra cone in an englacial lake. Jude-Eton et al. (2012) focused much of their work on understanding the dynamics and emplacement of tephra deposits formed during this early short-lived phreatomagmatic episode. They were able to match deposits with different characteristics (fall, surge, flow) to changes in seismicity and column heights over the course of the first 2 days, when most of the tephra formed (0.047 km^3 DRE). They were also able to estimate rates of ice cauldron development, which reached 65% of its final size within the first 12 hours, and 85% by 1600 hours on

2 November. Jude-Eton et al. (2012) reconstructed mass fluxes for the eruption (maximum of 218 m^3 s^{-1}), and suggested that early, high discharge rates did not produce a significant plume (<6 km) because much of the thermal energy being released by the fragmenting magma was used initially to melt ice and enlarge the cauldron. They also showed that particle aggregation was responsible for enhanced deposition of fine ash, in addition to ash scavenging by snowfall during the eruption. While the 2004 eruption was minor in volume compared to many other historical eruptions, the level of detail with which it was reported and interpreted by Jude-Eton et al. (2012), based on *in situ* measured deposits, was unprecedented for an explosive glaciovolcanic eruption. Their identification and documentation of simultaneous deposition by gravitational settling from the eruption column, by pyroclastic density currents, and by Surtseyan explosions gives important new insights into the complexity of vent-proximal processes that take place in many explosive glaciovolcanic eruptions.

While this large volcanic system is mostly covered by ice, at least one of its flank fissure systems extends beyond the southwestern part of Vatnajökull and it produced the infamous 1783–4 Lakigigar eruption. Lakigigar is the second largest historical effusive eruption on Earth and produced more than 14 km^3 of lava (Thordarson and Larsen, 2007).

3.3.12 Montagu Island 2001–2007

A long-lived small-volume eruption commenced on Montagu Island, the largest of the South Sandwich Islands, Antarctica, in September or October 2001 and lasted for 6 years (Patrick et al., 2005; Patrick and Smellie, 2013). The island is composed of basalts and basaltic andesites but the eruption itself was andesitic (Baker, 1990b; Pearce et al., 1995; and unpublished information of JLS). It is the only eruption on Montagu Island recorded in historical time and it was documented almost entirely using the global MODVOLC satellite monitoring system (Wright et al., 2002, 2004). The eruption was focused on Mt Belinda, a small pyroclastic cone with a crater 150 m wide, situated in the centre of an extensive ice-filled caldera 7 km in diameter that forms much of the summit of the island. It consisted mainly of persistent low-intensity pyroclastic activity characterised by a small plume, which distributed a thin ash drape on the summit ice in the northeast corner of the island. Numerous pits littered the snow-covered slopes of Mt Belinda and were melted by bombs ejected on ballistic trajectories. Three main effusive events were observed. It is unclear if any were associated with ponded meltwater (cf. lava-fed deltas) but none was observed on satellite images. The first effusive episode, in late 2001 or early 2002, created a lava 600 m long and 200 m wide. The second lava, emplaced at some time between March and August 2003, was larger and extended 2 km, swelling from c. 100 m-wide close to source, to 200 m at 1.5 km, ending in a fan-like shape 500–600 m across. It melted a deep gully in the surrounding ice and terminated against a rocky ridge forming the local caldera wall. The third and largest effusive event took place in September 2005. Lava flowed from Mt Belinda all the way to the sea at the north coast, a distance of c. 3.5 km. It formed a lava-fed delta with a steep polylobate terminus 800 m wide that extended the coastline out by 500 m. To reach the sea the lava (a block lava) flowed down steep cliffs at the margin of a glacier. Subglacial meltwater caused major disruption of the glacier, which surged a few hundred metres into the sea. Additional effusive events took place after September 2005 (e.g. relatively long lavas in November 2005 and March 2006; and several short lavas between August 2006 and July 2007) but all episodes were probably brief. Some of the later short lavas may be local breakouts caused by freezing at the lava front causing back pressure in the lava body. The presence of a prominent degassing plume and a short wavelength infrared anomaly throughout the eruption indicate that lava remained at a high level in the conduit and was probably maintained by magmatic convection. Intriguingly, a lava lake in the crater of ice-clad Mt Michael on Saunders Island, previously identified by satellite observations (Lachlan-Cope et al., 2001), was concurrently active during much of the eruption on Montagu Island. The Mt Belinda eruption ended around September 2007.

3.3.13 Eyjafjallajökull 2010

The Eyjafjallajökull eruption, Iceland, began on 20 March and continued until 22 May 2010 (Table 3.1; Fig. 3.9); Gudmundsson et al., 2012a; Edwards et al., 2015a). The main explosive episode is best known for the disruption the airborne ash caused to air traffic in Europe and across the Atlantic Ocean (see Chapter 15) but the eruption also displayed a variety of interactions between magma, snow and ice. Eyjafjallajökull is a 25 km-long, 15 km-wide and 1600 m-high stratovolcano (Loughlin, 2002). Its upper slopes are covered by a small ice cap 80 km^2 in area and 50–200 m thick. Similar to the Gjálp eruption (Section 3.3.10), no increase in geothermal activity was detected before the eruptions. However, repeated seismic swarms in the 20 years preceding the eruption were thought to be indications of the system becoming reinvigorated (Sturkell et al., 2003b). The most recent prior eruptions occurred in 1612 and 1821–3. The first phase was effusive and occurred from two short fissures, feeding small 'a'ā basaltic lava flows on top of a 1–5 m-thick seasonal snowpack on a relatively ice-free saddle between Eyjafjallajökull and Katla known as Fimmvörðuháls (Edwards et al., 2012). The lavas flowed predominantly on top of the snow, since the rate of snow melting by radiative heat transfer from the lava was too slow to remove the snow ahead of the advancing lava; much of the snowpack was also temporarily covered with tephra that acted to slow heat transfer (Edwards et al., 2012). Gradual conductive melting of the snow followed, extending over months. Three months after the end of this first phase of the eruption, only subtle evidence of lava–snow interactions were observed such as collapse features from melting of underlying snow, and anomalous tephra mounds formed by draping of the tephra blanket over snow-covered blocks. The main explosive eruption (bulk tephra volume of 0.27 km^3), which produced mostly very fine grained trachyandesite tephra, started in the shallow summit caldera on Eyjafjallajökull on 14 April and created three small craters, one of which eventually expanded to become the dominant vent.

Fig. 3.9 Aerial views of Eyjafjallajökull in southern Iceland, July 2010. (a) View looking to the west across the (inactive) Fimmvörðuháls eruption site (ridge in the foreground) in March 2010, towards the ash-covered summit ice cap of Eyjafjallajökull, with its steam plume. (b) View of the tephra deposits covering the summit glacier. Note the numerous bombs and blocks lying on the ash-covered ice and associated shallow depressions caused by melting. Tracks on left of the image are footprints in the tephra. (c) View looking to the south showing the Gigjökull lava flow and its canyon melted in the glacier on the northern flank of the volcano.

The eruption melted through 200 m of caldera ice in about 4 hours, and created a prominent plume that penetrated the stratosphere and transported ash over all of Europe apart from the Iberian Peninsula (Magnússon et al., 2012). Ash also fell on western Russia and the eastern seaboard of North America. In contrast to Gjálp, the Eyjafjallajökull eruption occurred through ice cauldrons with vertical walls (Magnússon et al., 2012). Due to the relative thinness of the surrounding ice, deformation of that ice was relatively minor outside the cauldrons. Ice melting was rapid in the first 24 hours (500–1000 m^3 s^{-1}), but declined to c. 200 m^3 s^{-1} on the second day. By day three, melting around the vent largely ceased as the surrounding ice cauldron had enlarged to c. 500 m, thus removing ice from the main vent locus, and tephra falling on the surrounding ice largely insulated it. During those first days, meltwater drained northward down the steep outlet glacier Gígjökull in several swift jökuhlaups with a discharge c. 1000–10 000 m^3 s^{-1}. The jökulhlaups drained subglacially for the first 1–1.5 km but then emerged on both sides of the glacier and flowed mainly supraglacially down the <100 m-thick Gígjökull. The meltwater carried with it large amounts of tephra, filling a proglacial lagoon at the margins of the sandur in the first 24 hours. At least one small lahar travelled to the south. On the

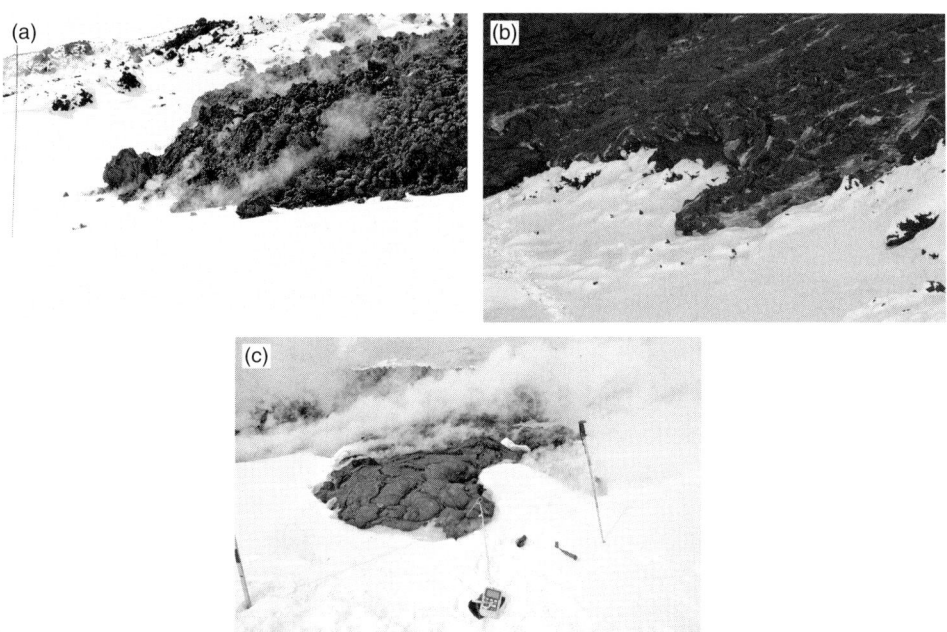

Fig. 3.10 Photographs of lavas flowing on snow during the 2012–13 Tolbachik eruption in Kamchatka, Russia. (a) 'A'ā lava flowing on snowpack; 2 m-high snow pole for scale. (b) 'A'ā sheet lava flowing over snowpack; footprints in snow in front of the lava for scale. (c) Pāhoehoe lava moving through snowpack. Trek pole is 1.5 m long.

eighth day of the eruption, lava started to flow subglacially towards the north, progressing very slowly for about a week. It gradually melted through Gígjökull and increased in thickness to 100 m. The lava advanced about 2 km in a few days as it progressed through 50–100 m-thick ice, forming a c. 20 m-thick lava flow by melting the ice above the advancing lava front. The lava stopped moving early in May as the eruption intensity increased and became fully explosive again. Magma–ice interaction and melting was minor after the lava stopped flowing. Tephra accumulated to a thickness of several metres on top of the glacier surrounding the craters and there were also minor base surges. Surface melting of the snow and ice was limited.

Like the 2004 Grímsvötn eruption, the 2010 Eyjafjallajökull eruption produced a relatively small volume of tephra compared to larger historical Icelandic eruptions (e.g. 1783–4 Lakigigar, 1947 Hekla). However, it is one of the few recent Icelandic glaciovolcanic eruptions that produced both explosive and effusive behaviour. It is also one of few historical examples in which basaltic magma intruded a more chemically evolved magma system to produce an eruption (Sigmundsson et al., 2010b). Finally, the 2010 eruption also highlighted the potential for external forces (e.g. flow direction of the northern stratospheric jet stream) to dramatically change the area affected by a relatively small eruption. The jet stream at the start of the 14 April explosive phase was perfectly positioned to rapidly carry ash to the southeast, ensuring that it directly affected European airspace very soon after the eruption commenced. By contrast, an eruption less than one year later (2011) at Grímsvötn produced a larger volume of ash, but because the jet stream was flowing in a northerly direction, the larger eruption had a much smaller impact on air traffic.

3.3.14 Tolbachik 2012–13

Tolbachik volcano is a massif that comprises two overlapping composite volcanoes in addition to flank rift zones within the southern end of the Klyuchevskoy volcano group, in the Central Kamchatka Depression (Fig. 3.10; see Chapter 2). The southern fissure zone was the location of the Great Tolbachik Fissure Eruption in 1975–6, which produced the largest effusive eruption in Kamchatka during historical time. The 2012–13 eruption began on 27 November 2012 and continued until early September 2013. Thus, it extended throughout the winter in Kamchatka and presented one of the most important opportunities to make qualitative and quantitative observations of interactions between a variety of lava flow types and seasonal snowpack. While initial snow cover at the elevation of the main eruptive fissures (Menyailov vents at 1900 m a.s.l. and Naboko vents at 1650 m a.s.l.) was probably less than 1 m, during the course of the subsequent 5 months the snowpack deepened until it was greater than 4 m locally by early April. High initial effusion rates (c. 440 m^3 s^{-1}; Gordeev et al., 2013; Dvigalo et al., 2014; Belousov et al., 2015) produced lava flows that travelled more than 9 km during the first 24 hours. However, these rates decreased through January until they reached a longer-term steady-state flux of c. 18–19 m^3 s^{-1} (Dvigalo et al., 2014; Belousov et al., 2015). Ground and air-based images from the first 6 months of the eruption commonly show voluminous production of steam during interactions between lava flows and the snowpack (Fig. 3.10). Edwards et al. (2014a) gave a brief account of these interactions focused specifically on the influence of lava effusion style. They found that two of the three dominant types of lavas ('a'ā and sheet lava) advanced rapidly enough to travel across snowpack before the snow completely melted. This behaviour was consistent with observations of lavas during the first phase of the 2010 Eyjafjallajökull eruption (at Fimmvörðuháls; Edwards et al., 2012). However, more slowly advancing pāhoehoe lava migrated through the snowpack by a combination of melting and physical displacement of the snow (Fig. 3.10). The sub-snow emplacement produced widely visible steam plumes, and the resulting water-enhanced cooling of the flows produced distinctive bulbous vitric textures indicative of emplacement within a water-rich environment. Documentation of these textures is the first direct field-based indication that lavas can preserve evidence of snow during emplacement (Edwards et al., 2015b; see also Mee et al., 2006). The field results confirm hypotheses developed from large-scale melt–ice experiments that suggested lava–ice/snow interactions should produce distinctive textures (Fig. 3.10; cf. Mee et al., 2006). Pits dug into the snowpack in front of advancing 'a'ā and pāhoehoe lavas allowed meltwater temperatures to be measured directly and also observations of the processes by which the lava flows interacted with snowpack. Limited observations were also made on localised phreatic and phreatomagmatic behaviour generated by collapse of 'a'ā lavas into water-saturated snowpack.

3.4 Important lessons learned and observations needed for future eruptions

3.4.1 Lessons learned

The historical eruptions described in this chapter have provided significant insights for interpreting ancient deposits over a wide range of environmental conditions (see Chapter 13) and for better understanding likely hazards from eruptions (see Chapter 15). While no historical eruption has produced a complete 'classic' tuya sequence, the 1996 Gjálp eruption demonstrated processes characteristic of a tindar eruption (Gudmundsson et al., 1997). They provide constraints on how fast (minutes to hours) eruptions can melt through thick (Gjálp: 600–700 m) or thin (Grímsvötn: 200 m) ice. They have also given us insights into both proximal coherent (Veniaminof) and fragmental (Grímsvötn) deposits, as well as information on how far small (2010 Eyjafjallajökull) and large (1918 Katla, 1991 Hudson, 1989/2009 Redoubt) mass flows may travel and associated processes. A wide range of volcanic behaviour has also been observed, including very local lava–snowpack interactions (1947 Hekla, 2010 Eyjafjalajökull (Fimmvörðuháls), 2012–13 Tolbachik); lava flows travelling through and beneath ice (1984 Villarrica, 1991 Hudson, 1994 Llaima); and a wide range of pyroclastic density currents (e.g. surge, flow) and fall deposits interacting with ice and snow (1985 Nevado del Ruiz). Many of these processes leave identifiable records in modern deposits that record interactions between magma and the cryosphere that can be applied to ancient systems and thus facilitate increasingly detailed palaeoenvironmental reconstructions (Chapter 13), and their identification also makes it possible to significantly improve hazard assessments for all ice-clad volcanoes.

Other important lessons learned include:

- While not all deposits produce clear records of interaction with ice, many do.
- Interactions can produced enhanced explosivity which may facilitate longer-distance ash transport; however, wet eruption columns may also be more susceptible to aggregation processes to help scrub ash from the atmosphere.
- Melting rates are controlled by the speed with which magma can intermix with snow or ice and the ability to confine the resulting meltwater; rapid draining due to pre-existing drainages, fractures caused by the activity, or presence of high permeability snow will slow heat transfer.
- Low-permeability barriers (ice, pre-existing topography) will facilitate magma–water heat transfer, and impoundment can lead to higher potential for larger, more damaging floods.
- Deposits directly related to melting of snow or ice can be ephemeral and need to be documented, sampled and measured as soon as safely possible after the eruption.
- The presence of a snow or ice cover can contribute to early warnings of impending activity via visible fractures to help measure ground surface motion and fumaroles for estimating heat fluxes.

3.4.2 *Observations needed during future eruptions*

While better deployment of monitoring equipment facilitates improved correlations between changes in energy releases and deposit formation (e.g. Jude-Eton et al., 2012), many of the world's ice-clad volcanoes are poorly monitored or even lack monitoring equipment. Having local seismic networks will be critical for better constraining conditions during eruption initiation – i.e. whether they are explosive or effusive – and may need augmentation from infrasound methods. Better measurements of particle densities in tephra plumes are needed to determine how and when transitions occur between 'wet' and 'dry' eruptive phases related to when an eruption no longer involves interaction with ice, and also for accurately determining magma fluxes and energy transfers. Our understanding of the role of external water in the formation, transport, deposition and even alteration of pyroclastic deposits of all types is lacking as well. Also needed are better studies of the cooling of lava flows after emplacement, and investigations of the mode of formation of cooling fractures (perhaps using infrasound techniques) including determining the precise influence and timing of external water on fracture speeds and fracture spacing.

4

Physical properties of ice important for glaciovolcanic eruptions

4.1 Introduction

A principal effect of an ice mass is to impound an erupting volcano, which at least temporarily confines most of the erupted products and any meltwater produced during the eruption. Only tephra generally escapes the confines of ice, and in some glaciovolcanic eruptions even tephra can be largely confined (Stevenson et al., 2011; Jude-Eton et al., 2012). Thereafter, in addition to the thickness of that ice, it is the physical properties of the ice (i.e. its structure (proportion of ice, firn and snow, as layers), basal thermal regime, rheology and hydraulics) that determine the evolution of the erupting system, the specific types of lithofacies formed and the sequence architecture (Smellie, 2000, 2009; Russell et al., 2014). Thus, the presence of the surrounding ice determines why glaciovolcanic eruptions and their products are different from their submarine or non-glacial lacustrine equivalents and how they can generally be distinguished. Those ice properties of particular importance to glaciovolcanic eruptions are reviewed in this chapter.

4.2 Physical structure of an ice mass

The physical structure of an ice mass can be characterised at the dimensions of individual grains of ice (millimetres) as well as the overall dimensions of the entire ice mass (kilometres thick; 10^6 km^2). At the microscale, when snow accumulates, it slowly transforms to ice by compaction. The process involves the crushing and sintering of ice crystals and the movement of ice particles to stronger more stable packings, resulting in a decrease in volume and increase in density (Herron and Langway, 1980). The transformation from snow to ice is greatly facilitated in a warm and wet environment such as in the accumulation zones of warm glaciers where water percolates down through the snow and freezes at depth. Dry snow has a density of c. 400 kg m^{-3}, transforming to firn at a density of c. 600 kg m^{-3}, and finally becoming ice at a density of c. 920 kg m^{-3}. The rapidity with which these changes take place depends on the temperature profile of the ice mass. In warm ('temperate'; see next section) glaciers, the firn–ice transition can occur at depths of only a few tens of metres (e.g. <20 m in the Upper Seward Glacier, Alaska) whereas it may be more than 100 m in cold ('polar') glaciers (e.g. 115 m at South Pole; Herron and Langway, 1980; Paterson, 1994). This leads to

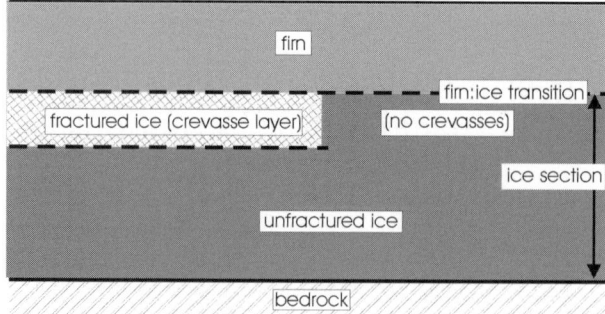

Fig. 4.1 Empirical cross section through an idealised glacier showing the division into (1) firn (and snow); (2) unfractured and (3) fractured ice (from Smellie, 2001). Firn, snow and fractured ice are permeable and can act as aquifers, unlike unfractured ice which is an aquiclude, at least on the timescale of most eruptions. The same structure is present in wet-based and cold-based glaciers. Not to scale.

glaciers being stratified in their accumulation areas (in ablation areas, snow and firn are removed annually by melting, leaving ice at the surface), with a layer of snow overlying firn then ice (Fig. 4.1). The thickness of snow and firn is greatest on cold and polythermal ice.

The hydraulics of an ice mass are controlled by its physical structure (Section 4.5), which strongly impacts glaciovolcanic eruptions (Smellie, 2001, 2006). Snow and firn are porous whilst ice (in the absence of fractures) is essentially non-porous. Thus, cavities filled by meltwater created during a glaciovolcanic eruption may overflow through the porous surface layers above the firn–ice transition (Smellie, 2006, and see Section 4.5). Whilst water can also be transmitted through (warm) ice along the crystal boundaries, it is extremely slow (Nye and Frank, 1973) and, on the timescale for most volcanic eruptions (days to a few decades), it is too small to have a significant effect. It can thus be ignored compared with water flow in fractures (crevasses and pervasive fine-scale connected fractures; Fountain et al., 2005), tunnels (including moulins within the ice and Röthlisberger and Nye channels at the base of the ice), and at the basal ice–substrate boundary.

At the largest spatial scale, Nye (1951, 1952a,b) proposed a general relationship between the thickness of ice at the centre of an ice sheet and the overall dimensions of the ice sheet:

$$H^2 = \frac{2\,\tau_0}{\rho\,g}[L - x] \qquad (4.1)$$

where H is the maximum thickness of the ice, τ_0 is the driving stress (assumed to be equal to the basal shear stress for ice that is perfectly plastic, typically ~100 kPa), ρ is the density of ice, L is the overall distance from the position of H to the edge of the ice sheet and x is a specific location along that distance transect (from Cuffey and Paterson, 2010). The equation is that of a parabola, which is a good first approximation to the profile shape for larger ice bodies. As ice thickness plays a critical role in determining the dynamics at the start of an eruption, the relationship described by equation (4.1) is a starting point for trying to predict ice thicknesses beneath an ice mass. Alternatively, given some minimum constraints on ice thickness extant during an eruption (e.g. Chapter 13), it is possible to estimate empirically the dimensions of an ice body associated with an erupting glaciovolcano by solving the equation for *L* (e.g. Edwards et al., 2011):

$$L = H^2 \frac{g\,\rho}{2\,\tau_0} \tag{4.2}$$

The ability to estimate first-order dimensions of the enclosing ice body during an ancient glaciovolcanic eruption provides some constraints on whether the eruption occurred through a local or regional-scale ice body, with the attendant implications for palaeoclimatic reconstructions.

4.3 Thermal regime

The temperatures of glaciers are not uniform. Whether basal ice is at the melting point or below has profound implications for glacier motion: efficient motion (and associated processes of glacial erosion, deposition of tills and fluvial sediments) can only occur if ice is at or close to its melting point. Thus basal ice temperature is an important guide as to whether an ice mass is dynamic over time and able to contribute significantly to fluctuating global sea levels, or is more stable with a slower, more paced impact on sea level. In addition, the thermal regime is critically important with regard to the role of meltwater (i.e. the hydraulics; Section 4.5). The basal condition of an ice mass also influences subglacial drainage, the permafrost thermal regime and ground ice conditions.

Factors that affect the temperature profile through ice include its thickness, precipitation, ambient temperature, geothermal heat flux and frictional heat generated by flow, which vary spatially and temporally. This has led to a simple threefold classification of glaciers in terms of their thermal characteristics, usually referred to as basal thermal regime. Using glaciovolcanic sequences to determine basal thermal regime is thus an important contribution to understanding palaeoenvironmental evolution and long-term changes to global climate (Chapter 13).

Warm ice, also called temperate or thawed-bed, is at the pressure melting point everywhere except in a thin surface layer during winter. Movement is predominantly by basal sliding, which produces the most efficient erosion. It is associated with a wide range of glacial landforms and abundant meltwater that variably reworks any tills present. **Polythermal ice**, also called sub-polar, is only above the pressure melting point in a thin layer at the bed. It is below the pressure melting temperature elsewhere. Thus, it also has a thawed bed and moves by sliding but meltwater and fluvial activity are not well developed. Polythermal valley glaciers are typically frozen to their bed at their thin margins and have a thawed bed in their interiors. By contrast, ice sheets may have a basal patchwork or mosaic of frozen-bed and thawed-bed ice (Siegert and Dowdeswell, 1996; Pattyn, 2010), although wet-based conditions are probably dominated by flow in ice streams and outlet glaciers (Golledge et al., 2013). Debris entrainment and deposition can be substantial at the junction between the two thermal states, where the sliding thawed-bed ice is advected onto the frozen-bed ice (Hambrey and Glasser, 2012). Warm and polythermal ice masses are also known collectively as wet-based. Finally, **cold ice**, also known as polar, dry-based or frozen-bed, is below the pressure melting point throughout. Because it is frozen to its bed, the ice moves principally by internal deformation. It thus largely protects the underlying landscape (Gellatly et al., 1988; Fabel et al., 2002; Goodfellow, 2007). Although some erosion can occur, especially in steeper-gradient areas with higher driving stresses and basal velocities (Golledge et al., 2013), it is typically minor and has distinctive characteristics unlikely to be confused with erosion by wet-based ice (cf. Atkins et al., 2002; Lloyd Davies et al., 2009). A major consequence of thermal regime is that colder glaciers require significantly more energy to melt compared with warmer glaciers since energy has to be used simply to raise the temperature to the melting point. For example, 209 kJ of energy are required to raise the temperature of 1 kg of ice at -50 °C to the freezing point, whereas it takes 333 kJ to melt 1 kg of ice at 0 °C (Wilson et al., 2013); the presence of impurities in ice, such as air, salts and CO_2, will lower the melting point relative to pure water ice.

With the significantly lower albedo of dirt compared with ice, dirty ice (e.g. ice with tephra particles) will also absorb more heat from sunlight. This may raise the surface temperature of the dirt above the melting point of ice. As the heat is transmitted by conduction to the ice, the ice should melt more rapidly than ice that is dirt-free (Wilson et al., 2013). Conversely, a thick layer of tephra will protect underlying ice from melting (Wilson et al., 2013). Soluble and insoluble impurities can also greatly increase the effective specific heat capacity of ice.

4.4 Rheology

How a glacier deforms and at what rate depends sensitively on its temperature and the temperature profile throughout the ice mass. The relationship is not linear. For example, under a given stress, ice at 0 °C deforms at a rate 100 times faster than ice at -20 °C (Hambrey, 1994), whilst cooling ice from -10 °C to -25 °C reduces the deformation rate by a factor of only five (Paterson, 1994). Rheology is also affected by the presence of solid (non-volatile) impurities such as bedrock-derived dirt (including aerosols) or tephra. Their influence on ice rheology is relatively predictable and they cause ice strength and thus strain response time to increase. Ice strength generally increases rapidly as the proportion of dirt increases. However, once again the response is not linear. The effect of particulate concentrations less than 10% is negligible whereas greater proportions of dirt significantly strengthen ice. By contrast, ice containing proportions of dirt exceeding 85% is weaker than pure ice, probably because of a loss of cohesive strength (Benn and Evans, 1998). Because volcanoes often produce tephra, it is likely that ice in glaciated volcanic regions will be stronger than in non-volcanic regions but it may also show strength variations depending on local ash thicknesses.

One of the important principles for understanding how ice around a cavity created by an eruption will respond depends on the rate at which the ice can deform. Glen (1952) conducted a series of steady-state compression experiments to determine the rate at which ice will deform by plastic creep, and found that, to a good approximation, rates of ice creep follow a power law with the following form:

$$\dot{\epsilon} = A\tau^n \tag{4.3}$$

where $\dot{\epsilon}$ is the shear stress, τ is the rate of strain, and n and A are constants. Nye (1953) derived equations for calculating the rate of deformation of ice above cylindrical and spherical cavities (equations 4.4 and 4.5) in order to investigate rates at which ice cavities would close at the base of a glacier, i.e.:

$$\dot{R} = r\left(\frac{P}{nA}\right)^n \tag{4.4}$$

$$\dot{R} = 3r\left(\frac{P}{2nA}\right)^n \tag{4.5}$$

In these equations $\dot{R}$ is the rate of cavity collapse, r is the radius of the cylinder/sphere, and P is the pressure at the cavity roof due to the overlying ice. Walder (1986) used a value of n equal to 3, and a value of A equal to 5.14×10^7 Pa $s^{1/3}$ to investigate water flow in subglacial cavities. Cuffey and Paterson (2010) have reviewed the factors that control values of n and A. They found that, while estimated values of n range from 1.5 to 4.2, a value of 3 was in most cases consistent with field data. However, values of A are complex functions of temperature, pressure and several other variables (Cuffey and Paterson, 2010). Tuffen (2007) applied the same equations to investigate rates of cavity growth during the onset of subglacial eruptions.

4.5 Hydraulics

A basic understanding of the hydraulics of a glacier system is important for explaining many aspects of glaciovolcanic eruptions. Ice can be cold, warm or polythermal. Because cold ice is below the pressure melting point everywhere, there is no free water and any created by melting during eruptions can only escape upward and away by overflowing or via ice fractures (Smellie, 2009, 2013). Conversely, warm ice is at the pressure melting point everywhere and will transmit water readily, principally via fractures and by basal escape at the glacier bed in thin sheets, linked cavities or tunnels (Hooke, 1989; Fountain and Walder, 1998; Fountain et al., 2005). Finally, polythermal ice is below the pressure melting point except at its base, but the wet-based zones are geographically limited in their extent, with frozen-bed conditions present either in thin marginal zones (glaciers) or as an irregular mosaic or patchwork (ice sheets; Pattyn, 2010; Hambrey and Glasser, 2012). Thus, basal meltwater moving below wet-based ice in polythermal ice masses may get trapped by freezing in contact with cold ice.

4.5.1 Water flow in snow, firn and ice

As explained in Section 4.2, snow and firn are porous whereas ice is non-porous. This also affects permeabilities. Although water can percolate through warm ice along crystal and grain boundaries, especially at three-grain intersections (Shreve, 1972; Nye and Frank, 1973), unfractured ice is effectively impermeable on the timescale of most eruptions. The permeability of water in snow is typically 10^{-5} m^2, but only c. 10^{-10} m^2 in firn (Colbeck and Anderson, 1982; Fountain, 1989), equivalent to a moderately permeable medium such as sand. They are thus aquifers and aquitards, respectively, whereas ice is an aquiclude. Thus, ice will impound meltwater. Conversely, if crevasses are present they will dominate water flow. Since most dry crevasses pinch out downward (typically at depths of c. 20 m in warm ice, deeper in cold ice; Glen, 1954), a surface layer of crevasses will channel away any overflowing meltwater. Crevasses will form only in ice undergoing extension and are not present everywhere.

However, even the permeabilities of snow and firn are very low and it is not immediately obvious how meltwater can escape rapidly enough to be significant (cf. Ambach et al., 1981). Therefore, the possibility of a water-filled vault (as may form in many glaciovolcanic eruptions) overflowing through a supraglacial firn or snow layer might be viewed as an unlikely process. A comprehensive understanding of snow and firn hydraulics is also hampered because of the unique combination of the melting properties of snow and ice and the processes involved, including liquid infiltration, heat conduction and refreezing, all of which are difficult to model (Illangasekare et al., 1990). Moreover, because of the heat transferred to the snow and firn particles, phase changes will occur continuously that will alter the hydraulic properties of the porous medium and greatly increase its ability to transmit the water.

The problem of permeability and supraglacial water escape was examined by Smellie (2006), who suggested that the effective permeabilities of snow and firn can be substantially enhanced during glaciovolcanic eruptions. For example, during a volcanic eruption flooding by warm water is likely to be very different from 'normal' meteoric wetting with cold water. Even seasonal (non-volcanically heated) water at ambient temperatures penetrates firn down to and along the firn–ice boundary (Gore, 1992; Fig. 4.2) and seepage rates will increase if patches of dry firn are present, i.e. firn with linked air-filled pores. The penetration rate will be even more enhanced when the water is heated since the water melts space very quickly, as is observed by the extensive and rapid losses of hot water used when drilling through firn (personal communication, K. Nicholls, British Antarctic Survey). Meltwater warmed by glaciovolcanic eruptions may be at 20 °C or more (Björnsson, 1992; Gudmundsson, 2003; Edwards et al., 2014a) and instantaneous melting will occur on contact with ice grains in firn. The melting will simultaneously

Fig. 4.2 View of englacial meltwater tunnel formed along a firn–ice boundary (arrows) and showing characteristic 'keyhole' shape caused by downcutting due to thermal erosion by the meltwater (image: Damian Gore).

increase the porosity and permeability, whilst cooling the inflowing water. The less dense firn also melts more easily than ice, requiring c. 66% of the energy required to melt a similar volume of pure ice, and dirty firn melts even more easily (firn with 20% rock particles (e.g. dispersed ash grains) requires just 77% of the energy to melt clean firn). Smellie (2006) suggested that, as a consequence of these characteristics, a percolation or melting zone develops behind the advancing wetting front. Behind that front the firn is fully melted or disaggregated (Fig. 4.3). The width of the melting zone (Δx in Fig. 4.3) depends on the melting rate, which is influenced by the pressure head and water temperature. In firn, Δx will be much shorter than in a non-melting aquifer (in which Δx will become longer with time as water penetrates further out, and $\Delta P/\Delta x$ will diminish with time). Δx will remain short as the melting zone migrates because all of the ice particles on the water-filled ice cavity side melt completely or are advected away (and melt completely) by turbulence. Because of the large volume of water in the cavity, ΔP is buffered. Thus, with Δx more or less constant at small values, $\Delta P/\Delta x$ will also be maintained and help to drive the lateral flow of water.

Finally, overflowing will also be facilitated in eruptions through a glacier with a sloping surface. This is because of two principal and related effects. (1) Because of progressive compaction-related variations in firn density, the meltwater will encounter less and less dense, more permeable firn at the same elevation making the migration easier. The seepage path will therefore develop a shallower gradient compared with the glacier surface and, once it exits onto the latter, will then develop rapid pipe flow (Fig. 4.3). (2) The transfer of heat by viscous dissipation during pipe flow will cause rapid downcutting of the underlying substrate (firn or ice), thus increasing the discharge of outflowing water. This has been observed in overflowing glacial lakes and results in distinctive 'keyhole'-shaped tunnels

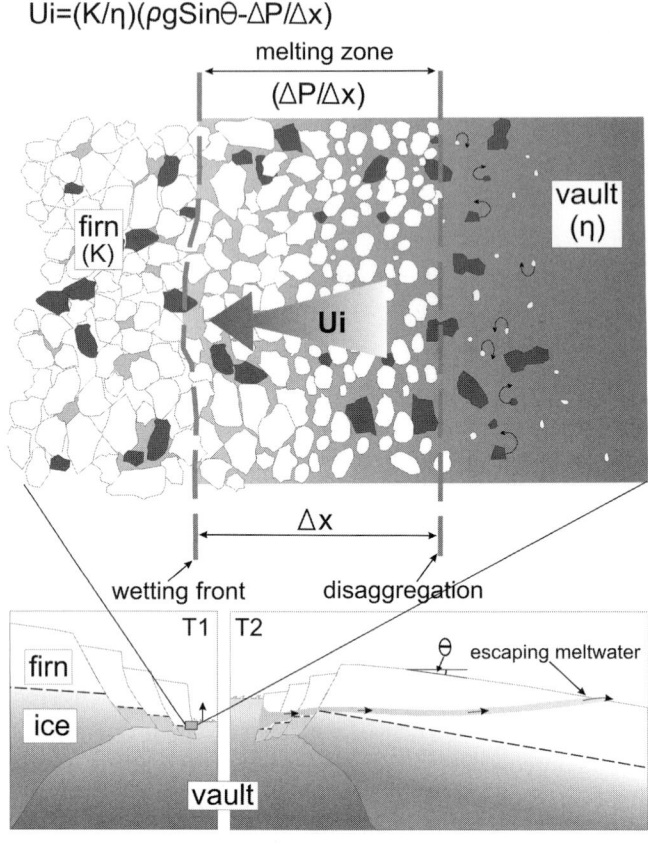

$$U_i = (K/\eta)(\rho g \sin\theta - \Delta P/\Delta x)$$

Not to scale

Fig. 4.3 Cartoon illustrating hypothetical melting processes and migration of meltwater in a narrow melting zone in firn at the margins of a meltwater-filled cavity during a glaciovolcanic eruption (see Smellie, 2006, for explanation of symbols and further details). The presence of warm water rapidly increases the permeability (by melting ice grains), possibly enhanced by eruption-induced seismicity causing turbulence, and ultimately leads to supraglacial meltwater outflow if local ice surface gradients are suitable (see right side of lower diagram).

and channels (Gore, 1992; Raymond and Nolan, 2002; Fig. 4.2). Supporting mathematical arguments can also be advanced based on Darcian flow of meltwater through a porous medium but it is by no means clear that most of the flow is Darcian and the reasoning will not be reproduced here (see Smellie, 2006). However, taken together, each of the effects outlined above will enhance the flux of meltwater overflowing through firn.

4.5.2 Hydraulic potential

The flow direction of water within and below a warm ice mass is controlled by the gradient of the hydraulic or fluid potential (ϕ_b; expressed as pressure, J m^{-3}), defined by

$$\phi_b = \underset{[A]}{\rho_w g z_b} + \underset{[B]}{P_w} \tag{4.6}$$

$$P_w = k_p \rho_i g (z_s - z_b) \tag{4.7}$$

where ρ_w and ρ_i are the densities of water and ice, respectively; g is acceleration due to gravity; z_b is the elevation of the point in question above a datum (normally the level of bedrock at the glacier snout); z_s is the ice surface elevation above the point in question, also relative to the same datum; P_w is water pressure; and k_p represents the range of subglacial water pressures, from atmospheric ($k_p = 0$) to full ice overburden pressure ($k_p = 1$) (Shreve, 1972; Björnsson, 1988).

If $k_p = 1$, combining equations (4.6) and (4.7) yields (Gudmundson et al., 2004)

$$\phi_b = \rho_i g \left[z_s + \frac{\rho_w - \rho_i}{\rho_w} z_b \right] \tag{4.8}$$

Equation (4.6) has two major parts: a gravitational potential for the water due to the relative elevation of the calculated point [A], and a pressure potential due to the thickness of overlying ice [B]. Water will flow from regions of high hydraulic potential to regions of low hydraulic potential. This equation governs how water films will move in warm ice and how water flows at the glacier bed. It is worth stating that the water does not move because of the high hydrostatic pressure per se, nor a gradient in that pressure (e.g. a deep body of standing water has both a high hydrostatic pressure and a pressure gradient yet it may be motionless). Rather, the water flow is caused by the gradient of the excess pressure over the hydrostatic pressure (Shreve, 1972). The surfaces with equal values of φ_b are an exaggerated mirror image of the surface of the ice mass (Fig. 4.4). This is important as it explains graphically how the surface slope of the ice mass is predicted to be about 10 or 11 times more effective than the slope of the bedrock in directing the water flow along the glacier bed (Björnsson, 1988; Gudmundsson et al., 2004). In other words, to overcome the fluid flow determined by the hydraulic gradient requires the bedrock gradient to be more than ten times the ice surface gradient. This property of meltwater hydraulics is responsible for the ability of subglacial water flow to flow upslope *against* the local bedrock gradient. Note that for flow in a partially full subglacial conduit connected to the atmosphere (i.e. $k_p = 0$) the water will simply flow down the local bedrock gradient.

For glaciovolcanic eruptions, an important consequence of water flow under a hydraulic potential has been highlighted by Björnsson (1988). During an eruption beneath a warm ice mass composed entirely of unfractured pure ice, the basal melting will rapidly cause a cauldron (depression) to form in the ice surface because of the lower density of ice compared with water. The equipotential surfaces therefore become distorted to mirror inversely the changes in ice surface. As can be seen in Fig. 4.4, meltwater becomes directed inwards and accumulates within an englacial cavity or vault. The surrounding ice forms an effective seal or barrier to any meltwater escape. In this scenario, the water accumulates

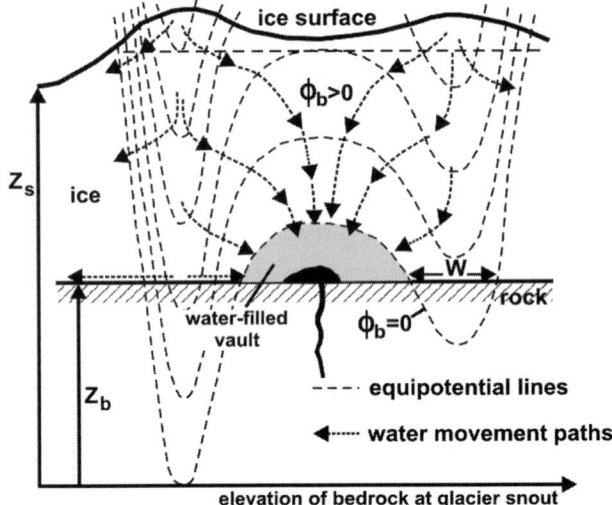

Fig. 4.4 Diagram illustrating the distribution and shape of equipotential surfaces, and englacial and subglacial water flow directions. Labelled lines are explained in the text. (Modified after Björnsson, 1988.) W: width of ice barrier that, with a perfect basal seal (as shown but probably only rarely achieved under wet-based ices, which are 'leaky' systems; see Fig. 3.8) will prevent water escaping from the vault. The sketch depicts a water-filled ice vault and overlying ice cauldron that might form in the early stages of a subglacial eruption (magma shown in black). If the eruption progresses, the vault will enlarge until an englacial lake develops with a surface open to the air.

until its depth reaches approximately 90% of the thickness of the ice barrier, whereupon the basal water pressure is able to float the surrounding ice. The escaping water, discharged in a sudden short-lived and often violently destructive event, is known as a jökulhlaup or glacier outburst flood (Björnsson, 1992, 2002). This model for subglacial eruptions was used to explain the construction of tuyas, with their encircling lava-fed deltas (Björnsson, 1988). However, it is deficient in three major respects: (1) it relies on an assumption that the overlying mass is composed solely of ice (i.e. the presence of firn and snow is ignored); (2) in a system sealed against basal water escape, water can only accumulate; and (3) the accumulation of volcanic rocks within the cavity in addition to meltwater should lead to the vault filling with water very quickly. Thus, in the Björnsson model, the basal water pressures in the vault will rapidly reach those of the surrounding ice barrier, lift it by flotation, and result in a jökulhlaup long before the volcanic edifice becomes subaerial. These conditions are *never* suitable for the evolution of lava-fed deltas, since the vault will drain long before a delta can form (Smellie, 2000, 2001, 2006).

The conundrum over how lava-fed deltas can form was partially resolved by Smellie (2001), who suggested that the ice mass was stratified, as explained in Section 4.2, with the presence and additional weight of a surface layer of permeable snow, firn and/or fractured ice capable of holding down the ice barrier until the rising meltwater surface was able to escape through the permeable surface layer (as described in Section 4.5.1). Thus, the supraglacial escape of meltwater from the vault enabled lava-fed deltas to prograde without the surrounding ice barrier being lifted (see also Section 8.2.2), which is impossible in a glacier composed wholly of ice. Of course, the thickness of the permeable surface layers has a maximum limit (c. 150–200 m; Smellie, 2000, 2001) and, because the thickness of an ice sheet is not similarly limited, the 'pinning-down' effects of the weight of permeable surface layers ceases once that ice sheet reaches thicknesses of 750–1000 m (Smellie, 2009).

Fig. 4.5 (a) View of anastomosing Nye channels carved by subglacial meltwater in limestone bedrock, Haute d'Arolla, Switzerland. The ice axe is c. 80 cm long. (b) Prominent Nye channels exposed at the base of Gigjökull on the northern flank of Eyjafjallajökull. The channels are c. 17–27 m in width.

The effects of basal escape of meltwater were not included in the explanation by Smellie (2001), although it unequivocally occurred during the 1996 tindar eruption of Gjálp, Iceland (Gudmundsson et al., 1997; Fig. 3.8). Thus, the next major development in understanding the hydraulics of eruptions with lava-fed deltas came with the recognition by Smellie (2006) that the elevation of the water in the meltwater lake was a compromise between the relative rates of subglacial versus supraglacial discharge. In that study, it was postulated that wet-based ice masses are intrinsically leaky systems, with large volumes of meltwater escaping via subglacial channels, fractures and water films. Although Röthlisberger and Nye channels might both be involved, only Nye channels (excavated into the bedrock) leave a clear record of their presence. They are developed particularly conspicuously on limestone bedrock (Fig. 4.5). Because these channels are related to the 'normal' (i.e. pre-existing, non-volcanic) seasonal hydraulics of the system, they are likely to be running full prior to a glaciovolcanic eruption. Any eruption-related, water-filled cavity will be superimposed on those channels and, as they are in hydraulic continuity with the rest of the glacier outside of the 'sealed' zone (*sensu* Björnsson, 1988), water is able to escape basally during the eruption. Moreover, as the water in the vault is volcanically heated (Gudmundsson, 2003; Jarosch et al., 2008), it will enlarge any channels by melting and because of the hydraulic head driving the meltwater out of the vault. Thus, the elevation reached by the water in a glaciovolcanic vault or lake is a balance between basal and supraglacial escape. Both overflowing and basal drainage are probably unstable over periods of days and weeks (Smellie, 2009). As represented by passage zones in lava-fed deltas, the water level in a vault is a highly dynamic feature during glaciovolcanic eruptions and passage zone elevations (i.e. fossil water levels; Section 13.4.1) may vary by several tens of metres. Examples of elevation changes up to 170 m are well documented (e.g. Fig. 13.3; Smellie, 2006).

5

Chemical and physical properties important to glaciovolcanic lavas

5.1 Introduction

Modern studies of glaciovolcanic eruptions and ancient deposits require quantitative knowledge of the chemical and physical properties for a wide compositional range of magma types in order to better predict likely hazards during an eruption (e.g. for calculating the degree and rate of meltwater production), or to extract important palaeoclimate proxy information from ancient deposits (e.g. estimating palaeo-ice thicknesses). While several studies have used detailed calculations in order to model aspects of the physics and chemistry of glaciovolcanic eruptions (e.g. Höskuldsson and Sparks, 1997; Edwards et al., 2002; Kelman et al., 2002; Gudmundsson, 2003; Tuffen, 2007), more frequently, published studies have incorporated minor modelling components that utilise a variety of thermodynamic parameters (e.g. Allen, 1980; Tuffen et al., 2002b; Edwards et al., 2009b). Here we review the relevant chemical and physical parameters needed to understand glaciovolcanic eruptions from a magmatic viewpoint. Our goal is to provide enough background information to facilitate the expansion of future studies that explore quantitatively the processes controlling volcano–ice interactions. A more detailed overview of physical processes and modelling can be found in Chapter 6. We note that the ability to fully characterise glaciovolcanic deposits requires access to a range of analytical equipment and techniques (partly reviewed in Chapter 7), which will not be discussed in detail here.

5.2 Compositional range

Almost the entire compositional range for igneous rocks is represented in known glaciovolcanic deposits (Fig. 5.1; Table 5.1). Calc-alkaline subduction zones such as the Cascades, the Andes, Kamchatka and the Aleutian arcs all have documented examples of glaciovolcanism that span the compositional spectrum from basalt to rhyolite. Tholeiitic, extension-driven magmatism in Iceland has produced glaciovolcanic deposits ranging from picritic basalt to high-silica rhyolite (e.g. Jakobsson et al., 2008; Jakobsson and Johnson, 2012), and hotspot volcanoes on the flanks of the Icelandic rift zones erupted transitional to mildly alkaline magmas with basalt–trachyte–rhyolite compositions (e.g. Hardarson and Fitton, 1991; Jakobsson et al., 2008). Finally, areas of continental extension, especially the

Table 5.1 *Bulk rock chemical compositions for representative glaciovolcanic rocks*

Source[a]	(1)	(2)	(3)	(4)	(5)	(6)	(7)	(8)	(9)	(10)
Sample #	90102	T5.4.6	SH9354	S-08–04	JS290	94BRE91	T5.34.2	BYE99	T5.13.2	AR2884
[TAS name]	Ol neph	Basanite	AOB	Tholeiite	Andesite	Trachyte	Trachyte	Dacite	Rhyolite	Rhyolite
SiO_2	42.7	45.24	46.44	48.91	58.99	59.60	63.74	64.89	70.28	77.5
TiO_2	2.76	3.42	2.83	1.5	1.548	0.33	0.32	0.47	0.24	0.32
Al_2O_3	11.72	14.27	16.97	15.53	14.14	16.50	15.3	16.54	13.88	8.4
Fe_2O_3[b]	2.3	2.12	2.03	1.86	1.91	1.54	1.13	0.73	0.85	0.86
FeO[b]	10.56	11.1	10.48	9.52	10.52	5.84	4.4	3.34	3.39	3.52
MnO	0.17	0.19	0.18	0.19	0.21	0.18	0.18	0.083	0.14	0.07
MgO	11.26	9.04	6.56	7.89	2.4	0.27	0.10	2.4	0.06	0.3
CaO	10.03	11.1	8.91	12.12	5.5	1.80	1.4	4.69	0.44	0.3
Na_2O	4.61	2.84	4	2.13	3.66	7.50	6.31	4.34	5.27	5.4
K_2O	1.5	0.87	1.08	0.19	1.53	4.78	5.29	1.71	5.24	3.1
P_2O_5	1.12	0.67	0.45	0.13	0.428	0.04	0.05	0.173	0.01	0
Sum	98.77	100.86	99.93	99.97	100.84	98.38	98.22	99.37	99.80	99.77

[a] (1) Francis and Ludden (1990), (2) J.L. Smellie and S. Rocchi (unpub.), (3) Russell and Hauksdóttir (2001), (4) Pollock et al. (2014), (5) Stevenson et al. (2006), (6) Edwards et al. (2002), (7) J.L. Smellie and S. Rocchi (unpub.), (8) Bye et al. (1999), (9) J.L. Smellie and S. Rocchi (unpub.), (10) Souther (1992).

[b] Calculated to be in equilibrium with the Qtz-Fay-Mt oxygen fugacity buffer.

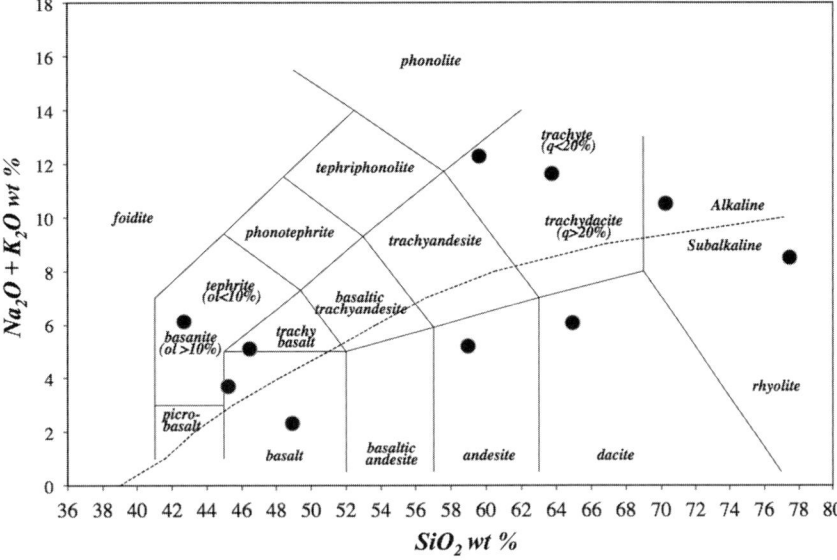

Fig. 5.1 Total alkalis versus silica diagram showing the compositions of selected rocks formed during glaciovolcanic eruptions and used as examples in this chapter (q: quartz, ol: olivine).

West Antarctic Rift, the East African Rift and the Northern Cordilleran Volcanic Province of western Canada, contain the largest compositional range of all, varying from nepheli-nites, basanites and alkali basalts to peralkaline phonolites and rhyolites (e.g. Francis and Ludden, 1990; Souther, 1992; Panter et al., 1997; Edwards and Russell, 2000; LeMasurier et al., 2003; Nardini et al., 2003; Furman et al., 2004). The only igneous compositions not represented in glaciovolcanic suites are ultramafic (komatiite and kimberlite families) and carbonatitic.

5.2.1 Why is it important?

The chemical composition of a magmatic system not only profoundly influences the physical properties that control how the magma and derivative lavas behave but it can also be a sensitive indication of eruption conditions and environments (e.g. formation of vesicles due to reduction in pressure). While major elements (with concentrations greater than about 1.0 wt %) control the most important physical properties (e.g. eruption tem-perature, viscosity and other thermodynamic properties), minor and trace elements (with concentrations between about 1.0 and 0.1 wt %, and less than 0.1 wt %, respectively) can be critical tools for unravelling the stratigraphy of complex deposits formed by multiple eruptions and for linking glaciovolcanism to broader climate–magmatism feed-backs (e.g. Jull and McKenzie, 1996; Edwards et al., 2002; Maclennan et al., 2002; Nyland et al., 2013). Amounts of volatile components, which can vary from major to trace concentrations, not only play a dominant role in determining basic eruption styles (effusive versus explosive), but can also be used to estimate syn-eruption pressures, and hence infer minimum ice thicknesses or englacial lake depths (e.g. Tuffen et al., 2010; see Section 13.5.1). Likewise, stable isotopes can be used to investigate interactions with non-magmatic fluids either during eruption or during subsequent alteration of the eruption products (e.g. Bindeman et al., 2008; Antibus et al., 2014).

5.2.2 Major, minor and trace elements

The concentrations of major elements fundamentally control the physical properties of the magmatic system (see below for detail). For example, the concentration of silica is one of the main controls on eruption temperatures, viscosities and glass transition temperatures. Significant shifts in major element concentrations during glaciovolcanic eruptions have only been documented for the 1985 Nevado del Ruiz (Gourgard and Thouret, 1990), the 1991 Hudson (Kratzmann et al., 2009), and the 2010 Eyjafjallajökull (Sigmarsson et al., 2011) eruptions, where eruption products varied from basalt almost to rhyolite in the fragmental deposits. More subtle shifts in major element concentrations have been reported for some basaltic centres in Iceland (e.g. Moore and Calk, 1991; Pollock et al., 2014) and in British Columbia (Moore et al., 1995; Edwards et al., 2009b, 2011). Some of these shifts may simply result from variations in crystal sorting or fractionation (e.g. Mathews Tuya, Pillow Ridge), while at other locations, different magma storage areas may have been involved during the eruption. A more detailed examination of case studies is reserved for Chapter 7.

The major element concentrations of a glaciovolcanic system also exert some control on features of the deposits. For example, only relatively fluid magmas (usually mafic) will form elongate and ellipsoidal pillow lavas that are generally less than 1 m in diameter. By comparison, more evolved magmas tend to form large lava lobes up to several metres in diameter that have also been interpreted as pillows (often called megapillows; Walker, 1992). However, at present it is unclear if the silicic lava 'pillows' are always diagnostic of subaqueous conditions because they have few modern analogues.

The concentrations of minor and trace elements are frequently determined by source region processes, and thus, just as glaciovolcanism spans a wide range of major element compositions, the same is true for elements in lower concentrations. However, minor and trace elements are less likely to exert strong controls on physical processes due to their low concentrations. The most important exception to this is volatile species, which are discussed separately in the following section. Minor and trace elements can be used to identify stratigraphical boundaries within deposits (Pollock et al., 2014), to correlate between physically separated deposits (McGarvie et al., 2007), to indicate changes in magma storage areas (Pollock et al., 2014), or even to identify changes in source region conditions (Jull and McKenzie, 1996; Maclennan et al., 2002). Several of these studies are discussed in more detail in Chapter 7.

5.2.3 Volatile contents

Compounds considered to be 'volatile' are frequently found as gas phases in equilibrium with magmas at very low pressures. These include H_2O, CO_2, SO_2, HCl, HF and a few other gaseous species. The amounts of dissolved H_2O and CO_2 in particular are critical for determining the likely explosivity of an eruption – i.e. whether or not the eruption will be initiated by magmatic explosions (Edwards et al., 2015a). If a separate gas phase has formed before the magma reaches the ice interface, it can increase the likelihood of an explosive eruption, which has significant consequences for how rapidly the overlying ice melts (Gudmundsson, 2003; Gudmundsson et al., 2004). A magma that has reached volatile saturation will produce a lava–gas mixture with a lower bulk density, and potentially enhance fragmentation even if the fragmentation is mainly caused by phreatomagmatic explosions from interacting with external water.

A number of models have been derived from a relatively small number of experiments aimed at understanding the relationships between pressure and volatile solubility in silicate melts (e.g. Dixon et al., 1995; Papale et al., 2006). These and other datasets have been exploited by several workers to create models to predict equilibrium volatile saturation conditions for H_2O with or without CO_2 (Newman and Lowenstern, 2002; Papale et al., 2006); recently published models can also account for some S species. Many of these models are freely available via the internet (e.g. VolatileCalc; http://volcanoes.usgs.gov /observatories/yvo/jlowenstern/other/software_jbl.html) and others are available as online calculators (e.g. the OFM model of Papale et al., 2006; http://ctserver.ofm-research.org /Papale/Papale.php). Most of the models estimate the solubility for one or a pair of volatile phases (most commonly H_2O and CO_2) as a function of pressure, bulk melt composition and temperature. Parfitt and Wilson (2011) give simplified equations for making order of magnitude H_2O solubility estimates for basaltic and rhyolitic melts:

$$n_{H_2O, \text{ bas}} = 0.1078 P^{0.7} \tag{5.1a}$$

$$n_{H_2O, \text{ rhy}} = 0.4111 P^{0.5} \tag{5.1b}$$

Following their lead, we can also suggest a simplified approximation for making order of magnitude CO_2 solubility estimates for basaltic and rhyolitic melts, which with slight modification gives reasonable estimates of low pressure (<300 MPa) CO_2 solubility in basaltic and rhyolitic melts compared to more complex models (e.g. Newman and Lowenstern, 2002):

$$n_{CO_2, \text{ bas}} = 0.00049 P \tag{5.2a}$$

$$n_{CO_2, \text{ rhy}} = 0.00071 P \tag{5.2b}$$

where n_{CO_2} is the wt % of CO_2 in the melt (bas: basalt, rhy: rhyolite) and the pressure (P) is in MPa. These simplified expressions compare reasonably well to more complicated expressions (Fig. 5.2a,b), but do not take in to account variations in temperature or melt

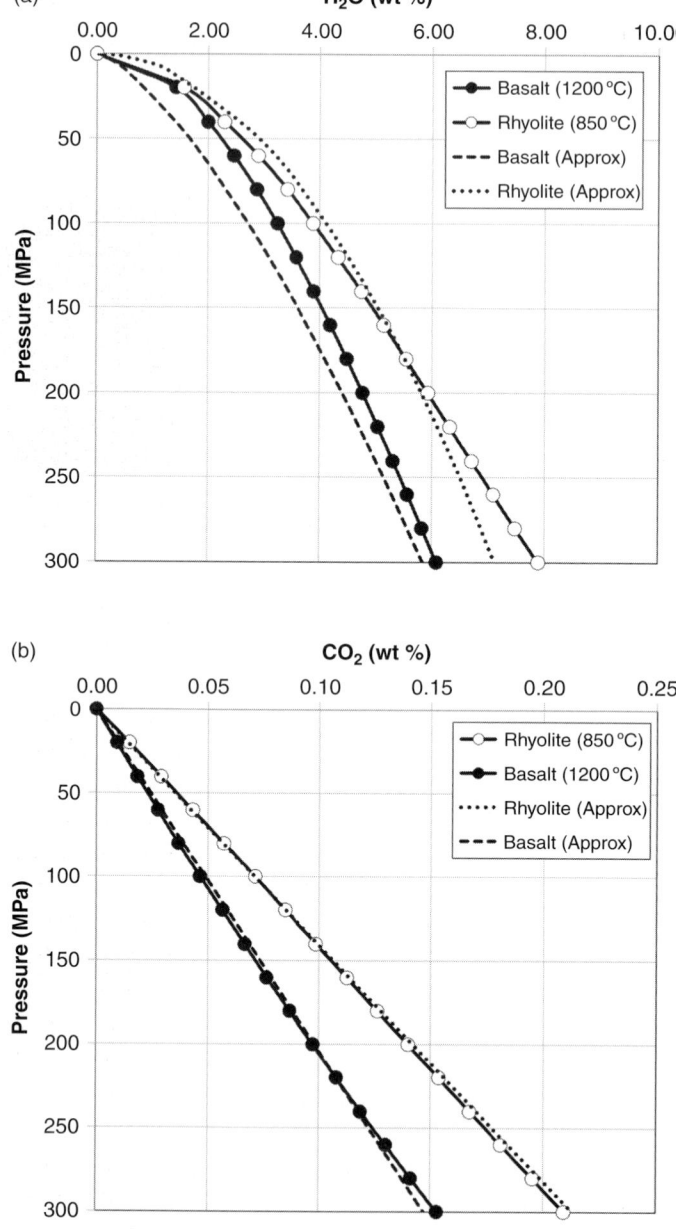

Fig. 5.2 Calculated volatile solubility for basalt and rhyolite as a function of P, H_2O and CO_2 concentrations. (a) Solubilities for H_2O in basaltic and rhyolitic melts. (b) Solubilities of CO_2 in basaltic and rhyolitic melts. See text for details of calculations.

compositions. They are also only for solubilities of either pure H_2O or pure CO_2, not for mixed volatile solutions. The model of Papale et al. (2006) is presently the one that can be applied to the widest range of melt compositions, temperatures and pressures.

The solubilities for all volatile species are dependent on pressure, temperature and the overall composition of the system. Basalt at its liquidus temperature (c. 1200 °C) in general has a lower solubility for H_2O than rhyolite at its liquidus temperature (c. 850 °C) by approximately 20 relative per cent (e.g. for a pressure of 100 MPa, basalt is saturated with c. 3.23 wt % H_2O while rhyolite is saturated with 3.88 wt %). For CO_2 the relative differences are similar (e.g. for a pressure of 100 MPa, basalt is saturated with c. 0.04 wt % CO_2 while rhyolite is saturated with 0.07 wt %). However, the solubilities of both species are highly dependent on each other, so that when mixed they significantly affect each other's solubilities.

Models for predicting the distribution of volatile species between a silicate melt and a separate fluid phase can be used to reconstruct the conditions extant during glaciovolcanic eruptions. Input parameters are the measured volatile concentration from the glass, the estimated temperature, and the bulk composition of the glass, with the latter assumed to be the same composition as the melt. The models then calculate the equilibrium distribution of the volatile species between the melt phase and a separate gas phase (assuming the melt is oversaturated for the specified conditions). Temperatures can be estimated either from mineral–melt equilibria thermometers (e.g. plag–melt, cpx–melt or ol–melt using models described in Putirka, 2008) or by calculating the presumed liquidus temperature using software such as MELTS, which can be accessed online (http://melts.ofm-research.org) or by downloading executable code (Ghiorso and Sack, 1995; Gualda et al., 2012). When volatile contents of volcanic glasses can be measured (e.g. vitric rims on pillow lava), and eruption temperatures can be estimated independently, it is possible to use the saturation models to constrain palaeo-pressures extant during eruption.

A number of studies have applied this technique to constrain palaeo-ice thicknesses including Dixon et al. (2002; tuyas in British Columbia); Schopka et al. (2006; Helgafell (a tindar) in Iceland); Edwards et al. (2009b; Pillow Ridge, Mt Edziza, British Columbia); Hungerford et al. (2014; Tenenna Cone); and Stevenson et al. (2009; Kerlingarfjöll, an intermediate-composition centre in Iceland). Icelandic magmas are generally considered to be relatively 'dry' and magmatic H_2O concentrations are quite low (less than 0.6 wt % for basalt; Nichols et al., 2002). Conversely, examples from British Columbia have relatively high H_2O concentrations as expected for more alkaline systems (up to 1 wt %; Edwards et al., 2009b; Hungerford et al., 2014). The results of these specific studies are discussed in detail in Chapter 7 (also Section 13.5.1.)

The main limitation with the technique is that CO_2 concentrations are typically at or below standard Fourier transform infrared spectroscopy (FTIR) detection limits (25 ppm), and H_2O solubility is very sensitive to CO_2 concentrations (Dixon et al., 2002; Edwards et al., 2009b; Tuffen et al., 2010). Thus, the difference in estimated pressure for a given value of H_2O (e.g. 0.8 wt %) associated with CO_2 contents of either 25 or 0 ppm results in ice thickness estimates that vary between 1000 m and 600 m, respectively, which can lead to substantially different environmental interpretations, e.g. whether the syneruption ice cover was a local ice cap or regional ice sheet (cf. interpretations for Pillow Ridge: Edwards et al., 2009b; see Section 13.5.1). Future refinements to current solubility models will need to account for the full contingent of volatile species, and will probably require a broad range of new experiments at relatively low pressures (5–10 MPa) to simulate relatively thin ice/ shallow water eruption conditions.

5.2.4 Stable isotopes

In volcanic systems, stable isotopes (i.e. those that are not products of radioactive decay) have at least three main uses: (1) to characterise magma source regions and signals from possible open-system processes (i.e. assimilation of crustal material, magma mixing, magma recharge and replenishment); (2) as geothermometers; and (3) to identify possible interactions between magmas and external fluid sources. The two most commonly used isotope families are hydrogen (^1H, ^2H, ^3H; reported as δD ‰ because ^2H is also known as deuterium) and oxygen (^{16}O, ^{17}O, ^{18}O; reported as $\delta^{18}O$ ‰). Values for most stable isotope systems are reported in units of per mil (‰), and are specified relative to defined standards (δ notation; see White, 2013 for detailed discussion). Typical values of δD for pristine mantle melts range from −40 to −90‰. Mantle values of $\delta^{18}O$ range from +4.5 to +8‰, although typically the 'mantle' range is +5.7 to +0.5‰ (Ito et al., 1987; Dorendorf et al., 2000). The most likely potential additional source of $\delta^{18}O$ ('contaminant') in glaciovolcanic systems is water. Although meteoric water can have values of $\delta^{18}O$ and δD that, taken individually, can overlap with pristine mantle values, in combination these two isotopes define a 'meteoric water' line that tightly controls variations in the isotopes based on latitudinal position where the magma erupted, and which is distinct from the values for mantle melts (Fig. 5.3). Although currently not much utilised to characterise glaciovolcanic deposits, future studies using these isotopes and others (e.g. Li, B, S) may provide another means of recognising changes in source regions during construction of larger volcanoes, as well as constraining areas of magma generation and storage. This may be particularly important as more detailed studies try to better understand how waxing and waning of ice sheets might affect the tapping of magmas stored at different levels of the crust and mantle. Magmas with long residence times in the crust are more likely to record evidence of the storage by having non-mantle values for stable isotopes. A few preliminary studies (Bindeman et al., 2004, 2008) show that some Russian and Icelandic volcanic systems have been contaminated by a low $\delta^{18}O$ source, potentially palagonitised volcaniclastic deposits or altered upper crustal deposits, either while transiting the crust or during storage there.

Because mass fractionation of stable isotopes is highly dependent on temperature, stable isotopes have long been used as geothermometers in fluid-dominated systems (White, 2013). While underutilised in glaciovolcanic systems, application of O isotopic geothermometers may provide an alternative means for estimating magma temperatures (e.g. using

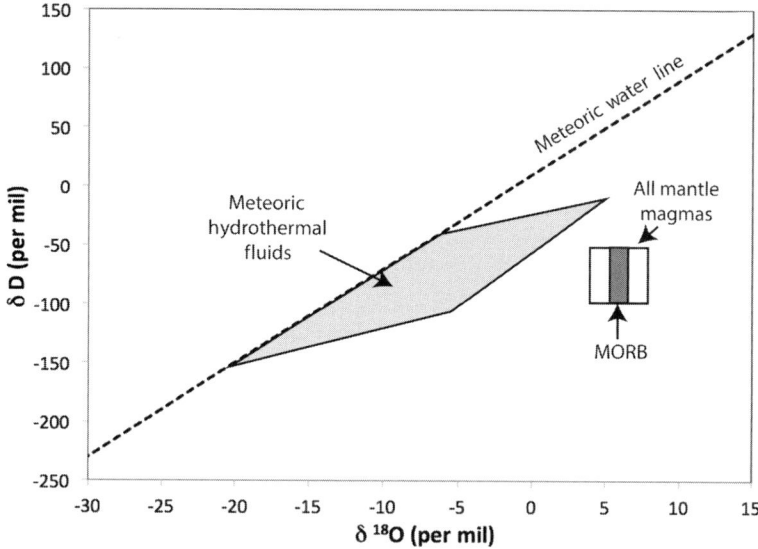

Fig. 5.3 Graph showing values of $\delta^{18}O$ versus δD for the Meteoric Water Line, representative samples of mantle melts and MORB, and hydrothermal fluids. See White (2013) for compilation of data sources.

plagioclase–olivine thermometry) or for identifying geothermal fluids that cause palagonitisation of vitric fragmental deposits.

Additionally, stable isotope systems may ultimately be important for helping to confirm the eruptive environments of ancient glaciovolcanic deposits. Edwards et al. (2013) showed that, even on the timescales of minutes, O isotopes can be exchanged between cooling basaltic melts and external water. In environments where putative glaciovolcanic deposits might have origins that are ambiguous between glacial or marine settings, differences in marine versus glacial water isotopes could be a critical discriminant (e.g. value of $\delta^{18}O$ for all marine water is defined as 0‰, so values for water from glacial settings should always be <0‰ or 'lighter'; Mortimer et al., 2008; Antibus et al., 2014).

5.3 Lava temperatures

Temperatures of materials erupted during glaciovolcanism can range from liquidus conditions for mafic lavas (c. >1200 °C) to below the glass transition temperature for rhyolitic lavas (<700 °C; Table 5.2; Giordano et al., 2005, 2008; Fig. 5.4). Knowledge of eruption temperatures is ultimately critical for constraining potential rates of melting and meltwater generation for hazard evaluation of ice/snow-clad volcanoes, and for reconstruction of heat transfer at ancient deposits. Both can be based on thermodynamic calculations (e.g. MELTS), a variety of geothermometers, experimental measurements, and field measurements during modern eruptions.

Table 5.2 *Calculated properties for representative glaciovolcanic rocks*

	(1)	(2)	(3)	(4)	(5)	(6)	(7)	(8)	(9)	(10)
Source[a]										
Sample #	90102	T5.4.6	SH9354	S-08-04	JS290	94BRE91	T5.34.2	BYE99	T5.13.2	AR2884
TAS name	Ol neph	Basanite	AOB	Tholeiite	Andesite	Trachyte	Trachyte	Dacite	Rhyolite	Rhyolite
$\rho\,(T_{liq})$[b]	2742	2764	2715	2715	2594	2459	2417	2448	2379	2355
Silica activity[b]	0.269	0.376	0.375	0.438	0.576	0.437	0.512	0.598	0.618	0.737
T_{liq} (anhy)[b]	1298	1213	1174	1212	1111	1090	1072	1148	1039	1018
T (>90% sol)	968	1078	1054	1087	925	883	892	920	889	751
ΔT (liq-sol)	330	135	120	125	186	207	180	228	150	267
T_g[c]	644	673	672	668	681	637	657	709	682	691
C_p (liq) J/kg K[b]	1530	1500	1450	1500	1360	1350	1330	1370	1310	1290
C_p (sol)[b]	1150	1180	1190	1210	1170	1200	1200	1210	1198	1190
$\Delta T^* C_p$ (kJ/kg) (liq-sol)	442	182	158	169	235	264	228	294	188	331
$\Delta T^* C_p$ (kJ/kg) (total)	1556	1454	1413	1485	1317	1323	1298	1407	1253	1225
ΔH_f (kJ/kg)	358	350	348	304	359	251	208	211	190	153
Total sensible + ΔH_f	1914	1804	1761	1789	1676	1574	1506	1618	1443	1378
Percentage sensible	81.3	80.6	80.2	83.0	78.6	84.1	86.2	87.0	86.8	88.9
Percentage ΔH_f	18.7	19.4	19.8	17.0	21.4	15.9	13.8	13.0	13.2	11.1
$\log \eta$ Pa s (T_{liq})[c]	0.5	1.3	1.8	1.4	3.9	4.42	5.11	4.49	6.07	6.64
η (Pa s)	3.0E+00	1.9E+01	6.9E+01	2.5E+01	7.8E+03	2.6E+04	1.3E+05	3.1E+04	1.2E+06	4.4E+06
η (15% crystals)[d]	6.0E+01	4.0E+02	1.4E+03	5.1E+02	1.6E+05	5.3E+05	2.6E+06	6.3E+05	2.4E+07	8.9E+07
η (1% H_2O)[c]	1.1E+00	4.9E+00	1.2E+01	5.9E+00	5.5E+02	4.9E+03	1.3E+03	1.7E+03	3.1E+04	1.3E+03

[a] See Table 5.1 for list of sources of bulk rock compositions

[b] Thermodynamic properties calculated using Rhyolite-MELTS (Gualda et al., 2012)

[c] Viscosities calculated from Giordano et al. (2006)

[d] Estimates for bulk viscosity calculated using liquid viscosities from Giordano et al. (2006) and the Einstein–Roscoe equation

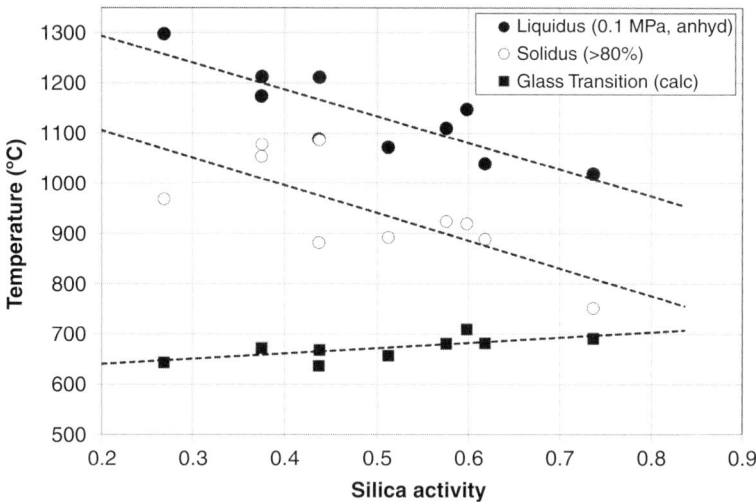

Fig. 5.4 Values of liquidus temperatures, solidus temperatures, and T_g's calculated using Rhyolite-MELTS and Giordano et al. (2006) for a range of magma compositions.

5.3.1 Why is it important?

The defining characteristic of glaciovolcanism is the interaction between magma/lava and ice/snow/water. While some physical interactions occur, such as fracturing of ice or compaction of snow, the most important interaction is transfer of heat from the magmatic system to the cryospheric system. The total amount of heat transferred is ultimately tied to the eruption temperature of the magmatic system, as well as the ability of the magma to transfer its heat to the environment. The eruption temperature marks the starting point for estimates of the total amount of *sensible heat* that can be released during an eruption (see below for calculations of sensible heat). Rapid heat transfer will produce the most efficient melting of the enclosing ice, and hence for a specific eruption will maximise the amount of work done by the magma on the cryosphere. This can translate into more rapid and severe hazards (e.g. jökulhlaups), as well as stronger fragmentation of the magma and production of volcanic deposits with a higher vitric component that are more susceptible to rapid alteration, cementation, and ultimately deposits that are more likely to be preserved in the geological record.

5.3.2 Estimates from field measurements

While it might seem as though measuring lava temperatures during an eruption would be commonplace, it is rarely done, especially for glaciovolcanic eruptions. In part this is due to the intrinsic dangers associated with most glaciovolcanic eruptions, and in part due to their remoteness. At present, the only glaciovolcanic eruption with a significant number of field temperature measurements is the 2012–13 eruption at Tolbachik volcano, in

Kamchatka, Russia (Edwards et al., 2014a; see Chapter 3). This Hawaiian-style eruption was dominated by effusion of fluid lava flows, many of which interacted with the local snowpack. During several stages of the eruption lava temperatures were measured by thermal probes and by forward-looking infrared radar (FLIR) cameras (Fig. 5.5). Thermal probes were inserted into actively flowing lava, and at Tolbachik they recorded temperatures up to 1353 K (1080 °C). FLIR measurements are carried out remotely and are thus safer, but they only map surface temperatures of lava flows. At Tolbachik the maximum temperatures recorded by FLIR were c. 1323 K (1050 °C), which is close to but less than the maximum internal temperatures of the lava flows. An advantage of FLIR measurements is that they can be used to obtain temperatures where the lavas are in contact with ice. Several satellites are also capable of collecting radiation signals in the infrared thermal spectrum to give broad-scale estimates of temperatures during eruptions that are either too remote or too dangerous for ground-based measurements (i.e. MODIS and SEVIRI; Ganci et al., 2012).

5.3.3 Estimates from magma/melt compositions

Thermodynamic models derived from experimental data have been used for many years to depict the crystallisation history of silicate melts. Many now are freely available on the internet as executable files (see Section 5.2.3 above). These programs allow the calculation of temperatures and pressures of crystallisation based on assumed liquid composition. It is important to remember that the whole-rock composition of a lava with crystals may not represent the composition of the actual liquid (i.e. 'melt') because of the presence of the crystals. However, if the mineral compositions and proportions are also known, they can be used to back-calculate the crystal-free magmatic compositions. One way to do this is by estimating liquidus temperatures from Rhyolite-MELTS (Gualda et al., 2012), or by using glass compositions and MELTS or bulk rock compositions. The majority of all volcanic rocks are porphyritic and, with the notable exception of a few lavas with rhyolitic composi-tions, few volcanoes erupt lavas that are completely crystal-free.

5.3.4 Estimates from geothermometers

The use of geothermometers to reconstruct equilibrium temperatures between crystalline phases or between crystals and melt is well established (Putirka, 2008; spreadsheets to be used for clinopyroxene and feldspar geothermometry are available online from http://www.fresnostate.edu/csm/ees/faculty-staff/putirka.html). For lavas with glass in the ground-mass, it is possible to estimate melt temperatures to within 10–20 °C using the compositions of plagioclase microlites or clinopyroxene microphenocrysts and melt compositions, all of which are generally measured using an electron probe microanalyser. In the absence of direct measurements taken from a lava during its emplacement, these methods may provide the best estimates of lava temperatures when the lava makes its first contact with ice.

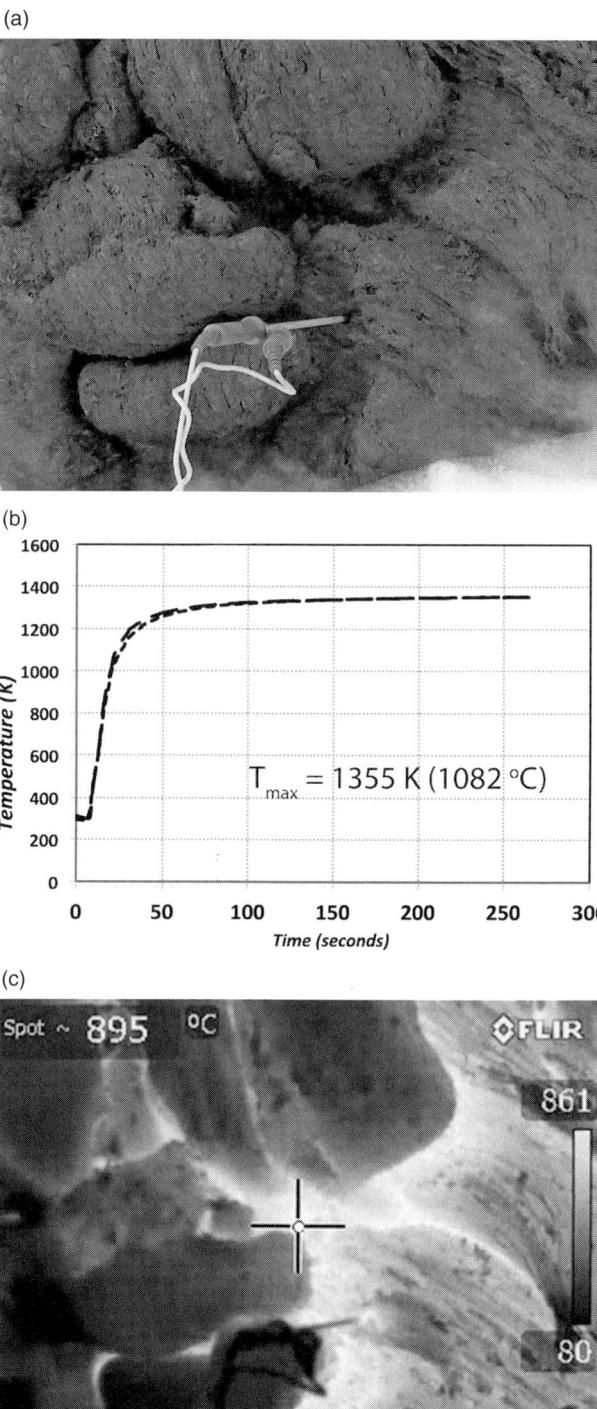

Fig. 5.5 Measuring temperatures in active Tolbachik lava flows. (a) Pāhoehoe lava lobe with thermal probe inserted into radiant crack to measure internal temperatures. (b) Graph of temperature versus time for measurement of pāhoehoe shown in (a). (c) Forward-looking infrared (FLIR) image of the surface of pāhoehoe lobe in (a). (A black and white version of this figure will appear in some formats. For the colour version, please refer to the plate section.)

5.3.5 Estimates from experiments

With suitable facilities, analogue experiments can also be conducted on volcanic materials and with appropriate scaling they may approximate the conditions extant during an eruption (Edwards et al., 2013). For example, materials can be heated to the point where glassy particles relax and become liquid (at the glass transition temperature), thus determining a lower limit for melt temperatures. Alternatively, crystalline samples can be heated to their liquidus temperature (i.e. all the crystals are melted) to determine maximum eruption temperatures.

5.3.6 Glass transition temperatures (T_g)

The glass transition temperature (T_g) is another important temperature parameter to be constrained for glaciovolcanic deposits, as it can be a record of rates of cooling induced by rapid heat transfer in an ice/water-dominated eruption environment. It is defined as the temperature at which the viscosity of a liquid changes such that it effectively behaves like a solid (Giordano et al., 2005). The glass transition temperature (T_g) marks the time during a cooling process at which the viscosity of a melt becomes virtually infinite, which indicates that the melt can no longer instantaneously react to strain in the way that a liquid can (Dingwell and Webb, 1990). Thus, at its T_g a melt essentially becomes a solid and reacts to strain in a brittle way. Experiments have demonstrated that if rates of cooling are increased for a specific silicate liquid composition, the T_g will increase as well (e.g. a cooling rate of 20 degrees min^{-1} for a basaltic glass induces a T_g of 739 °C, while a cooling rate of 5 degrees min^{-1} induces a T_g of 716 °C; Giordano et al., 2004); addition of water to melts lowers values of T_g by up to 200 degrees or more. Because rates of cooling strongly affect T_g, it is not possible to calculate the minimum T_g for a given melt composition. However, it is also possible to estimate a 'maximum' value of T_g based on thermodynamics, and this can be done in conjunction with estimates of melt viscosities (Table 5.2; Fig. 5.4; see model described in Giordano et al., 2008).

Comparisons of measurements of T_g for a range of compositions demonstrate why mafic systems are frequently more sensitive recorders of glaciovolcanic environments of eruption. Whether calculated (Table 5.2) or measured (Giordano et al., 2005), values of T_g are similar for all silicate melts. For example, Giordano et al. (2005) reported values of nominally anhydrous T_g for a compositional spectrum from basalt to dacite (spanning a compositional range of 20 wt % SiO_2) which at a constant rate of cooling only varied by 120 degrees (values for two trachyte samples were 788 °C and 721 °C, value for phonolite was 670 °C, value for dacite was 745 °C, and for basalt 739 °C). Yet the estimated liquidus temperatures for these melts would be significantly different, with the basalt likely to be c. 1200 °C, and the other melts closer to 1000 °C. This implies that rapid heat extraction is essential for producing basaltic glass, as the melt must reach a temperature up to 300 degrees below the solidus temperature rapidly enough to inhibit full crystallisation. However, for the higher silica melts T_g's are much closer to their solidus temperatures (typically 100–150 degrees), and so much less drastic heat extraction is required for these melts to reach their T_g's. Increasingly, researchers are attempting to use the cooling-path-dependent T_g's to recover rates of cooling in natural samples (e.g. Gottsman and Dingwell, 2001; Wilding et al., 2004; see Chapter 7 for more studies). The T_g is important because once it is reached by a liquid, the ability of crystals to grow on the timescales of an eruption decreases significantly, as diffusion of the nutrients for crystal growth is a few to several orders of magnitude slower in solid glass than in a liquid (Brady, 1995). Suppression of crystallisation can result in a loss of up to 20% of the total available thermal energy of the magma that would otherwise be released as latent heat (see Section 5.5.3) to cause melting.

5.4 Viscosity (η)

Viscosity is the resistance to flow for a fluid. It is usually measured in units of pascal seconds (Pa s) and can range over at least five orders of magnitude in glaciovolcanic systems due simply to compositional variations (see Table 5.2; Fig. 5.6). Viscosity is dependent on the temperature and major element composition of the magma/lava as well as the abundance of bubbles or solids in the liquid (phenocrysts, xenocrysts, crystal aggregates); of course composition and temperature are also intrinsically linked, as mafic melts not only have fewer network-forming cations (e.g. Si), which results in lower viscosities, but they also generally exist at higher temperatures. This inherent trend for mafic melts having lower viscosities is only partly mitigated by the higher solubilities for H_2O in felsic melts, which to a limited degree acts to lower viscosities (Table 5.2). Magmas without crystals or bubbles can behave relatively simply, i.e. similar to but non-Newtonian in behaviour. A variety of models can be used to estimate viscosities of silicate melts (e.g. Giordano et al., 2008). Flow behaviour becomes very complex for melts with crystals and/or bubbles (cf. Harris, 2013a,b), in channelised or non-channelised systems, and under subaqueous conditions.

5.4.1 Why is it important?

Viscosity is critical for understanding glaciovolcanism for several reasons. Firstly, it is the physical property that exerts a primary control on rates of magma transport through the crust. A viscous magma will rise more slowly, which may increase the length of time for background heat fluxes and hence melting overlying glaciers. It is also more likely to degas ahead of eruption, and thus have a lower potential for volatile-driven explosivity. Conversely, lower-viscosity magma will rise more quickly, advecting heat to the surface more rapidly where it is used for melting ice, and retaining higher volatile contents, which will increase the potential for explosivity. Secondly, lava effusion rates will be higher for lower viscosity melts, also potentially leading to more rapid melting of overlying ice (because of the rates of melting are dependent in part on contact surface area), and potentially generating greater hazards in terms of meltwater discharge volumes available.

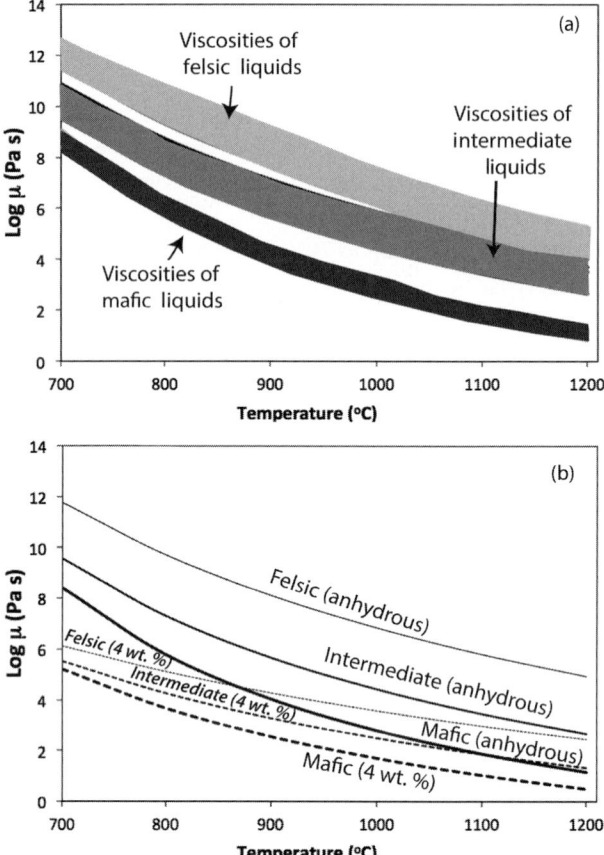

Fig. 5.6 Viscosities of a range of magma compositions. (a) Temperature versus viscosity. (b) H_2O concentrations versus viscosity. Note that viscosities of some magmas, e.g. phonolite, can extend between those of intermediate and felsic compositions.

Viscosity is also a primary control on the morphology of effusive deposits. Highly viscous lavas are more likely to form lava domes, or very thick, slow moving lava flows (e.g. Tuffen et al., 2013; see Chapter 8).

5.4.2 Models versus measurements

Accurate measurements of viscosities in the field are challenging (e.g. Cashman and Mangan, 2014), and at present no viscosity measurements have been published for glacio-volcanic eruptions. Comparisons of field measurements with those calculated from models are complicated because the thickness of flows can be difficult to measure in the field, the presence of crystals and bubbles can also have a significant effect on viscosities, and their abundances can be difficult to constrain in active flows. Jeffrey's equation provides a crude means for estimating viscosities of lava flows (η) in the field based solely on the slope of the surface down which the lava flows (α), its density (ρ), its thickness (d), and its velocity (V):

$$\eta = \frac{\sin{(\alpha)}g\rho d^2}{3V} \tag{5.3a}$$

The constant in the denominator can have a range of values, but Harris (2013a) suggests that a value of 3 is appropriate for channelised flows where the width of the channel is greater than its depth, and a value of 8 for channels with a semi-circular cross section. While the slope and velocity measurements are easily made in the field, the bulk density for the flow is more difficult to evaluate due to possible heterogeneities in the distribution of vesicles (thus collection of a sample from the surface of the lava flow might not be representative of the entire flow) as is the 'effective' thickness of the flow. The thickness estimate is particularly critical as it is squared in the equation. Thus, if all other measurements are accurate, but a flow is estimated to be 5 m thick yet has a true thickness of only 4 m, that can alter the viscosity estimate by half an order of magnitude. For 'a'ā lavas knowledge of the thickness of the basal autobreccia is also required, as is channel depth and thickness of the basal boundary layer for pāhoehoe lavas, which need to be subtracted from flow thickness estimates. MacDonald (1963) suggested that, in general, field measurements are likely to produce only crude estimates of true lava viscosities. Moore (1987) modified Jeffrey's equation by adding a damping term to account for the possibility of flows having yield strengths.

From the perspective of hazard mitigation, Jeffrey's equation can be rewritten to allow for the prediction of lava velocities on a given topographical gradient assuming that viscosities can be estimated a priori:

$$v = \frac{\sin{(\alpha)}g\rho d^2}{3\eta} \tag{5.3b}$$

This relationship can also be the starting point for modelling rates of lava movement across snow/ice surfaces in order to estimate potential rates of meltwater production from melting at the base of a lava flow.

5.4.3 Changes in viscosity during eruption

As viscosity is highly dependent on temperature, cooling of lava during effusion will produce significant changes, as will the loss of volatiles as the lava degasses. Decreasing temperatures will also lead to crystallisation, which will also increase lava viscosity. The Einstein–Roscoe relationship is frequently used as a first-order approximation to estimate the effects of the addition of solids to a liquid (cf. Harris, 2013b):

$$\eta_{\text{eff}} = \eta_{\text{liq}}(1 - R\Phi)^{-2.5} \tag{5.4}$$

where η_{eff} is the effective viscosity, η_{liq} is the viscosity of the melt (e.g. from Giordano et al., 2008 or similar), R is equal to $1/\phi_{\mathrm{max}}$ (where ϕ_{max} is the maximum crystal packing content, which is typically taken to be either 0.5 or 0.7; Marsh, 1981; Pinkerton and Stevenson, 1992), and Φ is the volume fraction of solids in the liquid. Based on this relationship, addition of 15 vol % crystals to a silicate melt will increase viscosities by about an order of magnitude (see Table 5.2); for example, the viscosity of a tholeiitic basalt would increase from 25 Pa s to 510 Pa s with the addition of 0.15 volume fraction crystals.

While the ubiquitous presence of bubbles must also have an effect, their role can be either to increase or to decrease viscosity (Llwewellin and Manga, 2005; Gonnermann and Manga, 2013; Harris, 2013b). In general, the presence of spherical bubbles is taken to mean that the bubbles are affecting the melt in a manner similar to crystals, but with a slightly different method of estimation:

$$\eta_{\mathrm{eff}} = \eta_{\mathrm{liq}}(1 - \omega)^{-1} \tag{5.5a}$$

where ω is the bubble content. If the bubbles are sheared (elliptical not spherical), that is an indication that they are helping to accommodate the applied stress and may actually be acting to reduce the effective viscosity. This effect can be estimating using:

$$\eta_{\mathrm{eff}} = \eta_{\mathrm{liq}}(1 - \omega)^{5/3} \tag{5.5b}$$

Even more complex formulations are being developed that can account for the presence of all three phases (liquid, crystals, bubbles) and take into account the relative sizes of crystals versus bubbles (cf. Harris, 2013b).

5.5 Other thermodynamic properties

Many fundamental properties of silicate melts can be calculated based on a variety of experimental results. Understanding these properties and their significance will facilitate the designing of more detailed investigations and development of quantitative models for reconstructing eruption dynamics of ancient glaciovolcanic deposits, or predicting the outcomes of future eruption scenarios. In addition to melt temperatures and viscosities (described above), the most important other thermodynamic parameters include the latent heats of fusion, heat capacities, glass transition temperatures, melt densities, thermal conductivities and thermal diffusivities.

5.5.1 Why are they important?

Thermodynamic properties are used in all quantitative modelling of magmatic and volcanic processes. The establishment of large databases of relevant parameters and their implementation into computer algorithms allows for easy estimation of many important conditions relevant to glaciovolcanism. For example, it is becoming more common for studies to estimate the total amount of thermal energy released during an eruption and the rate at which the energy was likely released, to gain insight into whether or not the eruption would have been able to melt through the overlying ice (e.g. Nielsen, 1937; Allen, 1980; Höskuldsson and Sparks, 1997; Kelman et al., 2002; Tuffen, 2007; Edwards et al., 2009b). Such a calculation requires knowledge of the sensible heat released by the magmatic system during the eruption (i.e. heat capacity × temperature range through which the magma cooled), the extent to which the lava crystallised once it made contact with the surrounding ice-dominated environment (i.e. volume proportion of specific mineral phases × their respective latent heats of fusion), and the rates at which heat was transferred from the magma to its surroundings (calculated using densities, thermal conductivities and thermal diffusivities). This type of estimation creates less speculative assessments of potential discharge rates and volumes of ancient jökulhlaups, and consequent changes to the enclosing ice. As has been speculated for the Pleistocene–Holocene transition, the increasing frequency of eruptions during the early stages of deglaciation may also have a number of feedback mechanisms that accelerate glacial–interglacial transitions (e.g. Huybers and Langmuir, 2009; see Chapter 14), including speeding glacial destruction by melting large volumes of ice, increasing rates of surface heating by deposition of tephra on ice surfaces (although this relationship is complex and can also lead to insulation; see Chapter 4), and facilitating faster ice flow via increased basal lubrication from syn- and post-eruption meltwater production.

5.5.2 Models versus measurements

Ultimately models that are used for estimating the thermodynamic properties of magmatic products are based on experimental measurements. While most experiments are not specifically designed to replicate glaciovolcanic conditions, large databases of compiled experimental results provide a useful and sometimes the sole means of constraining syn-eruption conditions.

5.5.3 Latent heats of fusion (ΔH_f or L)

The phase transition from liquid to solid (crystals) is accompanied by a release of energy, known as the latent heat of fusion (ΔH_f or L), which is the change in enthalpy associated with the change of state from solid to liquid or vice versa; it is the same amount of thermal energy needed to make a crystal melt. For glaciovolcanic eruptions, the latent heat of fusion is important as it can account for up to 20% of the total thermal energy available to be released during an eruption, and hence to do work on the surrounding cryosphere (Table 5.2). While values of L for common silicate minerals are generally within an order of magnitude (L for quartz is 157 kJ kg^{-1}, L for forsterite is 1010 kJ kg^{-1}; Spera 2000), they show enough variation that it can be important to use specific values corresponding to the

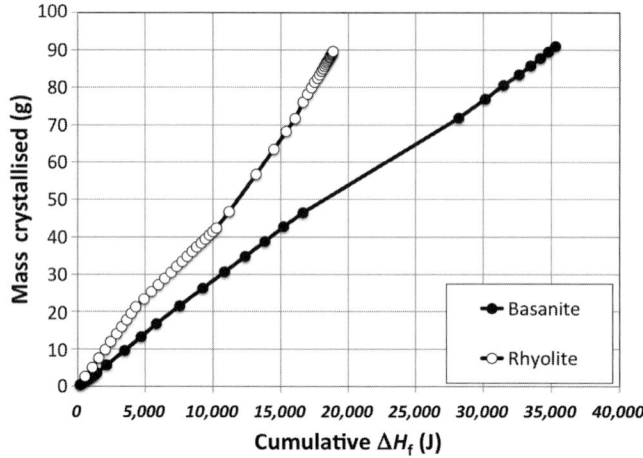

Fig. 5.7 Estimated cumulative latent heat (ΔH_f) production calculated using Rhyolite-MELTS for basanite and rhyolite (see Table 5.2 for compositions).

minerals present in a given magmatic system (Fig. 5.7; see Table 5.2 for compositions). While, strictly speaking, calculation of L for mineral compositions that are solid solutions should take into account potential enthalpies of mixing, estimates that are accurate to within about 10% approximate to simple mixtures of end-member mole fractions. For crystallising magmas, Rhyolite-MELTS can be used to estimate the overall ΔH_f for a given magma composition (Table 5.2); largely due to the almost order of magnitude difference in ΔH_f between olivine and quartz, mafic systems have higher proportions of energy to release via crystallisation (e.g. c. 350 kJ kg^{-1} for mafic magmas as opposed to <200 kJ kg^{-1} for felsic magmas; Fig. 5.7; Table 5.2).

5.5.4 Heat capacities of melts, crystals and magmas (C_p)

Heat capacity is a measure of the ability of a material to absorb thermal energy, and is defined as the amount of thermal energy required to raise the temperature of the material by one degree Celsius. It has units of J kg^{-1} K^{-1}. For most calculations important for under-standing glaciovolcanism, we are interested in the heat capacity at a constant pressure, or C_p. Values of C_p can be calculated for silicate melts and most common minerals from a variety of standard databases (e.g. Berman and Brown, 1985; Berman, 1988; Rhyolite-MELTS: Gualda et al., 2012). Typical values of C_p for silicate minerals and melts are approximately 1000–1400 J kg^{-1} K^{-1}, but are a function of temperature (e.g. the C_p for a tholeiitic basalt at 1200 °C is 1500 J kg^{-1} K^{-1} but decreases to 1200 J kg^{-1} K^{-1} when the basalt is at the solidus of 1087 °C; see Table 5.2 for other compositions). Heat capacities are particularly important for calculating the amount of heat released by simple cooling of a melt, which is also referred to as the 'sensible heat' of the system. This thermal energy accounts for 80–100%

of the potential thermal energy available during an eruption to do work on the surrounding ice/water.

5.5.5 Melt densities (ρ)

The densities of magma, lava and tephra are critical for understanding many aspects of glaciovolcanic eruptions. Magma densities play an important role (together with magma chamber overpressure, viscosity and volatile exsolution) in determining the buoyancy forces that drive ascent rates (see Chapter 6), and hence the degree to which an exsolved volatile phase will remain in equilibrium with the host melt during ascent. The density of a lava flow is also important for determining its thermal diffusivity (see Section 5.5.6), and hence how rapidly it can transfer heat to its surroundings. Tephra densities will influence settling rates for particles from a subaerial plume and strongly influence subaqueous transport and deposition. Edwards et al. (2002) suggested that changes in magma density during assimilation–fractional crystallisation (AFC) processes could make crustal magma storage areas more susceptible to 'glacial pumping' (see Chapter 14). They showed that changes in magma density during AFC processes coupled with ice thickness could control whether or not glaciovolcanic eruptions were more likely to be initiated by subglacial or supraglacial events.

Estimating magma densities can be made in a number of ways. Lange and Carmichael (1990) provided values for major oxide components from which densities for melts can be calculated as a function of pressure. Several thermodynamic databases (e.g. Berman and Brown, 1985) provide algorithms for calculating the densities of silicate minerals as a function of pressure and temperature. Alternatively, the Rhyolite-MELTS computational software can be used to estimate melt, crystal and bulk magma densities (melt plus crystals) also as a function of temperature and pressure (e.g. Ghiorso and Sack, 1995; Gualda et al., 2012). For lava and tephra, direct measurements from samples provide the best estimates, although densities can also be calculated given knowledge of vesicle volumes (i.e. the density of the sample is a product of the density of the non-vesicular portion times one minus the volume fraction of vesicles). Porosities can be estimated using image analysis techniques (e.g. Shea et al., 2009), Archimedes principle, or more advanced techniques such as X-ray microtomography (e.g. Song et al., 2001).

5.5.6 Thermal conductivity (λ) and diffusivity (κ)

Values for thermal conductivity (λ; measured in W m^{-1} K^{-1}) and thermal diffusivity (κ; calculated as $\lambda/(\rho\,C_p)$ with units of m^2 s^{-1}) are fundamentally important for estimating rates of heat transfer during glaciovolcanic eruptions (e.g. Gudmundsson, 2003). However, whereas values of thermal conductivity for minerals are relatively well known (e.g. Jaupart and Mareschal, 2010), values for silicate melts are not. Historically, values of κ are generally assumed to be c. 1×10^{-6} m^2 s^{-1} (e.g. Jaeger, 1968; Höskuldsson and Sparks,

1997), but recent measurements for silicate melts show that values of κ can be as low as $0.4 \times 10^{-6} \, \mathrm{m^2 \, s^{-1}}$, which would result in significantly longer times for conductive cooling of lava (e.g. Romine et al., 2012; Hofmeister et al., 2014). For example, using a simple standard equation that relates the timescales (τ) and length scales (l) of heat transfer to calculate the approximate time required to cool a 1 m-thick lava flow, i.e.

$$\tau = l^2 / \kappa \tag{5.6}$$

we see that using a value of κ of $1.0 \times 10^{-6} \, \mathrm{m^2 \, s^{-1}}$ predicts a cooling time of c. 280 hours, or between 11 and 12 days. However, using a value of κ of $0.4 \times 10^{-6} \, \mathrm{m^2 \, s^{-1}}$ predicts a cooling time of c. 695 hours, or between 29 and 30 days. While the above approximation could be refined for specific geometries and boundary conditions, it illustrates the point that when estimating rates of heat transfer by conduction, using the best available values is important to accurately test hypotheses related to interpreting ancient deposits, or to make predictions about the consequences of future eruptions.

5.6 Summary

Rapidly expanding abilities to exhaustively characterise the chemistry of magmatic systems are particularly critical to studies of glaciovolcanism. Understanding and quantifying how much energy is transferred between magmas and their ambient environments to do work on the surrounding ice is important for hazard assessments and for understanding the short- and long-term impacts of glaciovolcanism on the cryosphere. As measurement techniques, precision and accuracy continue to improve, we will also improve our ability to use glaciovolcanic deposits to reconstruct palaeo-ice thicknesses/palaeo-lake depths. Application of the techniques outlined here will greatly increase our ability to understand in greater detail the origins of modern and ancient glaciovolcanic deposits, and they are becoming an important part of publishable scientific studies.

6

Physics of glaciovolcanism

6.1 Introduction

This chapter reviews the most important aspects of the physical processes that govern glaciovolcanic eruptions. It is well beyond the scope of one chapter to describe in detail the physics and mathematical methods needed to accurately describe glaciovolcanic eruptions, so for more detailed examples and general information about the physics of volcanic phenomena, we refer interested readers to Fagents et al. (2013). Here we briefly review constraints on magma generation and migration, eruption rates, heat transfer, vapour saturation and bubble growth, processes related to rapid magma/lava fragmentation, modes of emplacement, and the formation of cooling fractures from the perspective of glaciovolcanism. These are all important for identifying and interpreting ancient glacio-volcanic sequences, for understanding the hazards posed by future volcano–ice interactions, and for understanding the impacts of eruptions on the cryosphere.

6.2 Magma generation

The physics of magma generation is a broad subject, which has seen revolutionary advances in our abilities to understand source region processes since the 1990s. High-pressure experiments (e.g. Duffy, 2008) and algorithms based on thermodynamics (e.g. p-MELTS: Ghiorso et al., 2002; Gualda et al., 2012) provide means of assessing the roles of source regions and variations in temperature and pressure conditions in produc-ing different compositions of basaltic magma, which are fundamentally important for understanding the heat and mass transfer processes during glaciovolcanism. Empirically, three different mechanisms can begin the process of magma production that may ultimately lead to magma–ice interaction: (1) increases in mantle temperature (e.g. mantle plumes/hotspots; e.g. Iceland, Hawaii); (2) decreases in pressure (e.g. divergent boundary magmatism in Iceland; continental rifting in West Antarctica and East Africa; transtension in western Canada); and (3) addition of volatile components (continental and oceanic arcs; e.g. Andes, Cascades, Aleutians, Kamchatka). While all three mechanisms can be important for glaciovolcanism, the last two are especially relevant. Some workers (e.g. Jull and McKenzie, 1996) have suggested that changes in lithostatic pressure due to deglaciation could be a possible trigger for enhanced rates of volcanism related to major climate change. In addition, production of 'wet' magmas with higher-than-normal dissolved volatile concentrations can lead to deeper and more extensive magma fragmentation, which is an important factor in controlling how rapidly heat is transferred to the surrounding ice to produce meltwater and potentially dangerous flooding.

6.2.1 Temperature increases (heat fluxes)

Whether convection in the mantle is driven mainly by thermal or density gradients, ultimately it facilitates advection of heat from the mesospheric mantle to the asthenospheric mantle and eventually to Earth's surface. While the existence and location of mantle plumes is unresolved, the identification of areas of anomalously high heat flow (hotspots) is unequivocal. The potential role of deep mantle upwellings may be key to driving continental rifting, and hence related to the ultimate driving forces for glaciovolcanism in the West Antarctic and East African rifts (see below). However, at least two hotspots have been directly linked to volcano–ice interactions: Iceland and Hawaii. While the typical temperature gradient (geotherm) for most parts of the mantle is below the solidus for peridotitic mineral assemblages, slight perturbations in temperature may be enough to start the melting process. Calculation of mantle 'potential temperatures' (T_p) has become a standard procedure for studies of basaltic magma systems and is used to infer minimum temperature conditions during the initiation of source region melting (e.g. McKenzie and Bickle, 1988). While the exact interpretation of T_p is still debated, many workers have shown that values for T_p in the source regions of Hawaiian and Icelandic ('plume'-related) basalts are c. 200 °C hotter than for the source regions of MORB (e.g. Putirka, 2008).

The Icelandic hotspot is presently thought to underlie the east-central part of Iceland, including the Vatnajökull ice sheet, which overlies several active volcanic centres (Grimsvötn, Bárdarbunga, Kverkfjöll; Thordarson and Höskuldsson, 2008). Broad-scale lithospheric doming from the hotspot likely plays a role in the enhanced base elevations of these volcanoes and potentially influences the stability of the ice cap. Of course, without the thermal heating from the hotspot and consequent additional magma input, Iceland and its glaciovolcanoes would not exist! Certainly, the extensive glaciovolcanic activity in this area is due at least in part to magma generation related to the hotspot. For example, Grimsvötn volcano, situated on the eastern flank of the active zone of rifting that forms the central tectonic graben in Iceland that demarcates the North American and Eurasia plate boundaries, has erupted three times between 1998 and 2011 (see Chapter 3).

While it seems unlikely, given their latitude, the reduction of atmospheric temperature with altitude (i.e. lapse rate) and significant elevations of Hawaiian volcanoes (>4000 m) results in the precipitation and accumulation of snow periodically. Mauna Kea and possibly

Mauna Loa have been glaciated at times during the Quaternary (Porter et al., 1977; Porter, 1979a,b,c, 1987). Like Iceland, the Hawaiian Islands would not exist without magma generated by the underlying hotspot. Porter (1987) described glaciovolcanic lava flows and intercalated glacial deposits near the summit of Mauna Kea (see Section 2.5.5). Given its similar elevation, it seems likely that Mauna Loa has also been glaciated during the Quaternary, but its much higher rate of lava production means that any deposits formed prior to 10 000 years ago are now inaccessible because they will have been buried by many generations of Holocene non-glacial lava flows.

6.2.2 Decompression

The most important, but least intuitive, method of melting in the mantle is by adiabatic decompression. It is the driving force for magma production at mid-oceanic ridges (MORB), where the highest melt production rates on Earth occur, and in which rising (i.e. decompressing) mantle intersects melting curves calculated for mantle compositions (Philpotts and Ague, 2009). Decompression melting is important in two different contexts for glaciovolcanism. The first is simply that the largest areas of volcano–ice interaction on Earth occur in extensional tectonic settings, where decompression melting is likely dominant: Iceland (divergent boundary between North American and Eurasian tectonic plates), West Antarctica (continental rift zone), and western Canada (transtensional boundary between the North American and Pacific tectonic plates). The second context is the inferred relationship between deglaciation and increased eruption rates (e.g. Jull and McKenzie, 1996; Slater et al., 1998; see Chapter 14). Some workers have suggested that the rapid removal of 2 or 3 km of superimposed ice at the end of glacial periods, equivalent to lowering subjacent pressures by c. 18–27 MPa, can result in increased rates of magma production in the source region, leading to increased eruption rates. Other workers have suggested that extension associated with isostatic rebound could trigger fracturing around extant magma bodies in the crust, also leading to more frequent eruptions (Gudmundsson, 1986).

6.2.3 Addition of volatiles

The third main trigger for mantle melting is the addition of magmatic volatiles via dehydration reactions in the upper part of subducting tectonic plates ('slabs'). This process likely contributes significantly to magma production in large areas of relatively unstudied glaciovolcanism at several of Earth's largest volcanic arcs, including the Aleutian, Cascade, Andean, Kamchatka and Scotia arcs. While 'wet' melting has significant consequences for the evolution of magmatic systems (e.g. Philpotts and Ague, 2009), it may be particularly important for glaciovolcanism in determining whether an eruption is effusive or explosive at initiation. The processes leading to magma fragmentation are complex (see Section 6.5, below), and they are also critical to understand because the efficiency of heat transfer for explosive eruptions is likely to be much higher than for effusive eruptions due to the greater rapidity of heat transfer to meltwater from small turbulently mixed tephra particles compared to large (and largely static) pillows (e.g. Gudmundsson, 2003; Woodcock et al., 2012). Thus 'wet' magmatic systems may have a higher likelihood of being explosive at onset, leading to more rapid melting of overlying ice.

6.3 Magma migration

Once melt forms, it is almost always less dense than the source material from which it was derived. The resulting density contrasts generate an 'excess pressure' (P_{ex}) leading to a buoyancy potential that drives magmas to leave their source regions (see Philpotts and Ague, 2009, pp. 28–32). The critical component of P_{ex} with respect to glaciovolcanism is that the uppermost part of the lithostatic column is made of ice, which has a much lower density (c. 920 kg m^{-3}) than average crustal rocks (c. 2400 kg m^{-3}) or sediments (c. 1500 km m^{-3}). Hence, from a simple, static approach, it is possible to calculate for a given magma located in the crust or lithosphere how much of the P_{ex} is due to ice loading of a local (c. 500 m thick) or regional (c. 3 km thick) ice cap:

$$P_{ex} = P_l - P_m \tag{6.1a}$$

$$P_m = g\rho_m h_m \tag{6.1b}$$

$$P_l = g\rho_l h_l \tag{6.1c1}$$

$$P_l = g(\rho_{lm} h_{lm} + \rho_c h_c + \rho_i h_i) \tag{6.1c2}$$

$$P_{ex} = g(\rho_{lm} h_{lm} + \rho_c h_c + \rho_i h_i - \rho_m h_m) \tag{6.1d}$$

where the densities of magma (ρ_m), the lithospheric mantle (ρ_{lm}), the crust (ρ_c), and ice (ρ_i) are known, as well as their absolute thicknesses (h_m, thickness of magma column; h_l, thickness of entire lithospheric column; h_{lm}, thickness of lithospheric mantle; h_c, thickness of crust; and h_i, thickness of ice). Assuming all other values are constant, and including the thickness of ice in the total lithospheric thickness, a regional ice sheet 3 km thick reduces P_{ex} by approximately 6 MPa in comparison to a local ice cap that is 500 m thick (Fig. 6.1). While it seems somewhat counterintuitive that more ice leads to a reduction in P_{ex}, this is a consequence of the ice density being much less than the density of the rock column. Thus, including ice in the rock column leads to an overall reduction in the average column density, thereby reducing P_{ex}. The key is that P_{ex} must exceed the strength of the overlying rock to cause brittle failure and generation of a dyke (see Wilson et al., 2013 for more detailed discussion). This is also one of the potential links between deglaciation and climate: melting of overlying ice sheets can increase P_{ex} and increase the likelihood of eruptions. Edwards et al. (2002) showed that the presence of ice could favour either subglacial or supraglacial eruptions, depending on magma densities and how they change during fractionation in magma storage areas.

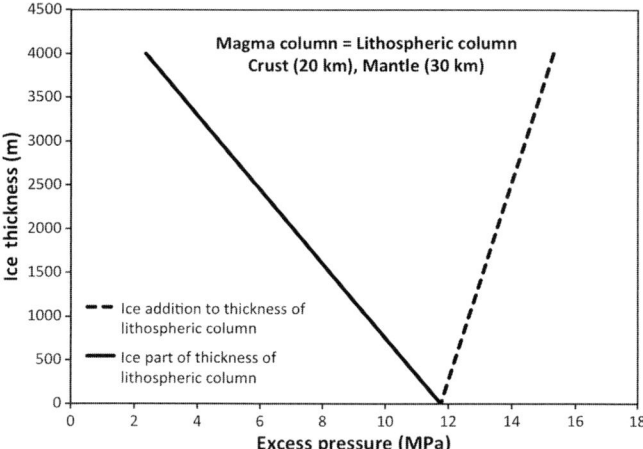

Fig. 6.1 Graph showing variations in excess pressure (P_{ex}) versus ice thickness. The solid line shows how increasing the thickness of ice that is part of the lithospheric column decreases P_{ex} for supraglacial eruptions. Conversely, the dashed line shows that adding an increasing amount of ice to the lithospheric column can lead to an increase in P_{ex} for subglacial eruptions.

6.4 Heat transfer

Heat transfer can be by any combination of three different processes: conduction, advection or radiation. Extensive books have been written about heat transfer by conduction (e.g. Carslaw and Jaeger, 1959), and modern engineering studies have gone to great lengths to produce models for a large diversity of systems to understand heat transfer (e.g. Incorpera et al., 2010). Glaciovolcanic eruptions represent more complex environments for heat transfer than either subaerial or subaqueous eruptions because of the presence of ice, which, upon melting, produces water and potentially steam, thus greatly complicating heat transfer physics. Nielsen (1937) was the first to make broad estimates of heat released during an eruption to assess whether or not the estimated volume of ice melted during the 1934 Grimsvötn eruption (c. 10 km^3) could have been driven by heat transfer from the erupted materials. His empirical calculations showed that basaltic magma could potentially melt about ten times its own volume of ice, which is approximately the same estimate that has been used by many subsequent workers (e.g. Allen, 1980; Höskuldsson and Sparks, 1997). However, Nielsen (1937) decided that another heat source must have been present as he did not feel that the 1934 eruption produced sufficient magma. The next iteration of thermal modelling was by Allen (1980), who used a simplified model starting with the emplacement of a single lava pillow into ice. His estimates of the rates of heat transfer and the heat flux showed that a single pillow could melt approximately four times its volume of ice. Almost two decades later, the more advanced study by Höskuldsson and Sparks (1997) of the thermodynamics of volcano–ice interactions suggested roughly equivalent volumes of melting for basalt as had been estimated by Allen (1980). Many subsequent workers have explored heat transfer from lava or tephra to ice (Kelman et al., 2002; Tuffen et al., 2002b;

Gudmundsson, 2003; Kelman, 2005; Tuffen, 2007; Wilson and Head, 2002, 2007b; Edwards et al., 2012), and it has become a routine procedure to estimate potential volumes of ice melted during the formation of glaciovolcanic edifices (e.g. Edwards et al., 2009b, 2011, 2013; Russell et al., 2013).

Heat conduction is one of the most important processes of heat transfer in most glaciovolcanic systems. It is the primary process that allows thermal energy to move across the lava boundary into the environment, at least at length scales ≤ 0.1 m. Advection encompasses all processes that physically transport material that has more thermal energy than its surroundings. For example, when magma moves from its source region, it transports heat to another part of the Earth. Lava flows move much more rapidly than heat can be conducted to their surroundings (typical velocities for basaltic lava flows range from 0.1 to 5 m s^{-1}, while thermal diffusivities are typically 10^{-6} to 10^{-7} m^2 s^{-1}). If this was not so, lavas would cool in place and would be unable to flow. When magma/lava is in contact with water, it is possible that convection in the water will generate a high heat flux from the magma/lava boundary (e.g. Höskuldsson and Sparks, 1997) and facilitate opening of a cavity in the ice. Finally, radiation of heat is the fastest means by which thermal energy can be transported, although it is mainly important at high temperatures.

6.4.1 Heat conduction

Critical to understanding how heat is lost by conduction is knowledge of thermal conductivity/thermal diffusivity (Chapter 5), and knowledge of the geometry of a cooling body. While conduction from the top and bottom of a lava flow can be modelled essentially as two infinite, parallel boundaries, when lava domes are emplaced beneath ice, or pillow lavas in water, their more irregular geometries require more complicated expressions to approximate cooling (e.g. Jaeger, 1968). For example, Wilson and Head (2007b) assumed a planar geometry to model the cooling and heat flux from a lava flow emplaced onto ice, and used the following analytical expression from Carslaw and Jaeger (1959):

$$T = T_{\mathrm{b}} + (T_{\mathrm{c}} - T_{\mathrm{b}}) \left(2\mathrm{erf}\, \frac{x}{2\sqrt{\kappa t}} - \mathrm{erf}\, \frac{x-d}{2\sqrt{\kappa t}} - \mathrm{erf}\, \frac{x+d}{2\sqrt{\kappa t}} \right) \qquad (6.2)$$

where T is the temperature of interest, T_{b} is temperature of the substrate, T_{c} is the initial temperature of the lava, x is the distance from the contact to a point inside the flow, d is the overall thickness of the flow, κ is the thermal diffusivity and t is time. This geometrical formulation is appropriate for a planar contact at the base of a lava flow. However, it would not be appropriate for modelling the emplacement of a lava pillow, where the cooling surface has a broadly cylindrical shape. To model cooling of a lava pillow, a more appropriate starting point would be an analytical solution for a cylindrical geometry, which is also given by Carslaw and Jaeger (1959) as:

$$T = T_{\mathrm{o}} + \left[(T_{\mathrm{s}} - T_{\mathrm{o}}) \left(1 + \frac{2}{x} \sum_{i=1}^{\infty} \frac{J_0(r\alpha_i)}{i J_1(x\alpha_i)} e^{-\left(\kappa \alpha_i^2 t\right)} \right) \right] \tag{6.3}$$

where T_{o} is the initial temperature, T_{s} is the temperature surrounding the system, $J_0(r\alpha_i)$ are the zero-order Bessel functions, $J_1(x\alpha_i)$ are the first-order Bessel functions, α_i are the roots of the Bessel functions, x is the diameter of the cylinder, and κ represents the thermal diffusivity specific to the material. Using this solution, it is useful to briefly explore how variations in pillow size and thermal diffusivity, the two most important controls on cooling, change the expected timescales of heat transfer from a basaltic lava pillow. A basaltic lava pillow 0.5 m in diameter that cools by conduction will be below its solidus temperature within approximately 2 hours, and will be cooled to ambient (i.e. 0 °C) temperature after approximately 10 hours, assuming a value for κ of 1×10^{-6} m^2 s^{-1}. However, simply changing the value of κ to 0.7×10^{-6} m^2 s^{-1} extends the cooling time to more than 3 hours to the solidus, and more than 20 hours to reach ambient temperature. The timescale differences are even more significant for larger pillows: increasing the diameter of the cooling pillow to 1 m changes timescales for cooling to the solidus to between 8 ($\kappa = 1 \times 10^{-6}$ m^2 s^{-1}) and 10 hours ($\kappa = 0.7 \times 10^{-6}$ m^2 s^{-1}), and the overall cooling time for the pillow to between 40 hours ($\kappa = 1 \times 10^{-6}$ m^2 s^{-1}) and 60 hours ($\kappa = 0.7 \times 10^{-6}$ m^2 s^{-1}).

Although analytical solutions have been formulated for many different types of boundary geometries and boundary conditions (e.g. constant boundary temperature; time-variant boundary temperature), numerical solutions (e.g. finite element; finite volume) are frequently used to allow for more exact controls on modelling specific geometries (e.g. Klingelhöfer et al., 1999), to allow more detailed exploration of variations in boundary conditions and to simulate moving boundary conditions (e.g. Stefan-like conditions; Carslaw and Jaeger, 1959; Manglik and Singh, 1995). To properly model heat transfer under conditions where ice is melting either above or below lava, and hence the lava–ice boundary is not stationary, the Stefan formalism is required. This type of problem only has analytical solutions for a very limited range of boundary conditions, and so it essentially requires numerical modelling techniques. For some larger lava bodies, heat conduction is slow enough that melting caused by emplacement of lava may have only a small effect on surrounding ice. The 2004–6 emplacement of the dacitic lava dome at Mt St Helens showed that if effusion rates are very slow, lava can physically push ice aside with negligible heat transfer (e.g. Fig. 3.4b; Walder et al., 2007, 2008).

6.4.2 Heat advection

Advection, or physical transport of thermal energy, can be important for several different components of heat transport during glaciovolcanic eruptions. From the perspective of magma/lava, movement of either is a form of heat advection. As magma emerges at the contact with ice either through a dyke or another exit (e.g. diatreme), it conveys a convective heat flux (q_{adv}) to the ice interface. Likewise, as magma moves along the ice–bedrock contact, in the case of a sill, or as lava flows supraglacially, each is advecting thermal energy simultaneously which can cause melting of ice or heating of surrounding water. A simple estimate of rates of heat advection can be gleaned from estimating heat fluxes as a function of volumetric eruption rates (q_m), which can vary over several orders of magnitude but generally fall within the range 0.1 to 500 m^3 s^{-1}, from

$$q_{adv} = \frac{q_m C_p^m \rho_m}{A} \tag{6.4}$$

where C_p^m is the heat capacity of the magma/lava in units of J kg^{-1} K^{-1}, ρ_m is the density of the magma/lava, and A is the cross-sectional area of the surface of interest in m^2.

Advection can also be important if the magma/lava is in contact with water that is undergoing natural convection (e.g. Höskuldsson and Sparks, 1997). In this case, the heat flux from the boundary contact between magma/lava and water due to natural convection is approximated by

$$q_{conv} = C_p^w \rho_w J (T - T_i)^{\frac{4}{3}} \tag{6.5}$$

where C_p^w is the heat capacity of water at T, ρ_w is the density of water at T, T is the water temperature, T_i is the temperature of the ice roof, and J is defined as

$$J = \gamma \left(\frac{\alpha g \kappa^2}{v} \right)^{\frac{1}{3}} \tag{6.6}$$

where γ is a dimensionless constant with a value of approximately 0.1, α is the coefficient of thermal expansion for pure water at T, g is the gravitational constant (9.8 m s^{-2}), κ is the thermal diffusivity of water at T, and v is the kinematic viscosity at T (from Huppert and Sparks, 1988). Data for the physical properties of water can be obtained from a number of sources (e.g. http://www.mhtl.uwaterloo.ca/old/onlinetools/airprop/airprop.html). Finally, this information can be used to calculate the rate of melting ($\dot{a}$) of the overlying ice using

$$\dot{a} = \frac{q_{conv}}{H_s} = A(T - T_i)^{\frac{4}{3}} \tag{6.7}$$

where

$$A = \frac{C_p^w \rho_w J}{H_s}, \quad H_s = \rho_i [c_i (T_m - T_\infty) + L_i] \tag{6.8}$$

and H_s is specific enthalpy of melting of the ice (originally described by Huppert and Sparks, 1988, and applied to ice melting by Höskuldsson and Sparks, 1997). Our recalculated one-dimensional estimated heat fluxes from convecting water using the above equations produce ice melting rates of 10–40 m per day (slightly higher than those reported by Höskuldsson and Sparks, 1997), assuming that the temperature of the convecting water remains between 30 and 100 °C. However, although Höskuldsson and Sparks (1997) suggested that these calculated rates are close to the lower limit of those occurring during eruptions in Iceland, the rates of melting actually observed were significantly faster. For example, the 1996 Gjálp eruption melted through 600–700 m of ice in c. 30 h, for an average melting rate of c. 20 m h^{-1} (Gudmundsson et al., 2004).

6.4.3 Thermal radiation

Thermal radiation is the most rapid type of heat transfer. It depends largely on the temperature difference between the object that is losing heat to its surroundings and the emissivity of the material (likely 0.8–0.95 for lava). The thermal radiation balance between two bodies at different temperatures is given by the Stefan–Boltzmann law (e.g. Ball and Pinkerton, 2006):

$$q_{rad} = \varepsilon\sigma\left(T_1^4 - T_o^4\right) \qquad (6.9)$$

where q_{rad} is the rate of heat transfer with units of W m^{-2}, ε is emissivity, taken as 0.9 for lava (Harris, 2013a,b), σ is Stefan's constant (5.67 × 10^{-8} W m^{-2} K^{-4}), T_1 is the surface temperature of the lava and T_o is the surface temperature of the body receiving the radiation (assumed to be 273 K), which during a glaciovolcanic eruption is ice. If all the radiated heat is used for melting the ice, and assuming a latent heat of fusion for ice of $L_i = 3.35 \times 10^5$ J K^{-1} with an assumed density of 920 kg m^3, the rate of melting of ice ($\Delta x/\Delta t$), where x is metres of ice and t is time, either in front of an advancing lava flow (e.g. Edwards et al., 2012) or beneath a supraglacial lava flow, is

$$\left(\frac{\Delta x}{\Delta t}\right) = \frac{q_{rad}}{\rho_i L_i} \qquad (6.10)$$

The maximum possible melting rate from radiation occurs where a wall of lava is close to a wall of snow or ice. This rate is calculated by substituting (6.9) into (6.10) for q_{rad}, yielding the maximum melting rate:

$$\left(\frac{\Delta x}{\Delta t}\right) = \frac{\varepsilon\sigma\left(T_1^4 - T_o^4\right)}{\rho_i L_i} \qquad (6.11)$$

Edwards et al. (2012) showed that when the melting rates of ice according to equation (6.11) over a range of values for T_1 (0–1473 K) are compared with the average advance rates of the lava flows during the first several days of the 2010 Fimmvörðuháls eruption, it is clear that thermal radiation is an order of magnitude too slow to melt a layer of ice ahead of rapidly advancing lava. Observations from large-scale experiments using basaltic melt and snow/ice (Edwards et al., 2013), as well as from the 2012–13 eruption at Tolbachik, Russia (Edwards et al., 2014a, 2015b), are also consistent with this theoretical limit of melting due to radiation. Thus, even radiation is probably too slow to generate enough heat to allow lava flows to melt impounding snow/ice unless the lava flows are travelling less than 5 m h^{-1} (Edwards et al., 2012, 2015b).

6.4.4 Efficiency of heat transfer

Most estimates of the volumes of ice melted during glaciovolcanic eruptions are theoretical limits based on relatively vague assumptions about how efficient heat transfer will be from the magma/lava to the surroundings. However, Gudmundsson (2003; see also Edwards et al., 2015a) explored in detail the likely efficiency of heat transfer, focusing on the end-member cases of effusive and explosive eruptions. During the early stages, a glaciovolcanic eruption is likely to be fully subglacial, so the heat released by rapid cooling of the erupted material is used almost exclusively for melting ice and heating up the resulting meltwater. During the early stages the eruption rate is also likely to be highest. Under these circumstances energy transfer between magma and ice can be described by simple calorimetric relationships. Because not all of the thermal energy carried by magma is released instantaneously, the temperature of the deposited tephra or lava is significantly higher than that of the surroundings. The energy used for ice melting is E_i, while the total potential thermal energy advected to the glacier base is E_m; for a specific magma composition E_m comprises the total 'sensible heat' released by cooling of the magma plus the potential latent heat that could be released by crystallisation (see Chapter 5), minus the heat absorbed by exsolution of the fluid phase (an endothermic process). The ratio of these two energies is the *eruption-timescale* ice-melting efficiency, f_e:

$$E_i = f_e E_m \tag{6.12}$$

For a fully realistic description of the physical problem, the efficiency should be evaluated as the ratio of the energy fluxes, meaning that E_i and E_m should denote energy per unit time. In practical terms changes in ice volume melted in an eruption are estimated over discrete time periods. The energy required to melt a given volume of ice (V_i) is calculated from

$$E_i = V_i \rho_i [C_i (T_0 - T_i) + L_i] \tag{6.13}$$

where V_i and ρ_i are, respectively, the volume and density of ice melted, T_i is the initial temperature of the ice, T_0 is its melting point (c. 0 °C), C_i the specific heat capacity, and L_i is the latent heat of fusion. The total thermal energy of the magma, E_m, is given by

$$E_m = V_m \rho_m [L_m + C_m (T_m - T_0)] \tag{6.14}$$

Here the subscript m refers to magma/lava/tephra properties, with V_m being the volume of volcanic material erupted and ρ_m being its density. In some eruptions crystals may be a significant fraction of the erupted volume, whereas fragmentation facilitates cooling and suppresses crystallisation by pushing the melt to its glass transition temperature (T_g) faster than crystal nucleation and growth (e.g. Kelman et al., 2002). In the former case the latent heat must be multiplied by the percentage of liquid magma. In the latter case no crystallisation occurs and thus no latent heat is released ($L_m = 0$). The efficiency value f_e is in general not easy to determine, but for eruptions where fragmentation appears to be dominant, $f_e = 0.5$–0.8 is appropriate (Gudmundsson, 2003; Edwards et al., 2015a). For effusive eruptions (e.g. pillow lavas), the value is much less, but over short timescales f_e is likely to be 0.1–0.3 (Gudmundsson 2003; Edwards et al., 2015a).

The density of lava pillows is considerably higher than that of tephra. If this is taken into account, efficiencies of 0.1–0.3 for basalts lead to melting of ice volumes of 1.2–5 times the lava volume. Corresponding values for fragmentation, having efficiencies of 0.5–0.8, allow for melting of ice volumes 4–7 times that of the tephra volume. Thus, the volume of ice melted is generally expected to be larger than the volume taken up by the eruption products. However, it is important to note that these values are much less than the maximum possible melting values that have been discussed in the literature (e.g. 10–14 times the volume of magma; e.g. Höskuldsson and Sparks, 1997). They also assume that all the available magmatic ('sensible') heat is extracted down to 0 °C, almost instantaneously ahead of emplacement of the magma, which is unrealistic. This is the so-called space problem highlighted by Smellie (2000, 2008). It is also important to understand that over longer timescales (years to decades) heat will continue to be released, but at a rate that will have much less direct impact than that which occurs on the timescale of the eruption. For example, while rapid cooling of the magma to temperatures below T_g on short timescales (seconds to minutes) 'loses' the initial potential contribution of the latent heat, over longer timescales volcanic glass is thermodynamically unstable, and will eventually crystallise, frequently driving processes such as palagonitisation, which is critical for consolidating deposits and increasing the likelihood of survival from subglacial erosion.

6.5 Fragmentation processes

At least four different processes can induce fragmentation during volcanic eruptions (see Kokelaar, 1986): (1) magmatic fragmentation driven by exsolution of magmatic gases and stresses induced by magma discharge; (2) 'quench' fragmentation driven by thermal shock as melts are rapidly cooled to their glass transition temperatures; (3) phreatomagmatic fragmentation driven by rapid expansion of external water that becomes enclosed by melt (also known as bulk interaction steam explosivity); and (4) collisional fragmentation produced by high-speed collisions between particles. The high likelihood that all four of these processes will be active during glaciovolcanic eruptions means that the componentry of the deposits produced by those eruptions can be extremely complex (cf. Heiken and Wohletz, 1985; Zimanowski et al., 1997; Murtagh and White, 2013; Liu et al., 2015). While the majority of historical glaciovolcanic eruptions have been dominated by highly explosive activity, abundant evidence from the geological record also shows that many glaciovolcanic eruptions have significant effusive episodes. As for all volcanic eruptions, one of the main controls on explosivity during glaciovolcanic eruptions is the relationship between external pressure and the solubility of volatile phases for a given magma (e.g. Sparks, 1978). In general, high glaciostatic/hydrostatic pressures and 'dry' magmas will favour effusive eruptions, while low pressure and 'wet' magmas will favour explosive eruptions. Edwards et al. (2015a) have mapped out for a 'typical' basaltic magma three regions in which different fragmentation mechanisms might be dominant as a function of the initial water content in the magma (Fig. 6.2). For high glaciostatic pressures and relatively low magmatic H_2O concentrations, volatile exsolution is minimised as is expansion of heated meltwater; in these environments fragmentation will be dominated by thermal stresses. However, as initial H_2O concentrations increase and/or glaciostatic pressure decreases, volatile exsolution and expansion become increasingly important, as does the potential for phreatomagmatic fragmentation. Once the volume of exsolved magmatic gases reaches 50% or higher, it is increasingly likely that magmatic processes will begin to dominate the disruption of the host melt, and by c. 75% many melts will transform from bubbly foams to jets of gas with suspended droplets of melt. Thus, even some 'thick' ice eruptions may initially be explosive, which favours high heat transfer efficiencies (see Section 6.4.4 above) and rapid melting. However, the extant pressure during glaciovolcanic eruptions is complicated by the properties of the ice overburden, which can melt during the course of an eruption, reducing ambient pressures and forming a partly to completely water-filled cavity within the ice. Pressures within partly filled vaults may be substantially less than glaciostatic, e.g. if the vaults are connected to the atmosphere at an ice front via incompletely filled subglacial tunnels (e.g. Tuffen, 2007). While the nucleation and growth of bubbles is a critical component of magmatic fragmentation, recent studies (summarised by Gonnermann and Manga, 2013) have suggested a number of potential criteria with which to assess the likelihood of magmatic fragmentation, including: (1) the presence of a critical volume fraction of bubbles; (2) a critical stress; (3) a critical strain rate; (4) a critical potential energy; or (v) a critical inertial state.

The presence of external water adds an additional complication to fragmentation at any stage of eruption, as it can cause high thermal stresses in the magma leading to 'quench' fragmentation, as well as phreatomagmatic explosions driven by shockwave breakup of the magma followed by rapid expansion of trapped liquid water being converted to steam (see Wohletz et al., 2013, for a thorough discussion). The relative importance of quench, phreatomagmatic, magmatic and collisional fragmentation will ultimately depend on the unique magmatic and glacial conditions present during an eruption.

Woodcock et al. (2012) investigated a number of different scenarios for heat transfer between individual fragments and water in a glaciovolcanic environment. They examined the effects of steam coatings slowing heat transfer between spherical particles and water. Their modelling allowed for a more accurate estimation of heat retention for

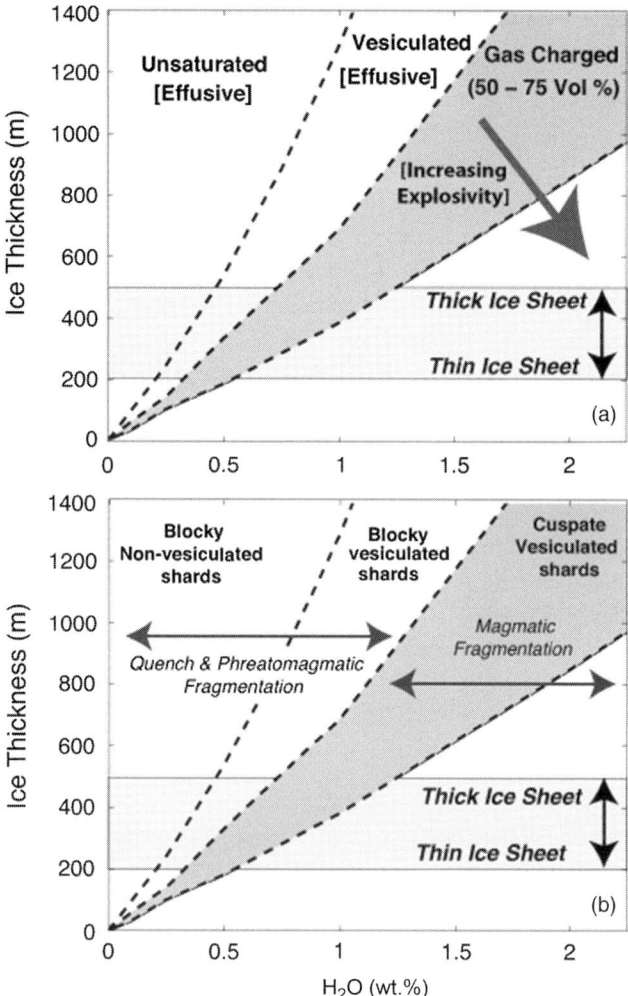

Fig. 6.2 Conditions favouring 'dry' versus 'wet' glaciovolcanic eruptions (from Edwards et al., 2015a). (a) A combination of initial magma water concentrations and ice thickness control whether eruptions are effusive without vesiculation, effusive with vesiculation, or likely to be gas-charged with higher initial explosivity. (b) Characteristics of fragmental deposits and fragmentation mechanisms as a function of initial magma water concentrations and ice thickness.

a measured grain size distribution from a sample from the 1996 Gjálp eruption, and concluded that 70–80% of the initial heat content of the grains would be transferred to the enclosing water while the grains were settling through the water column. These results are consistent with the long-term cooling of the Gjálp tephra mound, which released about two-thirds of the available heat during the first two weeks of the eruption (Jarosch et al., 2008).

6.6 Volatile saturation and vesiculation

The solubility of dissolved volatile phases (H_2O, CO_2, SO_2, HCl, HF, etc.) in silicate melts is a function of temperature, pressure and melt composition (e.g. Papale et al., 2006). Thus the potential for any magma to produce (exsolve) a separate gas phase increases as the magma migrates towards Earth's surface and lower lithostatic pressures. In glaciovolcanic settings this is important for several reasons. Firstly, the boundary between effusive and explosive eruptions is demarcated at least in part by the volatile-phase volume fraction that is in the melt. Explosive eruptions lead to more rapid ice melting, and more significant hazards (e.g. Gudmundsson, 2003). Secondly, the concentrations of dissolved volatiles remaining in the magma at eruption onset can be used to constrain the thickness of ice/depth of water present during the eruption (e.g. Dixon et al., 2002; Höskuldsson et al., 2006; Schopka et al., 2006; Edwards et al., 2009b; Stevenson et al., 2009; Tuffen et al., 2010; see Section 13.5.1). Finally, because englacial lakes formed during eruptions can be inherently unstable (Smellie, 2006), it is possible for sudden drainage events to cause vesiculation events that might be preserved in glaciovolcanic deposits (e.g. Höskuldsson et al., 2006; Graettinger et al., 2013c; Pollock et al., 2014), leading to enhanced recovery of palaeo-hydraulic information.

The key to understanding the role of pressure on gas expansion can be gleaned by examining the ideal gas law ($PV = nRT$), rearranged to emphasise the relationship between the volume (V) of a gas and its ambient pressure (P):

$$V = \frac{nRT}{P} \qquad\qquad (6.15)$$

where n is the number of moles of gas; R is the universal gas constant ($8.3145\ \mathrm{J\,K^{-1}\,mol^{-1}}$); and T is the temperature of the gas. Thus, as the pressure on a magma decreases during ascent, not only does the solubility of gases in the melt decrease, but once a separate fluid phase has formed its volume continues to increase due to the effects of decreasing pressure both on volume and on fluid saturation. While the ideal gas law is a legitimate starting point to explore V–P relationships, at present many different equations of state for H_2O–CO_2 mixtures are easily accessible to more accurately estimate the effects of decompression on fluid volumes (http://esdtools.lbl.gov/gaseos/gaseos.html). It has been recognised since the 1960s that the relative volume of fossilised gas bubbles (i.e. vesicles) could be used as a crude indicator of eruption pressures (Jones, 1969a). While present methods for using measured water contents in glasses from pillow rims have some intrinsic problems (see Chapters 7 and 13), an important criterion for applying the methodology is the presence of vesicles to demonstrate that indeed the melt was actually saturated with a fluid phase. If no vesicles are present, palaeo-hydrostatic pressure estimates can only indicate minimum environmental pressures.

6.7 Constraints on modes of emplacement

Modelling of modes of emplacement for glaciovolcanism have followed two main courses: the first depends largely on theoretical analysis of potential end-members (e.g. Allen, 1980;

Höskuldsson and Sparks, 1997; Tuffen, 2007; Wilson and Head, 2002, 2007b; Wilson et al., 2013), and the second focuses on models that tend to be less quantitative but which were derived to explain specific field observations (including landform analysis) and deposits (e.g. van Bemmelen and Rutten, 1955; Jones, 1969b; Allen, 1980; Smellie and Skilling, 1994; Hickson, 2000; Edwards and Russell, 2002; Kelman et al., 2002; Tuffen et al., 2002b; Schopka et al., 2006; Smellie, 2007, 2009, 2013; Edwards et al., 2009b; Hungerford et al., 2014; Pedersen and Grosse, 2014; Russell et al., 2014). Here we will focus primarily on the studies that have attempted to quantitatively constrain physical processes related to glacio-volcanic lava emplacement.

Wilson and Head (2002, 2007b) and Wilson et al. (2013) outlined in detail three general scenarios for the initial emplacement of magma/lava in a glaciovolcanic setting: (1) intrusion of a sill at the ice–bedrock interface (used to explain the formation of deposits along the southern coast of Iceland; Smellie, 2008); (2) intrusion of a dyke into overlying ice (proposed for deposits at Hoodoo Mountain; Edwards and Russell, 2002); and (3) lava flowing onto an ice surface from above (supraglacial; e.g. eruptions at Veniaminof volcano (Alaska), Fimmvörðuháls (Iceland, in 2010) and Tolbachik (Kamchatka, in 2012–13); Edwards et al., 2014a). Wilson and Head (2002) focused on physical constraints for (1) and (2), including descriptions of the pressure gradient driving the propagation of dykes into overlying ice, and exploring conditions that would allow for sill propagation at the bedrock–ice interface. The main result for the case of a dyke, originating from a crustal magma chamber at a depth of 3 km, is that the height to which the dyke can penetrate the overlying ice (h) can be estimated from

$$h = \frac{z(\rho_r - \rho_b)}{\rho_b} + y\frac{\rho_i}{\rho_b} + \frac{\Delta P + P_a - P_t}{g\rho_b} \tag{6.16}$$

where z is the distance from the magma reservoir to the rock–ice interface (3 km), ρ_r is the bulk density of crustal rocks overlying the magma reservoir (taken to be c. 2300 kg m^{-3} for crust composed of volcanic rocks), ρ_b is the bulk density of the magma in the dyke (slightly less than the bulk density of the magma (ρ_m) due to the presence of gas bubbles; c. 2500 kg m^{-3} is a reasonable value but this can be estimated for specific pressures and magmatic volatile concentrations), y is the thickness of the overlying ice, ρ_i is the density of the ice (920 kg m^{-3}), ΔP is the excess pressure of the magma in the magma reservoir (c. 5 MPa is a reasonable value), P_a is the atmospheric pressure (0.1 MPa), and P_t is the pressure in the dyke tip, which can be estimated from

$$P_t = \frac{n_t(1 - f)\rho_m R T_m}{mf(1 - n_t)} \tag{6.17}$$

where n_t is the mass fraction of volatiles exsolved from the melt (equal to 0.008 for a melt with initial volatile concentrations of 0.7 wt % H_2O and 900 ppm CO_2), f is the volume fraction of bubbles in the melt (taken to be c. 0.8 as a maximum value before the melt is

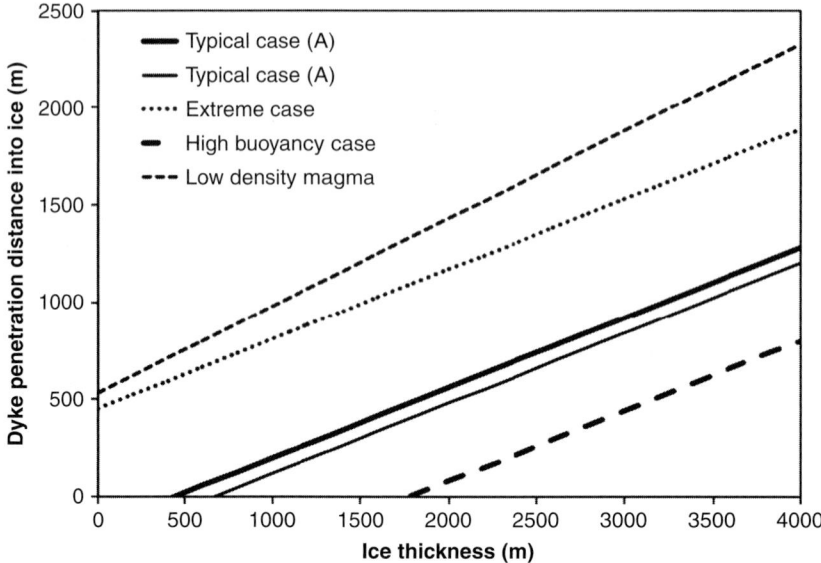

Fig. 6.3 Variations in parameters that control dyke penetration into overlying ice. Ice penetration distances are largely a function of assumptions about magma density (assumed to be 2500 kg m^{-3} for all cases except low density magma for which density is 2000 kg m^{-3}) and excess dyke tip pressure. Typical cases are for magmas in a crustal magma chamber, and high buoyancy case is for a deeper mantle-derived magma. Calculations are described in the text and follow the discussion in Wilson et al. (2013).

totally disrupted), ρ_m is the density of the melt (c. 2700 kg m^{-3} for a typical basalt), R is the universal gas constant (8.3145 J K^{-1} mol^{-1}), m is the average molecular weight of the gas phase, and T_m is the temperature of the melt (c. 1473 K for basalt). For the specified parameters Wilson et al. (2013) estimated that a reasonable P_t would be c. 3.2 MPa. P_t values can easily be recalculated for different magma compositions, dissolved volatile contents and exsolved volatile mass fractions. For the scenario above, Wilson et al. (2013) showed that for ice thicknesses (y) greater than 500 m, dykes should just be able to penetrate into the overlying ice by a few tens of metres, and for values of y of 1000 m, dykes could penetrate more than 200 m (see Fig. 6.3 for examples).

Wilson and Head (2002) also gave a detailed analysis of the physics of emplacement of a sill at the rock–ice interface. Their analysis showed that it is reasonable to predict that such subglacial sills can form, and also demonstrated that such a sill would produce a larger area of magma in contact with the overlying ice than would a lava flow moving on top of the ice. According to their results, after approximately 1 hour of intrusion the sill would be more than 1 m thick and have a lateral extent of at least 700 m on either side of the feeder dyke. While the complexities of this type of modelling make it difficult to attain anything more than order of magnitude estimates of intrusion rates, they at least confirm that under reasonable assumptions the processes are feasible and thus provide

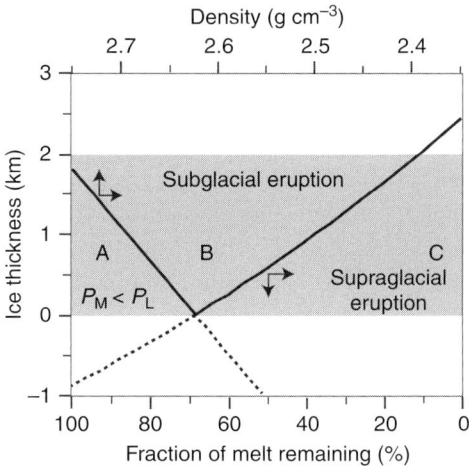

Fig. 6.4 Comparison of density changes to a crustal magma undergoing assimilation–fractional crystallisation and ice thickness (from Edwards et al., 2002). Three regions are identified: (A) eruption is not favoured; (B) subglacial eruptions are favoured; and (C) supraglacial eruptions are favoured.

some quantitative constraints for field-based observations (e.g. Smellie 2008; see Section 10.6).

Using a similar approach based on magmastatic driving pressures, Edwards et al. (2002) explored the potential for subglacial versus supraglacial lava emplacement based on changing magma densities during assimilation–fractional crystallisation at mid-crustal levels. This work was driven in part by attempts to understand the production of large volumes of phonolitic–trachytic magma at Hoodoo Mountain volcano (British Columbia), in a tectonic setting otherwise dominated by alkali olivine basalt. The formation of intermediate composition lavas in an extensional setting almost certainly required magma storage and differentiation. In a location such as western Canada, which has been repeatedly covered by ice sheets over the past 2 Ma, the waxing and waning of ice possibly as thick as 3 km can have a potential impact on lithostatic pressures ($\Delta P > 26$ MPa). Edwards et al. (2002) showed that, as a function of decreasing magma density during assimilation–fractionation processes, three broad regions could be defined, i.e. where (1) ice loading was insufficient to compensate for magma density; (2) the ice loading/magma density combination was sufficient to allow for sub- to intraglacial emplacement; and (3) a combination that could drive eruptions to the top of the ice surface (Fig. 6.4).

The formation of cavities inside overlying ice bodies has been explored with increasingly sophisticated numerical models by a number of workers (Allen, 1980; Höskuldsson and Sparks, 1997; Kelman et al., 2002; Tuffen et al., 2002b; Kelman 2005; Tuffen, 2007). The models take into account basic thermal constraints, but some also examine the development of pressure within an ice cavity (Höskuldsson and Sparks, 1997; Kelman, 2005;

Tuffen et al., 2002b; Tuffen, 2007) as well as rates of ice deformation (Tuffen, 2007). The landmark contribution by Höskuldsson and Sparks (1997) covered a broad range of physical processes ranging from estimations of differing ice melting potentials between basalt and rhyolite, to rates of cavity growth, and estimations of the within-cavity pressure changes. They showed that it was feasible for an effusive eruption with discharge rates of $10^4 \text{ m}^3 \text{ s}^{-1}$ to melt through 500 m of ice within about 11 days (later work by Gudmundsson (2003) showed that melting rates can be much faster for explosive eruptions), which is on the general timescale of the duration of some historic glaciovolcanic eruptions but is far slower than the observed rates of melting during the Gjálp eruption in 1996 (cf. Gudmundsson et al., 1997, 2004). While later workers have refined many of the calculations of Höskuldsson and Sparks (1997), this work covered a large breadth of the important aspects of modelling glaciovolcanic eruptions.

Kelman et al. (2002) was one of the first contributions to address constraints on emplacement of more silica-rich melts. Their work focused on glaciovolcanism in the Garibaldi segment of the Cascade volcanic arc, which is dominated by magmas with andesitic and dacitic compositions. Their work was the first to explore the thermal consequences of the reduction in heat transfer from magma to ice due to melts being forced rapidly to their glass transition temperature (T_g). Kelman et al. (2002) pointed out that because the difference between liquidus temperatures and T_g for basalt was much greater (>400 °C) than that for dacitic magmas (c. 150 °C), basaltic magmas were more likely to release a larger fraction of their potential latent heat from crystallisation than were more silicic melt compositions. This resulted in the dacitic melt compositions only releasing 70–80% of the amount of thermal energy released by basaltic melts. At its crux, this issue also affects our understanding of the 'space problem' highlighted by Smellie (2000, 2008). The heat released down to T_g represents sensible heat that is instantaneously available for melting and thus able to physically create space for magma to occupy during emplacement. Because it is a small proportion of the total heat available in a cooling magmatic system (assumed, unrealistically, in all published models to cool down to 0 °C), it spotlights how little space can actually be melted by magma as it is emplaced (Smellie, 2008).

Tuffen (2007) presented one of the first models to combine thermal effects originating from the emplacement of basaltic melt at the base of an ice body with the physical laws of ice response (e.g. Nye's Law; see Chapter 4). The main results of this work provided constraints on intrusive versus explosive subglacial eruptions, mainly focusing on whether or not a cavity would develop that could facilitate explosive fragmentation. The calculations assume that cavity pressures of 5 MPa or greater will prevent magmatic fragmentation, and those less than 5 MPa will have magmatic fragmentation (see Section 6.4, above). These calculations showed that at higher rates of magma discharge ($100 \text{ m}^3 \text{ s}^{-1}$) melting of the overlying ice may open a large cavity between the magma and ice (80 m high) within the first half hour of eruption; for lower magma fluxes ($1 \text{ m}^3 \text{ s}^{-1}$) only smaller cavities developed and these closed as ice deformed in response to the eruption, leading to closure of the cavities and subsequent intrusion-dominated eruptions. However, it is still unlikely that this explanation resolves the 'space problem', since it relies on a greater volume flux of magma to provide enough heat to melt the ice and does not address how that space is instantaneously created ahead of the invading magma (Smellie, 2008).

Large-scale experiments by Edwards et al. (2013) have given new insight into supraglacial emplacement of basaltic lava flows, including observations on the melting rates for lava emplaced directly onto ice versus a boundary layer between ice and lava, formation of steam-buffering within the boundary layer, gas injection into lava, and lava-exploitation of pre-existing weaknesses in ice. Their large-scale experiments using up to 100 kg of remelted basaltic lava poured on top of ice, snow and ice/snow covered with differing thicknesses of sand showed that basaltic melts can quickly melt through any ice/snow and then exploit pre-existing gaps within the substrate to enlarge, occupy and travel through sub-ice drainage networks. However, the presence of a boundary layer separating the lava and ice/snow significantly reduces melting rates and can prevent lava from accessing the base of the frozen substrate.

6.8 Formation of cooling fractures

One of the final processes that occurs towards the waning stages of a glaciovolcanic eruption, and which can leave important clues for the environmental interpretation of the deposits, is the formation of fractures in lava caused by a build-up of thermal stresses (i.e. 'polygonal' and other joints). While this is very much an area of active research (e.g. Long and Wood, 1986; Aydin and DeGraff, 1988; DeGraff and Aydin, 1993; Lore et al., 2000; Goehring and Morris, 2008; Lodge and Lescinsky, 2009a,b; Mattsson et al., 2011; Forbes et al., 2012, 2013, 2014a; Hetenyi et al., 2012), it is worth reviewing here briefly the present state of terminology and knowledge. The features of cooling fractures are discussed more extensively in Section 9.3.7. The science of fractometry has long of been of interest to engineering sciences in order to better understand how materials fail (e.g. Hull, 1999), and recent work has focused on bringing concepts from this extensive literature into the study of geological materials. Ultimately, understanding the geometry and dimensions of such fractures may provide key clues to rates of heat extraction from lavas, to more confidently identify those that were emplaced into ice-rich environments.

Pioneering work by Long and Wood (1986) on Columbia River basalts first suggested that the diameter of individual columnar joints was an indication of rates of cooling. They showed that the large diameter joints in the upper and lower colonnades have larger average grain sizes, hence slower cooling, than for the interior of the lavas characterised by smaller diameter joints and smaller average grain sizes (called entablatures). They inferred that the increase in cooling rates for the interior of the lava could only happen by introduction of a coolant, inferred to be water emplaced by flooding caused when the lavas dammed coeval rivers. From this type of observation was formed a mythology about joint sizes and cooling rates that is pervasive in the glaciovolcanic literature. Studies by Aydin and DeGraff (1988), DeGraff et al.

(1989), and DeGraff and Aydin (1993) applied fractography (e.g. Hull, 1999) to the study of lava flow fractures in order to assess the direction of joint propagation. Recently, more quantitative studies of fracturing have tried to systematise the nomenclature of fracture patterns and model the build-up of thermal stresses in a cooling lava flow (Goehring and Morris, 2008; Lodge and Lescinsky, 2009a,b; Forbes et al., 2012, 2014a,b). In particular, the work by Forbes et al. (2012, 2014a,b) has expanded the use of fractometry terminology from the engineering literature to identify fractures with different fracture modes.

6.9 Summary

Whilst physical modelling of glaciovolcanic eruptions is still a young endeavour, much of the basic physics applicable to glaciovolcanic eruptions has been developed from eruptions in other volcanic environments and so can easily be back-applied. Predictions for eruption conditions need to take into account knowledge of the prevailing tectonic environment. Areas where magmas inherently have higher water contents are also likely to be more explosive, and higher percentages of magmatic fragmentation result in faster heat transfer, loss of latent heat of crystallisation, and different eruption products than in 'drier' systems. Most heat transfer models can be improved by using more realistic boundary conditions (e.g. Stefan problems), and more sophisticated models for boundary geometries. However, basic physical models explain the order-of-magnitude behaviours described from field-based observations, and some empirical physical models have been developed to incorporate thermal properties of the magma–ice system as well as the physical properties of the ice as it might respond over the timescales of a sustained eruption.

7

Analytical studies of glaciovolcanic materials

7.1 Introduction

The goal of this chapter is to review published studies that use a variety of analytical methods as a principal means of investigating glaciovolcanic deposits. While an increasingly complex array of analytical techniques is becoming important in modern studies of glaciovolcanism, ranging in scale from an entire edifice to individual analyses of elements with concentrations of parts per billion, many basic petrological techniques (e.g. petrography, major element geochemistry) are still the foundation for most analytical studies. A comprehensive evaluation of more advanced analytical procedures involved is beyond the scope of this book, but here we demonstrate the array of analytical techniques that has been used and highlight some of the most detailed published studies.

7.2 Morphometry

There are few studies of the morphology and morphometry of the different types of glaciovolcanic edifices (Hickson, 2000; Smellie, 2007, 2009, 2013; Russell et al., 2014). They can be used as an important proxy source of palaeo-environmental information (see Chapter 8), and are particularly important when trying to identify planetary glaciovolcanic edifices (e.g. on Mars; Chapter 16), for which information about deposits is limited to remotely sensed data including relatively high-resolution topographical information. The quantitative study of landform dimensions and characteristics has benefited tremendously from the advances in spatial analysis software, as well as considerably improved abilities for measuring positional information using the Global Positioning System (GPS) and Differential-GPS, referred to collectively as geographical information systems (GIS). The study of glaciovolcanoes should be particularly amenable as much of the early evidence used to identify subglacially erupted volcanoes in British Columbia, Iceland and on Mars was the distinctive shapes of the edifices. Attempts have been made to apply more rigorous methods, and morphometry (i.e. the statistical quantification of morphology). They have already been broadly applied to volcanic landforms (e.g. stratovolcanoes; Grosse et al., 2009) and are now being used to discriminate between different volcano sub-types (e.g. NETVOLC and MORPHVOLC, Grosse et al., 2014). In addition, the advent of satellite-based (Tandem-X) and ground-based (LiDAR) techniques for high-resolution topographical measurements is likely to drive a new generation of morphological studies of glaciovolcanic edifices (e.g. Pedersen and Grosse, 2014).

A discussion of morphometric data applied to glaciovolcanic landforms is presented in Chapter 8. Here, we briefly outline some of the new and emerging techniques available for quantitative analysis. Pederson and Grosse (2014) reported a routine that estimates passage zone elevations remotely for examples for which digital elevation models have sufficiently high resolution. They used the breaks in slope at the base of passage zones from volcanoes in southwestern Iceland to constrain passage zone elevations and volcano heights. The study also used morphometrical techniques to generate new insights into stratovolcano morphologies (cf. Grosse et al., 2009, 2012, 2014).

Recent work in Iceland and British Columbia has also used structure-from-motion techniques to create three-dimensional 'point clouds' of data that can be used to measure stratigraphical relationships in areas that are physically impossible to reach and to characterise the shapes of individual lava units (e.g. lava lobes; Fig. 7.1). Similarly, a combination of DGPS and laser range-finding is being tested on vertical quarry walls in Iceland, to facilitate more precise thickness measurements of inaccessible stratigraphical units (unpublished information of BRE).

7.3 Major, minor and trace element geochemistry

While most of the geochemical methods necessary for characterising glaciovolcanic deposits are routine (e.g. X-ray fluorescence (XRF) analysis of bulk rock samples for major and minor elements; electron microprobe analysis (EMPA) of major oxides in glass shards; inductively coupled plasma mass spectrometry (ICP-MS) analysis for trace elements), a relatively small number of published studies have combined geochemical modelling with physical volcanology (e.g. Edwards et al., 2002; Pollock et al., 2014). Systematic analysis and modelling can be very useful, however, for correlating volcanic units and thus infer their relative ages (e.g. McGarvie et al., 2007), to identify and investigate the significance of stratigraphical changes (Moore and Calk, 1991; Moore et al., 1995; Werner and Schmincke, 1999; Edwards et al., 2009b; Pollock et al., 2014), and to infer deeper-level processes by which fluctuations in overlying ice thicknesses affect magma storage areas (Edwards et al., 2002) or source region melting conditions (e.g. Jull and McKenzie, 1996; Maclennan et al., 2002).

7.3.1 Major element studies

Significant shifts in magma compositions during glaciovolcanic eruptions have been documented for the 2010 Eyjafjallajökull and 1991 Hudson eruptions, in which

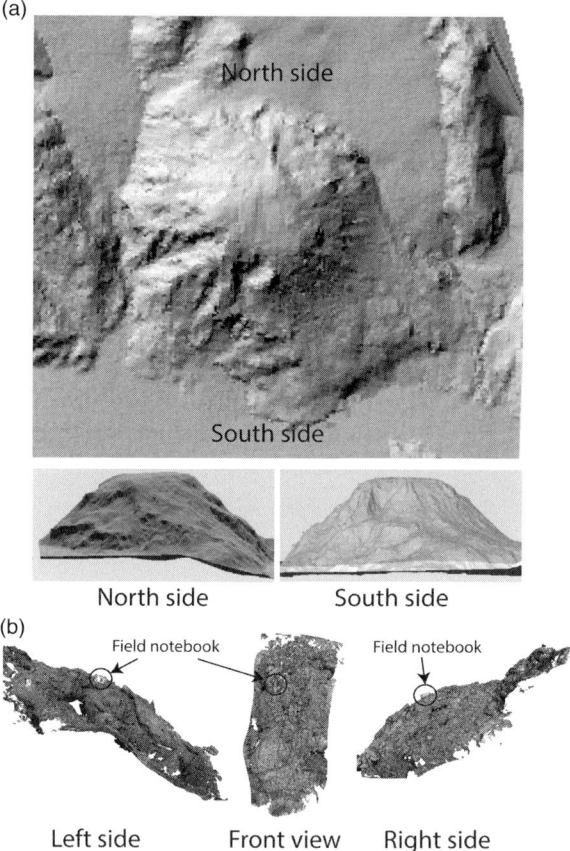

Fig. 7.1 Examples of morphometric analysis. (a) Digital elevation model of Hoodoo Mountain volcano with topographical profiles of the north and south sides showing its distinctive morphology. (b) Point clouds made from field photographs of an individual lava lobe emplaced within Gigjökull glacier during the 2010 Eyjafjallajökull eruption. Structure-from-motion produces three-dimensional renderings from photographs from which morphometric measurements can be made (point cloud courtesy of M. James).

compositions varied from mafic almost to felsic in the fragmental deposits. More subtle shifts in major element concentrations have been reported for some basaltic centres in Iceland (e.g. Moore and Calk, 1991; Pollock et al., 2014), British Columbia (Moore et al., 1995; Edwards et al., 2009b, 2011) and Antarctica (Smellie, 2001). Some of the compositional shifts may simply result from variations in crystal sorting or fractionation (e.g. Mathews tuya, Pillow Ridge), while at other locations, different magma storage areas may have been involved during the eruptions, or the eruptions may be polygenetic as opposed to monogenetic. For example, in the Tuya area of British Columbia, Moore et al. (1995) argued that a compositional shift from slightly tholeiitic to mildly alkaline bulk rock compositions recorded changing pressure conditions during glaciovolcanic eruptions, tapping first a deeper and then a shallower storage area. However, compositional shifts of very similar magnitude at Mathews tuya, which is adjacent to the volcanoes studied by Moore et al. (1995), were shown to be compatible with minor changes in extent of crystallisation consistent with eruption from a single magma storage area (Edwards et al., 2011). Moore and Calk (1991) analysed glasses from six tuyas and inferred minor fractionation on the timescales of edifice formation (years to decades) to explain small differences in glass compositions with height above the volcano base. On a broader scale, several studies in Iceland (see Chapter 14) have shown that within the same local magmatic system the presence or absence of a thick ice cover influences depths and extents of melting, suggesting that major element compositions alone might discriminate between interglacial and syn-glacial eruptions (Hardarson and Fitton, 1991; Slater et al., 1998; Maclennan et al., 2002; Sims et al., 2013).

7.3.2 Minor and trace element studies

Minor and trace elements have been used to identify cryptic stratigraphical boundaries within sequences (Pollock et al., 2014), to correlate between physically separated deposits (McGarvie et al., 2006), to indicate changes in magma storage areas (Pollock et al., 2014), and to identify changes in source region conditions (Jull and McKenzie, 1996; Maclennan et al., 2002). While many glaciovolcanic edifices have been assumed to be 'monogenetic', i.e. all of the deposits formed during a continuous eruption from the same magma batch, some studies have shown that even small edifices may have formed by multiple eruptions, over timescales of decades to thousands of years (e.g. Icefall Nunatak, Antarctica: cf. LeMasurier et al., 1994; Smellie, 2001). Pollock et al. (2014) identified compositionally different sets of deposits within a relatively small, basaltic pillow-dominated tindar in southwest Iceland. Abandoned quarry exposures showed four units of glaciovolcanic pillow lavas, two horizons of fragmental deposits, and multiple intrusions. While physical breaks distinguish at least temporary shifts in the centres of eruptive activity, analyses of rare earth elements (REE) showed that the deposits can be split into two geochemical groups with distinctive characteristics that are not easily explained by closed system fractionation. Pollock et al. (2014) used these observations to investigate several different models for the plumbing that supplied different magmas to the same fissure system.

Several small, physically separated rhyolitic tuyas have been identified by McGarvie et al. (2006) in the Torfajökull area of south central Iceland. They argued that many of the tuyas likely formed during the same event based on similarities in trace element abundances. Because of inaccuracies in the dating of young volcanic rocks, it was impossible to identify whether the centres were separated in time or were cogenetic and erupted simultaneously from different vents. Thus, geochemistry may provide the only evidence to test for consanguinity of spatially separated deposits (see also Flude et al., 2010). Demonstrating

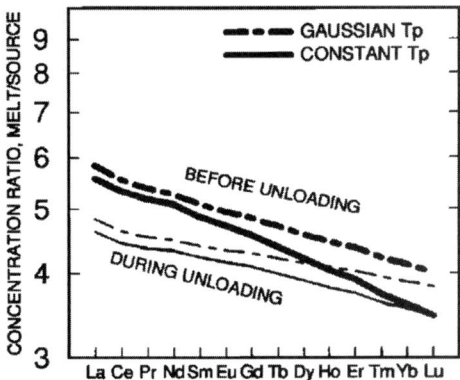

Fig. 7.2 Diagram showing possible geochemical effects of glacial unloading on rare earth element concentrations for mantle melts and two different mantle potential temperature assumptions (constant T_p and Gaussian T_p) (diagram is from Jull and McKenzie, 1996). Jull and McKenzie (1996) argued that unloading should have a distinctive geochemical signal indicative of source region melting changes due to isostatic rebound during deglaciation.

consanguinity will also affect estimates of the volumes of magma erupted, and thus estimates for the volume of ice that may be melted.

Nyland et al. (2013) used trace element compositions of tephras in cores of Antarctic shelf sediments to argue for linkages between deglaciation and volcanic activity within the West Antarctic Rift System. They found that the ashes had distinctive chemical compositions, and tephra layers appeared to be more abundant within sediments deposited during the transitions from glacial to interglacial stades, suggesting that variations in ice thickness may have acted as a trigger for the eruptions.

Trace element variations in lavas erupted in Iceland during glaciations and interglacials were also used by Jull and McKenzie (1996) and Maclennan et al. (2002) to argue that isostasy has an effect on the depths and/or degrees of melting in Iceland. They found that REE concentrations were distinctly different for lavas erupted during glaciations (Fig. 7.2; cf. Nyland et al., 2013). While many subsequent workers have suggested links between glaciation and volcanism, the two Icelandic studies were first to demonstrate quantitatively the consequences of ice thickness fluctuations on degrees of mantle melting based on trace element modelling.

7.4 Volatile elements

An increased understanding of the role of volatile compounds is critical for many aspects of glaciovolcanism, as discussed in Chapters 5 and 13 (Dixon et al., 2002; Tuffen et al., 2010). However, while several studies have included volatile analyses, many problems exist with the analytical techniques and in the models used to interpret the information derived from those elements. The volatile species mainly include H_2O, CO_2, SO_2, HCl and HF. Water and CO_2 contents are particularly important for determining the likely explosivity of an eruption, especially during the earliest stages. If degassing occurs before the magma reaches the ice interface, it can increase the likelihood of an explosive eruption onset, which has significant consequences for how rapidly the overlying ice melts (Gudmundsson et al., 2004). A pervasively degassing magma also has a lower bulk density that can potentially enhance fragmentation even if the fragmentation is mainly caused by phreatomagmatic explosions from interacting with external water. Sulphur is also a critical volatile component and several studies have used changes in S concentrations to constrain hydrostatic pressures (Moore and Calk, 1991; Moore et al., 1995; Dixon et al., 2002).

Volatile element concentrations are particularly important for documenting the eruptive environment (see Section 13.5.1). Jones (1969a) examined the sizes and volume percentages of bubbles in glaciovolcanic pillow lavas in southern Iceland to argue empirically that pillows with smaller bubbles erupted beneath deeper water (or greater ice thicknesses). Since the mid-1990s it has become common to measure concentrations of H_2O and CO_2 in the vitric rinds of pillow lavas in order to try to estimate the eruption pressures (e.g. Dixon et al., 2002; Höskuldsson et al., 2006; Schopka et al., 2006; Edwards et al., 2009b; Stevenson et al., 2009). Tuffen et al. (2010) outlined five essential criteria that need to be met in order for volatile studies to be able to reconstruct ice thicknesses: (1) the magma needs to be volatile saturated; (2) degassing should be in equilibrium; (3) there should be no post-quenching movement of the glassy material sampled; (4) sample volatile concentrations should be spatially homogeneous; and (5) there should have been no post-quenching hydration. These and a wide range of additional factors influencing ice thickness calculations were also discussed by Owen et al. (2012), including effects of magma density values selected, which absorption coefficient and solubility models were used, assumptions about magmatic temperature (water solubility is temperature dependent), and sensitivity of Fourier transform infrared spectroscopy (FTIR) analysis to sample thickness. For the first time in any study, each of these criteria was assessed as a possible influence on the spread of data. Some criteria can be hard to assess, for example the presence of vesicles used to indicate degassing from a volatile-saturated magma (Höskuldsson et al., 2006; Tuffen et al., 2010; Owen et al., 2012). Bubbles in felsic magmas also commonly collapse and heal by annealing (Tuffen and Castro, 2009; Owen et al., 2012). A lack of visible vesicles may thus not preclude the possibility of vesiculation. Another potentially serious problem lies in the determination of CO_2. The concentration of CO_2 has a large influence on the water solubility–pressure relationships in a melt and therefore affects the amount of water in the glass (Dixon et al., 2002; Stevenson et al., 2009; Tuffen et al., 2010). Because of the relatively high detection limits for CO_2 currently available in FTIR analysis (typically 30 ppm), a CO_2 peak is usually unresolved in the spectra. Since CO_2 is much less soluble than water, evidence for water degassing in the samples should indicate that all or nearly all of the CO_2 has escaped and volcanoes that degas at pressures below 5 MPa (i.e. equivalent to c. 550 m of ice overburden) will have virtually no CO_2 left in the magma (Tuffen et al., 2010).

Studies that have applied this technique to constrain palaeo-ice thicknesses have met with varying degrees of success (e.g. Moore et al. (1995; Tuya Butte, Ash Mountain, South Tuya), Dixon et al. (2002; Tanzilla Butte in British Columbia); Schopka et al. (2006; Helgafell (a tindar) in Iceland); Edwards et al. (2009b; Pillow Ridge, Mt Edziza, British Columbia); Stevenson et al. (2009; Kerlingarfjöll, an intermediate-composition volcano in Iceland) and Hungerford et al. (2014; Tennena Cone); see Chapters 10 and 11). Icelandic magmas are generally considered to be relatively 'dry' and magmatic H_2O concentrations are relatively low (less than 0.6 wt % for basalt from the main rift axis; Nichols et al., 2002), although possibly higher near the hotspot trace in east-central Iceland (0.85–1.04 wt % for quartz tholeiite from Kverkfjöll; Höskuldsson et al., 2006). Conversely, examples from British Columbia have consistently high H_2O concentrations, as expected for more alkaline systems (up to 1 wt %; Edwards et al., 2009b; Hungerford et al., 2014).

The most detailed studies of suites of volatile species in glaciovolcanic deposits are by Moore and Calk (1991), Moore et al. (1995) and Dixon et al. (2002). Moore and Calk (1991) analysed major and minor element compositions of sideromelane in pillow rims and glassy tephra from six glaciovolcanoes in Iceland (Herdubreid, Hlödufell, Raudafell, Estadalsfjall, Kalfstindar and Geitafell). Their glass analyses were obtained using an electron microprobe, and importantly they used established glass standards for the calibration of their analyses to avoid inaccuracies from matrix effects. The S concentrations of the glasses varied from values of <0.01 wt % for subaerial flows to >0.1 wt % for lava pillows at Estadalsfjall. For all of the edifices studied, subaerially erupted units (e.g. lavas and tephra) had the lowest S contents, and lava pillows had the highest values. Three of the volcanoes clearly showed trends of S concentrations decreasing with increasing elevation above the volcano base (Herdubreid, Raudafell, Kalfstindar), while results for the other three were ambiguous. Moore and Calk (1991) also noted that samples with higher values of S had spherules of sulphide minerals within vesicles, which they interpreted as an indication of S-degassing under pressure.

Moore et al. (1995) and Dixon et al. (2002) conducted similar studies of four glaciovolcanoes in British Columbia (Tuya Butte, South Tuya, Ash Mountain and Tanzilla Butte, all in the Tuya Volcanic Field; see Chapter 2) to document degassing histories as a function of elevation within the edifices during the course of eruptions, and to infer source region processes. Moore et al. (1995) analysed S, H_2O and Cl at Tuya Butte (number of samples (n) = 26), South Tuya (n = 32) and Ash Mountain (n = 57; Fig. 7.3). Chlorine and S concentrations were analysed using the same method as in Moore and Calk (1991), whereas the concentrations of H_2O were obtained by FTIR. They suggested that samples from lower in the stratigraphy at each volcanic centre were relatively 'undegassed', with c. 0.1 wt % S, c. 0.02 wt % Cl and c. 0.5 wt % H_2O. Degassed samples from the three centres showed greater variation in S and H_2O concentrations, but the concentrations were all lower than in the undegassed samples (Ash Mountain: 0.06 wt % S and 0.2 wt % H_2O; South Tuya: 0.07 wt % S, H_2O not measured; Tuya Butte: 0.03 wt % S, H_2O not measured). The behaviour of Cl proved more difficult to reconcile with a specific degassing history, as

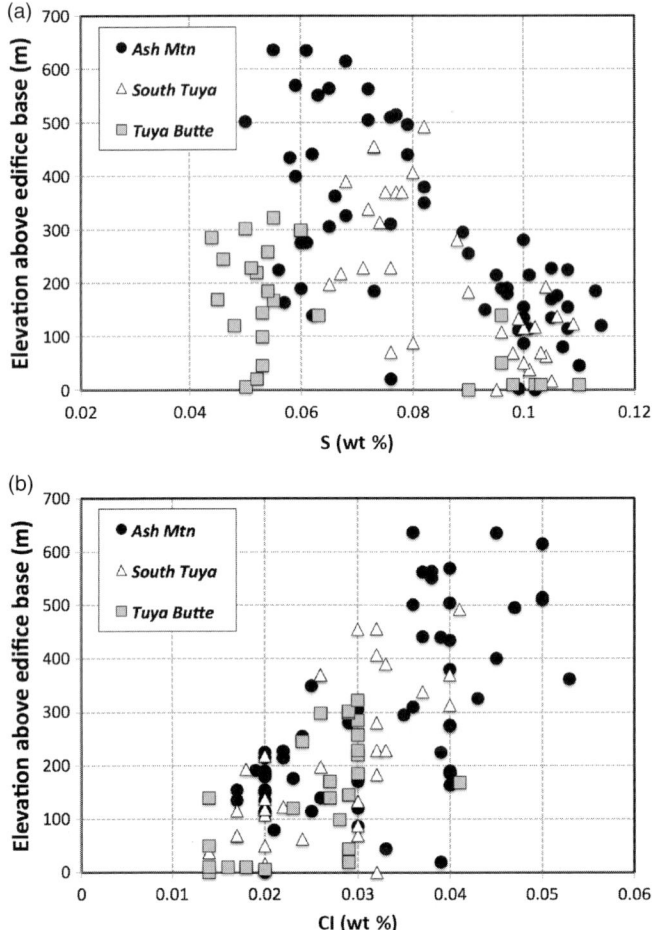

Fig. 7.3 Concentrations of (a) S and (b) Cl from Ash Mountain, South Tuya and Tuya Butte in northwestern British Columbia as a function of relative height above the edifice base. General decreases in volatile concentrations with edifice height (but not Cl) are consistent with eruption progressing into shallower water due to edifice building or lake level lowering (data are from Moore et al., 1995).

Cl concentrations in all three centres increased with stratigraphical height to values of c. 0.03 wt %.

Twenty-one samples from Tanzilla Butte were analysed by Dixon et al. (2002). They included FTIR analyses of CO_2. They found concentrations of H_2O between 0.5 and 0.92 wt %, S between 850 and 1100 ppm, Cl between 0.02 and 0.04 wt % and CO_2 less than the FTIR detection limit (c. 30 ppm). Assuming CO_2 concentrations of ≤ 30 ppm, they estimated minimum eruption hydrostatic pressures of 35.5 to 80.5 MPa (equivalent to ice thicknesses of 394 to 894 m). However, they did not find any consistent stratigraphical variations, despite sample elevations varying over c. 500 m vertically.

Several other studies have reported less detailed information on H_2O and CO_2 concentrations in glaciovolcanic glasses. For example, Schopka et al. (2006) measured water concentrations in basaltic samples using FTIR over a stratigraphical height of 210 m at Helgafell, in southwestern Iceland. A combination of vitric rinds from pillow lavas and glass fragments from hyaloclastite was sampled. A range of values from 0.26 to 0.37 wt % H_2O for eight samples was found, but no detectable CO_2. Stevenson et al. (2009) measured H_2O concentrations in glasses from three intermediate-composition samples from Kerlingarfjöll, in central Iceland, with values of 0.67 to 1.32 wt %. Höskuldsson et al. (2006) measured H_2O concentrations ranging from 0.85 to 1.04 wt % in quartz tholeiite pillow lavas in five separate pillow ridges at Kverkfjöll, on the northern margin of Vatnajökull. They found no CO_2 and on that basis assumed that maximum CO_2 concentrations were less than detection limits, which they reported as 30–40 ppm. Edwards et al. (2009b) measured H_2O, S and Cl concentrations from five separate basaltic pillow-rim glasses over a stratigraphical height of 400 m at Pillow Ridge in northern British Columbia with H_2O values ranging between 0.54 and 0.81 wt % H_2O. The concentrations did not vary systematically from base to summit, and this was interpreted to indicate that water levels must have fluctuated irregularly during the ridge-building eruptions. Values for S (0.2–0.23 wt %) and Cl (0.04–0.05 wt %) showed no systematic variations either. Hungerford et al. (2014) analysed H_2O, S and Cl (CO_2 was below detection limits of 25 ppm) in eight samples of basaltic pillow rims, and found values from 0.68 to 0.86 wt % for H_2O, 0.21 to 0.43 wt % for S, and 0.02 to 0.04 wt % for Cl. The samples were collected over an elevation range of 400 m within a lava flow interpreted to have been emplaced in a cavity beneath the ice, so the variations in elevation were not stratigraphically controlled.

From an analytical perspective, the main challenge with using unexsolved volatile species to reconstruct palaeo-hydrostatic/glaciostatic pressures is that CO_2 concentrations are typically at or below current FTIR detection limits (c. 30 ppm), H_2O solubility is very sensitive to CO_2 concentrations, and no models can currently use the full suite of volatile species (Edwards et al., 2009b; Tuffen et al., 2010; discussed more fully in Section 13.5.1). Thus, the difference in estimated pressure for a given value of H_2O (e.g. 0.8 wt %) associated with CO_2 contents of either 25 or 0 ppm results in ice thickness estimates that vary between 1000 m and 600 m, respectively, which can lead to substantially different environmental interpretations, i.e. whether the syn-eruption ice cover was a local ice cap or regional ice sheet (e.g. interpretations of ice cover during the formation of Pillow Ridge, British Columbia: Edwards et al., 2009b; interpretations of timing of eruptions relative to maximum ice thicknesses: Höskuldsson et al., 2006). Moreover, the criterion that all the CO_2 has ideally degassed (Tuffen et al., 2010) is probably satisfied for most of the samples from Bláhnúkur (Iceland) analysed by Owen et al. (2012), but it may not have been met for samples analysed from Tanzilla Mountain and Pillow Ridge (British Columbia), hence the large errors in ice thicknesses declared by Dixon et al. (2002) and Edwards et al. (2009b). Improvements in FTIR techniques that lower detection limits and increased access to ion microprobes (with detection limits of 1–2 ppm) will continue to help improve the utility of volatile species for palaeo-ice reconstructions.

7.5 Mineral, glass and palagonite compositions

While variations in magmatic compositions in glaciovolcanic deposits may record changes in magma chemistry related to syn- and post-glacial eruptive conditions (e.g. Maclennan et al., 2002), the most detailed studies have focused on the formation of authigenic minerals during hydrothermal alteration of volcanic glass (palagonitisation and associated zeolite formation; see Section 13.4.4). Several investigations have specifically examined the formation of palagonite in glaciovolcanic settings (Furnes, 1984; Jercinovic et al., 1990; Thorseth et al., 1991; Bishop et al., 2002), and others have compared the processes and products in different alteration environments (Stroncik and Schmincke, 2001, 2002; Pauly et al., 2011). Volcanic glass is inherently out of equilibrium with its surroundings in surficial conditions and is highly susceptible to 'palagonitisation', whereby mafic glass (sideromelane) reacts with available water to produce a mixture of clay minerals, zeolites and other quasi-crystalline solids. The process has a long history of textural studies (e.g. Peacock, 1926), and the 'Palagonite Formation' in Iceland was one of the earliest-recognised regionally extensive glaciovolcanic units. There have been many studies of the processes involved in glaciovolcanic and subaqueous settings, as well as comparisons with low-temperature pedogenic reactions that convert volcanic glass to mainly clay minerals (e.g. Furnes, 1984; Jercinovic et al., 1990; Schiffman et al., 2002; Pauly et al., 2011; Thorseth et al., 1991). Jarosch et al. (2008) also monitored the slow cooling of the tephra pile produced during the 1996 Gjálp eruption over a timespan of several years. These studies are important for understanding the evolution of porosity and permeability in glaciovolcanic edifices, which can in turn affect longer-term rates of heat transfer (e.g. cooling history of tephra at Gjálp: Jarosch et al., 2008; palagonite formation at Llangorse Mountain, British Columbia: Harder and Russell, 2007). Other authors have investigated terrestrial deposits in order to gain insights into weathering and alteration products on Mars (Bishop et al., 2002).

Studies by Jercinovic et al. (1990) and Thorseth et al. (1991) focused specifically on palagonitisation in glaciovolcanic outcrops. Jercinovic et al. (1990) analysed deposits at Ash Mountain, South Tuya and Tuya Butte in north-central British Columbia. They suggested that the overall stoichiometry of palagonite most generally approximated that of smectite $[(Na,Ca)_{0.3}(Mg,Fe)_{2-3}((Si,Al)_4)O_{10}(OH)_2 \cdot n(H_2O)]$, and that the paragenesis for minerals was typically phillipsite, chabazite, analcite and calcite (from earliest to latest formed). Observed rinds varied in thickness but generally showed compositional zoning. Depending on local environmental parameters, including pH, palagonite can be either low-Al (Ni, Co, Cr are also depleted) or high-Al (Ni, Co, Cr are retained). Jercinovic et al. (1990) also argued that the process was isovolumetrical, and that the degree of heterogeneity was partly due to stronger controls at a microenvironmental level rather than macroenvironmental. Thorseth et al. (1991) analysed glaciovolcanic basaltic and basaltic andesite palagonite from Iceland. They found that reaction fronts varied from sharp to diffusional. At the microscale, palagonite growth commences as granules that eventually form chains and ultimately develop a sponge-like texture. They also suggested that element

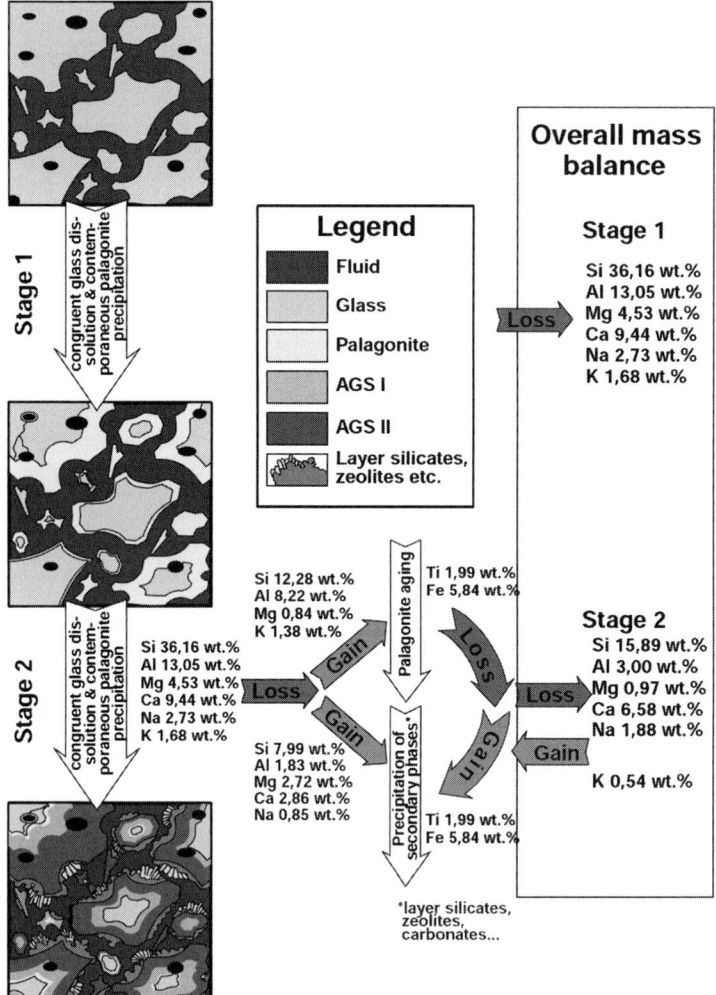

Fig. 7.4 Schematic diagram summarising the compositional effects (mass gains and losses) during progressive abiotic alteration of mafic glass to palagonite in a variety of environments (subglacial, meteoric, marine; from Stroncik and Schmincke, 2001). The process is envisaged taking place in two important stages. Stage 1 consists of palagonite formation by congruent dissolution of sideromelane glass, characterised by partial loss of Si, Al, Mg, Ca, Na and K. Stage 2 consists of the previously precipitated unstable palagonite taking up Si, Al, Mg and Ca and releasing Ti, Fe, Ca and Na to become more stable smectite. (A black and white version of this figure will appear in some formats. For the colour version, please refer to the plate section.)

mobility was selective, and varied as a function of pH, with higher pH values (>3) favouring precipitation of Fe, Al, Ti hydroxides in pores spaces. Stroncik and Schmincke (2001, 2002) and Pauly et al. (2011) all presented summaries of the broad compositional changes (Figs. 7.4, 7.5).

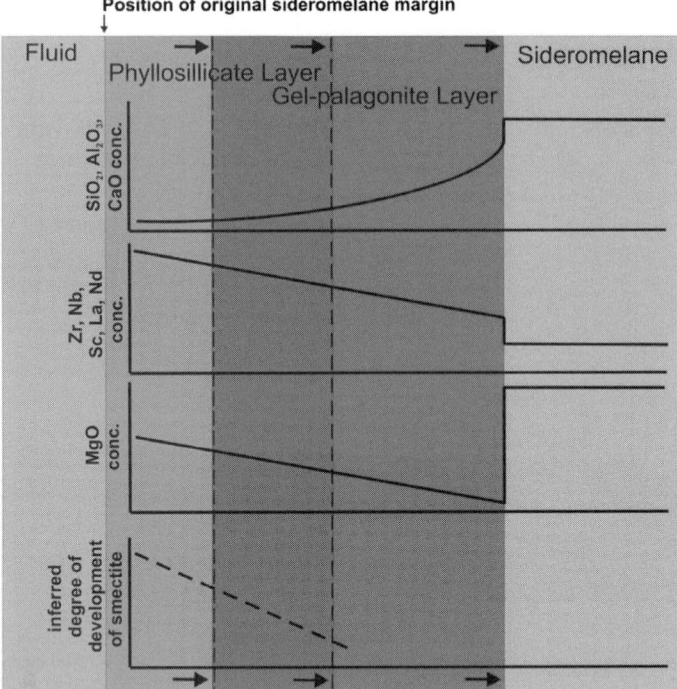

Fig. 7.5 Summary diagram illustrating elemental changes during progressive abiotic palagonitisation, also in a variety of alteration environments, envisaged by Pauly et al. (2011). The arrows indicate the evolution direction of gel palagonite to crystalline smectite with time, which proceeds from the outer surface of the rind (most altered) and migrates inward; thus, smectite growth is greatest at the fluid boundary. In this study, it was suggested that the mass changes are not isovolumetric (contrary to Jercinovic et al., 1990, and Stroncik and Schmincke, 2001) and original glass composition is also important. The rate and duration of the alteration may also be important (e.g. burial versus hydrothermal). The study also assessed changes in trace element concentrations.

Johnson and Smellie (2007) examined zeolite minerals in Antarctic lava-fed deltas in order to distinguish between deposits formed in glaciovolcanic or submarine settings. They used variations in chabazite ($Ca_2[Al_4Si_8O_{12}] \cdot 12H_2O$) and phillipsite ($K_2(Ca_{0.5},Na)_4[Al_6 Si_{10}O_{32}] \cdot 12H_2O$) compositions to investigate the potential of zeolite minerals as environmental discriminants. They found that a ratio of (Na + K)/Ca ratios in the two zeolite minerals could discriminate between those formed in freshwater (glaciovolcanic) and seawater. Approximate values for phillipsite of less than 3.0 and for chabazite of less than 1.0 were thought to be indicative of freshwater formation, but the continuum of compositions observed implies that the discriminant values used for either phase probably comprise a range rather than a discrete number. There are at least two problems in the use and interpretation of zeolite compositions to infer eruptive palaeoenvironment: (1) zeolites are susceptible to becoming overprinted by the compositions of later fluids (e.g. seawater) flushing through the rock sequence, e.g. along microfractures (Johnson and Smellie, 2007);

and (2) the zeolite compositions may differ according to whether the local alteration system was closed or open, irrespective of the eruptive setting (Antibus et al., 2014).

Spectroscopic studies of minerals from terrestrial samples are also becoming common in characterising styles of alteration on Mars (e.g. Bishop et al., 2002). Bishop et al. (2002) analysed samples from two Icelandic edifices, Hlödufell and Thorolfsfell, for major element compositions and used visible–infrared reflectance to measure spectra for possible comparison with data from the Mars Pathfinder lander.

7.6 Stable isotope studies

There have been few stable isotope studies of glaciovolcanic deposits. The main processes investigated have been at broad magmatic scales, including studies in Kamchatka (Bindeman et al., 2004, 2010), Alaska (Bindeman et al., 2001) and in Iceland (Bindeman et al., 2008). Mortimer et al. (2008) determined stable isotopes (O and C) in carbonates, and Antibus et al. (2014) also determined O and H isotopes in silicates in veins and breccias associated with volcaniclastic rocks in Antarctica as an aid to inferring eruptive palaeoenvironment. Finally, a pilot study has also investigated rapid isotopic exchange between glasses and meltwater using large-scale volcano–ice interaction experiments (Edwards et al., 2013).

The studies by Bindeman et al. (2001, 2004, 2008, 2010), in Kamchatka and Iceland, focused on explaining apparently anomalously low $\delta^{18}O$ values in volcanic rocks by assimilation of crustal rocks whose isotopic compositions reflected interaction with fluids derived from melted glacial ice. The signal of 'low $\delta^{18}O$' tephra and its potential role in identifying interaction between hydrothermal fluids with glacier signatures and volcanism was first described at Fisher Caldera on Unimak Island, Alaska (Bindeman et al., 2001). Magmatic phenocrysts and bulk rock samples from the 9100 ka climactic eruption at the caldera have values of $\delta^{18}O$ that are anomalously low ($<+5.5‰$) compared to ranges for typical volcanic materials ($+5.5–6.4‰$). Their model for the origin of the low $\delta^{18}O$ values included assimilation of hydrothermally altered crustal materials accompanying differentiation of a basaltic magma; it was suggested that the hydrothermal fluids were formed by incorporation of meteoric water depleted in $\delta^{18}O$ by contributions from glacially derived water, which is typically 'lighter' (i.e. with lower values of $\delta^{18}O$) than meteoric waters from interglacial periods. In Kamchatka, Bindeman et al. (2004, 2010) found that Pleistocene–Holocene silicic ignimbrites have values of $\delta^{18}O$ that are demonstrably lower ($+4–5‰$) than values for 'normal' volcanic rocks, or for 'high $\delta^{18}O$' volcanic rocks ($+6.0‰$ for basalts and $6.4‰$ for rhyolites). They suggested that a possible explanation for the anomalously low values comprised melting or assimilation of crustal material that was hydrothermally altered by meteoric water whose values of $\delta^{18}O$ recorded the periodic presence of glaciers covering parts of Kamchatka. Extensive numerical modelling of assimilation–fractionation by Bindeman et al. (2008) showed that it was feasible to produce at least some of the anomalously low $\delta^{18}O$ values found in Icelandic basalts (with values of $\delta^{18}O$ ranging from +3.1 to 4.5‰ in groundmass glass) by contamination with glaciovolcanic hyaloclastite and altered crustal rocks. Bindeman and Serebryakov (2011) applied similar arguments to recognise glaciation at 2.4 Ga in Karelia, Russia, based on extremely low values of $\delta^{18}O$ and δD in metamorphic rocks. While there is no indication that the rocks were originally glaciovolcanic, their study demonstrates that the same principle can be applied to ancient deposits thought to have formed in glaciovolcanic environments.

The study by Antibus et al. (2014) was able to demonstrate that the alteration fluid was probably freshwater (i.e. glacial), with alteration temperatures ranging between 5 and 100 °C. They also made the first attempt to use stable isotopes in authigenic carbonate in volcani-clastic rocks as a proxy for palaeoclimate in the region that is independent of other climate proxies, although because of the assumptions that were necessary, the results were not definitive. By contrast, the study by Mortimer et al. (2008) showed that the altering fluid was probably marine and they used the result to infer significant uplift since the rocks formed.

Edwards et al. (2013) conducted large-scale pilot study experiments using basaltic melt to investigate rates of melting and textures produced when the melts are poured on top of ice. The preliminary data showed that, even on short timescales (minutes), O isotopes exchanged measurably with the basaltic melt, and thus $\delta^{18}O$ values could be used to identify volcanic rocks that were emplaced in water-rich environments.

7.7 Geochronometric studies

One of the most important aspects of glaciovolcanic deposits is that, once their environment of formation has been identified, they constrain the location of previously existing ice masses. Additionally, geochronometry allows the deposits to be correlated with global climatic changes, such as those recorded by the marine record which reflect how ocean volumes vary in response to changing volumes of ice. A broad range of techniques have been applied to constraining the eruption ages, including: K–Ar, 'high-precision' Ar–Ar, magnetostratigraphy, U/Th–He, and cosmogenic nuclides.

For older sequences (pre-Pliocene), standard K–Ar geochronology is sufficient to provide general ages for older ice masses (e.g. LeMasurier and Rex, 1982). However, the resolution of ages for younger sequences is more problematical (e.g. Souther et al., 1984; Komatsu et al., 2007a,b). Standard K–Ar geochronology requires samples with high concentrations of K to produce ages with precision of c. 40–50 k.y., which is a similar timescale for entire glacial–interglacial Milankovich cycles. Hypotheses testing potential feedbacks between ice sheet fluctuations and changes in volcanic mass eruption rates and eruption frequencies need much greater precision (preferably of a few k.y.) in order to link eruptions directly to the global record of ice fluctuations. A new unspiked method of K–Ar dating can produce ages with greater precision (<10 k.y.; Guillou et al., 2010). Precision in Ar–Ar geochronology is also typically 40–60 k.y. for rocks of Neogene age (Smellie et al., 2008; Mahood et al., 2010). However, precisions for rocks with intermediate to felsic compositions can now be considerably better (few k.y.: McGarvie et al., 2006; Flude et al., 2008, 2010; Singer et al., 2008). Some investigations have also tried using K–Ar geochronology on partly melted xenoliths to take advantage of higher K concentrations, with limited success (e.g. Souther et al., 1984).

Blondes et al. (2007) used U/Th–He geochronometry on zircons within a crustal xenolith from a glaciovolcanic deposit (informally referred to as 'Little Bear Mountain') located in western British Columbia. This technique is based on the observations that He is produced by the radioactive decay of U and Th, but it can be retained in the host mineral as long as the mineral is not exposed to temperatures greater than c. 100 °C. Once that temperature is reached, the He rapidly diffuses out of the crystal, and the geochronometer is 'reset'. Blondes et al. (2007) showed that, given the measured sizes of the xenoliths studies, residence in the host magma for more than 1 hour was sufficient to flush all of the He out of host crystals, effectively resetting the system. In the case of the mafic Little Bear Mountain deposits, this produced a precise age of 157 ka ± 3.5 k.y.

Efforts to find more accurate means of measurements have attempted several new routes. Liccardi et al. (2007) used surface exposure geochronometry to determine exposure ages for a series of subaerial capping lavas on tuyas from northern Iceland. This technique is based on the production of ^{3}He isotopes by cosmic-ray bombardment of minerals (olivine in the study by Licciardi et al., 2007). They found a consistent pattern of ages and tuya summit elevations across Iceland and interpreted the exposure ages as eruption ages. The analytical and calculated errors for this technique are strikingly low; Licciardi et al. (2007) reported one-sigma errors ranging from 200 years to 2200 years for eruptions between 20 and 8.5 ka. However, the technique has several limitations. For example, determining the amount of exposure time of glaciovolcanic edifices is not straightforward. Deep snow cover can limit exposure to cosmogenic radiation for long periods during a year, and the duration of those periods may change significantly over time. Licciardi et al. (2007) calculated that snow cover of 10 cm for 8 months annually (assuming a snow density of 200 kg m^{-3}) would only affect the exposure ages by 0.8%. However, snow cover of 100 cm for 8 months annually (assuming a greater snow density of 500 kg m^{-3}) would result in ages varying by c. 18%. They suggested that the consistency of their dataset argued against snow shielding as a problem, but that some of the age differences identified could be caused by local variations in snow cover. This technique is likely to be only applicable to rocks formed in the waning stages of the Last Glacial Maximum (<10 000 y BP), as those formed during the last glacial would have been covered by ice of indeterminate thickness, which would have completely blocked the production of cosmogenic radionuclides.

7.8 Quantitative analysis of grain sizes

Many glaciovolcanic studies have reported grain size ranges for indurated lithofacies, but few have quantified the grain size distributions in unconsolidated materials. Jude-

Eton et al. (2012) examined deposits from the 2004 Grímsvötn eruption, and Gudmundsson et al. (2010) and Taddeucci et al. (2011) described the ash fall patterns and influence of ash aggregates from the 2010 Eyjafjallajökull. The most extensive studies of grain size distributions are from mass flows of several types, including lahars (many eruptions of Cotopaxi, for the 1991 Hudson eruption, and from Popocatepetl), jökulhlaup deposits in Iceland, and ice-slurry deposits from Redoubt and Ruapehu (see Chapters 3 and 15).

The complexities of proximal tephra distribution from a glaciovolcanic eruption were documented by Jude-Eton et al. (2012) for the 2004 Grímsvötn eruption. They conducted granulometric analysis of 230 tephra samples from the eruption and estimated that approximately half (2.4×10^7 m^3) of the total estimated volume of tephra (c. 4.7×10^7 m^3) was deposited outside of the main ice cauldrons. Using grain size and deposit characteristics they identified three distinct eruption phases that occurred over a period of 36 hours. Three different tephra deposition processes were identified: air fall, pyroclastic density currents, and discrete phreatomagmatic explosions (minor).

Capra et al. (2004) published analyses of sediment grain sizes and shapes from lahars at Popocatepetl volcano in Mexico that were generated by melting of glaciers on top of the volcano during eruptions (1997) and by remobilisation of pumice flows (2001). They used a combination of field and laboratory analyses of grain sizes to compare and contrast characteristics of the lahars. They also measured grain shapes and calculated eigenvectors to characterise the three-dimensional fabrics of the deposits. The 1997 deposits are thought to be have been generated directly by melting of summit ice and snow during explosive eruptions and erosion during transport. Because of the relatively high volume of meltwater generated by the 1997 eruption (10^7 m^3), the lahar transformed downstream into a sediment-laden stream flow. By contrast, the lahar generated during the 2001 eruption contained a much smaller volume of water (0.5×10^5 m^3). It travelled a similar runout distance to that in 1997 but was emplaced as a debris flow, without transformation to a stream flow.

7.9 Summary

It is likely that future investigations of glaciovolcanism will not simply be field-based but will increasingly include a range of types of laboratory analyses. Geochemical studies are already becoming used more routinely for the compositional characterisation of geological units, but they can also be used to identify and unravel stratigraphical complexities, often not obvious in the field, that will improve our understanding of the connections between magmatic processes and eruption dynamics. A better understanding of volatile behaviour is also required and the use of stable isotopes is likely to become more common. Both have the potential to add critical information for unravelling the complex details of glaciovolcanic processes, as well as for improving palaeoenvironmental reconstructions. Analytical uncertainties in determining CO_2 concentrations remain problematical as is the precision of many geochronological techniques. Many new techniques are highly promising and the more frequent implementation of 'standard' volcanological techniques on unconsolidated materials (e.g. grain size analysis and grain morphometry; Jude-Eton et al., 2012) will lead to a greater understanding of the eruptive origins of fragmental deposits in glaciovolcanic sequences.

8

Landform classification and morphometry
of glaciovolcanic centres

8.1 Introduction

This chapter describes the classification and morphometry of glaciovolcanic landforms that formed by subglacial eruptions. Thus, it includes all subglacially erupted volcanoes (Smellie, 2013) but excludes glaciovolcanic features that formed simply by chilling of lava against a downslope ice barrier (i.e. ice-impounded lavas; Mathews, 1952a; Lescinsky and Sisson, 1998; Harder and Russell, 2007; Kelman, 2005). There are no known distinctive landforms that formed during interaction with snow or the cryolithosphere. Few publications exist that discuss glaciovolcanic morphometry (Hickson, 2000; Kelman et al., 2002; Komatsu et al., 2007a; Smellie, 2007, 2009, 2013; Jakobsson and Gudmundsson, 2008; Jakobsson and Johnson, 2012; Pedersen and Grosse, 2014; Russell et al., 2014). The subject is important, however, particularly for planetary research in which volcano edifice shape is a major guide to past glacial environments and for assessing planetary water inventories (see Chapter 16). Moreover, even when only the landforms are recognised, they are not a simple Boolean climate signal. They can yield a range of palaeoenvironmental parameters such as ice thickness, its surface elevation and, potentially, thermal regime (see also Chapter 13). For simplicity, in this chapter a simple distinction is made between mafic and felsic rocks, with the division made at 57% SiO_2.

8.2 Classification of glaciovolcanic landforms

The published descriptions of glaciovolcanic landforms and their sequences are dominated by examples of relatively short-lived monogenetic eruptions beneath wet-based (i.e. warm or polythermal) ice (e.g. Mathews, 1947, 1951; Jones, 1969b, 1970; Smellie et al., 1993; Bergh and Sigvaldason, 1991; Smellie and Hole, 1997; Tuffen et al., 2001, 2002a; McGarvie et al., 2006; Schopka et al., 2006; Edwards et al., 2011; Russell et al., 2014). There are no published examples of monogenetic volcanic landforms formed under cold-based ice although the resulting shapes of the glaciovolcanic edifices have been postulated (Smellie, 2007, 2009, 2013; see Section 8.3). There are few descriptions of the much longer-lived (polygenetic) subglacially erupted central volcanoes, but examples have been described that were erupted under warm, polythermal and cold-based ice (e.g. LeMasurier et al., 1994; Smellie, 2000; Edwards and Russell, 2002; Wilch and McIntosh, 2002; Stevenson et al., 2006, 2011; Smellie et al., 2008, 2011a,b, 2013a). The polygenetic volcanoes have a compound morphometry that may be very different from that of the monogenetic landforms (Section 8.2.1). This chapter is based principally on the study by Smellie (2013) in his review of mafic and felsic glaciovolcanic landforms, updated with new morphometric data.

Early studies of glaciovolcanism noted the peculiar shapes of some edifices, mainly the distinctive flat-topped steep-sided mafic volcanoes now referred to as mafic and felsic tuyas (Mathews, 1947, 1951; also van Bemmelen and Rutten, 1955; Jones, 1969b, 1970). However, most studies have focused on the internal structure, lithofacies and generalised models for (mainly mafic magma) emplacement rather than on volcano morphology and morphometry and only tuyas were recognised as a discrete glaciovolcanic landform. Since then, several morphologically distinct glaciovolcanic landform types have been described. Hickson (2000) divided mafic centres in British Columbia into four broad types (subglacial mounds incorporating a basal pillow pile overlain by explosively generated tephra; tuyas or stapi (stapar); tindars; and subglacial stratovolcanoes). She also compared subglacial lavas with distinctive cooling joints and elongate morphologies with eskers (see also Smellie and Skilling, 1994). Kelman et al. (2002) expanded Hickson's study to encompass magmas with mafic and felsic compositions and described tuyas (which they called flow-dominated for felsic magmas), subglacial domes and subglacial flows, the latter also called 'esker-like'. Kelman (2005) also tabulated nine glaciovolcanic landforms, comprising tuya, subglacial mound, tindar, pillow sheet, esker-like flow, remnant cliff after supraglacial eruption, subglacial dome, flow-dominated tuya and impoundment feature. Jakobsson and Gudmundsson (2008) and Jakobsson and Johnson (2012) divided mafic glaciovolcanic landforms in Iceland into tuyas, tindars and pillow ridges and mounds. They used basic morphometry (width, length and volume) to illustrate some fundamental differences, principally that tuyas are wider and have greater volumes than most tindars (also Smellie, 2007). However, the most comprehensive discussion of glaciovolcanic landforms and their morphometry is by Smellie (2007, 2009, 2013) using a much expanded dataset, in which at least nine morphotypes were documented. They comprise mafic tuyas; felsic tuyas (two types – tephra-dominated and lava-flow dominated); felsic domes and lobes; tephra mounds and ridges; pillow mounds, ridges and sheets; and sheet-like sequences (two categories: Dalsheidi-type and Mt Pinafore-type). Of these, all are volcanic edifices except the sheet-like sequences. The classification of glaciovolcanic landforms used in this book is shown in Fig. 8.1. Jakobsson and Johnson (2012) and Smellie (2013) also published morphometric datasets for the glaciovolcanic edifices they considered. Finally, Russell et al. (2014) reclassified the glaciovolcanic landforms in a significant departure from all previous classifications, in which they included many of the landforms as different types of tuya (e.g. conical tuyas, linear tuyas, complex tuyas; see Section 1.4.2 for discussion); not all glaciovolcanic landforms were included in their classification. The classification of Smellie (2007, 2009, 2013) is used as the basis of this chapter.

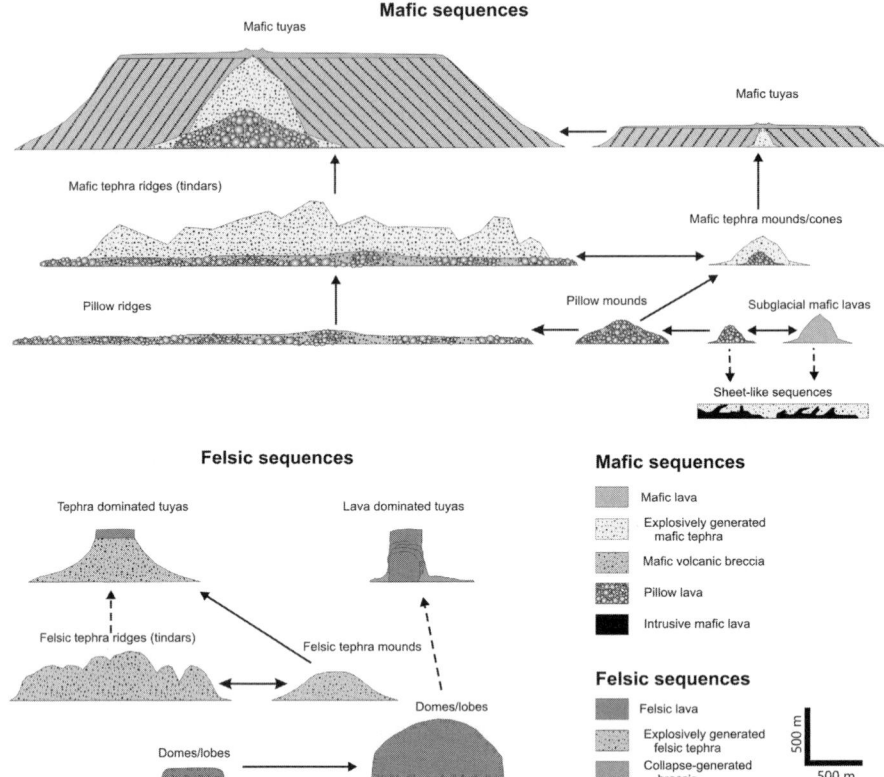

Fig. 8.1 Classification and hierarchy of glaciovolcanic landforms and their principal constituent lithofacies (modified after Smellie, 2013). The arrows indicate the evolutionary progression of landforms, e.g. pillow mounds evolving up into tindars, then into tuyas. The scale is only indicative. (A black and white version of this figure will appear in some formats. For the colour version, please refer to the plate section.)

The large number of glaciovolcanic landform types is a function of the interaction between two primary factors: (1) the physical conditions (location, amount, thickness and thermal regime of confining ice; the meltwater hydraulics (i.e. location and quantity of trapped water and its migration); and localised versus fissure eruption); and (2) the magmatic conditions (chemical composition of the erupting magma; effusion rate; and total erupted volume) (Hickson, 2000; Smellie, 2007, 2009, 2013). The simple presence and confining effects of surrounding ice have also resulted in glaciovolcanic landforms being generally taller than subaerially formed volcanic landforms, with steeper slopes (Fig. 8.4b). There is also a clear unidirectional evolutional hierarchy amongst the volcanic landforms (Fig. 8.1). For example, mafic pillow mounds and ridges frequently evolve up into tephra mounds or tindars (tephra ridges), which themselves may develop into tuyas. Similarly, felsic tephra mounds and ridges may evolve into tephra-dominated felsic tuyas whilst felsic and inter-mediate domes may transform upwards into lava-dominated felsic tuyas (Kelman, 2005).

By contrast, sheet-like sequences (which are not edifices) do not change up into any other glaciovolcanic landform. These gradual transformations, together with the occurrence of intermediate-composition edifices with transitional rheologies (Kelman et al., 2002;Kelman, 2005; Stevenson et al., 2009) have resulted in overlapping morphometric parameters for the diverse landforms, although overall there are also significant differences (see below).

The morphometric parameters used here are height above volcano base, basal length or width, volume and aspect ratio. Aspect ratio is defined as the ratio of maximum edifice height to width. It is a particularly crude parameter when applied to features such as tuyas with contrasting surface slopes (e.g. gently dipping summit regions versus steep-dipping perimeter cliffs). It is less accurate than calculating average slope based on mapping the varied slope angles across an individual edifice (Pedersen and Grosse, 2014) but is simpler to measure without requiring use of photogrammetric methods (e.g. DEMs). Because sheet-like sequences are not edifices, deposit thickness is used rather than height. Summit craters are also invisible using aspect ratio as a parameter but it may not be much of a problem since craters, formed mainly of poorly consolidated tephra, are typically eroded rapidly and are not a long-lasting feature anyway, and their original form may be masked by lava lake development during the final eruptive stage (Rossi, 1996). Pedersen and Grosse (2014) used additional parameters such as summit width, summit area and mean slope angle, and derived indices such as ellipticity to quantify edifice elongation (equivalent to basal length to width ratio of Smellie, 2013) and irregularity, which quantifies edifice complexity. Their measurements, based on 20 m resolution DEM photogrammetry of aerial images, are more accurate than those used by Smellie (2013).

8.2.1 Polygenetic glaciovolcanic edifices

At present it is thought that most polygenetic glaciovolcanic edifices do not create distinctive landforms, with the possible exceptions of Hoodoo Mountain volcano in British Columbia (Edwards and Russell, 2002; Edwards et al., 2002) and Undirhlíðar pillow-dominated tindar in Iceland (Pollock et al., 2014). Other polygenetic edifices occur as stratovolcanoes and shield volcanoes with morphologies not diagnostic of eruption in association with ice. The shield volcano profile, with its low flank gradients, is particularly common in effusive mafic centres owing to lava-fed deltas, whose surfaces mirror the gradients of the draping coeval ice sheet. Some edifices have high cliff margins owing to the buttressing by former ice (e.g. Hoodoo Mountain, British Columbia; Edwards and Russell, 2002; Edwards et al., 2002). However, few large polygenetic glaciovolcanic centres have been investigated. The largest one currently known is Mt Haddington (James Ross Island, Antarctica), which erupted from Middle Miocene times ($\leq$c. 12 Ma; Smellie et al., 2008; Marenssi et al., 2010). Mount Haddington has a shield-like profile rising to 1.6 km and a basal diameter of 60 km (Nelson, 1975; Smellie et al., 2008, 2013b). It might be referred to informally as a tuya volcano because it erupted subglacially and has relatively gentle surface slopes ($<10°$)

and prominent peripheral cliffs constructed mainly from pāhoehoe lava-fed deltas (Skilling, 2002; Smellie, 2006). However, unlike the smaller monogenetic tuyas, in which the steep peripheral margins are a primary constructional feature related to steep-dipping hyaloclastite foreset bedding, the cliffs on James Ross Island are destructive features mainly formed by repeated gravitational collapses due to loading by the brittle volcanic edifice and the deformation of weak underlying layers of soft Cretaceous sediment (Oehler et al., 2005; Davies et al., 2013; Smellie et al., 2013b). Without those collapses, it is unlikely that it would display anything other than a shield profile undiagnostic of its glacial setting.

Other large polygenetic glaciovolcanic edifices investigated include several prominent Mio-Pliocene centres in Victoria Land, Antarctica. Although broadly similar in dimensions to Mt Haddington (summit elevations c. 2 km, basal diameters c. 20–40 km), they differ in being constructed from multiple 'a'ā lava-fed deltas and their bounding cliffs are eroded by glaciers and the sea (Smellie et al., 2011a, 2013a, 2014). The flow fronts of 'a'ā-fed deltas are much less pronounced than pāhoehoe deltas (Smellie et al., 2013a). Like James Ross Island, the Victoria Land edifices resemble non-glacial mafic shield volcanoes. Thus, because of the lack of a distinctive tuya morphology (i.e. more or less flat top with steep peripheral slopes), none of these large polygenetic subglacially erupted volcanoes should be called tuyas.

Hoodoo Mountain in British Columbia is a Quaternary phonolite and trachyte polygenetic stratiovolcano 1850 m in height and c. 6 km wide at its base (Fig. 8.2). It has

Fig. 8.2 View looking to the northwest of Hoodoo Mountain volcano in western British Columbia. Elevation change from the Iskut River (foreground) to the top of Hoodoo Mountain is 1900 m (image: D. Huisman).

Fig. 8.3 (a) View of Herdubreid, Iceland, a monogenetic tuya c. 600 m high. Note the domical shape of the summit, which is a small basaltic shield formed of subaerial lavas. (b) View (from left to right) of Thorisjökull, Geitlandsjökull and Eiriksjökull (Iceland's largest tuya). Thorisjökull is c. 8 m wide and rises c. 600 m. Larger tuyas are more flat-topped because the summit is dominated by the surfaces of lava-fed deltas rather than the pyroclastic cone around the vent.

a prominent flat top capped by an ice cap and prominent subvertical encircling cliffs. It thus has a broadly tuya-like appearance although it is a composite edifice, formed as a result of multiple eruptions over the past 85 ka (Edwards and Russell, 2002; Edwards et al., 2002). The steep flanks are a result of eruptions beneath and against ice and glacial erosion. Like the Antarctic polygenetic centres, the shape of Hoodoo Mountain reflects a variety of influences, only some of which (e.g. 200 m high vertical cliffs) are diagnostic of glacial conditions.

Likewise, at least one elongate pillow ridge or pillow-dominated tindar is known to have a polygenetic origin: Undirhlíðar, in southwestern Iceland. Recent studies at its northern end have demonstrated that it is polygenetic (Pollock et al., 2014).

In summary, there are currently no clear diagnostic morphological parameters for large, long-lived polygenetic volcanoes that can be used to interpret them unambiguously as glaciovolcanic. Even the presence of a prominent flat summit region and steep margins (as at Hoodoo Mountain) require ground studies of the constituent lithofacies to prove a glaciovolcanic origin. However, petrogenetic and geochronometric studies of large supposedly monogenetic tuyas and tindars may identify more edifices with complex origins.

8.2.2 Mafic tuyas

Of all the glaciovolcanic landforms, mafic tuyas are the most distinctive. Also known non-genetically as table mountains and mesas (although not all table mountains and mesas are tuyas), they were amongst the first glaciovolcanic landforms to be recognised and described. In plan view, tuyas are oval to straight-sided (van Bemmelen and Rutten, 1955; Komatsu et al., 2007a; Jakobsson and Johnson, 2012). The combination of flat to gently dipping summits and steep-sided flanks is essentially diagnostic (Fig. 8.3), although it also occurs in shoaling to emergent marine Surtseyan edifices (e.g. Thordarson and Sigmarsson, 2009) and tuyas have also been described as Surtseyan in terms of their constructive processes (Smellie and Hole, 1997). Landform analysis alone cannot distinguish marine Surtseyan edifices from glaciovolcanic tuyas. Identification needs to be based on differences in the internal architecture of the constituent lithofacies, particularly evidence for variable water levels during the eruptive period (Smellie, 2006). Most tuyas are monogenetic but detailed lithofacies analysis has revealed that even some small glaciovolcanic edifices with seemingly simple tuya-like morphologies are polygenetic (Wörner and Viereck, 1987; Smellie, 2001; Smellie et al., 2006b; Skilling, 2009; Fig. 8.4).

Monogenetic tuyas are defined by the presence of lava-fed deltas, which give the edifices their distinctive gently sloping tops and steep flanks. Ponded water, in this case a meltwater lake, is required in order that deltas can form and they relate to a late subaerial episode of lateral progradation formed after the edifice has penetrated the enclosing ice sheet (Smellie, 2000, 2001, 2006; Skilling, 2002). As suggested above, two types are known depending on the nature of the feeding lava: pāhoehoe and 'a'ā, with different implications for the constituent lithofacies and architecture (Smellie et al., 2013a). Pāhoehoe deltas can have substantial delta faces rising to 1 km, at least, depending on the thickness of ice that they were emplaced in (Smellie, 2007, 2013; Smellie et al., 2008, 2009; Jakobsson and Johnson, 2012). 'A'ā deltas of similar size have not been described. Compared with pāhoehoe deltas, individual 'a'ā deltas have much less steep terminations suggesting that they will form less distinctive landforms. However, thus far only 'a'ā deltas emplaced in thin ice (<300 m)

Fig. 8.4 (a) View of Shield Nunatak, a small tuya in Victoria Land, Antarctica. Although probably monogenetic and just 2 km in diameter, rising c. 300 m, the edifice was constructed in at least three main episodes during periods in which the surface elevation of the coeval meltwater lake varied (see Worner and Viereck, 1987). (b) View of Hlödufell, Iceland (c. 4 km long and c. 600 m high). The two prominent flat surfaces in the summit region were caused by two episodes of lava-fed delta progradation in meltwater lakes during periods of different water levels in the surrounding englacial lake (see Skilling, 2009). Contrast the flat-topped steep-sided morphology with the subaerial basaltic shield of Skjaldbreidur, at left behind Hlödufell.

have been described and it is possible that 'a'ā deltas emplaced in much thicker ice might develop the steeper delta faces more akin to pāhoehoe deltas (Smellie et al., 2013a).

All tuyas commence as either pillow mounds or tephra mounds (tuff cones or tindars). However, the morphology of the earlier-formed constructs is largely or completely obscured by the subsequent draping lava-fed deltas and does not influence the final tuya morphology. The principal limitation on the lateral extent of a tuya is the volume of magma discharged. By contrast, the height of a tuya is determined principally by the thickness of the surrounding ice. For eruptions in wet-based ice, a theoretical height limit of c. 750–1000 m has been suggested based on the principle that a relatively stable meltwater lake is required for lava-fed delta progradation, whereas the meltwater-filled englacial vault developed by eruptions in much thicker ice sheets will always float the surrounding ice, thus permitting the water to escape. However, this suggestion is based on the depth of an englacial vault being determined by overflowing through a supraglacial permeable layer composed of snow, firn or fractured (crevassed) ice (Smellie, 2006, 2009). Since there are effective limits to the thickness of the permeable layer, water accumulating in very thick ice (>750–1000 m) will be able to lift its encircling ice barrier and escape (see Chapter 6). Qualitative support for this suggestion is provided by the observation that the tallest tuyas documented so far are c. 1000 m high (Fig. 8.5). However, if the surface of an englacial lake is not fixed by overflowing but is dominated by efficient subglacial drainage, the limitation on tuya height is removed since the pressure at the base of the lake may never exceed that exerted by the surrounding ice barrier. This height limitation does not apply to polygenetic glaciovolcanic centres since ice reforms and cloaks an edifice after each eruption. Each successive eruption thereafter simply records the thickness of the draping coeval ice it encounters. Thus, polygenetic volcanoes can grow to heights unconstrained by regional ice thicknesses. A height limitation also does not apply to monogenetic tuyas prograding in cold-based ice. Since cold ice is frozen to its bed, the meltwater lake cannot lift the encircling ice (see Section 8.3).

Mafic tuyas have the greatest volumes of any glaciovolcanic landforms, the data defining a well-defined trend with the highest ratio of volume:area (Fig. 8.5). The trend is curved (concave-up) and fits a power law or polynomial distribution. On a plot of basal length (*maximum* horizontal extent) versus height, tuyas form two groups corresponding to edifices erupted from a point source (or from a fissure that subsequently focused on a point source), which have relatively symmetrical outcrops, and those erupted from fissures, with asymmetrical outcrops (Smellie, 2013). The result is not unexpected since the length of a volcanic outcrop will be strongly influenced by the mode of eruption, and fissure length is a tectonic rather than magmatically controlled variable. However, local bedrock topography can also affect edifice morphology since the volcanic products might be limited laterally by a steep surface (Hickson, 2000), or else flow preferentially down-hill such that the edifice becomes distorted. Two discrete fields are also defined when aspect ratio is plotted against width (i.e. *minimum* horizontal extent). Because tectonic effects are removed in this plot, the prominent field of tuyas with greater widths is most simply explained as those edifices that were constructed during unusually voluminous eruptions, i.e. when the constituent lava-fed deltas prograded to much greater distances than deltas formed in less voluminous eruptions. In this case there is no a-priori reason why two discrete fields should be present and it is likely that these fields will merge as more data become available.

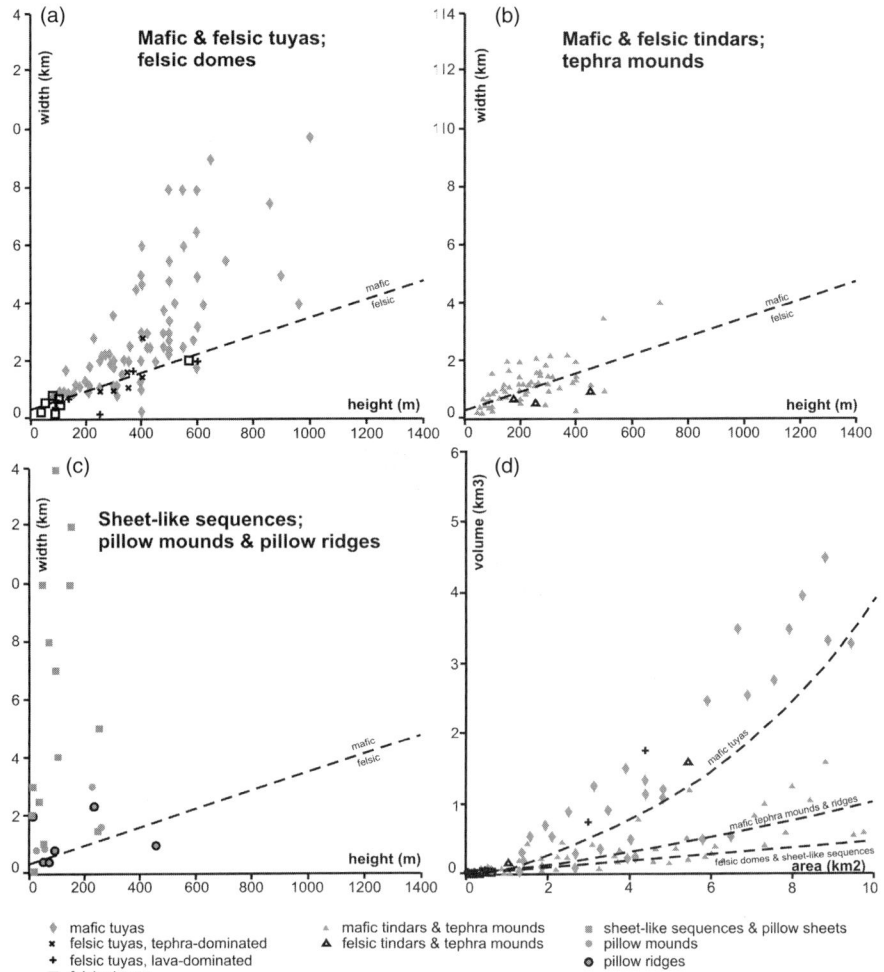

Fig. 8.5 Plots of (a)–(c) width versus height and (d) volume versus area for glaciovolcanic landforms (modified after Smellie, 2013).

8.2.3 Felsic tuyas

Far fewer felsic tuyas have been described compared with mafic tuyas, reflecting their greater scarcity. Most crop out in Iceland (McGarvie, 2009) but a few have also been described in British Columbia (Kelman et al., 2002; Kelman, 2005). They differ in many respects from their mafic counterparts. Whilst they share the flat-topped, steep-sided landform diagnostic of tuyas (Fig. 8.6), they differ in their constituent lithofacies (Chapter 12). There are also two types based on differences in their gross internal structure: (1) tephra dominated and (2) lava dominated. As their names suggest,

Fig. 8.6 Laufafell: a rhyolite tuya in the Torfajökull volcano complex, Iceland. The tuya is c. 4 km long and rises c. 400 m above the surrounding ground.

tephra-dominated tuyas have extensive cores (typically a few hundred metres thick) of poorly exposed fine-grained tephra overlain by a relatively thin (tens of metres) caprock of flat-lying subaerial felsic lava (Tuffen et al., 2002a; Stevenson et al., 2011). By contrast, lava-dominated felsic tuyas form very distinctive flat-topped columns or short blade-shaped ridges with near-vertical sides formed almost entirely of flat-lying, ice-constrained subaerial lavas, some of which may plaster the flanks (Mathews, 1951). Fragmental deposits are either absent or scarce. In all but one case, the lava caprocks formed within a cylindrical tunnel connecting the englacial ice cavity to the atmosphere (cf. Gudmundsson et al., 1997). The exception, an unusual lava-dominated edifice at Prestahnúkur (Iceland), was interpreted as a felsic tuya by McGarvie et al. (2007), essentially based on the presence of a poorly developed flat lava caprock. Its aspect ratio is also comparable with felsic tuyas (Smellie, 2007, 2013). Conversely, it has a 300 m-thick basal pedestal of perlitised lobe-bearing breccias dissimilar to other lava-dominated felsic tuyas and also unlike subglacial domes (see Chapters 11 and 12). Finally, it may have been formed entirely subglacially whereas tuyas are defined by having capping lavas erupted subaerially. These conflicting characteristics illustrate that not all glaciovolcanic landforms can be neatly pigeonholed.

Morphometric data for felsic tuyas are still scarce, such that it is difficult to be definitive about diagnostic characteristics. However, the data currently available suggest that all felsic tuyas have high volume:area ratios comparable to mafic tuyas (Fig. 8.5). However, they have slightly to markedly higher aspect ratios compared with mafic tuyas (Fig. 8.5). In part this is probably due to a combination of (1) higher viscosities, (2) lower intrinsic heat, and (3) an absence of lava-fed deltas. Thus, the magma simply piles up over the vent and, buttressed by the confining walls of ice, grows up more rapidly than it grows out. Felsic tuyas also have the highest aspect ratios of any glaciovolcanic landforms measured so far, overlapping significantly only with those for subglacial felsic domes.

8.2.4 Tephra mounds and ridges

After tuyas, tephra mounds and ridges are the most commonly described glaciovolcanic landforms. Tephra mounds are cone-like edifices, i.e. tuff cones (the conical tuyas of Russell et al., 2014), whereas tephra ridges are also called tindars (Jones, 1969b; Smellie, 2000, 2013; the linear tuyas of Russell et al., 2014). Like tuyas, different types can be distinguished depending on whether the magma is mafic or felsic, although the morphological distinctions are less obvious. All form due to the explosive interaction of magma and meltwater, i.e. they are hydrovolcanic. The tephras correspond to an early 'tindar phase' of glaciovolcanic eruptions and formed prior to any development of tuya capping units (Smellie, 2000, 2013). The structure and lithofacies of mafic tephra mounds and ridges are relatively well described although there are comparatively few published examples (Skilling, 1994; Smellie and Hole, 1997; Werner and Schmincke, 1999; Smellie, 2001; Schopka et al., 2006; Russell et al., 2013; Chapter 10). By contrast, the internal structure and lithofacies of felsic tephra mounds and ridges are less well known, with fewer published examples (all occur as pedestals to tephra-dominated felsic tuyas: Tuffen et al., 2002a; Stevenson et al., 2011). The tephra is usually rather poorly exposed beneath very mobile scree, probably as a consequence of felsic sideromelane being relatively resistant to post-eruptive clay alteration and the tephra pile remaining unconsolidated (Jakobsson and Gudmundsson, 2008).

All the tephra mounds have much lower volume:area ratios than for mafic and felsic tuyas (Fig. 8.5). Ratios are lower only for the sheet-like sequences and felsic domes. Aspect ratios also tend to be relatively high, but with considerable overlap (Fig. 8.5). Felsic tephra mounds mainly have higher aspect ratios than mafic examples, although there are still very few data available. Aspect ratios are similar to compositionally analogous tuyas, in other words, mafic tephra mounds/ridges with mafic tuyas and felsic mounds/ridges with felsic tuyas. The morphometry of glaciovolcanic tephra mounds has also been compared with that for subaqueous (non-glacial) tuff cones. There is complete overlap in aspect ratios between mafic examples from the two settings, whereas felsic glaciovolcanic examples have higher aspect ratios (and therefore steeper slopes) than non-glacial tephra mounds/ridges, although the datasets are still too small to be definitive. The result for mafic examples was a surprise as it had been predicted that higher ratios should characterise glaciovolcanic tephra mounds/ridges of all compositions (Smellie, 2009; Pedersen and Grosse, 2014). The higher ratios for mafic examples could be caused by some combination of (1) tephra being buttressed by the confining ice walls (Smellie, 2000, Fig. 5); (2) rapid early diagenesis and lithification of the warm wet tephra pile, as observed in the 1996 tindar eruption of Gjálp, Iceland (Gudmundsson, 2003; Gudmundsson et al., 2004; Jarosch et al., 2008); and (3) glacial erosion of the tindar flanks (Smellie, 2013, Fig. 12). However, the observation may be a result of a small and possibly unrepresentative dataset. For example, it may include glaciovolcanic examples in which a basal broad pedestal of pillow lava, with its much lower aspect ratio, was inadvertently included when measuring the landform, thus creating unrecognised composite edifices. This compound landform is well illustrated by Ash Mountain, a glaciovolcanic tephra mound constructed on a pillow mound formed during the same eruptive episode in British Columbia (Mathews, 1947; Allen et al., 1982; Moore et al., 1995; Hickson, 2000; Russell et al., 2014; Fig. 8.7).

8.2.5 Sheet-like sequences

Glaciovolcanic sheet-like sequences are relatively thin deposits (few tens of metres) with narrow, sinuous sheet, ribbon-like or lobate outcrops that may extend 10 or more kilometres down-dip, some of which are described as esker-like (Smellie and Skilling, 1994; but see Section 8.2.6). Although not edifices, the outcrops are a distinctive sequence and glaciovolcanic landform type. The edifices coeval with sheet-like sequences are either unknown or undiagnostic and are thought to be simple pyroclastic cones (tuff cones and scoria cones), fissure-erupted tindars or even pillow ridges (Smellie, 2008).

Two types of sheet-like sequences have been defined and called Dalsheidi-type and Mt Pinafore-type after the original localities where they were first described (Smellie, 2008). Despite substantial macroscopic similarities, the two types have somewhat subtle differences in their lithofacies and, particularly, the order in which the lithofacies occur (but see Section 10.6). However, the differences do not affect their morphometry and the two types are considered here together. Compositions are mainly mafic, but examples with intermediate and even felsic compositions (trachyte, phonolite and rhyolite) are also known but rare (Edwards and Russell, 2002; Smellie et al., 2011a). Because sheet-like sequences are so thin, they are affected to a greater extent by the morphology of the underlying bedrock topography than other glaciovolcanic landforms and many are confined laterally by incised valleys (Walker and Blake, 1966; Smellie et al., 1993; Smellie and Skilling, 1994; Loughlin, 2002; Smellie 2008). This will also affect their morphometry.

Examples of glaciovolcanic sheet-like sequences vary in thickness from a few tens of metres to a few hundred metres. Widths are similarly variable, ranging from several tens of metres to kilometres. In Iceland they have been traced up to 14 km down-dip (Bergh, 1985; Bergh and Sigvaldason, 1991). They often occur in stacked units within large stratovolcanoes but are sufficiently undiagnostic in their overall appearance that they can only be interpreted as glaciovolcanic after detailed lithofacies studies. Moreover, erosive effects of wet-based ice can also significantly modify outcrop dimensions and morphology between eruptions (Fig. 8.8). Sheet-like sequences include the thinnest glaciovolcanic units known, combined with a disproportionately large areas and low total volumes. There is substantial overlap with area:volume ratios for subglacial domes and lobes.

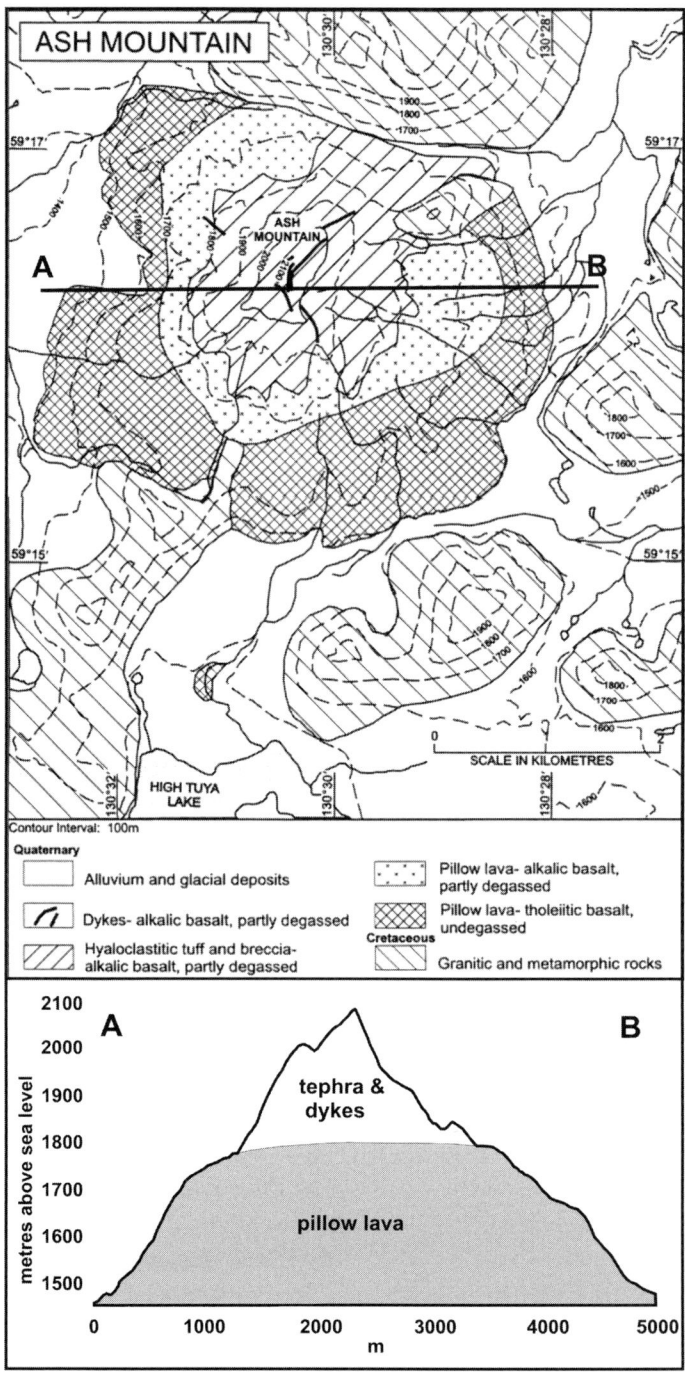

Fig. 8.7 Map (from Hickson, 2000) and cross section of Ash Mountain, British Columbia. The edifice consists of a tuff cone built on top of a lower-profile pillow volcano (mound). Note the different profiles of the two stages of coeval edifice construction, which can create problems when gathering morphometric data.

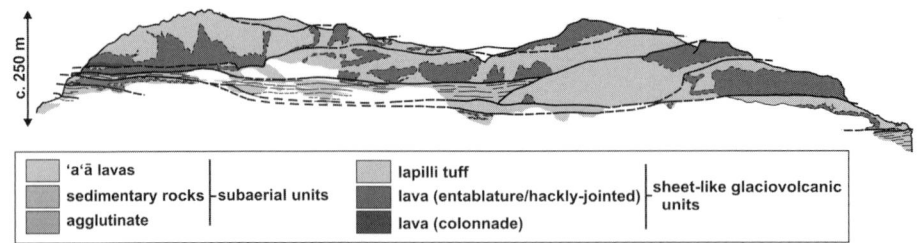

Fig. 8.8 Field sketch of alternating glaciovolcanic sheet-like sequences and subaerial lavas; Eyjafjallajökull, south Iceland (JLS, unpublished). Each of the sequences is separated by a prominent glacial erosional unconformity. The view is an oblique strike section. (A black and white version of this figure will appear in some formats. For the colour version, please refer to the plate section.)

There is also complete overlap in length, width and thickness with parameters for pillow sheets (see Sections 8.2.7 and 10.5).

8.2.6 Esker-like sequences

Some glaciovolcanic sequences have very narrow, sinuous morphologies that have been inferred to form by emplacement within esker-shaped cavities (tunnels) at the base of ice masses (Mathews, 1958; Hickson 2000; Lescinsky and Fink, 2000; Kelman et al., 2002; Hungerford et al., 2014). The sequences are dominated by either massive jointed lava (e.g. examples from the Garibaldi belt in southwestern British Columbia; Mt Rainier, USA) or pillow lava (distal parts of Tennena Cone, northwestern British Columbia). Fragmental volcanic material and glaciofluvial diamict are present locally beneath the lavas. There is a broad resemblance to some lava-dominated sheet-like sequences (cf. Walker and Blake, 1966; Smellie et al., 1993) and the distinction into a separate category of esker-like sequences is not yet firmly established.

8.2.7 Pillow mounds, ridges and sheets

All the described examples of pillow-dominated glaciovolcanic landforms are basaltic. More evolved magmas also form pillows, but only one glaciovolcanic example is known, in Iceland, of dacitic composition (Stevenson et al., 2009). Glaciovolcanic pillow mounds (often called pillow volcanoes) and pillow ridges are common (Edwards et al., 2006, 2009b; Höskuldsson et al., 2006; Carrivick et al., 2009; Jakobsson and Johnson, 2012; Pollock et al., 2014). They also typically form the cores of many tindars and tuyas (Jones, 1969a,b, 1970; Werner and Schmincke, 1999), where they and their landforms are often largely obscured from view by the overlying tephra pile. Pillow lava-dominated landforms range from low mounds with relatively steep sides and smooth profiles (Fig. 8.7) to subdued ridges with gentle slopes that were erupted from fissures. A third category comprises extensive thin pillow sheets. These are composed entirely of pillow lava and are only a few tens of metres thick but are up to 3 km wide and 15 km or more in length (Snorrason and Vilmundardóttir, 2000; Vilmundardóttir et al., 2000). Pillow sheets have only been identified in Iceland. Their lithofacies and internal structure have not yet been described and their status as a discrete glaciovolcanc landform and their mode of formation are currently uncertain (see Section 10.5).

Few morphometric data exist for glaciovolcanic pillow-dominated landforms. Aspect ratios are very low, comparable only to subglacial sheet-like sequences with which they also may be correlated (Chapter 10). They also have amongst the lowest volume:area ratios of any glaciovolcanic landforms, although the data are strongly biased towards supposed pillow sheets and thus may be unrepresentative of pillow mounds and ridges. Pedersen and Grosse (2014), on the basis of a very small dataset, did not observe any clear morphometric differences between pillow ridges and tindars, which conflicts with the observation by Smellie (2013, Figs. 7 and 12) that pillow mounds typically have low profiles whereas tuff cones/ridges (tephra-dominated) have steeper slopes probably mainly due to buttressing by enclosing ice.

8.2.8 Felsic domes and lobes

The morphometry of felsic domes and lobes is not well known. The greatest number of felsic domes has been described by Kelman (2005) in British Columbia. Compositionally they range from andesite to rhyolite and trachyte–phonolite but andesite examples are most common (Edwards and Russell, 2002; Kelman et al., 2002; Tuffen et al., 2001, 2002a; Kelman, 2005; Section 11.3). Because they have undergone variable glacial erosion, the original dome shapes are generally inferred from the pattern of radiating cooling joints and associated glassy margins, interpreted as ice-contact features within subglacial cavities (Tuffen et al., 2002b; Kelman, 2005). The outcrops are dominated by felsic lava and fragmental rocks are typically scarce to absent. The felsic domes are amongst the smallest glaciovolcanic landforms, with widths much less than 1 km and heights over 300 m, overlapping in height only with sheet-like sequences and pillow sheets. However, they are distinguished easily from these by the very small areas of the felsic domes. Unsurprisingly, they generally have the smallest volumes and they also have somewhat higher aspect ratios than the mafic landforms whilst overlapping with most other felsic landform types.

8.3 Glaciovolcanic landforms constructed under cold-based ice

All of the above is based on monogenetic volcanic landforms constructed in association with wet-based (warm or polythermal) ice, in which the hydraulics of basal escaping meltwater exerts a major influence on edifice morphometry. By contrast, cold ice is frozen to its bed. In addition, the rheology, thermal requirements of melting and the meltwater hydraulics are very different (Chapter 4). Since those properties will strongly influence subglacial volcano construction (Smellie, 2009), it is worthwhile considering whether the glaciovolcanic landforms will be significantly different. The only published examples of glaciovolcanic edifices constructed under cold ice are several large polygenetic central volcanoes in Antarctica (Wilch and McIntosh, 2002; Smellie et al., 2011a,b, 2013a; Section 8.2.1, above). They were constructed beneath a thin draping ice cover that resulted in shield-like profiles. No monogenetic edifices have been described. The following description is for monogenetic edifices only.

Cold ice is below the pressure melting point and is frozen to its bed. Thus, no basal meltwater is present and any generated during eruptions is unable to escape except by overflowing as it is displaced upward by the addition of the newly formed volcanic products. Moreover, because of the increased energy required to melt cold ice (Section 4.3), the walls of the englacial cavity will melt back at a slower rate. The combination of these characteristics will result in the cavity rapidly becoming filled by volcanic materials, with the water displaced up and overflowing supraglacially. Lava-fed deltas emplaced in cold ice may overrun onto the ice surface because of the (faster) rates of lava advance relative to the (slower) rate of ice melting (Chapter 6). After eruptions cease, normal ice flow will advect away the supraglacially deposited lithofacies or else the deposits will collapse to form a pile of rubble if the ice melts due to climatic factors

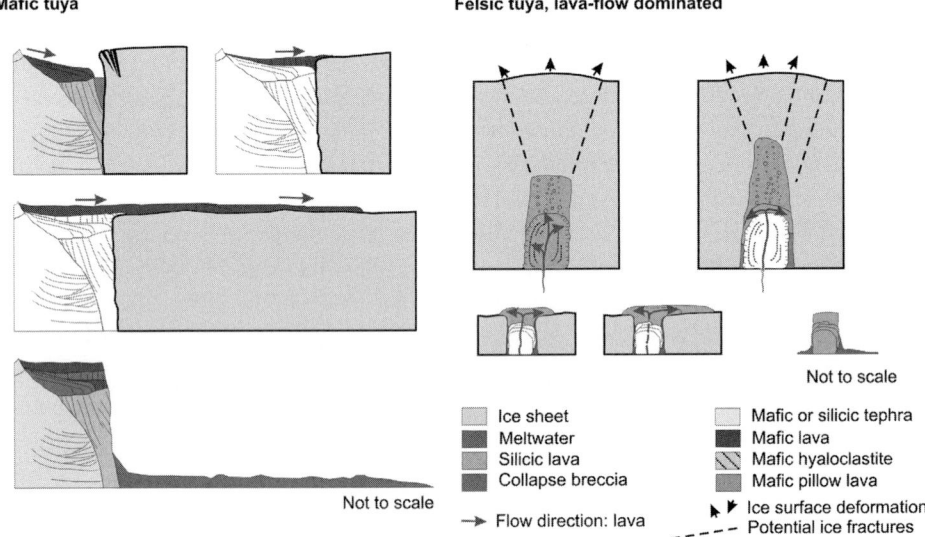

Fig. 8.9 Sketches illustrating postulated glaciovolcanic landforms and principal constituent lithofacies for eruptions associated with cold-based ice (modified after Smellie, 2013). The major differences from comparable landforms formed under wet-based ice are that lava-fed deltas will probably be shorter and the edifices taller. (A black and white version of this figure will appear in some formats. For the colour version, please refer to the plate section.)

(perhaps less likely in view of the timescales involved for climate change compared with eruption durations). Thus, the volcanic edifices will be tall and narrow, i.e. aspect ratios are likely to be high irrespective of the mode of formation of these edifices (i.e. whether as domes, tuyas, tindars or pillow mounds). The ratios will probably be amongst the highest of any glaciovolcanic landform and similar to or greater than those for felsic domes and tuyas irrespective of magma composition (Fig. 8.9) although they will get modified by post-glacial gravitational collapse. Sheet-like sequences cannot form in cold ice.

9

Lithofacies in glaciovolcanic sequences

9.1 Introduction

Understanding glaciovolcanic systems relies on a detailed knowledge of the spatial and temporal distribution of the volcanic lithofacies produced and their likely origins. Without that knowledge, our ability to interpret correctly both the eruptive record and the preserved history of former glaciers and palaeoclimate is severely impaired (see Chapter 13). A consistent approach using rigorous lithofacies criteria is also important to enable direct comparisons between field sites by different workers. The objective of this chapter is to provide basic descriptions and interpretations of the major volcanogenic lithofacies that together make up glaciovolcanic sequences.

9.2 Terminology

A lithofacies is a deposit that has some combination of distinctive characteristics (e.g. stratification; grain types, sizes and shapes; fabric; sorting; fossils; jointing; colour; and composition) that taken together bear a direct relationship to the depositional processes that produced them and help to interpret the depositional environment. Lithofacies can be broken down into smaller and smaller units depending on the purposes of the study, and it is certain that authors will modify and further subdivide the lithofacies described in this chapter. The range of lithofacies included here is not intended to be exhaustive, nor is it prescriptive. However, we describe the major volcanogenic lithofacies likely to be encountered in glaciovolcanic sequences and provide a list of informal codes that can be used as a practical aid to rapid logging in the field (Table 9.1). In the code notations, the rock type (e.g. lapilli tuff, breccia) is given greatest prominence using capital letters. Following common notational procedures (e.g. Branney and Kokelaar, 2002), the capital letter is preceded by the most significant distinguishing characteristic (e.g. graded (g), diffuse-bedded (db)) using lower-case letters. If desired, further distinguishing characteristics, such as presence of abundant obsidian clasts or pillows or style of grading (inverse, normal), can be added as a subscript in parentheses. Thus, $gLT_{(p)}$ would denote a graded lapilli tuff with dispersed pillows. If the same lithofacies occurs repeatedly within a stratigraphical sequence, numerals can also follow the distinguishing features to denote the relative position of a lithofacies (e.g. $gLT_{(p)}1$ versus $gLT_{(p)}2$; e.g. Edwards et al., 2009b).

Table 9.1 *Primary volcanic lithofacies found in glaciovolcanic successions and suggested informal codes that can be used as a practical aid to rapid logging in the field*

1 Coherent lithofacies			
Lithofacies	Code	Comments	Interpretation
Pillow lava	pL	Mafic(-intermediate) compositions	Subaqueous lava effusion
Compound lava (pāhoehoe)	cL	Mafic(-intermediate) compositions	Subaerial (air-cooled) lava; low gradients and discharge rates
Sheet lava, coarsely jointed	sL	Sheet- or ribbon-like; mafic(-intermediate) compositions	Subaerial (air-cooled) lava; steeper gradients and higher discharge rates than cL
	dL	Lava dome; (intermediate-)felsic compositions	Distinguished by roughly equant dome shape
	cL	Coulée lava; (intermediate-) felsic compositions	Short stubby lava flow more elongate than a dome
Sheet lava, blocky or curvicolumnar jointed	jL	Mafic(-intermediate) compositions	Water-cooled lava
Felsic lava	fL	Type of sheet lava but texturally distinguishable from mafic and most intermediate lavas; includes some intermediate-composition lavas	Subaerial and subglacial lava effusion
Lava, clastogenic (agglutinate)	aL	Mafic(-intermediate) compositions	Mainly vent-proximal products of dry (subaerial) magmatic eruptions mainly of Hawaiian type, including fire-fountaining; found in crater walls
Intrusive rocks	dL	Dyke	Hypabyssal intrusions with different shapes emplaced in the volcanic edifice
	sL	Sill	
	iL	Irregular intrusion	

2 Fragmental (volcaniclastic) lithofacies				
Lithofacies[a]	Code	Description	Comments	Interpretation
Hyaloclastite	H	Massive	Mainly glassy clasts, typically non- or poorly vesicular and blocky	Found close to host lava; generated *in situ* by quench fracturing and spalling; essentially an autobreccia formed by water-chilled lava (cf. cmB)

Table 9.1　(*cont.*)

2 Fragmental (volcaniclastic) lithofacies				
Lithofacies	Code	Description	Comments	Interpretation
Breccia/tuff breccia	cB	Clast-supported, massive	Mainly lithic clasts	Autobreccias formed by mechanical disruption of chilled crust during flow of subaerial lava (cf. H)
	B, TB	Matrix-supported, massive	Mainly lithic clasts	Variety of mass flow deposits generated by slumping and transported in concentrated and hyperconcentrated density currents; includes foreset beds with glassy matrices in lava-fed deltas
	clTB	Clast-supported, laminated in pore spaces	Mainly lithic clasts	Avalanched or slumped deposits with clasts sourced in pillow pile form the breccia framework; laminated tuffs present in pore spaces are sieve deposits
	gB, gTB	Graded	Mainly lithic clasts	Generated by avalanching and slumping; transported in concentrated and hyperconcentrated density currents (e.g. delta foreset beds)
Lapilli tuff [Lapilli ash]	LT	Massive		Eruption-fed concentrated density current deposits
	bLT	Thin-bedded	Individual beds < 10 cm thick	Eruption-fed concentrated density current deposits
	dbLT	Diffuse-bedded		Eruption-fed concentrated density current deposits
	dsLT	Diffuse-stratified		Eruption-fed concentrated density current deposits
	glLT	Graded, laminated	May contain slabby or contorted intra-clasts of bT	Eruption-fed concentrated density current deposits
	xLT	Cross-stratified		(1) Low-angle foresets: traction deposits associated with subaerial pyroclastic density currents; (2) high-angle foresets: traction deposits associated with subaqueous concentrated density currents or aqueous current reworking

Table 9.1 (*cont.*)

		2 Fragmental (volcaniclastic) lithofacies		
Lithofacies	Code	Description	Comments	Interpretation
Tuff [Ash]	T	Massive		Eruption-fed concentrated density current deposits and turbidites
	bT	Thin-bedded	Individual beds < 10 cm thick	Eruption-fed concentrated density current deposits and turbidites
	dbT	Diffuse-bedded		Eruption-fed concentrated density current deposits
	dsT	Diffuse-stratified		Eruption-fed concentrated density current deposits
	glT	Graded, laminated	Mainly fine–medium tuffs	Turbidites
	xT	Cross-stratified		(1) Low-angle foresets: traction deposits associated with subaerial pyroclastic density currents; (2) high-angle foresets: traction deposits associated with subaqueous concentrated density currents or aqueous current reworking
Lapillistone [Lapilli]	LP	Massive	May be intercalated with agglutinate and clastogenic lava	Vent-proximal fall deposits found mainly in the crater walls and on flanks, the products of dry magmatic eruptions mainly of Strombolian type
	gLP	Graded	As for LP; inverse grading	As for LP, modified by post-depositional runout on steep cone flanks

Other lithofacies modifiers:

  (dw) Dewatering structures
  (p) Lava pillows
  (n) Normal grading
  (i) Inverse grading
  (b) Bombs
  (o) Obsidian
[a] unconsolidated equivalents in parentheses

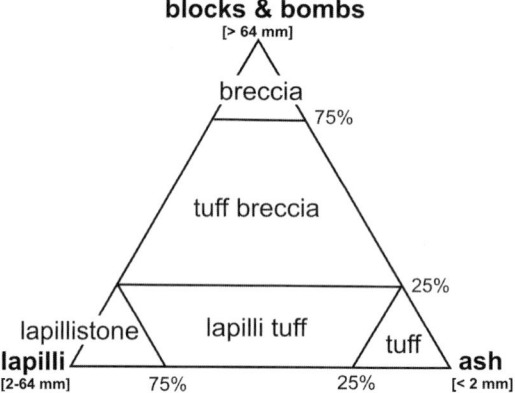

Fig. 9.1 Grain-size ternary diagram illustrating nomenclature used in this book for primary volcaniclastic rocks (modified after White and Houghton, 2006).

The terminology adopted is intended to be non-genetic and not inextricably linked to a local stratigraphy (e.g. móberg; see Glossary). We have divided the volcanic products of glaciovolcanic eruptions into two primary lithofacies groups: coherent and fragmental. *Coherent* lithofacies consist of lavas and hypabyssal intrusions, whilst *fragmental* lithofacies comprise the wide range of fragmental products created during explosive or effusive eruptions. The latter are also called volcaniclastic and we adopt the criteria and classification of White and Houghton (2006) for distinguishing volcaniclastic from epiclastic lithofacies. The grain size classification of White and Houghton (2006), with its additional category of extremely fine ash (<1/16 mm) corresponding to sedimentary mud (i.e. silt and clay), is also used here. However, we differ in one respect from White and Houghton (2006) in that we regard lapillistone as a valid and useful grain size category for volcaniclastic deposits formed of more than 75% lapilli (Fisher, 1961; Fig. 9.1). Lithofacies descriptors for tuff and lapilli tuff are mainly after Branney and Kokelaar (2002).

Epiclastic lithofacies are sedimentary deposits that form by the physical weathering and erosion of pre-existing rocks. Many texts describe the physical characteristics used in their recognition, for interpreting their mode of origin, and in their use for reconstructing past glacial environments. The epiclastic lithofacies most relevant to glaciovolcanism are those associated with glacierised terrains (e.g. Menzies, 1995, 1996; Hambrey and Glasser, 2003; Evans and Benn, 2004). The reader is referred to those publications for relevant details.

9.3 Primary coherent lithofacies

9.3.1 Pillow lava

Pillow lava (lithofacies pL; Table 9.1) is an important coherent glaciovolcanic lithofacies as it is a standalone indicator of the presence of water during lava effusion.

It has been recognised as a distinct lava form since the early 1900s, and was critical to interpretations of glaciovolcanic origins for early studies in Iceland (Peacock, 1926) and British Columbia (Mathews, 1947). Walker (1992) presented a summary of pillow lava textures for submarine and subglacial pillow lavas, and subsequent studies have described the characteristics of glaciovolcanic pillow lava textures, their stratigraphy and geochemistry (Höskuldsson et al., 2006; Edwards et al., 2009b; Pollock et al., 2014). A lava pillow has a tube shape (ovoid (pillow-like) in transverse section) and is emplaced beneath water. It can be very similar in appearance to a subaerial pāhoehoe lava lobe, but pillow lavas generally have a greater height to width ratio due to the buoyancy effects of the surrounding water (Fig. 9.2a; see Section 13.4.3). Individual lava pillows can range from about 1 m to 10 m in length, and from 0.2 m to 10 m in diameter, depending on their bulk composition (Walker, 1992; Fig. 9.2b). Pillow lavas generally have a vitric rim between 1 and 3 cm in thickness that forms by rapid cooling of the lava in water. They can show distinctive patterns of vesicle distributions, including concentric layers of vesicles or pipe vesicles. While a majority of glacio-volcanic pillow lavas are basaltic, there are also andesitic (Stevenson et al., 2011) and trachytic (LeMasurier, 2002) occurrences. However, because pillow lavas can form wherever subaerial lava flows enter rivers or lakes, and they are ubiquitous in marine

Fig. 9.2 Characteristics of glaciovolcanic pillow lavas. (a) Sketch and cross section of a pillow lava tube. (b) Glaciovolcanic lava pillows showing a broad range of sizes; Undirhlíðar quarry, Reykjanes Peninsula, Iceland. Note the thick chilled margins. The hammer is c. 40 cm long. (c) Photograph of a lava pillow with pipe vesicles; Reykjanes Peninsula, Iceland.

settings, independent proof of a glaciovolcanic emplacement is required (see Chapter 13). In at least one location (Hlödufell; Skilling 2009), pillows with flattened deformed shapes have been interpreted as indications of direct contact with ice. Unusual vertical pillow lava lobes have also been identified but their origin is less clear: they may form by emplacement into ice cavities, or they may simply be a response to emplacement on vertical faces within water (Hungerford et al., 2014).

Pillow lava is found in several different parts of glaciovolcanic sequences. Most general models for the evolution of glaciovolcanoes suggest that eruptions in thick ice will form pillow lavas initially (see Chapter 1), and most tindars and tuyas appear to have a basal pillow lava mound (e.g. Jones, 1969a,b; Skilling, 1994, 2009; Werner and Schmincke, 1999; Edwards et al., 2009b; Fig. 8.7). Recent work (Pollock et al., 2014) has shown that thick sequences of stacked pillow lavas can be separated into units that are geochemically distinct, indicating that they were likely emplaced during discrete eruption episodes. However, it is difficult to estimate the elapsed time between eruptions. Individual pillows and small pillow lobes are also found in the upper sections of subaqueous fragmental rocks (e.g. in tindars). The origin of those pillows is unclear. They could form by short-lived eruptions of degassed lava from the summit crater or perhaps dyke-fed onto the edifice flanks, or they could form as intrusive piles within the fragmental deposits. Lava pillows can also form at the subaerial–subaqueous transition (passage zone) where largely degassed subaerial lava flows enter water. Theoretically, it should be possible to distinguish between pillows extruded subaqueously from those that entered water subaerially by measuring the H_2O content of the glassy pillow rims (see Chapters 5–7).

Studies of submarine lava flows have provided important insights applicable to glaciovolcanic pillow lava formation, with the slope and effusion rate regarded as likely key parameters for determining submarine lava flow morphology (Bonatti and Harrison, 1988; Gregg and Fink, 1995, 2000). Recent work on glaciovolcanic examples in British Columbia (Hungerford et al., 2014) and Iceland (Pollock et al., 2014) has also identified multiple pillow lithofacies with morphological variations that may be related to slope and effusion rates. In the submarine environment, lava flow transitions between sheet, lobate and pillowed morphologies are thought to be linked to variable effusion rates (e.g. Griffiths and Fink, 1992) and similar reasoning may apply to glaciovolcanic examples. Early workers looking at submarine pillows also noted that the sizes of vesicles approximated empirically with inferred emplacement depth (Jones, 1969a). Changes in vesiculation patterns within glaciovolcanic pillow lavas may also record sudden, large pressure changes related to rapid water-drainage events (Höskuldsson et al., 2006).

9.3.2 *Compound lava (pāhoehoe)*

Compound lava (lithofacies cL; also called pāhoehoe; Walker, 1970) is a prominent lithofacies in glaciovolcanic sequences as it is a major component of the subaerial capping lavas at most mafic tuyas (see Chapter 10). It forms by subaerial effusion of mainly basaltic lava, but some more evolved lavas occasionally form pāhoehoe (e.g. phonolite on Mt Erebus, Antarctica; trachyte at Hoodoo Mountain, British Columbia). Pāhoehoe lava flows mainly form tabular sheets with thicknesses of several metres increasing (in glacio-volcanic successions) to over 150 m close to the eruptive source (Smellie et al., 2008). Some volcanic shields, formed entirely of pāhoehoe, are monogenetic and may exceed 600 m in thickness (e.g. Skjaldbreidur, Iceland; Walker, 1970). Pāhoehoe also includes the very voluminous and thick compound lavas that characterise flood basalt provinces (e.g. Self et al., 1998; Thordarson and Self, 1998). Pāhoehoe is constructed from multiple overlapping coeval flow units called lobes that, in cross section, have pillow-like shapes often called pāhoehoe toes (Fig. 9.3); they are collectively called lobes here. The lobes are finely to coarsely crystalline but typically have a thin ($\leq$ 1 cm) glassy surface rind. The glassy rinds are locally incomplete probably due to differences in cooling rates around the circumference during lobe effusion. The surfaces of individual lobes are smooth or less commonly ropy and they may show red oxidation. Vesicles are usually abundant, including pipe vesicles at lobe bases. The vesicles are characteristically spherical or ovoid, unlike vesicles in ʻaʻā lava, which generally have stretched out, irregular shapes. Vesicles in pāhoehoe may also occur in discrete zones a few centimetres to a metre thick related to flow inflation (Thordarson and Self, 1998; Umino et al., 2006). Despite broad similarities to pillow lava, pāhoehoe lobes are often noticeably more flattened than lava pillows probably owing to the absence of buoyancy effects associated with lava effusion in water (Smellie et al., 2013a; cf. Walker, 1970; cf. Figs. 9.2b and 9.3;

Fig. 9.3 View of pāhoehoe lava showing development of active flow lobes in a lava at Kilauea, Hawaii. Note the flattened profiles of the lobes, which contrasts with the more spherical profiles of lava pillows (cf. Fig. 9.2b).

Section 13.4.3) and the glassy rinds may be thinner and less complete. The diameters of pāhoehoe lobes are typically at least 1 m (< c. 10 cm in entrail pāhoehoe) but, with inflation, can be much greater (>50 m; Thordarson and Self, 1998). The thicker lava lobes may be coarsely columnar jointed and have several distinct platy jointed horizons individually up to 1 m thick. Entablature is absent except where lavas have been flooded by water (Long and Wood, 1986). Apart from rare rubbly zones (flow tops) known as slabby pāhoehoe formed of ≤1 m-long slabs of pāhoehoe crust, individual pāhoehoe lava lobes are not associated with other lithofacies.

9.3.3 Sheet lava

Sheet lavas are simple lavas, i.e. they are not divisible into multiple flow units derived from a single eruptive event. They have low aspect ratios (i.e. thickness to width or length) and are tabular or ribbon-like in form (collectively called sheet-like). The most common sheet lavas are ʻaʻā lava although some ʻaʻā lavas comprise more than one superimposed flow unit and are thus compound (Walker, 1970). Thicknesses are typically a few metres to a few tens of metres, but they increase from mafic to more evolved lavas. Lengths are a few hundred metres to a few tens of kilometres. Two broad types of sheet lava are identified here: coarsely jointed and blocky- or curvicolumnar-jointed.

Coarsely jointed sheet lava (lithofacies sL) encompasses most ʻaʻā and block lavas. ʻAʻā lavas are mainly mafic and are distinguished by prominent scoriaceous autobreccia surfaces, often red coloured due to subaerial oxidation, that alternate with sheets of massive lava, usually coloured grey or grey-brown (Fig. 9.4). Block lavas are principally andesite and dacite in composition and are internally structured similar to ʻaʻā lavas, comprising a core of platy or columnar-jointed massive lava encased in coarse rubble autobreccia formed of angular, usually dense lava blocks. The lavas are variably fine grained to coarsely crystalline due to relatively slow cooling in air, although intermediate lavas can be aphanitic or even glassy due to the shorter interval between the liquidus and glass transition temperatures and the greater ease with which they form glass (Chapter 5). Unlike compound (pāhoehoe) lavas, vesicles in ʻaʻā lavas are generally stretched by internal flow, becoming irregular and elongate in the process. Mafic and many inter-mediate lavas are characterised by irregular, steeply dipping joints and columnar jointing of colonnade type, both types spaced on a metre scale (Fig. 9.7a). Intermediate lavas commonly show curviplanar sheet-like joints (known as platy joints) parallel to the lava surfaces that curve up particularly at the flow margins (ramp structure) and are seen on the flow top as prominent jagged ridges separated by zones of clinkers. Several different types of mafic ʻaʻā lava have been described based on their surface textures. They include slabby (proximal), scoriaceous, clinker and blocky (distal) ʻaʻā, whose textures form progressively downslope as a result of cooling and degassing in the main lava stream (Lipman and Banks, 1987). Slabby ʻaʻā, formed of disrupted slabs of pāhoehoe crust, is transitional to pāhoehoe and is also called slabby pāhoehoe. It is created where the flow

Fig. 9.4 Subaerial 'a'ā lava flow migrating over the surface of snow in April 2013 during the Tolbachik eruption, Kamchatka, Russia. The lava is covered by a thick layer of autobreccia that encloses a massive (molten) core. The height of the flow front is c. 4 m.

rate of the lava is too great for the cooled crust to deform plastically and it is fractured and disrupted into thick slabby plates. Scoriaceous 'a'ā comprises abundant highly vesicular and often oxidised scoria separated by smoother crust caused by stretched and upwelling lava. Clinkery 'a'ā lacks the patches of smooth crust present in slabby 'a'ā and is covered by lava clinkers that are less vesicular and oxidised (mainly grey in colour) than in scoriaceous 'a'ā. Finally, blocky 'a'ā resembles block lava although the surfaces of the blocks are less smooth due to vesiculation.

Blocky- or curvicolumnar-jointed sheet lava (lithofacies jL) differs from subaerial sheet lava in being mainly aphanitic to fine grained due to more rapid cooling and glassy rims are common. In addition, it is characterised by a variety of much more closely spaced cooling fractures such as blocky or hackly joints, spectacular narrow (typically 20–30 cm diameter) curvicolumnar joints known as entablature, pseudopillow joints and small column-on-column joints (Figs. 9.7b, c; e.g. Lescinsky and Fink, 2000; Lodge and Lescinsky, 2009a; see Glossary for a description of these joint types, and Section 9.3.7, below). Rapid chilling by water or ice is indicated by these and other features (see also Chapter 13). The lava may also have a basal zone of colonnade prisms that are much wider (c. 1 m diameter) but the zone is always suppressed in thickness relative to the entablature and may be absent (Long and Wood, 1986). Lavas that flow into water are also often thicker and poorly to non-vesicular compared with their subaerial equivalents due to the buoyancy and pressure effects of the overlying water causing magmatic volatiles to shrink or stay in solution (Smellie et al., 2013a; see Section 13.4.3). They may also be associated with poorly formed lava pillows and hyaloclastite at the lava margins.

9.3.4 High-viscosity lava (mainly felsic)

Whereas lavas with broadly felsic compositions can also form sheet lavas, they show several prominent textural features that distinguish them from mafic and intermediate lavas. In addition they tend to form domes (dL) and short stubby lavas known as coulées (cL). Felsic lavas have a simple textural stratigraphy controlled principally by some combination of cooling, fracturing and migration of gases exsolved during crystallisation. A typical subaerial rhyolite lava comprises, from base up: a basal breccia zone, either hyaloclastic or lithic; a central zone of crystalline lava flanked by obsidian layers; and a complex upper zone of coarse and fine pumice alternating with obsidian (Fink and Manley, 1987). The upper pumice breccia may be oxidised. Lithophysae and spherulites may be abundant especially in the obsidian layers. Lava thicknesses vary from less than ten metres to a few hundred metres (generally <200 m) with lateral extents of a few tens of kilometres (<50 km; Bonnichsen and Kauffman, 1987). The lava cores are typically 25–150 m thick and microcrystalline with a sugary texture related to devitrification. Widely spaced columnar cooling joints are crudely developed in the crystalline interior flanked by subhorizontal platy joints, whereas well-formed columnar jointing is conspicuous in marginal obsidian, occurring as fanning and horizontal columns sometimes associated with jigsaw fractures (Fig. 9.5). The lavas commonly show flow bands, which are particularly well seen in the obsidian zones. The flow banding is usually subhorizontal and similar to the lava orientation but it may steepen up through the lava and at steep lava margins, similar to platy joints. It may also be tightly folded or even contorted. Wedges of breccia encased by steep planar joints are also sometimes present in the basal parts and resemble spiracles (i.e. structures filled by fragmental material injected into a lava flow due to the explosive expansion of water to steam; Smellie et al., 2011a). They are a product of bulk interaction steam explosivity (Kokelaar, 1986). The combination of features suggests that many lavas were emplaced subaerially (i.e. those with oxidised autobreccia) but they commonly flowed on a wet substrate and were confined by ice surfaces. Other than oxidation of the upper pumiceous zone (Bonnichsen and Kauffman, 1987) in subaerial felsic lavas, few textures are unique to subaerial and subglacial examples. The presence of horizontal columnar joints signifying cooling against a steep ice surface and association with coeval till or a glacially modified substrate are probably the best criteria for glacial emplacement (Edwards et al., 2002; Tuffen et al., 2002a,b; Stevenson et al., 2006; McGarvie et al., 2007; Smellie et al., 2011a).

9.3.5 Clastogenic lava (agglutinate)

Clastogenic lava and closely related agglutinate (collectively grouped as lithofacies aL; cf. Skilling, 2009) are uncommon lithofacies in glaciovolcanic successions. They are restricted to subaerial vent-proximal locations where they occur as agglutinate containing thin (<10 cm to few metres thick) impersistent lenses and short dribbles of clastogenic lava showing faint ghost-like outlines of spatter clasts (Skilling, 2009). Reddening due to oxidation is common. Moderately dipping agglutinate has also been described on top of

Fig. 9.5 View of shallow-dipping columnar joints in the surface of a rhyolite lava intruded subglacially at Godafjall, Öraefajökull, Iceland.

a 25 m-thick trachydacite glaciovolcanic lava (Stevenson et al., 2006). The agglutinate is situated directly above a basal zone that shows evidence for local interaction of the lava with underlying pumiceous ash, with the ash locally back-injected upward a few metres into the lava. Formation of the agglutinate was interpreted to be a result of basal water, either in moist ash or melted snow patches, flashing to steam and explosively expanding into the overlying thick trachydacite, fragmenting and ejecting the molten lava interior in a manner akin to short-lived rootless cones in mafic lavas (cf. Fagents et al., 2002).

9.3.6 Hypabyssal intrusions

A wide range of intrusions has been described within glaciovolcanic sequences (e.g. Smellie, 2008; Edwards et al., 2009b; Pollock et al., 2014), but, apart from that by Smellie (2008), no

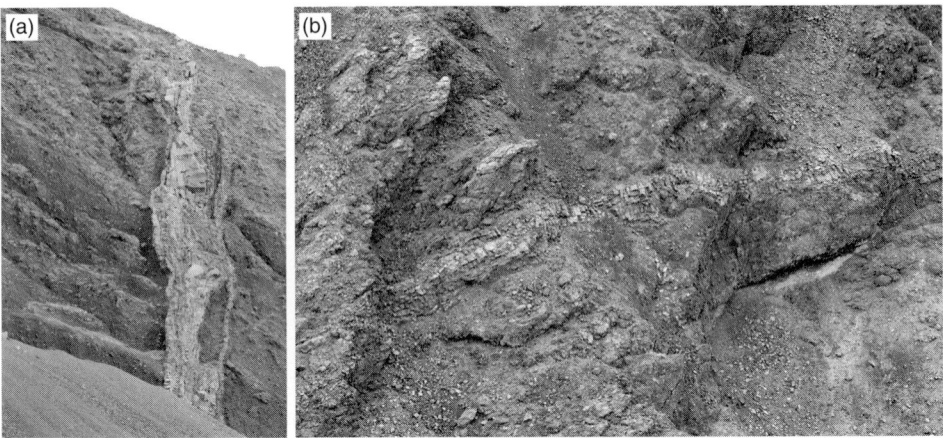

Fig. 9.6 Dykes and sills within glaciovolcanic deposits. (a) Dykes intruding tuff breccia at Lambafell quarry, southwestern Iceland. The dyke zone is 1.5 m wide. (b) Two metre-thick sill cutting through tuff breccia at Pillow Ridge, northern British Columbia, Canada.

systematic studies have been published that document how they formed. However, shallow-level intrusions that have been documented include dykes (dL), sills (sL), irregular intrusive masses (iL), and intrusive pillow lavas (Fig. 9.6). Intrusions can provide clues to the spatial construction of a glaciovolcanic edifice, the state of water saturation or strength of intruded fragmental material, and the local or regional stress field present. Glaciovolcanic intrusions are generally considered to be syn-eruptive features, as demonstrated by features such as marginal disruption into hyaloclastite and bulbous or pillow-like margins, signifying intrusion into a soft (unconsolidated) host. Intrusions are characteristically strongly dissected by cooling joints (see Section 9.3.7 below), and typically vary from 0.5 to 2 m in spacing. Where the host material is semi-consolidated or under lithostatic load, contacts are more likely to be linear (Fig. 9.6a). However, they are more likely to be wavy or contorted in a water-saturated or unconsolidated host (Edwards et al., 2009b; Graettinger et al., 2013c). The intrusion margins are characteristically finer grained, either aphanitic or vitric, and the host material locally shows thermal effects from the emplacement and cooling of the intrusion. At Pillow Ridge (British Columbia; Edwards et al., 2009b), a wide variety of intrusions were identified including dykes that are parallel to the ridge axis (probably parallel to the erupting fissure) and dykes that are essentially perpendicular. Some lava-fed deltas are also associated with an unusual form of dyke known as surface-fed dykes (Behncke, 2004). These dykes are injected down into the foreset breccias from the subaerial capping lavas, along fractures in the delta created during its advance. They terminate at the base of the delta without intruding the local basement (volcanic or otherwise).

Irregularly shaped intrusions are also common in glaciovolcanoes, but their origins are often less obvious. Most are mafic but irregular intrusions are also common in intermediate and felsic glaciovolcanoes (Tuffen et al., 2001; Stevenson et al., 2009, 2011). Massive columnar-jointed mafic intrusions were described in a quarry along Undirhlíðar ridge in

Iceland that may be the crystallised remnants of the within-ridge magma transport system (Pollock et al., 2014). Areally extensive intrusions are also ubiquitous in widespread mafic deposits dominated by fragmental lithofacies along the southern coast of Iceland, variously interpreted as glaciomarine (Bergh and Sigvaldasson, 1991) or subglacial sills (Smellie, 2008). Although sill-like in overall form, their upper surfaces are also characterised by multiple irregular apophyses (see Sections 10.6.2–10.6.4).

9.3.7 Cooling joints in lavas

The fracture patterns formed by thermal contraction in lavas ('cooling joints', columnar joints, polygonal joints, etc.) are amongst the most important features in glaciovolcanic successions. Their presence is frequently a diagnostic or common feature used to distinguish lava lithofacies and their eruptive environment. It is important to be able to recognise and interpret the origins of the different types present. Joints are environmentally diagnostic and can be used to infer either slow subaerial cooling or rapid chilling by meltwater or abutting against ice. In subaerial lavas, cooling joints are generally more widely spaced (usually >0.5 m apart, and up to 3 m; Long and Wood, 1986; Spörli and Rowland, 2006) than in water-cooled lavas (cf. Fig. 9.7). They range from irregular to well-formed prisms or columns, which are mainly hexagonal but can also be pentagonal or heptagonal. When the joints are orientated perpendicular to the presumed cooling surface and are generally parallel to each other they are frequently referred to as a colonnade, and they are inferred to form by the ordered expansion of cracks during slow conductive cooling of lava in air on a dry substrate (Grossenbacher and McDuffie, 1995; Goehring and Morris, 2008). The column surfaces show features related to the intermittent growth of the joints, mainly sets of chisel-like marks orientated perpendicular to the columns known as striae. Striae are several millimetres to a few centimetres in width and widths may vary with column diameter. For example, Lescinsky and Fink (2000) described columns 4 cm in diameter with striae 0.3–1 cm wide, and columns 30 cm across with striae 2.5–3 cm in width (also Goehring and Morris, 2008). Crack propagation also creates faint plumose structures on the striae surfaces whose asymmetrical shapes enable a joint propagation direction to be deduced (DeGraff and Aydin, 1987; Goehring and Morris 2008). The lava in which the columns are present is also typically fully crystalline and it may be quite coarse grained if it is a thick lava, consistent with slow cooling.

By contrast, lavas that are rapidly chilled by interaction with water (including meltwater) or by abutting against ice show a wide variety of cooling fractures, including curvicolumnar (also called entablature or kubbaberg), sheet-like, hackly, blocky and pseudopillow joints (Saemundsson, 1970; Long and Wood, 1986; Lescinsky and Fink, 2000; Lodge and Lescinsky, 2009a; Forbes et al., 2012; Fig. 9.7). They are described in the Glossary. Pseudopillow and sheet-like fractures were collectively called linear fractures by Lodge and Lescinsky (2009a). Entablature is the best known feature and it forms where water floods a lava surface or where lava directly abuts confining ice. The surfaces of the columns

Fig. 9.7 (a) Widely spaced, vertical columnar jointing in subaerial lava flow (sL), Gardner, Montana, USA. The columns are c. 0.8–1 m wide. (b) Blocky joints in basalt at Vermahlid, Eyjafjallajökull, Iceland. The notebook is 17 cm long. (c) Spectacular entablature joints and minor colonnade (by figure) in a basalt sill ('interface sill') at Seljaland, south Iceland (image: Anna Nelson). (d) Pseudopillow fractures in andesite lava, Cerro Blanco, Nevados de Chillán Volcanic Complex, Chile. The spectacles are c. 15 cm across (image: Katy Mee).

in entablature are typically aphanitic to glassy. Mathews (1951) was the first to describe glassy lava margins with subhorizontal columnar joints, which he ascribed to the lava quenching against an ice surface. The same feature has also been noted and similarly interpreted by many other authors, especially for felsic glaciovolcanic sequences (e.g. Tuffen et al., 2001; Edwards et al., 2002; Kelman, 2005; Stevenson et al., 2006; McGarvie et al., 2007). However, Harder and Russell (2007) suggested that a layer of rubble may intervene in some cases (also Kelman, 2005, Fig. 4.15), and Lescinsky and Fink (2000) documented narrow zones of hyaloclastite and hackly and sheet-like fractures outboard of any polygonal joints. It is not yet known whether these additional fracture zones and lithofacies are ubiquitous or only locally developed, as the evidence is often removed by erosion (Lodge and Lescinsky, 2009a). Long and Wood (1986) were amongst the first to suggest that entablature jointing was evidence for more rapid cooling, and they attributed its formation to infiltration by water. They noted that average sizes of groundmass crystals were smaller in lava entablatures and coarser in colonnades, which was also consistent with entablature joints having formed in a more rapidly cooled environment.

The sequence of fracture morphologies that develops around lava chilled by ice or meltwater was investigated by Lescinsky and Fink (2000), mainly for intermediate-composition lavas. In addition to locally developed hyaloclastite (which they described as shard-like fractures) sometimes associated with intact or broken lava pillows, they described a gradational sequence that includes hackly, sheet-like, pseudopillows and coarse and narrow columns (Fig. 9.8). Intersecting irregularly orientated chaotic fractures at the lava margins combine to create an appearance of irregular blocks or splinters on a rock surface and are grouped as hackly fractures. They are often removed by glacial erosion (Fig. 9.8a). The fracture spacing varies from sub-centimetre to tens of centimetres over short distances (<1 m). They are also prominent radiating away from small voids (e.g. ice-block meltout cavities; Figs. 9.8b, 9.9). Further into the lava, sheet-like fractures succeed the hackly fractures. They are parallel sets of long fractures spaced at c. 10 cm apart and up to c. 5 m in traceable length that are crossed by perpendicular fractures and produce a rectangular column-like effect. The sheet-like joints are succeeded by pseudopillows, which are arcuate fractures that intersect to yield a pillow-like appearance, may extend for several metres, and sometimes form sub-parallel sets (Forbes et al., 2012). The sheet-like fractures give way inward to polygonal columns. The columns are orientated perpendicular to the lava surface in an outermost zone in which they are 4–8 cm in diameter for rhyolitic–rhyodacitic lavas, 5–10 cm in diameter for andesitic–dacitic lavas, and up to 20 cm in basaltic lavas. Column diameter increases inward, sometimes by up to 3 times in less than 50 cm, by a process of fracture termination until columns are 30–50 cm across in andesite to rhyolite lavas and 40–60 cm in basalt. Fan-like and curved patterns are common in the zone of narrow columns (cf. entablature). Further into the lava interior, the crystal-linity increases and the fracture pattern is dominated by (1) larger-diameter (>50 cm) polygonal fractures; and (2) platy fractures with an average length of c. 80 cm, parallel to the lava margins and spaced c. 0.5–4 cm apart. In general, column fracture spacing is controlled by the physical properties of magma that vary with composition, properties such as Young's modulus, Poisson's ratio and thermal expansivity. Thus, smaller columns are found in felsic lavas and larger columns in mafic lavas (Lescinsky and Fink, 2000). For any composition, columns present in lava that has been water chilled are thinner than those formed in subaerial settings.

Lescinsky and Fink (2000) interpreted the sequence of fracture patterns as a reflection of the cooling history of lavas that grew during their emplacement rather than as a post-emplacement phenomenon. The prominent fracture orientation perpendicular to the lava margins (sheet-like and hackly fractures) indicates that the growth of those followed the solidification front, whereas it was suggested that fractures in the crystalline lava interior (coarse polygonal and platy fractures) probably grew after and during flow cessation. The regular inward progression of fracture types, from more closely spaced and irregular

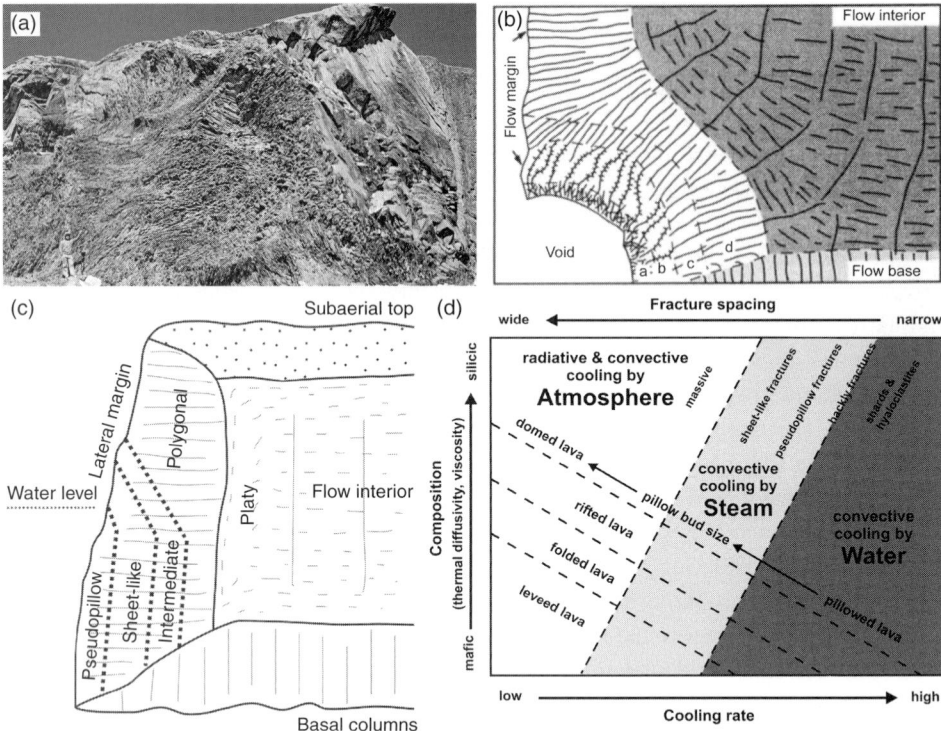

Fig. 9.8 (a) Complex cooling fractures developed in the base of an intermediate-composition lava at Nevada Coropuna, Peru. (b) Diagram depicting the progression in fracture types formed from a void at the base of a lava. Key – a: hackly, b: pseudopillow, c: sheet-like, d: polygonal (from Lescinsky and Fink, 2000). Irregular broader polygonal fractures and platy fractures are present in the lava interior. All transitions are gradational. (c) Schematic cross section of a lava at an ice-contact, showing the distribution of the major fracture types (from Lodge and Lescinsky, 2009a). (d) Qualitative depiction of lava fracture types in relation to cooling rate, cooling mechanism and lava composition (from Lescinsky and Fink, 2000). Cooling rates are sensitive to both cooling mechanism (in air (no shading), steam (mid grey) or water (dark grey)) and lava composition, with cooling largely a function of thermal diffusivity. Since felsic lavas have higher thermal diffusivities, they can cool more rapidly than mafic lavas. Fracture spacing is dependent on material properties such as Young's modulus, Poisson's ratio and thermal expansivity, which are composition dependent. Thus, fractures in felsic lavas are more closely spaced than in mafic lavas.

at flow margins (sheet-like and hackly fractures) to more regular and coarsely spaced towards the interior (platy and polygonal fractures), together with an increase in crystal-linity, are consistent with a decrease in cooling rates (possibly related to the amount of water available) away from the lava margins (Lodge and Lescinsky, 2009a,b).

Finally, Spörli and Rowland (2006) described distinctive small (few centimetres long) secondary polygonal fractures spaced at 5–10 cm that formed on the sides of primary cooling columns 0.5–3 m thick in an andesite lava in New Zealand. They were called

Fig. 9.9 (a) Ice-block meltout structure in a basal pillow lava sequence at Hlödufell, Iceland (see Skilling, 2009). Note the zone of fine-scale blocky joints surrounding the cavity. The notebook is 17 cm long. (b) Large ice-block meltout cavity at the base of an intermediate-composition lava flow at Coropuna volcano in Peru. The cavity was formed when the lava overran a large block of ice, which subsequently melted out.

'column on column' structures. Similar but more closely spaced small structures also occur as secondary fractures associated with pseudopillows (Lescinsky and Fink, 2000; Forbes et al., 2012). The structures described by Spörli and Rowland (2006) occur only in zones behind c. 10 m-wide outcrops of subhorizontal small columns developed at the margins of an andesite lava lobe. The marginal zone of small columns is backed by a zone of platy jointing (sheet-like joints?) that separates the small columns from lava with pervasive larger columns. The columns on columns affect only the larger-diameter columns. Unlike the quenched columns in entablatures, the small secondary columns are not glassy and they were interpreted to have formed by a second strong cooling event caused by a late influx of water when the lava had already cooled and crystallised. The lava has oxidised autobreccia and was subaerially emplaced, with no evidence for interaction with water or ice, such as hyaloclastite. However, the subhorizontal attitude of the small columns on the margins of the lava lobe together with the presence of the column on column structures were used to suggest that the lava advanced into snow and ice during a colder climate interval 40–32 or 25.5–12 k.y. ago.

9.4 Primary fragmental lithofacies

9.4.1 Hyaloclastite

We do not use hyaloclastite as a catch-all for any glassy volcaniclastic rock (Fisher and Schmincke, 1984; Schopka et al., 2006). Under the classification of White and Houghton (2006), the 'hyaloclastites' of Fisher and Schmincke are divided into hyaloclastite (*sensu stricto*) and lapilli tuff, similar to the distinction made by Honnorez and Kirst (1975) in which lapilli tuffs were called hyalotuffs. Hyaloclastite is described by White and Houghton (2006) as a discrete volcaniclastic deposit formed during effusive volcanism when lava is quenched and fragmented on contact with water. It is thus massive and formed essentially *in situ*, with no post-depositional downslope movement. Although the prominent large-scale homoclinal beds of breccia formed as foresets in lava-fed deltas have commonly been called hyaloclastite, under the White and Houghton (2006) classification they are breccias and tuff breccias (see Section 9.4.2, below).

Hyaloclastite (lithofacies H; Table 9.1) is most commonly and abundantly found in association with mafic magmas but it is also developed locally at the margins of intermediate and felsic lavas (e.g. Furnes et al., 1980; Tuffen et al., 2001; Edwards et al., 2002; McGarvie et al., 2007; Edwards et al., 2009b; Stevenson et al., 2009; Smellie et al., 2011a; Hungerford et al., 2014; Pollock et al., 2014). Hyaloclastite has no grain size connotation and may be ash, lapilli or breccia in mean grain size (e.g. Skilling, 1994), although breccias are volumetrically more common and ash-grade hyaloclastite is volumetrically minor. Hyaloclastite created this way will accumulate close to the source lava. The deposits are formed of angular ash-, lapilli- and block-sized, mainly poorly to non-vesicular blocky clasts and are generally framework supported (Fig. 9.10). Mafic hyaloclastite is usually dominated by lapilli-size clasts (>1 cm) whilst clasts in hyaloclastite associated with more evolved lavas or domes are often block sized. Both types contain little ash-grade matrix. The larger lapilli and blocks are generally aphanitic or finely crystalline, whereas finer clasts, especially those ≤ 1 cm, are predominantly glassy.

9.4.2 Breccia/tuff breccia

A wide variety of breccias and tuff breccias (lithofacies B and TB; collectively called breccia below) are found in glaciovolcanic successions. Hyaloclastite is a specific type of quenched-lava autobreccia that has already been described (lithofacies H). Many air-cooled lavas form massive fines-poor autobreccias (cB). They are ubiquitous in the subaerial 'a'ā lava sections of many glaciovolcanic successions but are seldom described as a discrete lithofacies (e.g. Smellie et al., 2011a). 'A'ā autobreccia is formed of clinkers, which are scoriaceous fragments of aphanitic chilled lava crust (Loock et al., 2010; Fig. 9.11). The clinkers are mainly lapilli and block sized, and ash-size fragments are minor. They form clast-supported deposits typically a few decimetres to several metres in thickness enclosing massive sheet lava (lithofacies sL; Fig. 9.4). Attrition of clinkers due to jostling during advection on the flowing lava causes variable rounding of sharp edges and generates minor ash-size matrix. In 'a'ā lavas of lava-fed deltas, steam effects at the passage zone can cause localised yellow coloration of clinkers signifying palagonite alteration (Smellie et al., 2011a; Fig. 9.12); it is an environmentally distinctive feature (Smellie et al., 2013a).

Other breccias are mainly tuff breccias, formed of lithic lapilli and blocks derived principally by (1) fragmentation of pillow lava during high-level hydromagmatic

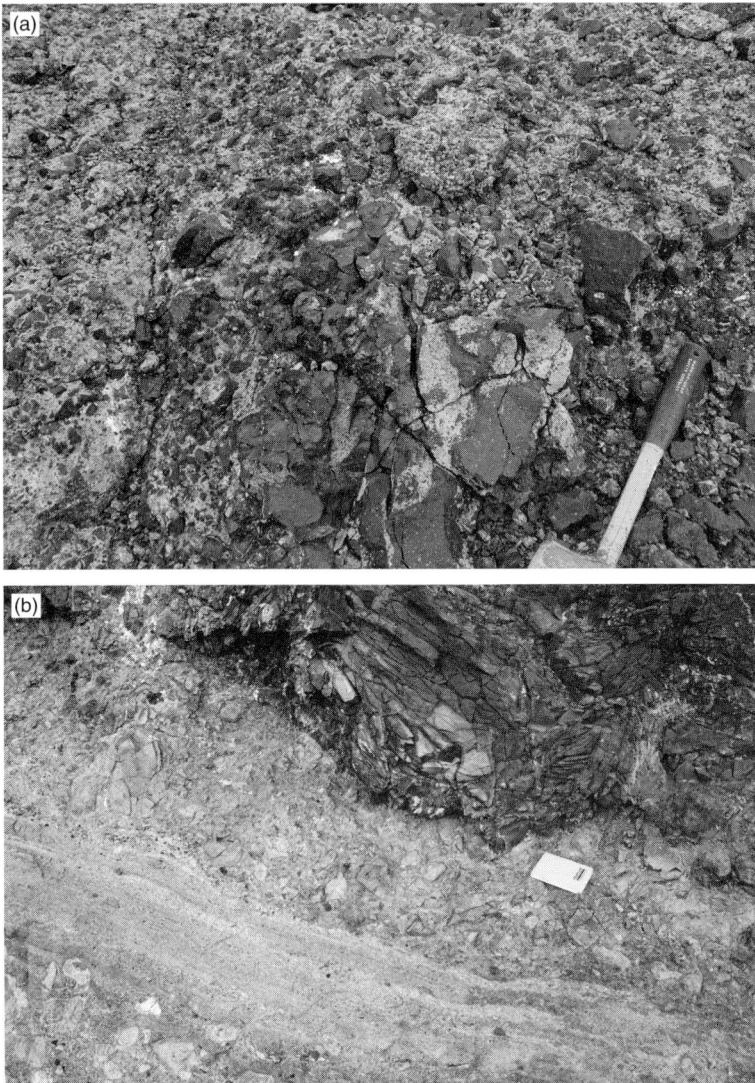

Fig. 9.10 (a) Fractured water-cooled lava surrounded by cogenetic hyaloclastite breccia. Minna Hook, Mt Discovery, Antarctica. The hammer is 35 cm long. (b) Rhyolite hyaloclastite breccia resting on subglacially deposited volcaniclastic beds, below subglacially emplaced, closely jointed rhyolite lava at Mandible Cirque, Daniell Peninsula, Antarctica. The notebook is 17 cm long. (A black and white version of this figure will appear in some formats. For the colour version, please refer to the plate section.)

Fig. 9.11 Oxidised autobreccia associated with subaerial 'a'ā lava at Cape Klovstad, Victoria Land, Antarctica. The ice axe is 60 cm long.

Fig. 9.12 Yellow (palagonite altered) finer clasts in autobreccia associated with subaerial 'a'ā lava at Coulman Island, Victoria Land, Antarctica. The alteration is attributed to localised steam activity associated with progradation of the lava as a lava-fed delta (see Smellie et al., 2011a). The pencil is 15 cm long. (A black and white version of this figure will appear in some formats. For the colour version, please refer to the plate section.)

Fig. 9.13 Close view of typical tuff breccia (TB) from lava-fed delta at Kima'Kho tuya, British Columbia. Note prominent lava pillow (c. 0.5 m in diameter). Compare with Fig. 9.10a (lithofacies H, formed *in situ*).

explosions, (2) slumping of volcanic piles of interbedded lapilli tuff and pillow lava, and (3) by quench and mechanical fragmentation of lava entering water at the brinkpoints of lava-fed deltas followed by avalanching of the accumulated debris down the delta front (Fig. 9.13; e.g. Edwards et al., 2009b; Skilling, 2009; Stevenson et al., 2009). Although the second category of breccia is probably epiclastic under the classification of White and Houghton (2006), it is entirely volcanic, monogenetic in composition and syn-eruptive, with a short residence time between initial deposition and redeposition. Breccia in category 3 has commonly been called hyaloclastite, but it is reclassified here following White and Houghton (2006).

The breccias in many lava-fed deltas form prominent crude large-scale foreset beds that are either massive or show inverse or normal grading (lithofacies $gB_{(i)}$, $gB_{(n)}$, $gTB_{(i)}$, $gTB_{(n)}$; Fig. 9.14; see also Fig. 10.12); the grading is often restricted to bed tops or bases. The foreset beds are not dominated by sideromelane. They typically have a preponderance of aphanitic and crystalline lithic clasts derived either from the incorporation of fragmented subaerial capping lavas following terrace collapse events (Mattox and Mangan, 1997) or from disintegrated lava pillows. Ash-size matrix is characteristically minor although clast size sorting during transport may generate tuff deposits distally, known as 'toeset beds'. As in hyaloclastite (lithofacies H), most clasts finer than c. 1 cm are glassy, and a population of tachylite grains derived from the margins of lava pillows is usually present. The lava-fed delta breccias also include pillow breccia or pillow-fragment breccia (e.g. lithofacies $cB_{(p)}$, $TB_{(p)}$; Jones, 1970; Skilling, 2009; Edwards et al., 2011) for breccias in which the proportion of fragmented and intact

Fig. 9.14 (a) Cross section through a pāhoehoe lava-fed delta showing well-developed, homoclinal tuff breccia foreset beds overlain by horizontal subaerial capping lavas. The cliff is 200 m high. Vega Island, Antarctica. (b) Oblique section across a pāhoehoe lava-fed delta at Kima'Kho tuya, British Columbia (Ryane et al., 2011; Russell et al., 2013). The section shown is c. 25 m high.

pillows is unusually high (i.e. volumetrically dominating). Lava pillows are relatively common in delta foreset breccias sourced in pāhoehoe. They are much less common, usually much larger and less spherical in breccia sourced in 'a'ā lava. 'A'ā-sourced breccia may also contain a small but distinctive proportion of vesicular and oxidised clinkers that

Fig. 9.15 Close view of numerous thin far-travelled lava stringers that flowed down tuff breccia foreset beds in a pāhoehoe lava-fed delta. James Ross Island, Antarctica. The view is c. 30 m high (image: Ian Skilling).

were formed as subaerial autobreccia (lithofacies cB) and advected by the lava into the meltwater lake (Smellie et al., 2011a, 2013a). The breccia may be interbedded with thin far-travelled pāhoehoe-sourced sheets that may extend a few tens of metres below the passage zone (Fig 9.15), or with thicker and often highly irregularly shaped lobes and sheet-like masses of massive lava in 'a'ā-fed deltas (Fig. 9.16). The combination of breccias and lava lobes in the latter has a distinctively chaotic appearance and the lithofacies was called lobe hyaloclastite by Smellie et al. (2011a); because of the paucity of hyaloclastite in the lithofacies, a better name may be lava lobe-bearing breccia (cf. lobe-bearing breccia in glaciovolcanic rhyolite described by McGarvie et al., 2007).

The individual breccia beds in lava-fed deltas are relatively thick (typically 1–5 m), steep-dipping (c. 35° diminishing down-dip to <10°; Skilling, 1994) and poorly sorted. Bedding in 'a'ā-fed deltas is much less well developed and most deposits are massive. However, where present, dips are generally less steep (i.e. c. <25°) than in pāhoehoe deltas (Smellie et al., 2011a, 2013a). The dip of the underlying pre-eruption substrate can exert an important influence and may increase bedding dips in 'a'ā-sourced breccias (Bosman et al., 2014). The beds have a wedge-like cross-sectional morphology overall as they thin and become finer grained down-slope, and onlap their substratum tangentially where tuff-rich toesets are sometimes (rarely) developed. Bedding discordances known as reactivation surfaces are formed when avulsion of the feeding lava results in the delivery of lava debris to a different delta front location, which then onlaps the previously deposited breccia (Fig. 9.17). Although the mode of formation of breccia in lava-fed deltas is commonly envisaged as 'avalanching', the deposits generally form as a result of repeated delta front collapses and the grains are transported mainly as sediment

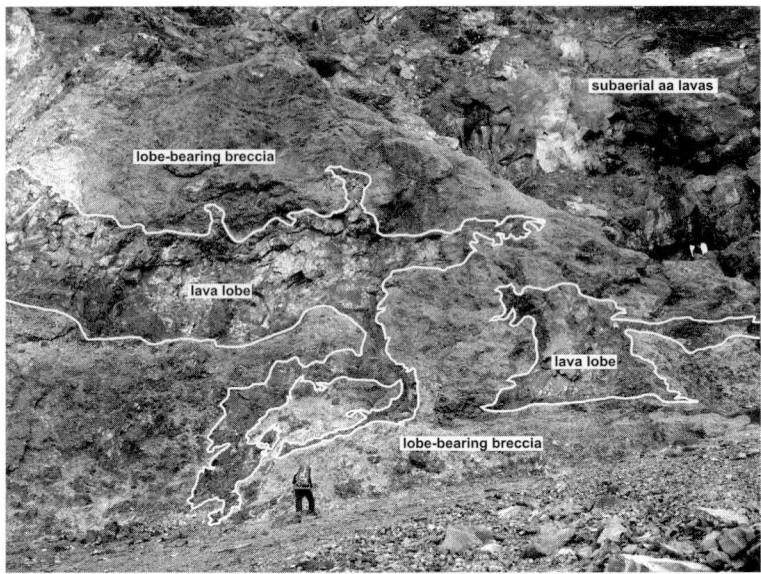

Fig. 9.16 Typical chaotic lobe hyaloclastite (lava lobe-bearing breccia) lithofacies formed of irregular water-cooled lava masses and breccia in an ʻaʻā lava-fed delta. Hallett Peninsula, Antarctica. The lavas at the top of the image are part of the subaerial lava capping unit of the delta. Note the striking contrast with the much better bedded subaqueous tuff breccia in pāhoehoe-fed deltas (Figs. 9.14a, 9.15, 9.17, 10.12; image: Sergio Rocchi; see also Smellie et al., 2011a, Fig. 3b).

Fig. 9.17 Prominent reactivation surface in a pāhoehoe lava-fed delta on James Ross Island, Antarctica. The cliffs are 200 m high (image: Ian Skilling).

Fig. 9.18 (a) Matrix-rich tuff breccia within a mafic tindar. Kalfstindar, Iceland. The notebook is 16 cm long. (b) Blocky breccia formed during gravity-driven avalanches derived from proximal pillow lavas. Pillow Ridge, British Columbia (see Edwards et al., 2009b).

gravity flows (concentrated density flows; Mulder and Alexander, 2001). They cease their movement by frictional freezing.

Breccias in other types of glaciovolcanic sequences consist of angular lava blocks either clast supported or more commonly matrix supported in massive yellow tuff or lapilli tuff matrix (lithofacies TB, cTB; e.g. Edwards et al., 2011, Fig. 4f; Fig. 9.18). The debris is mainly transported in concentrated and hyperconcentrated density flows. In mafic and intermediate-composition tindars the lithofacies is commonly encountered close to (above) the transition between the basal pillow mound and overlying hydromagmatically

Fig. 9.19 (a) Tuff breccia interpreted as a hot avalanche deposit, formed of irregular andesitic lava pillows dispersed in massive breccia formed from poorly vesicular pillow fragments. The amoeboid shape of the lava pillows indicates that they were hot and deformed plastically during transport downslope. Tindur, Kerlingarfjöll, Iceland (see Stevenson et al., 2009). The pencil is c. 15 cm long. (b) Hot avalanche deposit (breccia) with charred wood (black) resulting from the collapse of a lava dam at Mt Meager, British Columbia (Stewart et al., 2003).

generated lapilli tuffs. Similar lithofacies are also described in felsic and intermediate-composition tindars as both hot and cold avalanche deposits (e.g. Tuffen et al., 2001; Stevenson et al., 2009). That described by Stevenson et al. (2009; their 'contorted pillow fragment breccia'; Fig. 9.19a) consists of a pile of irregularly shaped, amoeboid and contorted poorly vesicular andesitic lava pillows up to 4 m in diameter and their fragments. They suggested that the lithofacies formed by collapsing or slumping of a freshly forming pillow lava pile, which caused extensive fragmentation and deformation of any still-molten pillow cores. The association with slope-parallel bulbous lava lobes with radial columnar joints indicates that the collapse(s) occurred whilst pillow lava effusion was taking place. Other deposits have blocks of wood carbonised during incorporation in the hot collapsing mass flow, as at Mt Meager, British Columbia (Fig. 9.19b).

Some tuff breccias contain conspicuous blocks of stratified and massive lithified lapilli tuff and tuff in massive tuff matrix (lithofacies TB; e.g. Smellie, 2001; Skilling, 2009). A few drape shallow-dipping slump-scar surfaces with which they are probably cogenetic, i.e. related to sector collapse. Others occur in subvertical zones cross-cutting bedding and spatially associated with syn-eruptive deformation such as faults and large collapsed blocks. The matrix-rich parts show irregular ghost-like domains of contrasting grain size indicating slumping and mingling of weakly consolidated strata. The coherent tuff clasts were presumably slightly more consolidated and more resistant to disaggregation. Narrow (<1 cm) dark-coloured alteration rims on some suggest peripheral interaction with hydrothermal fluids. The relationships of the lithofacies occupying steeply cross-cutting zones suggest that disaggregation and fluidisation has taken place during focused syn-eruptive dewatering of poorly or unconsolidated tuff and lapilli tuff deposits. Finally, laminated tuff breccia (clTB) is another uncommon but highly distinctive coarse lithofacies variant composed of clast-supported angular lithic lava blocks with yellow-coloured laminated tuff in the pore spaces. It formed from the gravitational collapse of pillow lava creating a subaqueous talus deposit with the laminated tuff back-filling the framework as a sieve deposit (cf. SHm lithofacies of Skilling, 1994).

9.4.3 Tuff

Tuff is a comparatively minor lithofacies in glaciovolcanic sequences (Smellie, 2001; Skilling, 2009). This has been explained as a consequence of most ash being transported in sediment density currents that bypass the edifice and being deposited further out, where it presumably is more prone to erosion and removal by ice (Cas et al., 1989; Skilling, 1994; Smellie and Hole, 1997; Smellie, 2001). However, there may be an additional factor involved. Since most glaciovolcanic eruptions occur through lakes created by melting holes through ice sheets, much of the ash from a subaerial column will fall onto the surrounding ice surface and will be advected away after the eruption ends. The area of the lake surrounding an erupting glaciovolcanic tuff cone is quite small (Figs. 1.1a, 3.7b and 9.20) and the coarser detritus (lapilli, blocks, bombs) falling in an annular zone around the collapsing column will be preferentially deposited in the lake. Thus, the effect of an

Fig. 9.20 Tephra cone formed during the 2004 eruption at Grimsvötn. Note the multiple strand lines (raised beaches) formed as the encircling meltwater lake subsided in stages (photo courtesy of B. Oddsson).

Fig. 9.21 Thin-bedded tuffs showing a variety of tractional structures (planar and ripple cross lamination) and normal grading. Icefall Nunatak, Marie Byrd Land, Antarctica. The pencil is c. 15 cm long.

eruption through a hole melted in an ice sheet may be self-collimating, resulting in a paucity of ash-size detritus. As this can only occur once a subaerial eruption column has formed, it will only affect the later-formed deposits draping the slopes of the tuff cone.

Tuff occurs in mafic successions in well-defined thin (generally 1–5 cm) planar beds of fine and very fine tuff (lithofacies T, bT, glT) which typically form thin sediment packages up to c. 2 m thick (Fig. 9.21). The sideromelane comprises poorly to non-vesicular blocky shards characteristic of formation during highly explosive hydromagmatic eruptions. The individual beds have sharp bases with some showing small-scale loading, normal grading and climbing ripple cross sets that pass up gradationally into very fine tuff. They correspond to T_{ae}, T_{be} and T_{ce} turbidites deposited in a meltwater lake. Some may be deposits of residual flows detached from concentrated or hyperconcentrated gravity flows, including in lava-fed deltas ('toeset' beds; Mulder and Alexander, 2001; Smellie, 2001). Another rarer tuff lithofacies consists of beds of coarse tuff that have subsided deeply into fine tuff, creating a conformable bed comprising mingled tuff showing dish and pillar structures (lithofacies $T_{(dw)}$). The effect is caused by dewatering associated with localised gravitational inversion (a form of Rayleigh–Taylor instability) triggered by the rapid deposition of coarser denser sediments on waterlogged finer sediments (Smellie, 2001) or a thixotropy caused by volcanic tremors.

Tuff is also found in felsic glaciovolcanic successions, in which it is subordinate to coarser volcaniclastic lithofacies. Stevenson et al. (2011) described massive tuff (T) erupted during the earliest stages of edifice formation at Kerlingarfjöll, Iceland. The ash contains <10% small pumice lapilli and <5% angular blocks of obsidian and flow-banded rhyolite; 72–98% of the ash is finer than 1 mm, some particles are aggregated, and matrix vesicles are present. The ash grains are non-vesicular to poorly vesicular and have a blocky morphology, consistent with a violently explosive hydro-magmatic origin of incipiently vesicular felsic magma. The deposits are comparable with products of phreato-Plinian eruptions and indicate abundant water interacting with magma at the vent whilst lacking evidence for meltwater ponding. The outsize lithic and obsidian clasts were probably picked up from underlying units. The massive texture of the deposits is conspicuous and, together with the other characteristics, implies essentially subaerial deposition from a collapsing column, possible containment within an ice cavity, and deposition from moist (<100 °C) pyroclastic density currents (cf. Branney and Kokelaar, 2002). Crudely bedded deposits of felsic ash with minor vesicle-poor obsidian and pumice were also described by Tuffen et al. (2008) at Dalakvísl, Iceland. The ash particles are broadly similar in appearance and origin to those described by Stevenson et al. (2011) but the deposits are crudely planar bedded, with beds 5–20 cm thick lacking erosive features and defined by subtle variations in grain size and sorting (i.e. dbT). Emplacement from high concentration aqueous density currents within a meltwater lake was suggested.

9.4.4 Lapilli tuff

Lapilli tuff is the commonest and most volumetrically dominant volcaniclastic lithofacies in mafic and intermediate-composition tindars and glaciovolcanic tuff cones (Smellie, 2001; Schopka et al., 2006; Edwards et al., 2009b; Stevenson et al., 2009) and their felsic equivalents (Tuffen et al., 2002a, 2008; Tuffen and Castro, 2009; Stevenson et al., 2011). In mafic and intermediate-composition sequences, lapilli tuff forms thick (tens of metres) eruption-generated sequences (*sensu* White, 2000, and Smellie, 2001) of laterally contin-uous planar beds with crudely defined bed surfaces (dbLT; Fig. 9.22). The sequences are bounded either by syn-sedimentary erosional surfaces (slump scars; see Section 9.5) or stratal packets formed of thin-bedded tuffs or lapilli tuffs (bT, bLT). The deposits are composed of variably vesicular blocky sideromelane and lesser tachylite and a variable proportion of outsize lithic clasts generally less than 25 cm in diameter but up to several decimetres. They were created during explosive hydromagmatic eruptions. Individual beds are a few centimetres to a few metres thick, poorly sorted and either massive (LT) or normal-graded ($gLT_{(n)}$). Erosive bases and amalgamation are common whereas reverse grading ($gLT_{(i)}$), internal planar stratification and basal loading are rare. Diffuse-stratified lapilli tuff (dsLT) is abundant in some outcrops (e.g. Helgafell and Kalfstindar, Iceland; Schopka et al., 2006; unpublished information of JLS). Outsize lithic clasts rarely have shallow impact structures. A few sequences have packets of diffuse-bedded lapilli tuffs (dbLT) that show upward thinning, fining and better-defined bedding that have been related to declining explosive activity and tephra input during continuous-uprush eruptive episodes (Smellie, 2000). In many sequences lapilli tuff also forms minor beds with sharp surfaces that are massive to normal graded, passing up through planar lamination into mudstone

Fig. 9.22 Diffuse-bedded lapilli tuffs at Icefall Nunatak, Marie Byrd Land, Antarctica. Ice axe (by rucksack) is 60 cm long.

(gLT; glLT). Some contain slabby to contorted tuff and lapilli tuff intraclasts. Their origin has been linked to ejection of discrete jets during intermittent jetting activity, the jets transforming to concentrated density currents in a water-filled ice vault (White, 2000; Smellie, 2001). They may also be products of local small-scale slumping events. Rare dune bedforms with steep lee-side laminations are either basal parts of concentrated density current deposits (cf. Lowe, 1982; Mulder and Alexander, 2001) or, when the bedforms are isolated, products of local traction reworking by aqueous currents (Smellie, 2001; Skilling, 2009). An unusual andesitic massive lapilli tuff lithofacies distinguished by the presence of c. 10% amoeboid, lensoid and flattened vesicular bomb-like clasts with glassy rims ($LT_{(b)}$) was described by Stevenson et al. (2009). It may represent the product of subaqueous fire-fountaining activity.

No unambiguous volcaniclastic criteria yet exist for recognising exactly when or where conditions in a subaqueous tuff cone or tindar sequence became subaerial, and thus recognition of 'pyroclastic passage zones' will be exceptionally difficult (cf. Russell et al., 2013). However, the following criteria are useful: accretionary and armoured lapilli tend to be more common in subaerial deposits although they can persist and even be reworked by traction currents following deposition underwater (Smellie, 1984); lapilli tuffs with low-angle ($\leq 10°$) cross stratification in sandwave structures of several types are an indication of subaerial conditions, as are antidune structures with their steep-dipping ($35°$) stoss-side laminae (e.g. Cole, 1991); and inflated pyroclastic density currents (subaerial 'surges') probably do not form ripples (personal communication, M. Branney), which conversely are commonly found in subaqueously deposited tuff. Fall beds, unrecognised so far in subaqueous glaciovolcanic tuff cones, should be present in the subaerial section but have yet to be unequivocally identified. They may also occur subaqueously, but abundant

Fig. 9.23 Massive felsic lapilli tuff at Kerlingarfjöll, Iceland (see Stevenson et al., 2011). Note poor bedding, grading and minor normal faults.

tephra dumped onto a surrounding englacial lake surface will probably overload the water column in most instances and sink as vertical sediment density flows rather than by fall settling of the individual particles (Manville and Wilson, 2004).

Because most felsic tuyas are lava dominated and the tephra sections are extensively obscured by surface wash and scree, few well-exposed felsic lapilli tuffs have been described (e.g. Edwards et al., 2002; Tuffen et al., 2002a). The most detailed study is by Stevenson et al. (2011) for rhyolite tephra-dominated tuyas at Kerlingarfjöll, Iceland. Unlike mafic–intermediate equivalents, which are orange-brown due to pervasive palagonite alteration, felsic lapilli tuffs are grey or white coloured. Those at Kerlingarfjöll are mainly varieties of massive lapilli tuff (LT; Fig. 9.23) with three lithofacies distinguished based on mean grain size contrasts and occurrence of cogenetic obsidian lapilli. The lithofacies individually form sections up to 50 m thick. The juvenile clasts are angular to sub-rounded, moderate to highly vesicular, white and grey pumice lapilli up to 5 cm across supported in a matrix of vesicular fine ash. Matrix vesicles are commonly present. The high vesicularity of the pumice lapilli and ash suggest magmatic fragmentation due to volatile exsolution but the abundance of fine ash is more consistent with significant hydromagmatic fragmentation. Ill-defined thick bedding is locally discernible suggesting that the deposits should be regarded as either LT or dbLT (Table 9.1). They were probably deposited from moist subaerial high-concentration pyroclastic density currents within an ice vault in which flow unsteadiness may be responsible for the poorly seen bedding rather than deposition of discrete beds linked to different eruptive episodes. The obsidian-bearing lapilli tuff ($LT_{(o)}$) occurs as domains within other LT lithofacies. The obsidian individually occurs as well-defined masses up to 3 m across with shattered cores and vesicular margins. The host lapilli tuff is poor in fine ash

Fig. 9.24 Diffusely stratified lapilli tuffs at Kalfstindar, Iceland.

and the pumice lapilli are relatively poorly vesicular (c. 30%). The lithofacies was interpreted as an unusual pyroclastic deposit fragmented during comparatively slow ascent of the magma causing some shear (rather than magmatic) fragmentation. It intruded and non-explosively mingled with other LT lithofacies during fluidisation by upward-streaming gases.

In addition, the Kerlingarfjöll felsic sequences contain a lithofacies of diffuse-stratified lapilli tuff (dsLT; cf. Fig. 9.24). The stratification is defined by subtle variations in the proportions of pumice and ash, forming layers 2–15 cm thick that are sub-parallel or exhibit low-angle truncations. It contains generally less fine ash than other LT lithofacies (<c. 50% compared with up to c. 70%) and much of the ash consists of bubble-wall fragments consistent with magmatic fragmentation under a lesser influence of magma–water interaction. Subaerial deposition from concentrated pyroclastic density currents is envisaged, with unsteady flow or episodic turbulence creating traction structures responsible for the stratification (Stevenson et al., 2011).

9.4.5 Lapillistone

Like agglutinate (Section 9.3.5, above), lapillistone (LP, gLP) is an uncommon glaciovolcanic lithofacies. All examples known to the authors are mafic in composition. The distribution of the lithofacies is largely confined to proximal (crater) locations surrounding the vent. It forms monomict red (oxidised) and grey or black poorly bedded deposits composed of coarse tachylitic scoria and bombs and is poor in ash-size clasts (Smellie, 2001; Smellie et al., 2011a). In bisected cones, the bedding dips radially outward. The deposits are the products of subaerial magmatic eruptions, forming scoria cones. Palagonite alteration locally affects lapillistone and is either caused by steam action associated with dyke activity affecting the

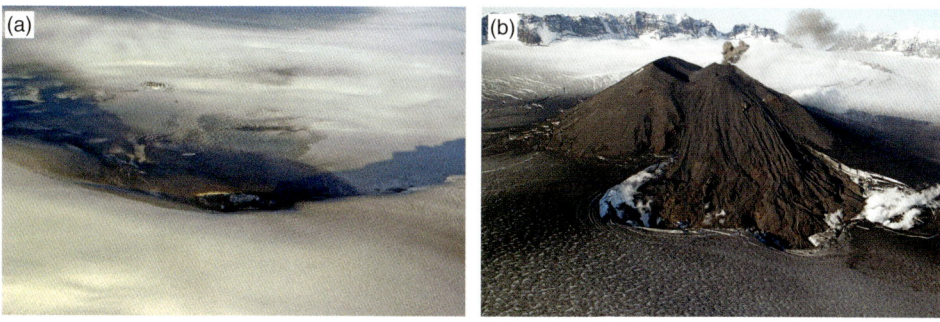

Fig. 1.1 Examples of recent glaciovolcanic eruptions. (a) Aerial view of a crater formed during the 2004 Grimsvötn eruption within the Vatnajökull ice cap, Iceland. The ice cauldron is c. 0.7 km in diameter and is flanked by ash formed during explosive activity. (b) Aerial view of the southern flank of the intracaldera cone during the 2013 eruption of Mt Veniaminof, Alaska. This is a large stratovolcano with a subaerial tephra cone situated within its ice-filled summit caldera, whose margin is visible in the background. The tephra cone protrudes above the surrounding ice and it emitted several 'a'ā lavas that flowed downslope until they abutted against and/or flowed on top of the surrounding ice, but rates of melting and drainage were such that no lake formed, unlike in 1983–4. The cone is c. 300 m high. (c) Sheet lava flowing over snowpack during the 2013 Tolbachik eruption, Kamchatka. The foreground lava lobe is c. 5 m wide. Note how minimal snow melting has occurred below the moving lava and the lava (with a temperature of over 1000 °C) is flowing over the snow surface. A black and white version of this figure will appear in some formats.

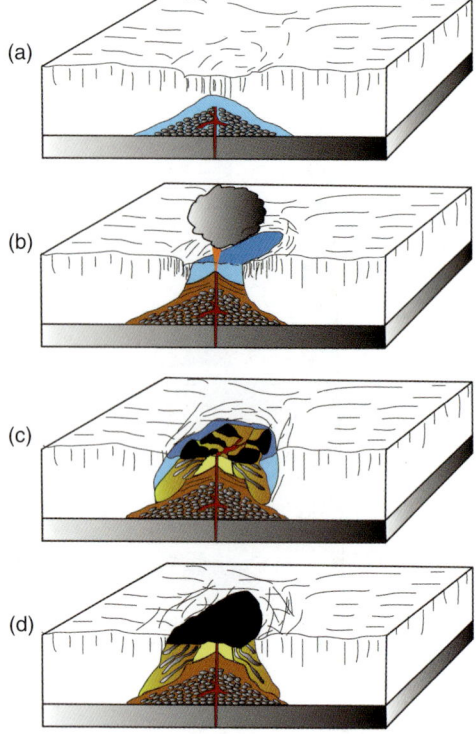

Fig. 1.3 Series of cartoons showing the sequential development of a tuya based on the 'classic' model described by Jones (1969b). See text for description of the constructional stages depicted. A black and white version of this figure will appear in some formats.

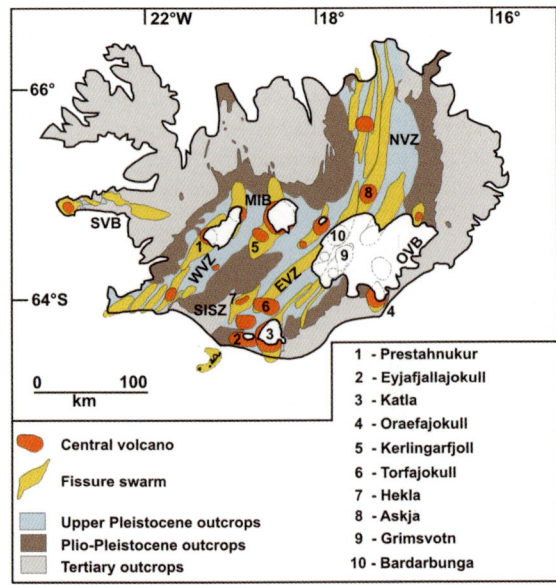

Fig. 2.5 Map of Iceland showing the volcanic zones and selected large volcanic centres (modified after Thordarson and Höskuldsson, 2008). Abbreviations: WVZ – Western Volcanic Zone; EVZ – Eastern Volcanic Zone; NVZ – Northern Volcanic Zone; MIB – Mid-Iceland Belt; SISZ – South Iceland Seismic Zone; OVB – Öraefajökull Volcanic Belt; SVB – Snæfellsnes Volcanic Belt. A black and white version of this figure will appear in some formats.

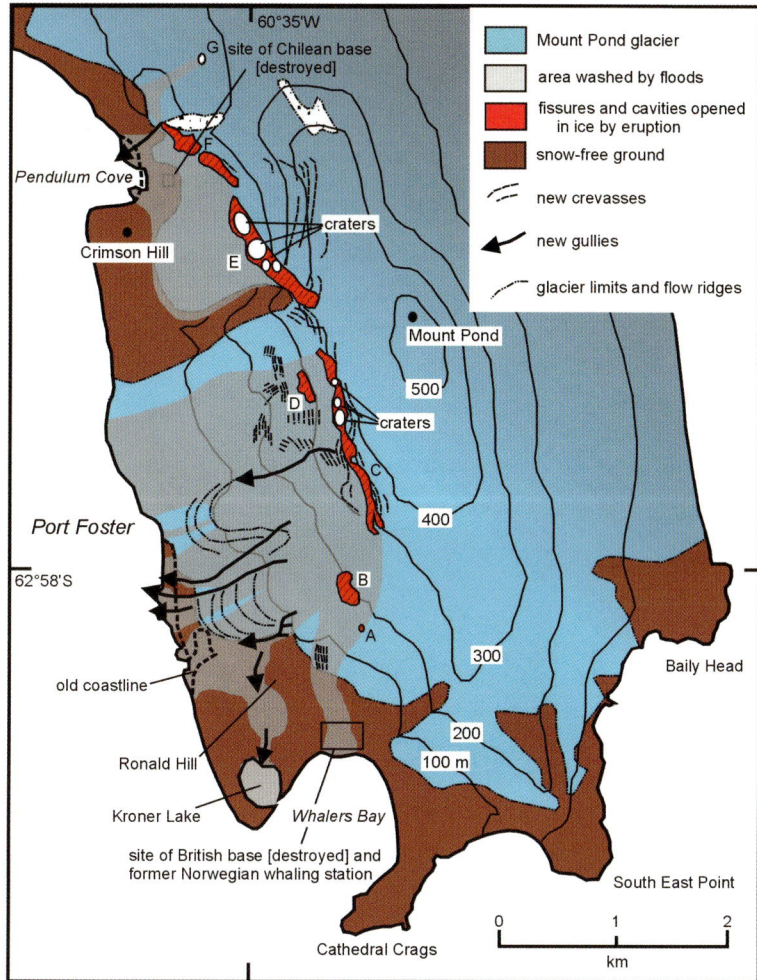

Fig. 3.3 Sketch map of eastern Deception Island showing the major features and effects of the 1969 subglacial eruption, including ice fissures and cavities opened up by the eruption, pyroclastic cones, the area washed by supraglacial meltwater flood, and the extended coastline in the floodplain below (from Smellie, 2002). A black and white version of this figure will appear in some formats.

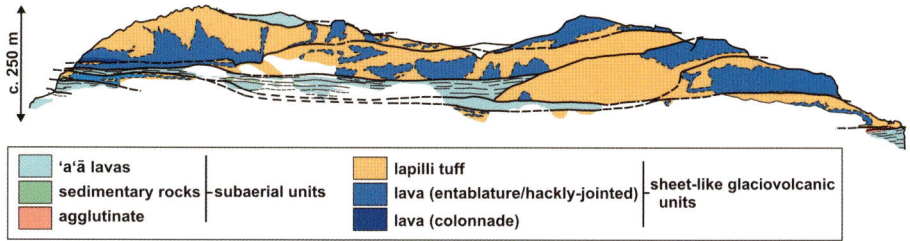

Fig. 8.8 Field sketch of alternating glaciovolcanic sheet-like sequences and subaerial lavas; Eyjafjallajökull, south Iceland (JLS, unpublished). Each of the sequences is separated by a prominent glacial erosional unconformity. The view is an oblique strike section. A black and white version of this figure will appear in some formats.

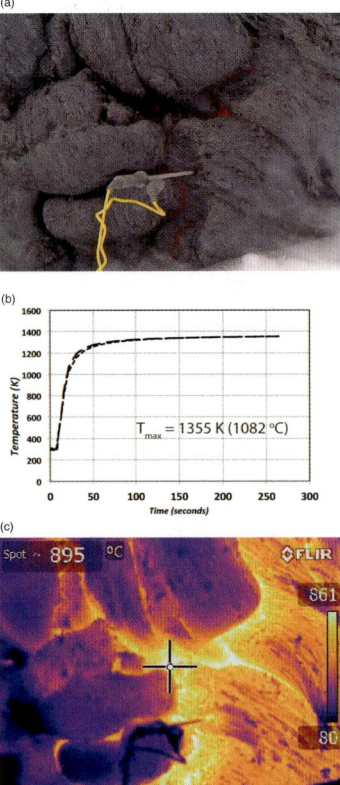

Fig. 5.5 Measuring temperatures in active Tolbachik lava flows. (a) Pāhoehoe lava lobe with thermal probe inserted into radiant crack to measure internal temperatures. (b) Graph of temperature versus time for measurement of pāhoehoe shown in (a). (c) Forward-looking infrared (FLIR) image of the surface of pāhoehoe lobe in (a). A black and white version of this figure will appear in some formats.

Fig. 9.12 Yellow (palagonite altered) finer clasts in autobreccia associated with subaerial 'a'ā lava at Coulman Island, Victoria Land, Antarctica. The alteration is attributed to localised steam activity associated with progradation of the lava as a lava-fed delta (see Smellie et al., 2011a). The pencil is 15 cm long. A black and white version of this figure will appear in some formats.

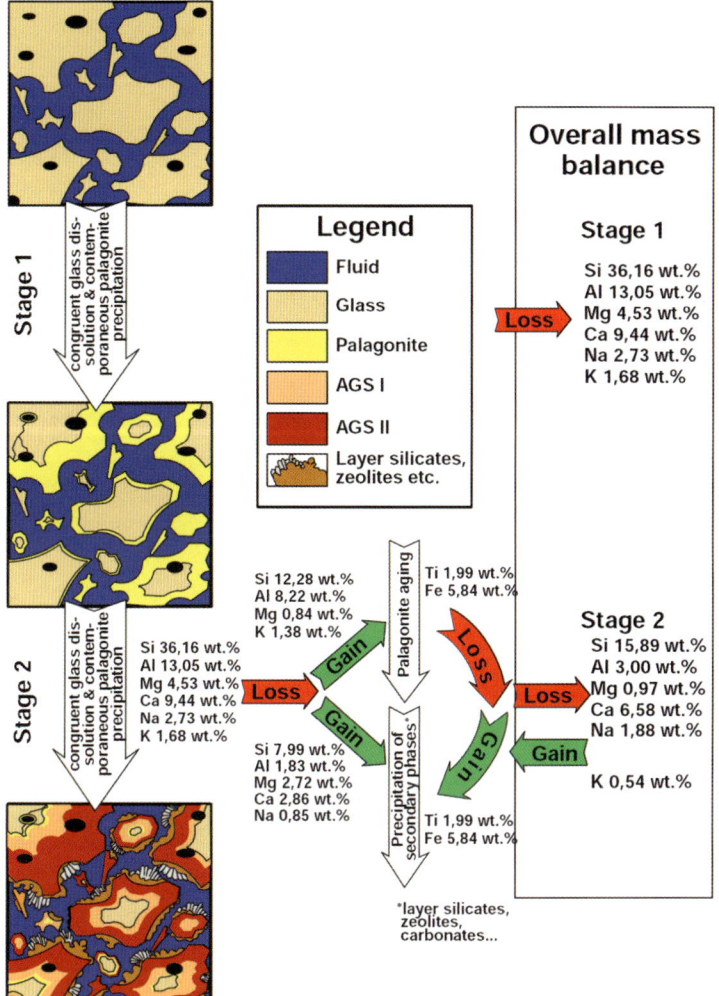

Fig. 7.4 Schematic diagram summarising the compositional effects (mass gains and losses) during progressive abiotic alteration of mafic glass to palagonite in a variety of environments (subglacial, meteoric, marine; from Stroncik and Schmincke, 2001). The process is envisaged taking place in two important stages. Stage 1 consists of palagonite formation by congruent dissolution of sideromelane glass, characterised by partial loss of Si, Al, Mg, Ca, Na and K. Stage 2 consists of the previously precipitated unstable palagonite taking up Si, Al, Mg and Ca and releasing Ti, Fe, Ca and Na to become more stable smectite. A black and white version of this figure will appear in some formats.

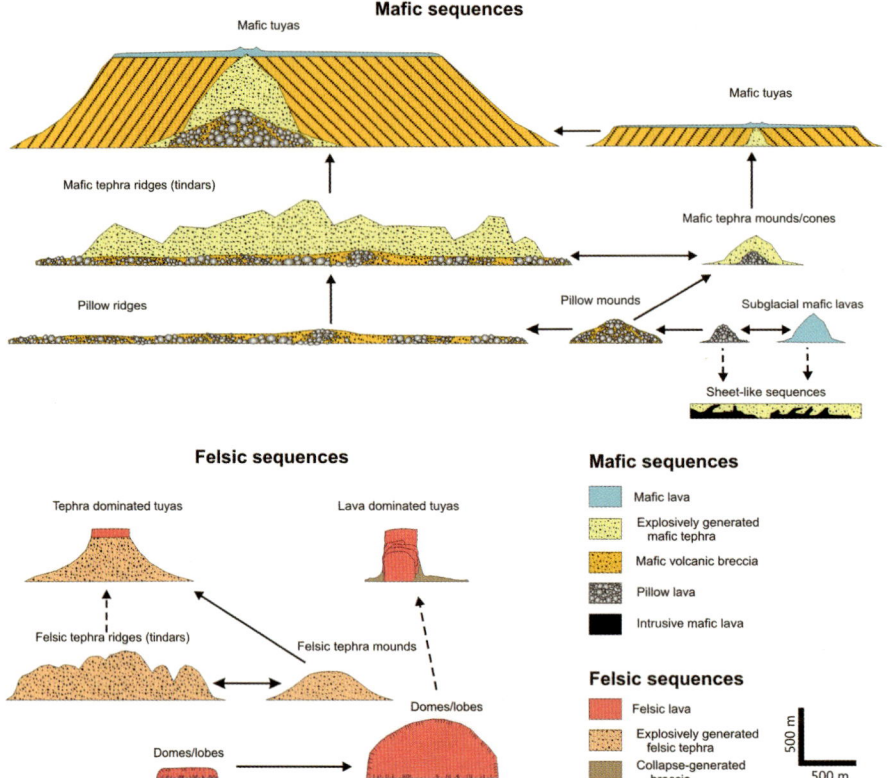

Fig. 8.1 Classification and hierarchy of glaciovolcanic landforms and their principal constituent lithofacies (modified after Smellie, 2013). The arrows indicate the evolutionary progression of landforms, e.g. pillow mounds evolving up into tindars, then into tuyas. The scale is only indicative. A black and white version of this figure will appear in some formats.

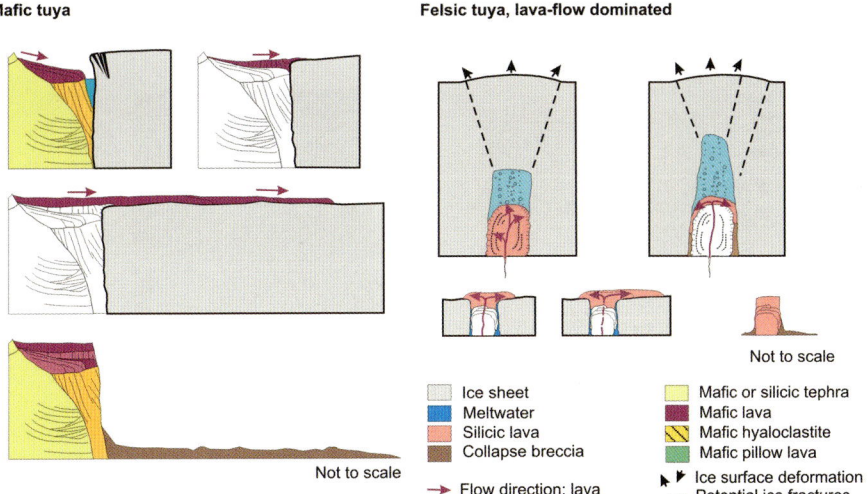

Mafic tuya

Felsic tuya, lava-flow dominated

Not to scale

Not to scale

Ice sheet	Mafic or silicic tephra
Meltwater	Mafic lava
Silicic lava	Mafic hyaloclastite
Collapse breccia	Mafic pillow lava

→ Flow direction: lava

▶ Ice surface deformation
--- Potential ice fractures

Fig. 8.9 Sketches illustrating postulated glaciovolcanic landforms and principal constituent lithofacies for eruptions associated with cold-based ice (modified after Smellie, 2013). The major differences from comparable landforms formed under wet-based ice are that lava-fed deltas will probably be shorter and the edifices taller. A black and white version of this figure will appear in some formats.

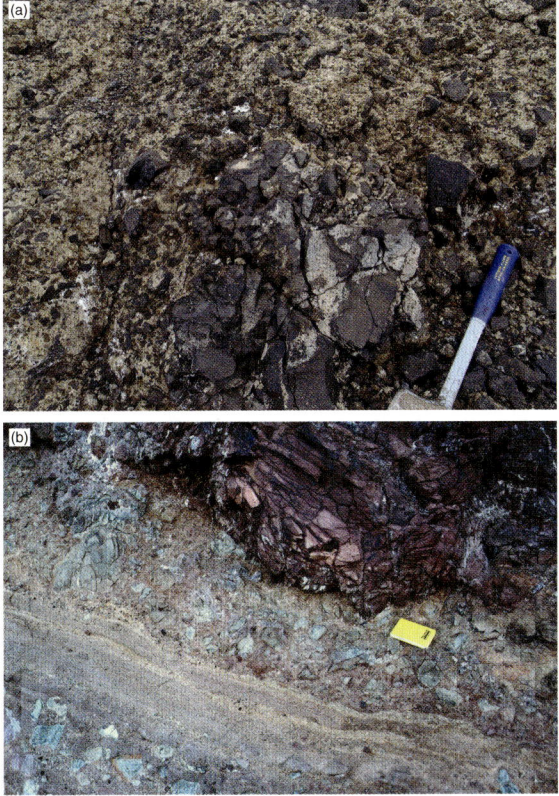

Fig. 9.10 (a) Fractured water-cooled lava surrounded by cogenetic hyaloclastite breccia. Minna Hook, Mt Discovery, Antarctica. The hammer is 35 cm long. (b) Rhyolite hyaloclastite breccia resting on subglacially deposited volcaniclastic beds, below subglacially emplaced, closely jointed rhyolite lava at Mandible Cirque, Daniell Peninsula, Antarctica. The notebook is 17 cm long. A black and white version of this figure will appear in some formats.

(a)

WEST

EAST

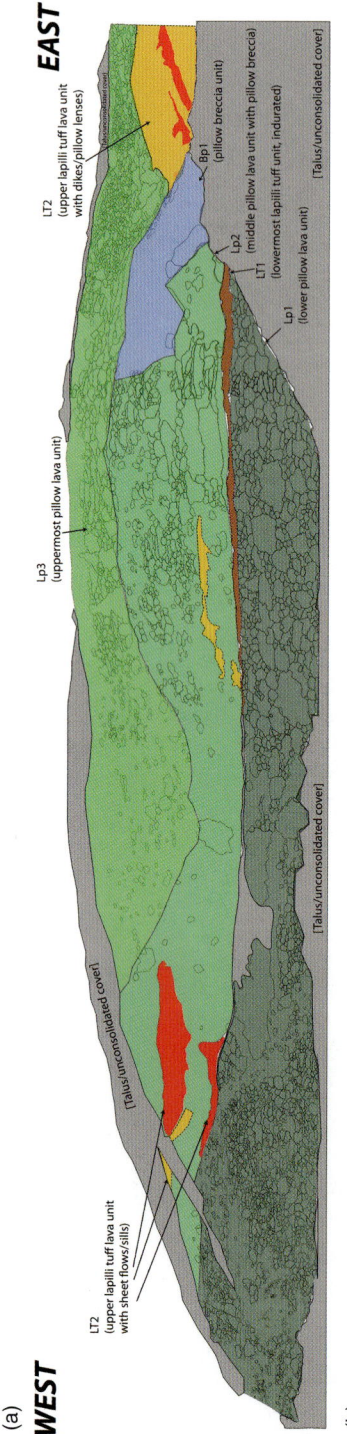

LT2
(upper lapilli tuff lava unit
with sheet flows/sills)

Lp3
(uppermost pillow lava unit)

LT2
(upper lapilli tuff lava unit
with dikes/pillow lenses)

Bp1
(pillow breccia unit)

Lp2
(middle pillow lava unit with pillow breccia)

LT1
(lowermost lapilli tuff unit, indurated)

Lp1
(lower pillow lava unit)

[Talus/unconsolidated cover]

[Talus/unconsolidated cover]

[Talus/unconsolidated cover]

(b)

Fig. 10.3 View and geological interpretation of a pillow ridge exposed in Undirhlíðar quarry, Reykjanes Peninsula, Iceland (from Pollock et al., 2014). A variety of evidence, including internal surfaces, some draped by lapilli tuff and tuff breccia, plus petrological differences, suggests that the edifice may be polygenetic. A black and white version of this figure will appear in some formats.

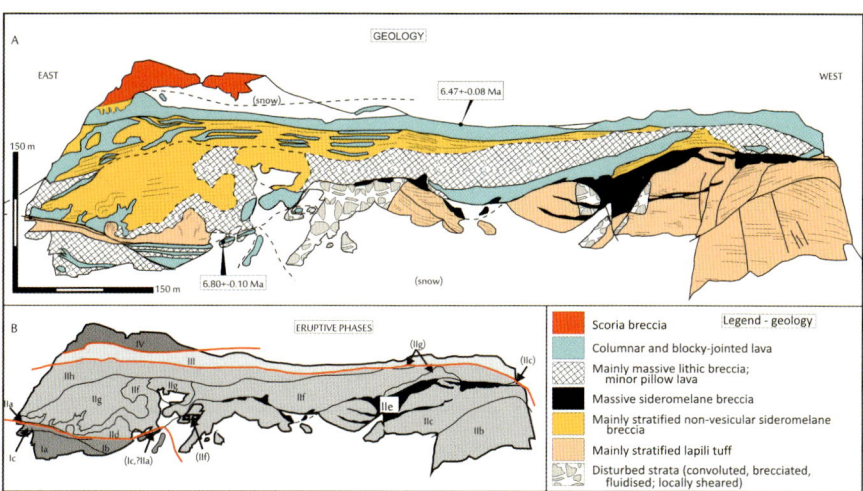

Fig. 10.4 Geological map of Icefall Nunatak, Marie Byrd Land, Antarctica (modified after Smellie, 2001). Although initially regarded as a monogenetic tuya, the outcrop is polygenetic and was constructed in at least four principal stages. A black and white version of this figure will appear in some formats.

Fig. 10.13 View of a passage zone in an 'a'ā lava-fed delta, Minna Hook, Mt Discovery, Antarctica. The passage zone is a very coarse feature that extends many metres. Note also that the subaerial capping lavas of the delta are markedly thinner and more evenly planar compared with those within the underlying passage zone, which are much thicker, have very irregular shapes and are enclosed in variably palagonite-altered breccia. The sharp top of the delta is glacially eroded (white line) and overlain by stratified sediments and a much thinner 'a'ā lava-fed delta with a very thin subaqueous section. Compare with Fig. 10.12. A black and white version of this figure will appear in some formats.

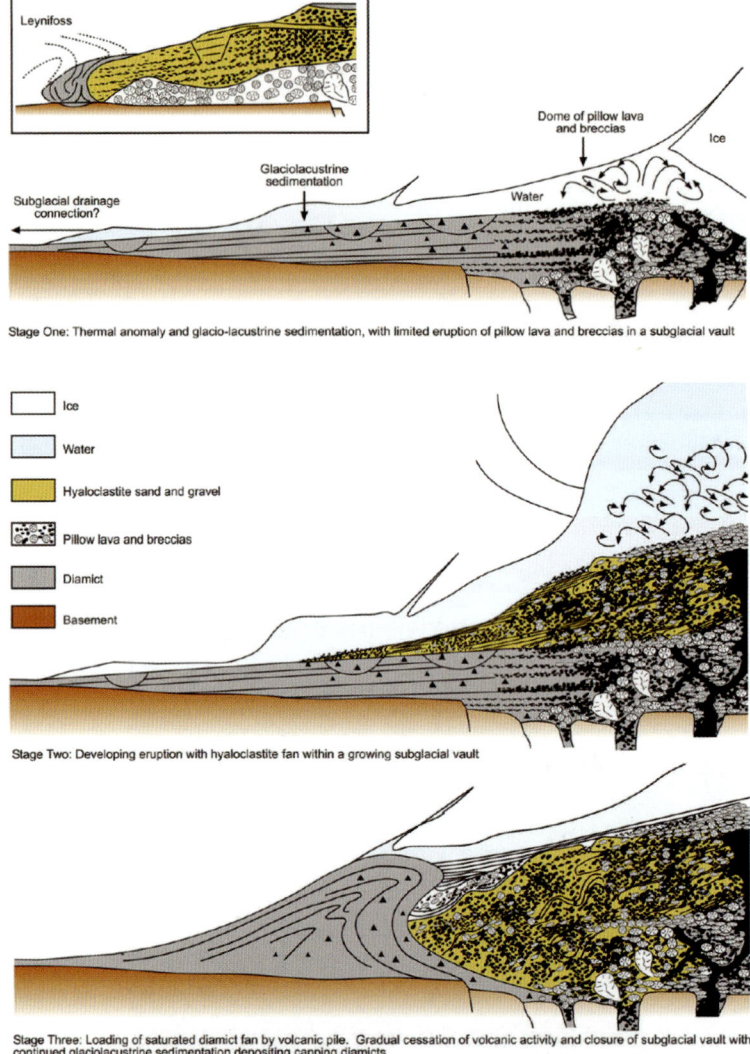

Fig. 10.9 Cartoons illustrating the formation of large-scale deformation structures by loading of lapilli tuffs into diamict and glaciolacustrine deposits flanking the Brekknafjöll tindar, Jarlhettur, Iceland (from Bennett et al., 2009). The section is sketched transverse to the tindar. Note that the authors' use of 'hyaloclastite' in the diagram is synonymous with lapilli tuff in this book. A black and white version of this figure will appear in some formats.

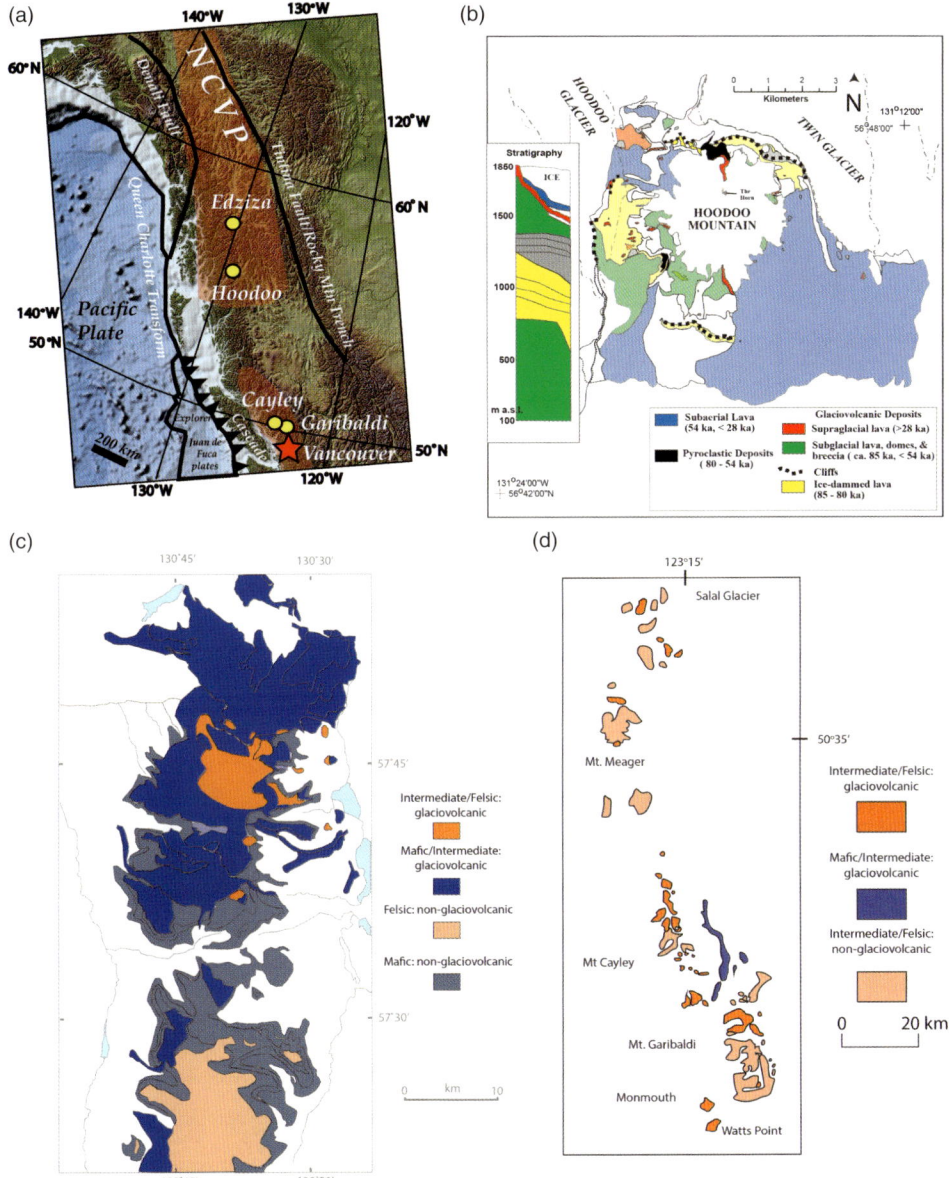

Fig. 11.1 Maps showing the main areas of intermediate glaciovolcanism in British Columbia. (a) Map of western Canada showing locations of detailed maps (b)–(d). (b) Schematic geological map of Hoodoo Mountain volcano showing glaciovolcanic deposits (after Edwards et al., 2000). (c) Generalised geological map of the Mt Edziza volcanic complex showing glaciovolcanic and non-glaciovolcanic deposits (after Souther, 1992). (d) Schematic map showing glaciovolcanic and non-glaciovolcanic deposits in the Garibaldi segment of the Cascade volcanic arc (after Kelman, 2005). A black and white version of this figure will appear in some formats.

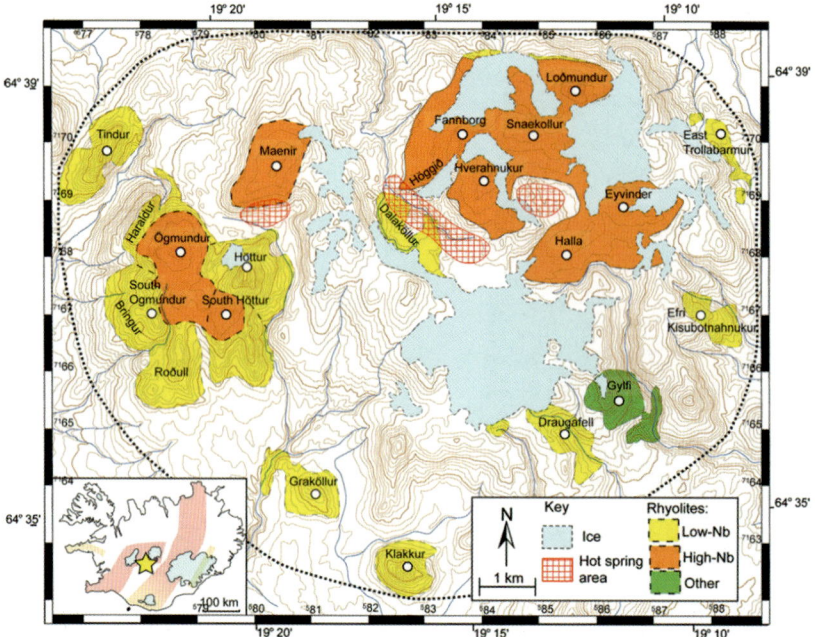

Fig. 12.8 Geological map showing the distribution of rhyolite volcanic centres in the Kerlingarfjöll volcanic complex, central Iceland (from Flude et al., 2010). The pecked line represents the approximate limit of the volcanic complex. A black and white version of this figure will appear in some formats.

Fig. 12.13 Sequence of rhyolite lava domes at Mandible Cirque, Victoria Land, Antarctica. The basal lava domes were subglacially emplaced and are separated by sharp uneven surfaces (red lines), at least some of which are glacial, overlain by diamict (including tillite) and stratified fluvial sediments. The rock face is c. 400 m high. A black and white version of this figure will appear in some formats.

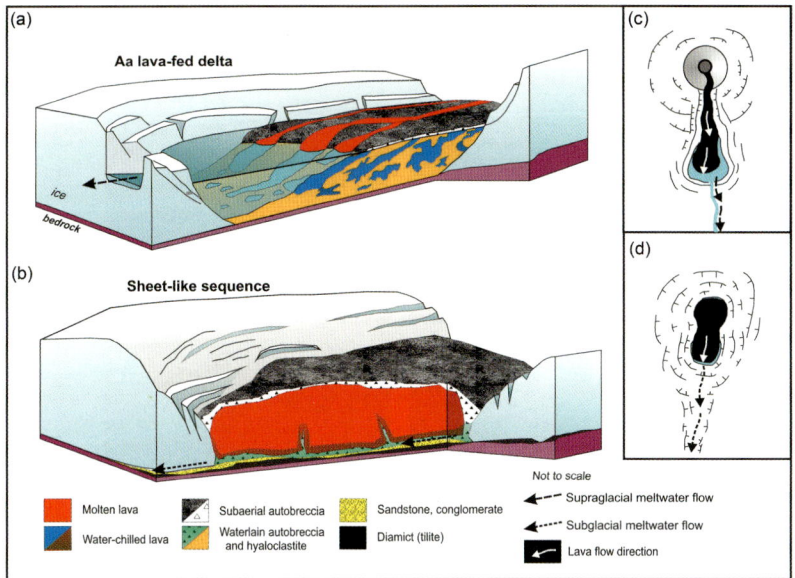

Fig. 12.14 Schematic perspective views depicting the major constituent lithofacies and inferred mode of emplacement of (a) 'a'ā lava-fed deltas and (b) felsic sheet-like sequences based on glaciovolcanic outcrops in northern Victoria Land, Antarctica (from Smellie et al., 2011b). A black and white version of this figure will appear in some formats.

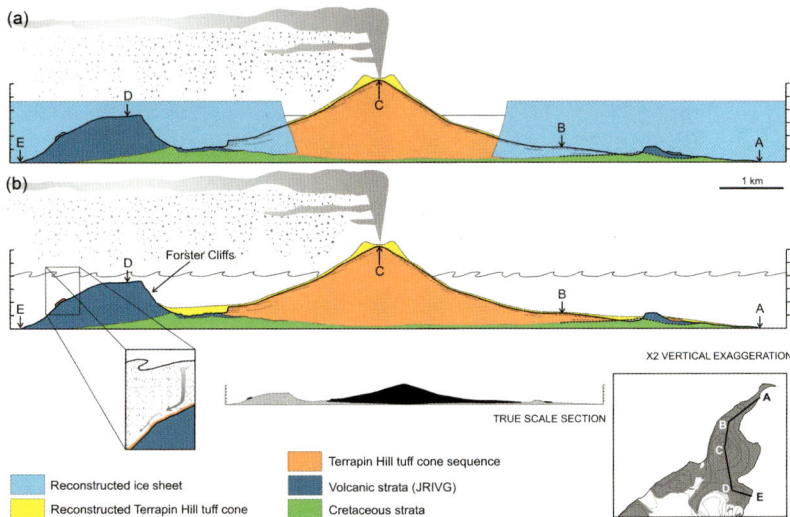

Fig. 13.16 Cartoons illustrating how volcano morphology can help to discriminate between eruptive environments. In (a), an ice sheet is reconstructed around a tuff cone (<660 ka) on James Ross Island, Antarctica. It is unlikely the tuff cone could have melted a hole in any overlying ice wider than the 3 km diameter hole sketched. The ice walls impinge unacceptably on the asymptotic tuff cone flanks. (b) A marine setting is thus favoured, helping to explain distal occurrences of tephra up to 5 km from the vent, some behind a prominent bedrock barrier (D), and the homoclinal dips of the constituent tuff beds, which are parallel to the surface slope all the way to the base of the cone (from Smellie et al., 2006a). A black and white version of this figure will appear in some formats.

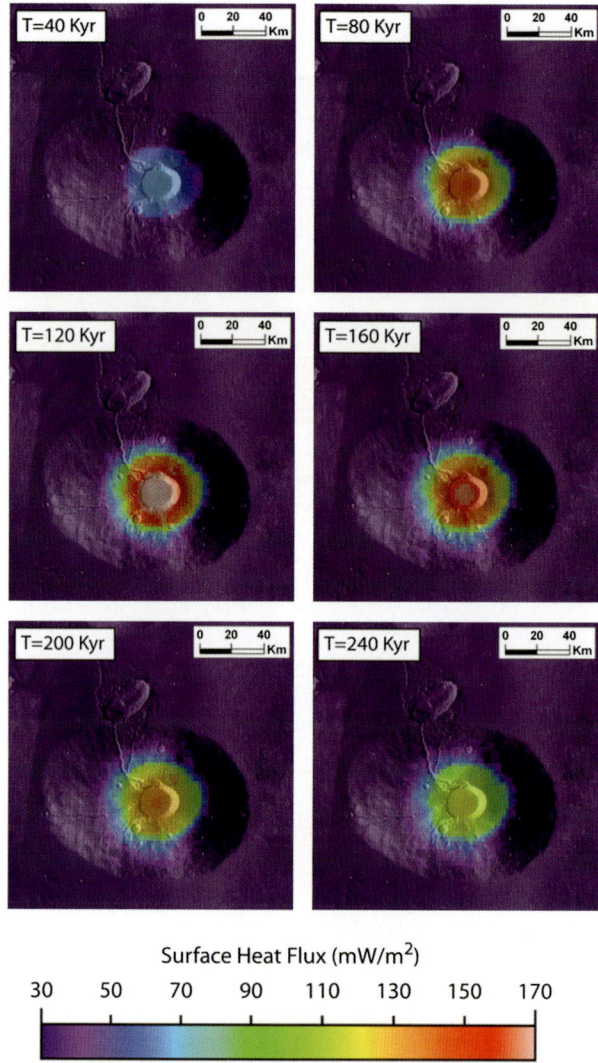

Fig. 13.14 Results of a conductive heating model on heat flow at the surface of a Mars volcano (from Fassett and Head, 2007). Note that the maximum lateral extent of the raised isotherms is about one caldera diameter on either side of the summit caldera and that most of the volcano surface area is unaffected. A black and white version of this figure will appear in some formats.

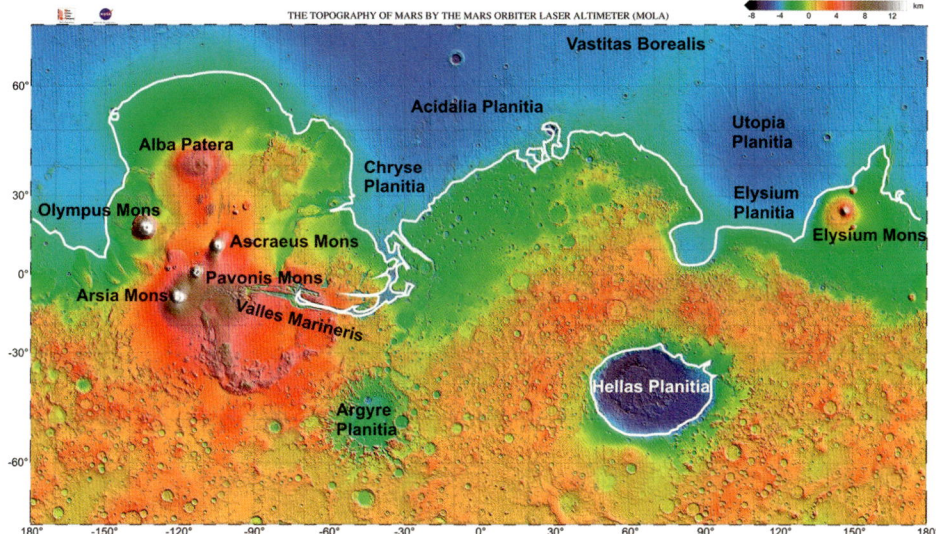

Fig. 16.1 Mars Orbiter laser altimeter (MOLA) map of the topography of Mars showing the pronounced separation into the two discrete physiographical provinces known as the Mars dichotomy. The white line is a depiction of a possible shoreline of a postulated Noachian ocean associated with the northern low-lying plains (shoreline from Clifford and Parker, 2001). A black and white version of this figure will appear in some formats.

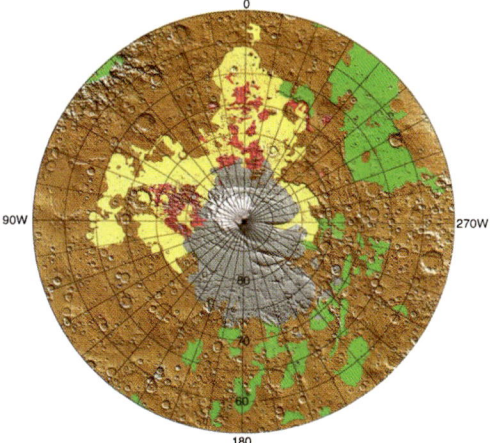

Fig. 16.5 Geological map of Mars' southern hemisphere showing the present ice cap (white), Amazonian polar layered terrain (grey), Hesperian–Noachian undivided (purple), and the Hesperian Dorsa Argentea Formation (yellow) (from Head and Pratt, 2001). The outcrop of the Dorsa Argentea Formation, a likely glacial deposit that contains numerous putative glaciovolcanic edifices (Fig. 16.10), extends far beyond the present polar ice cap and suggests the presence of a much more widespread ice sheet in Hesperian time. A black and white version of this figure will appear in some formats.

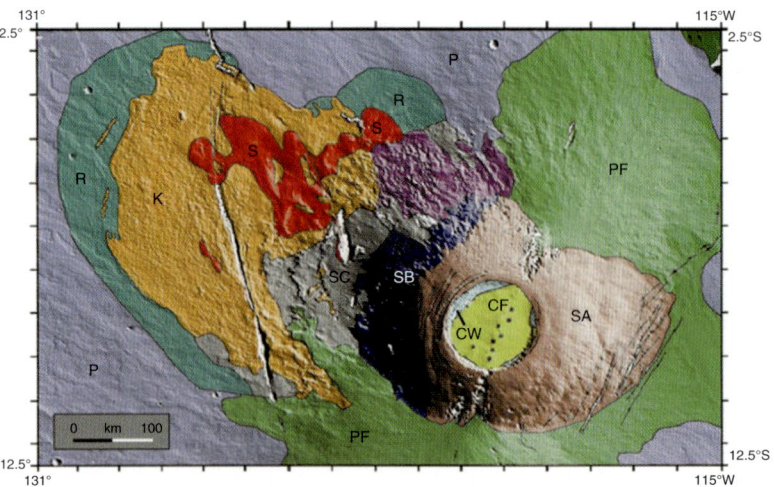

Fig. 16.6 Geological sketch map of Arsia Mons showing geological units believed to be glacial, including ridged (R), knobby (K) and smooth (S) deposits. They form a prominent fan-shaped deposit and are interpreted as drop moraines, sublimation till and rock glacier deposits, respectively (from Head and Marchant, 2003). A black and white version of this figure will appear in some formats.

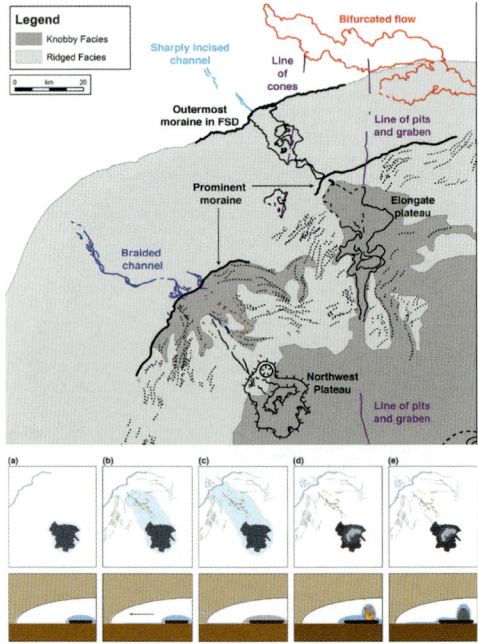

Fig. 16.12 Interpretation of the evolution of possible glaciovolcanic and associated features at Arsia Mons. (a) The Northwest Plateau (dark grey) may have been emplaced as a subglacial pillow sheet associated with a growing lens of meltwater. (b) The meltwater lens extends to the ice margin, causing the ice to stream (surge) and deposit moraines (thick black line in main map). (c) The lens drains catastrophically at the ice margin, forming jökulhlaup channels (blue lines in main map). (d) The eruption transforms from effusive to explosive, thermally eroding a tall englacial cavity above the vent. (e) Relict heat from the cooling volcanic products creates more meltwater that drains and transports debris in subglacial tunnels that is deposited as eskers (short olive green lines in main map) (from Scanlon et al., 2014). A black and white version of this figure will appear in some formats.

underlying water-saturated tephra pile (Smellie, 2001; cf. steam-driven palagonite alteration of subaerial 'a'ā autobreccia (see Section 9.4.2; Fig. 9.12); Smellie et al., 2013a) or it is possibly a later effect of weathering of the rock surfaces. Clastogenic lava and agglutinate may also be present. More rarely, scoria and bomb-bearing lapillistone is found in some lava-fed deltas where they are probably the localised product of short-lived littoral cone activity (Smellie et al., 2006c).

9.5 Coeval deformation features

Two broad types of syn-eruption deformation are present: (1) large-scale sector collapses of the volcanic edifice; and (2) small-scale folding and faulting. *Sector collapses* are major slumping events affecting many tens of metres of section that occur during the construction of tindar and subaqueous glaciovolcanic tuff cone edifices. They are probably responsible for the formation of sharp, convex-down curviplanar surfaces that are prominent in many tindars and interpreted as slump scars by Smellie (2000; Fig. 9.25). Although not unique to glaciovolcanic sequences, the slump scars are a particularly conspicuous feature. They may be more common in glaciovolcanic successions than in non-glacial equivalents. This is inferred because deposits constructing non-glacial tuff cones are unconstrained and beds are laid down on the cone flanks at or close to the stable angle of repose (Smellie et al., 2006a; Fig. 13.16); collapses may occur during or immediately following prolonged episodes of continuous-uprush activity, when the volcanic pile grows upward most rapidly and the supply of relatively coarse fragmental debris may pile up unstably surrounding the vent region. By contrast, deposits in glaciovolcanic tuff cones and tindars probably bank up

Fig. 9.25 Multiple large-scale convex-down arcuate slump-scar surfaces (white lines) in a mafic tindar. Kalfstindar, Iceland. The numerous surfaces attest to frequent collapses of the tindar flank during eruption. The rock face is c. 200 m high.

against enclosing ice walls and the unconsolidated debris will collapse readily when the supporting ice walls recede by melting during the eruption. In strike section the scars have shallow saucer shapes whereas in profile they are listric. They typically extend many tens to a few hundred metres laterally and dip outward away from the vent region(s). The deposits created by such collapses, consisting of massive tuff breccia (TB; Section 9.4.2, above), are not often preserved as they are generally advected to the base of the volcanic pile where they probably get removed by glacial erosion. In some instances very large detached blocks of coherent stratified lapilli tuff, individually up to a few tens of metres in diameter, are involved (Smellie, 2001, Fig. 6).

Small-scale *folding and faulting* are common in many tindars and subaqueous tuff cones (e.g. Smellie, 2001; Edwards et al., 2009b). The faults range in scale from those displacing beds by a few centimetres to metres, to rarer larger-scale faults offsetting tens of metres of strata. The latter are distinguished by their very steep inclination and planar morphology from the generally lower-angle listric (curved) slump-scar detachment surfaces. The structures vary from normal and reverse faults to thrust faults. The small-scale examples may also be listric and die out by becoming bed-parallel up- and down-section. Tuff matrix is commonly back-injected along the fault planes (Smellie and Hole, 1997; Smellie, 2001).

Deformation may also be suggested by the presence of packets of strata a few metres to a few tens of metres thick dipping at well beyond the angle of stable repose. These examples attest to slope instability affecting a more consolidated section and failure, with limited downslope displacement of the affected beds.

10

Mafic glaciovolcanic sequences

10.1 Introduction

Mafic volcanoes are the most abundant volcanic products on Earth and they also form the commonest glaciovolcanic sequences. They were the first glaciovolcanic sequence types to be described (see Chapters 1 and 2) and more mafic glaciovolcanic sequences have been described than any other. They include pillow volcanoes (mounds and ridges; possibly pillow sheets), tindars, tuyas, sheet-like sequences, ice-impounded lavas and polygenetic volcanoes.

10.2 Ice-impounded lavas

Published descriptions of mafic ice-impounded lavas are scarce and thus far are restricted to occurrences in British Columbia. Souther (1992) described examples in the Mt Edziza volcanic complex, a long-lived compound stratovolcano (see Section 10.7). Several valley-filling and subglacial basanite and nephelinite lavas crop out in the Llangorse Volcanic Field (Harder and Russell, 2007). That at Llangorse Mountain is a basanite 100 m thick and 350 m wide. It crops out in a prominent bluff that reveals its valley-confined nature (Fig. 10.1). The lava has well-developed columnar joints 1–2 m in diameter centrally with fanning orientations that define the orientation of the underlying valley floor. The lower lava margin has a prominent 30 cm-thick zone of quenched vesicular aphanitic basanite showing crude small (c. 1 cm diameter) polygonal joints resembling chaotic columnar or cube jointing and attributed to steam generation disrupting isotherms during water- or ice-mediated quenching (cf. Lescinsky and Fink, 2000). Unlike other ice-impounded sequences in British Columbia (see Section 11.2), the Llangorse lava conformably overlies 45–50 m of friable crudely stratified volcani-clastic sedimentary rocks, mainly diamictite, which may have provided the water that quenched the basanite. The diamicts are mainly composed of variably abraded basanite (including fluidal clasts and others with stretched vesicles indicative of pyroclasts) and palagonitised basanite glass together with clasts of peridotite and country-rock grano-diorite. They are believed to have been emplaced as debris flows (lahars) during floods that followed melting of snowpack or local ice during explosive eruptions, and were

Fig. 10.1 View of Llangorse Mountain lava flow showing overthickened north flank formed by lava that was impounded by former ice. The cliff face is 100 m high. (Geology after Harder and Russell, 2007.)

emplaced ahead of the arrival of basanite lava from the same eruption (Fig. 10.2). The lava may have provided significant heat (80–120 °C) which was maintained for several decades and which facilitated the *in situ* formation of palagonite in the diamictites. Harder and Russell (2007) noted that subaerial basanite lava flows have aspect ratios of c. 10^{-4} whereas the Llangorse lava has a ratio of 0.3, i.e. 3 orders of magnitude higher, indicating that it is overthickened (inflated), probably because of impoundment by a valley-wide barrier of ice. The Llangorse palaeotopography is thought to have been a valley drainage system blocked by ice at lower elevations.

10.3 Pillow mounds and ridges

Edifices composed mainly of basaltic pillows may be the commonest volcanic constructs on Earth (Walker, 1992), although most occur beneath Earth's oceans. However, glaciovolcanic examples are common and accessible in Iceland and British Columbia. Despite their abundance in many glaciovolcanic terrains (apparently not in Antarctica: LeMasurier, 2002), they are comparatively little studied. In his classic paper on pillow lava characteristics, Walker (1992) included measurements of glaciovolcanic pillow lavas in Iceland, but drew no distinction between the glaciovolcanic pillows and those erupted in other settings. The best-studied glaciovolcanic examples consist of ridges in British Columbia (Souther, 1992; Edwards et al., 2009b) and Iceland (Höskuldsson et al., 2006; Carrivick et al., 2009; Pollock et al., 2014).

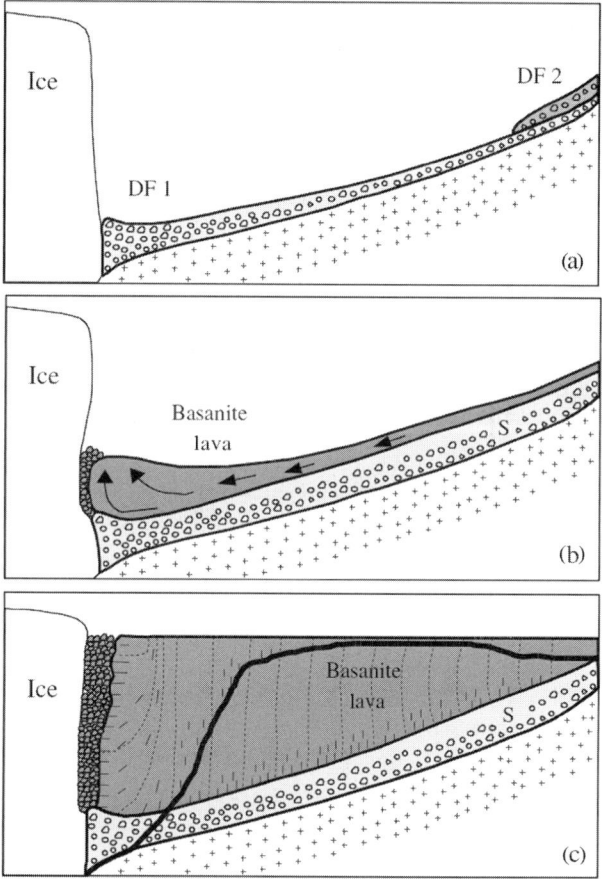

Fig. 10.2 Cartoon showing the consequences of ice impoundment on lava morphology; lava at Llangorse Mountain, British Columbia (from Harder and Russell, 2007). Vertical exaggeration > 15×. (a) Early stages, deposition of eruption-generated debris flows (DF) ponded at ice margin; (b) lava flow arrives and ponds against ice, inflating and overthickening at the flow front; (c) final post-eruption configuration, showing markedly inflated lava flow front. The light dashed lines represent columnar joints. The heavy solid line depicts present-day exposure level.

Pillow Ridge is part of the Edziza volcanic complex and is one of the most completely exposed glaciovolcanic ridges described so far. The ridge is c. 4 km long, 1 km wide, up to 450 m high, and has a present-day volume of c. 1.8 km^3. It formed from a fissure eruption on the northwestern flank of Mt Edziza (Souther, 1992; Edwards et al., 2009b). Pillow lava and fragmental lithofacies (tuff-breccia and lapilli tuff) are volumetrically sub-equal. The northern third of the ridge is dominated by five discrete sequences of pillow lavas. Lava pillows on the northeastern flank are orientated slope-parallel and may preserve the original shape of the edifice. However, the western side of the edifice has been dissected by an alpine glacier, providing an extensive ridge-parallel strike section.

It contains five stratigraphical packages of pillow lavas with surfaces mainly demarcated by coloration, although thin (<3 m) lenses of laminated fragmental rocks also occur locally. Lava pillow orientations (average plunges of 22°) consistently suggest downslope movement from south to north, turning west in the lowermost pillow unit at the north end of the ridge where it may follow a subglacial drainage pathway. Most of the associated dykes are ridge parallel and likely reflect the orientation of the eruptive fissure. Edwards et al. (2009b) showed that compositional changes occurred between the different pillow units consistent with formation as stages in an evolving magmatic system that was undergoing different degrees of fractionation, or eruptions of multiple magma batches. Souther (1992) proposed that the ridge began with explosive eruptions and construction of a subaqueous tuff cone, which was subsequently covered by a veneer of pillow lava fed from subaerial lava flows. However, Edwards et al. (2009b) discovered pillows at the base of the edifice, and H_2O concentrations in pillow rims are inconsistent with having been fed from subaerial lavas. They inferred that the eruption was initiated effusively in an englacial meltwater cavity and periodically experienced explosive episodes. They also used H_2O in pillow rim glasses to infer syn-eruption ice thicknesses of 300–1300 m (see Sections 6.6 and 13.5.1). Tsekone Ridge, another pillow ridge nearby, may represent a monogenetic eruptive unit (Souther, 1992; unpublished data of BRE). British Columbia also includes at least six other pillow ridges, but they have only been examined at reconnaissance level.

Pillow ridges are common throughout Iceland, but few have been described in detail. Höskuldsson et al. (2006) and Carrivick et al. (2009) described examples in the Kverkfjöll area of east-central Iceland, and Pollock et al. (2014) described another exposed in a quarry in southwestern Iceland. Höskuldsson et al. (2006) examined six separate ridges: Virkisfell, Kreppuhryggur, Lindarhryggur, Vegaskard, Karlshryggur and Karlsrani. The largest is Lindarhyrggur, which is 8 km long, 1 km wide and 0.3 km high with an estimated volume of 0.5 km^3, and the smallest is Karlsrani, only 0.25 km long, 0.1 km wide and 0.05 km high, with an estimated volume of 0.7×10^{-3} km^3. They focused mainly on the degassing history of the pillow lavas, although they also reported pillow dimensions. Many of the ridges have pillows with highly vesiculated cores, which were ascribed to rapid pressure release while the pillow cores were still molten. The pressure reduction may have been triggered by a jökulhlaup and was equivalent to the loss of c. 400 m of water in an englacial lake. Although Höskuldsson et al. (2006) did not provide detailed stratigraphical descriptions, Carrivick et al. (2009) suggested that many of the pillow ridges contain discrete eruptive units. The locus of the volcanism may also have moved from north to south, following the retreat of the northern edge of the Vatnajökull ice cap.

The most detailed published description of a pillow complex is that by Pollock et al. (2014). The quarry section they examined is cut into a glaciovolcanic ridge (Sveifuhals) that extends c. 7 km and has a maximum relief of 150 m (Fig. 10.3). The ridge is one of five fissure segments that together construct Reykjanes Peninsula in southwestern Iceland. Sveifuhals is composed of at least three discrete pillow lava units identified stratigraphically and compositionally. There are at least two distinct geochemical populations of pillow lavas and intrusions. Although the ridge is dominated by pillow lavas, like Pillow Ridge in

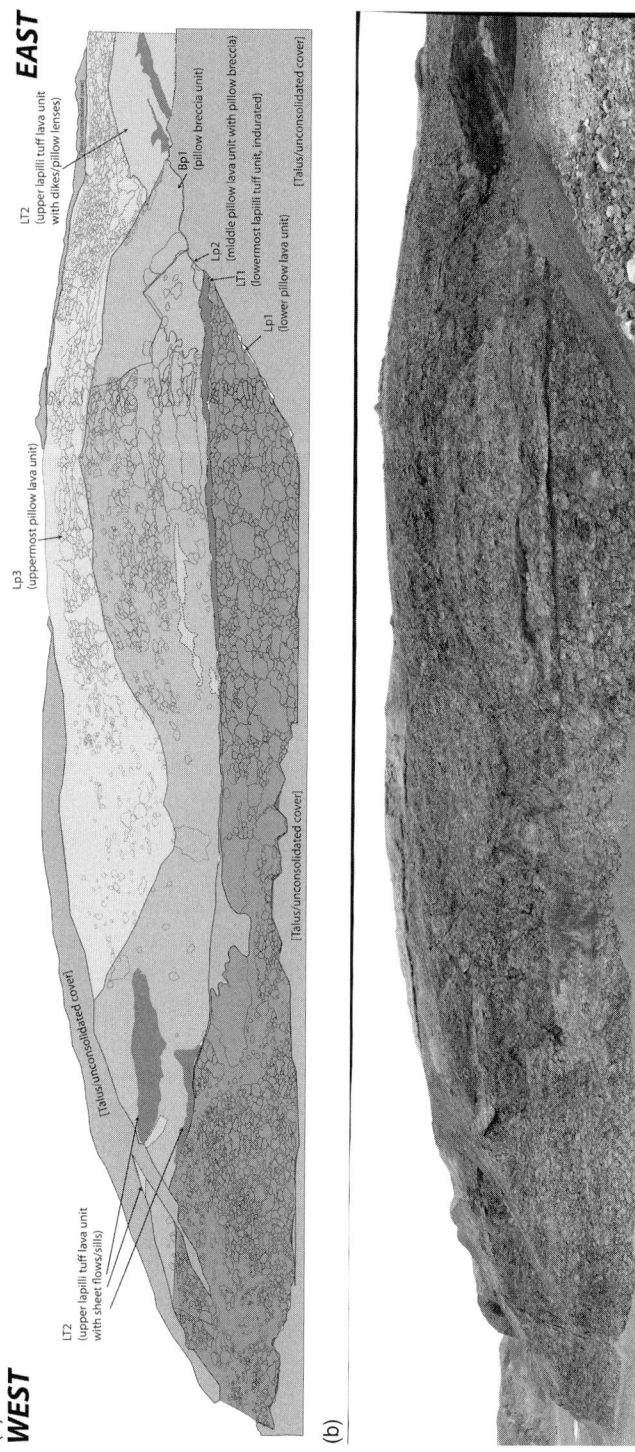

Fig. 10.3 View and geological interpretation of a pillow ridge exposed in Undirhlíðar quarry, Reykjanes Peninsula, Iceland (from Pollock et al., 2014). A variety of evidence, including internal surfaces, some draped by lapilli tuff and tuff breccia, plus petrological differences, suggests that the edifice may be polygenetic. (A black and white version of this figure will appear in some formats. For the colour version, please refer to the plate section.)

British Columbia (Edwards et al., 2009b), it also contains volumetrically and stratigraphically important units of tuff breccia and lapilli tuff, as well as intrusions.

While the study of glaciovolcanic pillow edifices is still in its initial stages, it is already
clear that detailed stratigraphical studies need to be coupled with geochemical investigations. Edwards et al. (2009b) and Pollock et al. (2014) showed that pillow ridges are more
than simple monogenetic piles. At both locations, sequential evolutionary relationships are
apparent involving glass-rich tuff breccia, cross-cutting dykes and pillow piles. Eruptions
may have commenced with volatile-driven subaqueous fire-fountaining, followed by effusion of pillows from a single feeder dyke. These studies are also beginning to unravel how
magma/lava is transported endogenously within growing glaciovolcanic pillow piles.
Whereas Kverkfjöll may be unique in the overwhelming preponderance of highly vesiculated cores within its lava pillows, variations in pillow lava vesicularity observed elsewhere
may provide important clues for understanding short-lived changes in eruptive environments (e.g. syn-eruption drainage events). Ultimately these processes may be critical for
understanding the initiation of many glaciovolcanic eruptions.

10.4 Tindars and tuyas

Mafic tindars and tuyas are regarded as monogenetic glaciovolcanic centres, although
diagnostic evidence for a monogenetic origin is generally assumed and seldom proven
(cf. Wörner and Viereck, 1987; Smellie, 2001; Section 10.4.1). They are common in Iceland
and British Columbia (e.g. Mathews, 1947; van Bemmelen and Rutten, 1955; Jones, 1969b,
1970; Hickson, 2000; Edwards et al., 2011). They also occur in Antarctica, although the
Antarctic glaciovolcanic province is dominated by large polygenetic shields and stratovolcanoes (Smellie et al., 1993; Skilling, 1994; Smellie and Hole, 1997; Smellie et al., 2006b,c,
2008, 2011a). Tindars and tuyas are described together here since tindars commonly evolve
up into tuyas and presumably most mafic tuyas have tindar (or tuff cone/tephra mound)
cores. Tindars consist of piles of mainly subaqueously erupted and deposited tephra. They
may be the commonest glaciovolcanic landform, even more numerous than tuyas
(Chapman et al., 2000; Schopka et al., 2006). Tindars usually rest on a coeval pillow
mound or ridge (Jones, 1969b, 1970; Werner and Schmincke, 1999; Skilling, 2009;
Edwards et al., 2009b; Section 10.3) but not all tindars have a pillow volcano stage
(LeMasurier, 2002; Schopka et al., 2006). Tindars represent an explosive eruptive phase
of relatively rapid vertical edifice aggradation that contrasts with the effusive phase of
lateral progradation characterised by most lava-fed deltas in tuyas (Smellie, 2000, 2009,
2013). Most tindars do not evolve into tuyas (e.g. Russell et al., 2014). Tindars differ from
glaciovolcanic tuff cones in that the latter are single centres whereas tindars are compound
edifices formed by multiple overlapping tuff cones erupted simultaneously along
a subglacial fissure (see Glossary). As a result, tindars can be defined on geomorphological
criteria (i.e. ridges with length to width ratio greater than 2:1; Jakobbson and
Gudmundsson, 2008; Russell et al., 2014). Together with glaciovolcanic tuff cones, they

can all be referred to informally as products of a 'tindar phase' of essentially subaqueous explosive hydromagmatic eruption (Smellie, 2000), although many tindars and glaciovol-canic tuff cones are subaerial in their final eruptive stages.

Intriguingly, neither of the namesakes for tuyas (Tuya Butte, British Columbia) nor tindars (Kalfstindar, Iceland) have been described in detail, although studies have been undertaken at both localities (Mathews, 1947; Allen et al., 1982; Moore et al., 1995; Jones, 1969b, 1970) and by the authors. Thus, we focus here on the results of four independent modern investigations of other tindars and tuyas. The outcrops are similar in overall evolution but each selected example shows different important aspects of tuya and tindar construction. Two of the edifices are tindars only (Helgafell and Breknafjöll (Jarlhettur)) whereas two evolved into tuyas (Icefall Nunatak and Hlödufell).

10.4.1 Icefall Nunatak, Antarctica

Icefall Nunatak is a small eroded satellite centre situated on the west side of Mt Murphy, a Late Miocene (9–8 Ma) alkaline basalt–trachyte stratovolcano in eastern Marie Byrd Land, Antarctica. It is constructed from the products of at least four eruptive episodes (stages I–IV) that took place between 6.80 and 6.47 (or 6.52) Ma (Smellie, 2001; Wilch and McIntosh, 2002). Although polygenetic, it is small (similar in size to many monogenetic tuyas). It also displays all the main stages of tuya construction, it is important for the unusually wide range of lithofacies present, especially those associated with edifice instability, and it was used to develop a significantly improved general model for tuya construction (specifically, one that identified meltwater overflowing from a glacial vault through a permeable supraglacial layer of snow and firn as a primary control on water levels in the vault, and the first documented evidence for basal meltwater escape; Smellie, 2001; also Smellie, 2006; see Section 10.4.5). Despite compositional similarities (all the magmas are alkali basalts), textural differences are evident between each eruptive stage, mainly differences in the types, proportions and sizes of phenocrysts. Each stage is also separated by a prominent unconformity. The nunatak is composed mainly of subaqueously emplaced lavas and volcaniclastic lithofacies. The latter are mainly lapilli tuffs dominated by juvenile glassy pyroclasts with blocky shapes and variable vesicularity characteristic of explosive hydromagmatic eruptions. Non-juvenile clasts are rare, which suggests that the fragmenta-tion of pyroclasts took place high in the vent, essentially within the tephra pile itself, and indicates that the vesiculating magma interacted with water-saturated tephra probably beneath surface water (a cavity-confined meltwater lake), *sensu* Kokelaar (1986, Fig. 2). The presence of blocky jointed (water cooled) and locally pillowed lava is consistent with the presence of a meltwater lake.

Basal exposures (Stage I of Smellie, 2001) consist of basalt lavas and breccias that were emplaced wholly subaqueously probably in a subglacial cavity. It may correspond to an early-formed pillow volcano stage of tuya construction, despite the paucity of pillows. Only 60 m of section are exposed (base unseen; Fig. 10.4). Stage I is dominated by faintly bedded

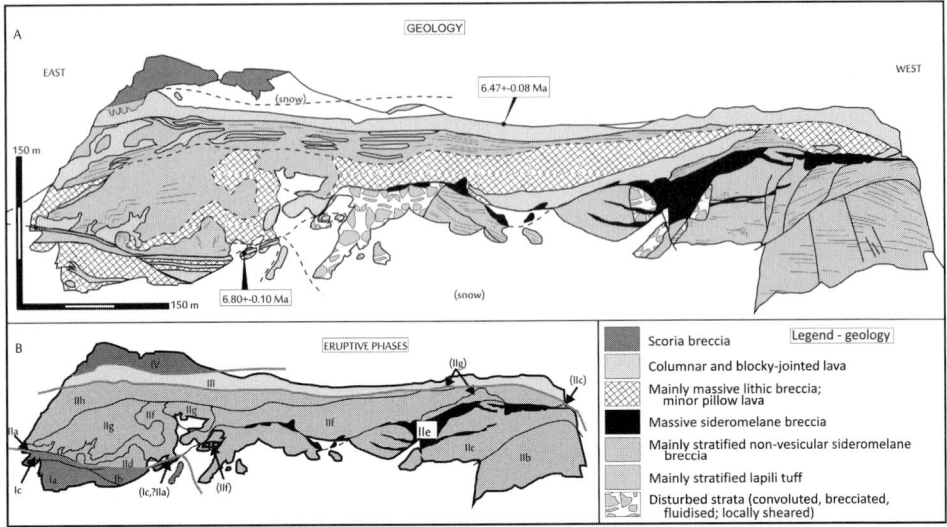

Fig. 10.4 Geological map of Icefall Nunatak, Marie Byrd Land, Antarctica (modified after Smellie, 2001). Although initially regarded as a monogenetic tuya, the outcrop is polygenetic and was constructed in at least four principal stages. (A black and white version of this figure will appear in some formats. For the colour version, please refer to the plate section.)

lithic orthobreccia with dispersed pillows and irregular lenses and lobes of blocky-jointed and pillowed lava a few metres thick (unit Ia). Sandy pockets of sideromelane are also present but larger clasts are finely crystalline. It is overlain by massive gravelly breccia with irregular lenses of scoriaceous and non-vesicular lava with lobes that intrude the surrounding breccia, and the deposit is overlain by two thin (1–6 m) scoriaceous sheet lavas separated by planar bedded gravelly breccias (unit Ib). The lavas may be fed by a vesicular dyke. Conformably overlying unit Ib is a wedge >30 m thick composed of crudely stratified gravelly and blocky breccia with dispersed pillows and continuous beds of graded gravelly sandstone with erosive bases (unit Ic). The Stage I breccias formed as joint-block deposits by mechanical breakage of subaqueously emplaced chilled lava, probably aided by hydrofracturing and steam explosions. The association with dispersed pillows and pillowed lenses suggests that the lavas were close to the transition between pillowed and blocky forms consistent with relatively high strain rates (favouring brecciation) and rapid magma discharge on steep (c. 30°) slopes. The presence of scoriaceous lava and scoria in unit Ic was ascribed to isolation of a steady-state subaqueous eruption column within a steam cupola (Kokelaar, 1983, 1986; Kokelaar and Busby, 1992) but it is also possible that the lavas (and unit I itself) are 'a'ā, and the scoria lava tops are subaerially formed clinkers that were advected underwater (cf. Smellie et al., 2011a, 2013a).

Stage II deposits dominate Icefall Nunatak. At the base is a thin (2 m) tabular deposit of interbedded sandstone and breccia showing grading and channelised unidirectional trough cross-stratification (unit IIa). It represents a period of traction sedimentation and basal meltwater escape from a subglacial cavity that had been postulated but not observed in an outcrop previously (cf. Smellie, 2000). The remainder of unit II comprises two main parts: a tindar phase (units IIb,c) and a tuya phase (units IIf–h) separated by a period of major edifice collapse(s) (units IId,e). The tindar phase is dominated by orange-yellow lapilli tuff (described by Smellie (2001) as gravelly sandstone in deference to the then-prevailing system of volcaniclastic nomenclature based on Cas and Wright, 1987). The lapilli tuffs are fine ash-poor and planar stratified, with diffuse bed boundaries and common amalgamation. They are gravity flow deposits, probably high-concentration density current deposits rather than turbidites (cf. Smellie, 2001; Fig. 9.22). They were emplaced in rapid succession during periods of sustained explosive eruptive activity known as continuous-uprush episodes, which provided a virtually continuous tephra supply, causing relatively rapid upward aggradation of a subaqueous tuff cone. There is also a much smaller proportion of thin-bedded coarse and fine tuffs showing better bed definition (sharp planar surfaces, locally erosive) and a variety of structures produced by loading and grading (Fig. 9.21). The tuff units are turbidites but they accumulated only sporadically, in more widely spaced pulses. They probably correspond to products of small vent-clearing explosions or discrete tephra-jetting activity associated with relatively low rates of magma supply during quiescent periods. Units IIb and IIc differ in colour, with unit IIb predominantly brown owing to the presence of significant fine ash, whereas most fine ash apparently bypassed the edifice to create the fines-poor lapilli tuffs of unit IIc. The two units are separated by a sharp, planar, discordant surface with local folding of adjacent beds that is the expression of coeval faulting. Eruption columns responsible for units IIb and IIc might have been entirely subaqueous, as suggested by the absence of accretionary lapilli.

Partial edifice collapse followed the construction of the subaqueous tuff cone and probably occurred in more than one episode. It is represented by several large blocks of lapilli tuff up to 20 m diameter with a variety of tilted attitudes (unit IId) that overlie the faulted or collapsed upper surface of unit IIc, and massive grey monomict orthobreccia (unit IIe) that unconformably drapes the fractured top of unit IIc and occupies fractures in the same. The latter is composed of blocky poorly vesicular lithic clasts and sideromelane, minor basalt pillows and <10% sideromelane ash matrix and may represent remobilised breccia associated with a tuya phase. Additional evidence for syn-eruptive instability consists of massive tuff composed of domains of heterogeneous mixtures of tuff blocks in tuff matrix and tuff-rich domains with ghost-like and swirling textures (lithofacies Z1), and slurry-like mixtures of tuff clasts dispersed in dark fine ash-rich tuff (lithofacies Z2). Z1 formed by the disaggregation and fluidisation of poorly consolidated and unconsolidated tuff during probable dewatering induced by earthquakes and phreatomagmatic detonation, whilst Z2 represents piecemeal sinking of rapidly deposited coarse ash into underlying fine ash along with concurrent dewatering of a further underlying coarser ash bed. Slumped folded lapilli tuff and tuff beds are also present, often present along an underlying decollement surface). There is a spatial association between the disturbed lithofacies (Z1 and Z2) and the faulted surface of unit IIc, suggesting that they may be related to collapse of the cone flank, including areas in which there was high water escape from coeval syn-sedimentary faults (lithofacies Z1). Fluidisation may be induced by momentary faulting-related pressure relief, causing pore water in the pile of tephra (overpressured due to rapid accumulation) to expand into the faults and triggering focused pore water discharge. In context, it is noteworthy that large-scale (outcrop-wide) syn-sedimentary slump scars caused by repeated flank collapses are a notable feature of tindars generally, and not just at Icefall Nunatak (Smellie, 2000). They attest to the instability of tall piles of poorly consolidated wet tephra buttressed by ice walls that melt back progressively and thus remove their support (Fig. 9.25).

At Icefall Nunatak, a tuya stage is represented by the final part of stage II (units f–h), and possibly also by stage III. In stage II, the tuya stage is composed of polyhedral blocky jointed and columnar lava and crudely stratified orthobreccia. The proportion of massive lava differs in the different units (greatest in unit IIf, least in IIg). The lavas are sheet lavas that are now known to be able to flow long distances underwater compared with pāhoehoe lavas (Mitchell et al., 2008; Tucker and Scott, 2009; Smellie et al., 2013a). The intrusive relationships between the lavas and associated breccias, including mushroom-like apophyses on the tops of some lavas, are features shown by subaqueously emplaced lavas in 'a'ā-fed deltas (Smellie et al., 2013a). By contrast, stages III and IV comprise a small scoria cone and laterally extensive water-cooled columnar lava fed from a neck. At the east end of the outcrop, the lava is associated with massive breccia, similar-looking to that in units IIf–h. The scoria cone is locally altered to palagonite, possibly related to local high temperature gradients set up by dyke intrusion in wet tephra underlying the dry scoria pile causing the upward transmission of steam. Stages III and IV evidently fed volcanism from at least two discrete vents, in different locations from those responsible for the earlier stages.

10.4.2 Hlödufell, Iceland

Hlödufell is a flat-topped basaltic massif within the Western Volcanic Zone south of Langjökull, southwest Iceland. It is 1186 m high, 2 km wide, 3 km long and rises about 700 m above the surrounding plain (Fig. 8.4b). The massif is a tuya with an evolution subdivided into four discrete growth stages on the basis of the lithofacies architecture (Skilling, 2009). Compared with Icefall Nunatak, there are no erosional or time-significant unconformities between the stages and no compositional variations that would imply a polygenetic origin, and the entire edifice is regarded as monogenetic although it was constructed from more than one vent. It has an age of 8.2 ± 1.0 ka based on cosmogenic ^{3}He dating (Licciardi et al., 2007). The tindar stage is less well exposed than at the other outcrops included in this section but a basal pillow pile and the lava-fed delta capping units are particularly well seen. Unlike Icefall Nunatak, the feeding lavas are pāhoehoe and two lava-fed deltas are superimposed, reflecting a rising water level during eruption.

Stage I comprises a subglacial lava complex, equivalent to a pillow volcano stage (Skilling, 1994; Smellie and Hole, 1997; Smellie, 2001, 2013; cf. basal pillow complex of Werner and Schmincke, 1999). It consists of ridges and mounds of pillow lava, with

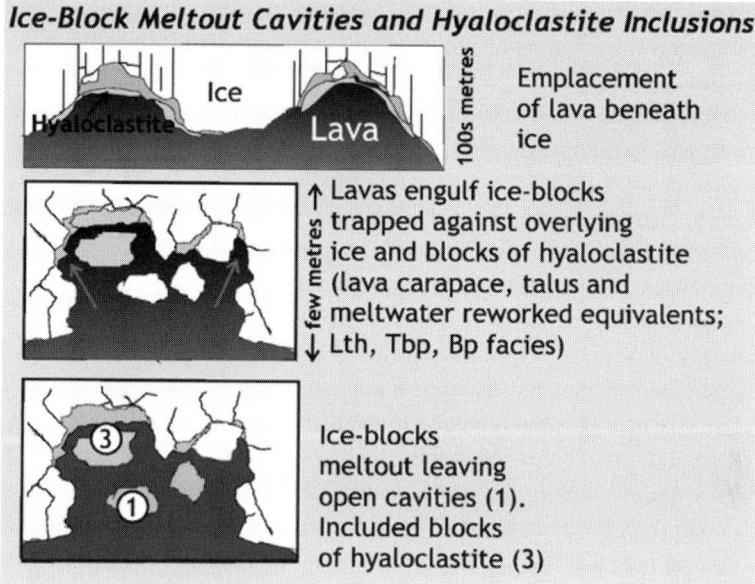

Fig. 10.5 Interpretation depicting the formation of ice-block meltout cavities on the surface of pillow lava at Hlödufell (from Skilling, 2009).

blocky and curvicolumnar-jointed tabular lavas more abundant than pillow lava, and associated with minor lava-derived lithic breccia and hyaloclastic lapilli tuff. The ridges and mounds are up to 350 m long and typically 30–70 m high, with steep margins and thicknesses up to 240 m. They occur up to an elevation of 750 m a.s.l., comparable with the summit of the isolated coeval(?) pillow mound of Thórólsfjell close by to the north, suggesting a common confining pressure control on the initiation of explosive hydromag-matic eruptions, i.e. minimum ice thicknesses of 250 m (summit plateau elevation of Hlödufell minus summit elevation of pillow mounds) required to suppress explosivity). The blocky jointed lavas are subaqueous sheet lavas emplaced at higher effusion rates than the associated pillows (Griffiths and Fink, 1992) and many were probably intruded endo-genously within the mound cores. Ice-block meltout cavities are common in the blocky jointed lavas, as are blocks of massive and bedded breccia (called 'hyaloclastite inclu-sions'). Interpretation of the ice-block meltout cavities suggests that they formed by brittle fracture of enclosing ice followed by displacement and rotation of ice blocks, which were trapped and engulfed by lava and melted (Fig. 10.5; cf. Fig. 9.9). This 'stoping' mechanism is another unique method by which subglacial lava can create space without direct melting of ice walls. Distinctive ice-contact surfaces are also locally well displayed, comprising pillows and lavas with common steep flat to mamillar glassy surfaces formed where they directly abutted ice. Mini pillow drips also occur on some of these surfaces and are interpreted as lava breakouts formed after the confining ice had melted back and created a narrow water-filled gap. Subhorizontal 'lava shelves' <1 m thick and >10 m long locally protrude from the steep ice-contact surfaces and may represent lava moulds of former ice fractures. They may represent an early stage in the 'stoping' process leading to the ice-block meltout cavities.

Stage II is represented by waterlain tephra that are products of hydromagmatic explosive activity, equivalent to the hyalotuff or tindar stages of Skilling (1994) and Smellie (2000, 2001; cf. tephra mound of Hickson, 2000, and Smellie, 2006). A dissected cone is preserved at one end of the massif summit and is probably the main vent for the tephra. The deposits have quaquaversal dips and comprise bedded lapilli tuffs together with various pillow-sourced breccias and pillow lava layers. Water depths greater than 200 m were suggested from the absence of bomb sags in the lapilli tuffs (cf. White, 2000). The breccias were probably mainly deposited from concentrated density currents whilst the thinly bedded lapilli tuffs may be subaqueous suspension deposits. Some lapilli tuffs are cross bedded, implying the existence of coeval traction currents and a vault that was actively draining, similar to interpretation of basal cross-bedded units at Icefall Nunatak described by Smellie (2001).

Stage III comprises the stratigraphically highest zone composed of subaerial pāhoehoe lava and a cogenetic lava-fed delta that drapes and forms the majority of the exposed rock on Hlödufell. The dominant lithofacies consist of steeply dipping tuff breccia and pillow breccia together with less common subaerial lava breccias. They are overlain across a horizontal passage zone by subaerial lava. The passage zone shows none of the prominent steps described by Smellie (2006), indicating a more or less stable surface to the associated meltwater lake (see Chapter 13). The subaerial lava breccias contain a variable proportion (mainly >75%) of crystalline subaerial lava clasts. They were deposited on the steep subaqueous delta slope following slope failure and bench collapse resulting in the disintegration of overlying pāhoehoe lava (Skilling, 1994, 2002).

Stage IV is the upper subaerial lava-fed delta sequence, fed by a well-preserved vent that now forms the summit of Hlödufell, although a second subsidiary cone is also present; both vents differ from those responsible for stages II and III products. The passage zone is also horizontal and the difference in elevation between stage III and IV passage zones is 150 m, indicating that coeval water levels rose between the two stages (cf. Smellie, 2006). This also indicates that the surface of the stage III meltwater lake was at least 150 m below the surface of the encircling ice. A series of prominent benches is present on the stage IV lava-fed delta and has a variety of possible origins, with glacial erosion considered the most likely.

10.4.3 Helgafell, Iceland

Helgafell is a well-preserved and well-exposed basaltic tindar in southwest Iceland, situated on the Reykjanes Peninsula south of Reykjavik (Fig. 10.6). It formed during an eruption under wet-based Pleistocene ice (probably Weichselian age; <110 ka) at least 500 m thick and possibly situated c. 15 km from the contemporaneous ice front (Schopka et al., 2006). Helgafell measures 2 km in length, 0.8 km wide, with a summit at 340 m a.s.l. that rises

Fig. 10.6 View, looking northwest, of Helgafell, a mafic tindar in southwestern Iceland. The hill rises c. 200 m above the surrounding post-glacial lavas (image: Mary Chapman).

200 m above the surrounding plain. Its total volume is c. 0.15–0.17 km^3, including an estimate for products currently buried by post-glacial lavas. Unlike construction of the tuyas at Icefall Nunatak and Hlödufell, Helgafell is solely a tindar formed of explosively erupted hydromagmatic tephra and associated intrusions. Gravity modelling indicates that the entire edifice has a low bulk density (1800 kg m^{-3}) suggesting that a basal pillow mound (bulk density 2300 kg m^{-3}) is absent; no subaerial lava capping unit is present. The feeding magma was an olivine tholeiite. There are no significant unconformities within the outcrop indicating that Helgafell is a monogenetic volcano.

Helgafell is dominated by lapilli tuff (described by Schopka et al., 2006, as hyaloclastite, *sensu* Fisher and Schmincke, 1984) but it also has isolated outcrops of pillow lava and intrusions, together amounting to just 1–2% by volume of the edifice. Two representative sections were described by Schopka et al. (2006). Section F, on the southeastern side of the edifice, consists of massive to poorly bedded lapilli tuff dominated by sideromelane (70–90% by volume) and including glassy juvenile scoria (bombs) with fluidal and ragged shapes. Lithic accessory clasts make up the remainder of the clasts and become somewhat more abundant up-section. Alignment bedding picked out by trails of larger clasts is the commonest bedding indicator but rare faint trough cross bedding and laterally discontinuous ash-grade laminae are also present, as are rare slide blocks of stratified lapilli tuff. The irregular basaltic intrusions present are coeval and commonly have peperitic and jigsaw-fit textures. They measure up to tens of metres thick and wide and have glassy margins. The lapilli tuffs were probably transported and deposited from eruption-fed gravity flows, probably hyperconcentrated density flows (Mulder and Alexander, 2001), during periods of high mass discharge that may correspond to continuous uprush events. A small proportion of the erupted fines were deposited from suspension during short quiescent periods yielding the fine laminae. In such a scenario of rapid aggradation, it was speculated that the deposits may have been dumped against an adjacent ice wall.

Section A, on the western side of the edifice, differs from section F in being composed of much better bedded lapilli tuffs. The beds in the basal third of the section are generally 0.5 m thick, with erosional bases. They comprise a basal massive layer that becomes normally graded upward, finally changing up into planar or cross-laminated fine ash. Loading, flame structures and clastic injections are also present. Other beds show normal and inverse grading and some are entirely massive. The remainder of the section consists of beds 0.1–2 m thick commonly showing alignment bedding together with finer ash layers showing planar or cross laminations. A 30 m-thick unit of convoluted layering is also present, and intact and fragmented pillows are common locally. Although the basal beds were interpreted as mainly high-density turbidites, the coarse grain size probably argues against such an origin (cf. Mulder and Alexander, 2001). They are probably better regarded as deposits of concentrated density flows. The overlying less well-bedded section was interpreted as a product of cohesionless mass flow deposits, possibly laid down from debris flows. However, their characteristics are also compatible with a density flow origin, probably as hyperconcentrated flow deposits similar to sediments in section F. The beds were said to be better sorted than in section F, a feature attributed to greater travel before deposition, in a setting where the confining ice walls were further away. The finest ash layers may have sedimented from suspension during short quiescent intervals (cf. Smellie and Hole, 1997). The convoluted strata attest to slope instability and slumping in which the sediment remained coherent, whereas the pillow-sourced beds may either be debris flow or coarse hyperconcentrated flow deposits resedimented following slope collapse, or else they represent magma injected into unconsolidated wet tephra with consequent instability and redeposition caused by associated steam explosions.

FTIR measurements of water contents showed no consistent variation with sample elevation (position in the edifice). Numerous variables affect volatile contents in magmas (reviewed by Tuffen et al., 2010, and Owen et al., 2012; see Sections 5.2.3 and 7.4) and their interpretation is often not straightforward (Section 13.5.1). For Helgafell, the lack of variation in water contents was attributed to quenching pressures that did not fall appreciably during construction of the edifice. Calculations by Schopka et al. (2006) suggested relatively low confining pressures of 0.8–1.6 MPa, equivalent to 80–160 m of water or 90–180 m of ice. By contrast, ice thicknesses of c. 500 m were estimated based on edifice height, the presence throughout of waterlain lithofacies and assumed ice surface sag of c. 150–200 m (from observations of subglacial eruptions; Gudmundsson et al., 1997; cf. Figs. 13.11 and 13.12). Such an ice thickness would create a minimum glaciostatic load of 4.5 MPa. Ice sag of c. 150–200 m would reduce that estimate to c. 3–3.5 MPa, although it assumes that basal water was able to drain away continuously to create the space for sagging to occupy. Although a reasonable assumption, there is no verifying evidence in the lithofacies exposed at Helgafell. A pressure reduction of 1–2 MPa in a subglacial cavity may be caused by partial support of the subsiding roof by the ice walls. Water escaping from the cavity in subglacial tunnels kept open by the volcanically warmed meltwater (at c. 10–20 °C; Gudmundsson et al.,

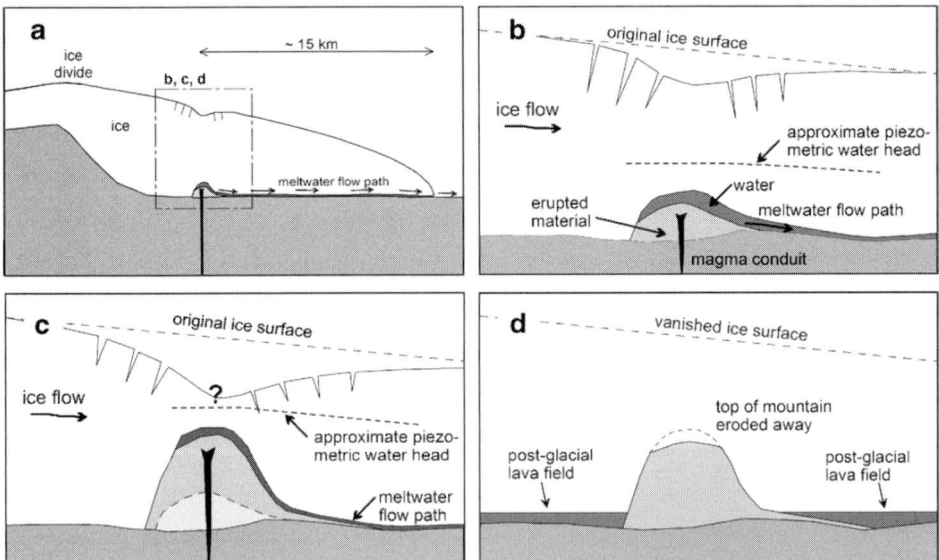

Fig. 10.7 Interpreted sequence of events leading to the construction of the tindar at Helgafell, southwestern Iceland (from Schopka et al., 2006). Pillow lava is minor and the eruption may have been explosive at the outset, possibly because the cavity was hydraulically connected with the glacier terminus.

1997; Höskuldsson and Sparks, 1997) will reduce the upwards floating force exerted by the water and will help to facilitate a roof collapsing into any void space created. The efficiency of drainage of meltwater from an eruption site is thus a key factor in determining cavity pressures. Moreover, if a hydraulic connection is established between the cavity and a low-pressure region, such as a glacier snout, the pressures over the vent will further reduce. This may have been the case at Helgafell, with a glacier terminus possibly situated <15 km away. Finally, the absence of a significant thickness of pillow lava at Helgafell is unusual (cf. Jones, 1969a,b, 1970; Smellie and Skilling, 1994; Werner and Schmincke, 1999). This might also be attributed to efficient basal meltwater drainage that lowered cavity pressures over the vent and facilitated the rapid and early transition to explosive hydro-magmatic activity (Fig. 10.7).

10.4.4 Breknafjöll, Iceland

The description by Bennett et al. (2009) of glaciovolcanic and associated glacial sedimentary lithofacies at Breknafjöll, central Iceland, is the only attempt so far to place glaciovolcanism within a landsystems concept (Benn and Evans, 1998). Jarlhettur is a prominent double volcanic ridge over 10 km in length within Iceland's Western Volcanic Zone, and Breknafjöll is the western ridge. It is an extension of the basaltic Kalfstindar–Raudafell–Hlödufell series of ridge-forming tindars and tuyas described by Jones (1969b,

Fig. 10.8 View of Jarlhettur, a double tindar ridge emerging from the south flank of Langjökull, Iceland. The ridges have the characteristic beaded appearance indicating eruption from multiple contemporaneous vents.

1970) and Skilling (2009). The Jarlhettur ridges have a serrated or beaded appearance (Fig. 10.8), which reflects their formation as a series of overlapping centres erupted from two parallel fissures, although it is unknown if they were all active simultaneously. The individual ridge segments vary between 1 and 1.6 km in length and are 0.3 to 0.8 km wide. It is assumed that the centres erupted during the last glacial cycle. Three depositional environments have been described and designated (1) pillow lava dome (i.e. pillow volcano); (2) hyaloclastite fan (i.e. tindar); and (3) glaciolacustrine fan.

Six lithofacies assemblages are recognised, comprising: stratified and laminated diamict with minor gravel and laminated silts (Assemblage 1); planar sheets and troughs of yellow hyaloclastite sand with gravel (Assemblage 2); massive or horizontally stratified pillow breccias with hyaloclastite sand (Assemblage 3); planar sheets and channel fills of hyalo-clastite breccia and sand sheets (Assemblage 4); pillows, disarticulated pillows, breccias and lava sheets/lobes (Assemblage 5); and subaerial lava (Assemblage 6). The terminology used by Bennett et al. (2009) is frequently at odds with that used in this book. In particular, they use words such as sand and silt for lithified rocks, and the terminology is fundamentally epiclastic whereas the bulk of the succession is volcanic and comprises syn-eruptively redeposited primary volcanic material, for which a volcaniclastic terminology is more appropriate (White and Houghton, 2006). Thus, whilst Assemblage 1 is undoubtedly epiclastic and its terminology is retained here, volcaniclastic terminology is used hereafter for Assemblages 2–4; Assemblages 5 and 6 are unambiguously volcanic and the terminology is unchanged.

The core of the Breknafjöll ridge is composed of Assemblage 5, which corresponds to a series of pillow mounds or ridges. The lithofacies also occur less commonly as isolated small lensoid bodies at higher positions in the ridge. Important characteristics include an association of the lava pillows with breccias interpreted as subaqueous talus deposits of hyaloclastite, some of which shows signs of welding thought to be caused by juxtaposition against a growing dome or pillow lava mound, and as lava fragmented by localised small steam explosions (cf. Staudigel and Schmincke, 1984), presumably in cases where water becomes trapped beneath the advancing pillows. Lobate lava masses within the pillow lavas have distinctive margin-parallel joint sets and cauliflower shapes interpreted to represent cooling of lava in ice cavities similar to features of glaciovolcanic rhyolites described by Tuffen et al. (2001, 2002a). Like the rhyolite examples, the ice cavities were assumed to be 'dry' (i.e. air- or steam-filled), although the supporting evidence is weak apart possibly for late-formed examples overlying lapilli tuffs at the summit of Fagradalsfjall (Bennett et al., 2009, Fig. 13). However, the usual occurrence of the lobes in intimate association with lava pillow mounds probably implies magma cooling within the pillow mounds, which were water-flooded, and they may be intrusions (acknowledged by Bennett et al., 2009; cf. Skilling, 2009). Some lava sheets also drape the pillow piles.

The next stage of tindar construction comprises lithofacies in Assemblages 2–4. These are dominated by lapilli tuffs that form several overlapping fans (called 'hyaloclastite fans' by Bennett et al., 2009). Most of the lithofacies were probably emplaced from sediment density flows with characteristics of hyperconcentrated and concentrated flows (Assemblage 2), probably during periods of high rates of magma discharge corresponding to continuous uprush hydromagmatic events involving a water-saturated tephra slurry interacting with magma within the erupting vent (Kokelaar, 1983, 1986; Smellie, 2000, 2001). It probably had a low, relatively dense eruption column, which was entirely subaqueous for much of the construction of the ridge. This stage also includes a major fining-upward succession of lapilli tuff and gravel breccias interdigitated with laminated tuff forming planar sheets, lenses and channel fills (Assemblage 4). The upward fining may be connected to an eruptive pulse of waning intensity (cf. Smellie, 2000, 2001). Horizontally bedded gravel breccias and pillow breccias in Assemblage 3 may be the products of explosive hydromagmatic fragmentation comparable with lapilli tuffs in Assemblage 2 or they were deposited following collapses due to flank instability. Because the tephra in the Breknafjöll ridge was erupted from a series of point sources (vents), the density flows have created a ridge formed of overlapping tephra fans similar in transport and depositional characteristics to many submarine sediment fans (Walker, 1975, 1978). Bedding attitudes generally decrease upslope. They are essentially slope-parallel at lower elevations whereas the ridge slopes higher up truncate shallower-dipping bedding consistent with deposition of that tephra against a former buttressing wall of ice (Smellie, 2000). Evidence for slope instability comprises displaced near-vertical cross-cutting lapilli tuff beds, and a chaotic group of mega-blocks, each with tilted bedding and interstices filled by massive lapilli tuff (cf. Smellie, 2001; see Section 10.4.1).

Jarlhettur has no well-developed episode of subaerial lava construction but irregular terraces of massive lava overlying lapilli tuff and breccia are present at the summits of Fagradalsfjall and Brekknafjöll. The lavas show no obvious signs of water chilling and they contain widely spaced subvertical cooling joints that are characteristic of slow conductive cooling. Their presence suggests that the uppermost parts of the tindar succession were subaerially exposed prior to a brief period of lava effusion.

By contrast, Assemblage 1 is epiclastic. It is draped on the Breknafjöll ridge, including summit locations, and also occurs at distal locations situated off the ridge. The assemblage is composed of combinations of massive, stratified and laminated diamictites, matrix-supported gravel breccias and laminated siltstones. The basal layer is generally a sheared diamictite that overlies a striated ice-modified bedrock surface of older subaerial lava. The laminated silt contains numerous lonestones whose proportion increases until the deposit become laminated diamictite. Small current scours and draped laminations are also present in the finer units and there are numerous cross-cutting channels filled by diamictite. The assemblage is interpreted as that of a glacial lake with varying amounts of debris rainout, current reworking and erosional scouring, with relatively rare evidence for quiet water conditions mainly affecting distal locations (Eyles et al., 1983, 1985). The distribution of the lithofacies, extending up to 0.75 km from the ridge axis, suggests that the lake was at times much more extensive than the ridge itself. The numerous channels are filled by remobilised diamictites and the contemporaneous water flow might have been relatively strong. The diamictite lithofacies draping the ridge were probably deposited by downslope remobilisation as debris flows on a low-angled fan, while the gravel breccias may represent deposits from hyperconcentrated, concentrated or perhaps turbulent flows (Postma, 1986; Mulder and Alexander, 2001; Bennett et al., 2006). The fan may have been associated with inflowing debris-charged meltwater from sub-ice portals, perhaps linked to coeval eruptions. Coeval deformation of the epiclastic sequences is present, comprising large diapiric and recumbent folds of diamictite and gravel breccia. They appear to represent deformation of glaciolacustrine fan deposits by slightly younger but essentially coeval volcanic products, which probably migrated laterally during edifice construction (pillow lavas and breccias), to overlie the unconsolidated water-saturated glacial materials (Fig. 10.9). Relationships such as these have not been described anywhere else, although Bennett et al. (2006) described similar glaciolacustrine diamictite fans on the down-ice (lee) side of tuyas in the Laugarvatn area (Iceland) that post-date the tuyas and were deposited in caverns created by enhanced geothermal heat from the cooling tuyas.

The Breknafjöll sequence contains large-scale unconformities, e.g. lapilli tuffs draping a glacially(?) eroded surface cut in underlying pillow lava, which suggest that the edifice is compound (polygenetic). However, other cited evidence for significant time breaks is less convincing, e.g. a surface separating two lapilli tuff fans and the presence of a 'mega-channel' 30 m thick and >50 m wide. At least some of those features may be syn-eruptive, including slump-scar surfaces, which are common and prominent in many tindars (Smellie, 2000).

10.4.5 A general model for construction of tindars and tuyas

Most tindars and tuyas commence with a pillow volcano stage in which mounds or hummocky ridges are constructed (Fig. 1.3). The processes and lithofacies involved in their construction are described in Section 10.3. Whether or not a pillow volcano stage

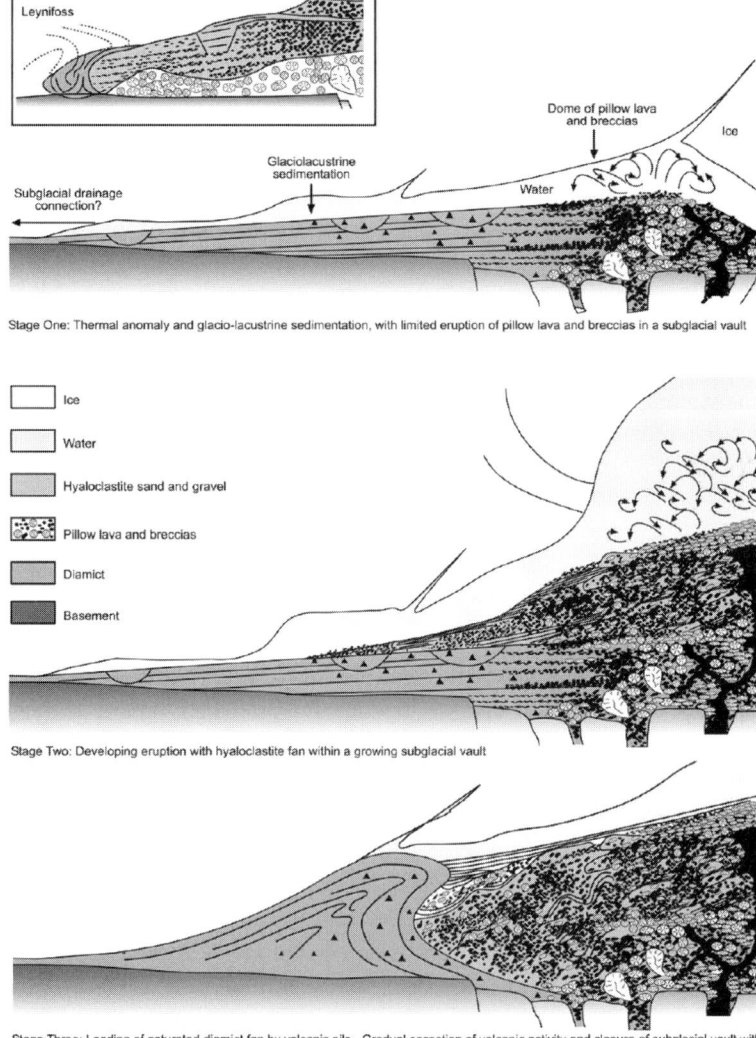

Fig. 10.9 Cartoons illustrating the formation of large-scale deformation structures by loading of lapilli tuffs into diamict and glaciolacustrine deposits flanking the Brekknafjöll tindar, Jarlhettur, Iceland (from Bennett et al., 2009). The section is sketched transverse to the tindar. Note that the authors' use of 'hyaloclastite' in the diagram is synonymous with lapilli tuff in this book. (A black and white version of this figure will appear in some formats. For the colour version, please refer to the plate section.)

takes place is probably largely determined by the volatile content of the magmas, with relatively volatile-rich magmas (e.g. alkalic basalts compared with tholeiitic basalts) more likely to erupt explosively. During the pillow volcano stage, the ambient pressure of overlying ice and meltwater may suppress vesiculation to a variable degree.

However, as the ice thins (assuming that meltwater also escapes) and the volcanic pile grows upward, so the ambient pressure over the vent diminishes. This permits greater degrees of vesiculation which, because of the expansion of gases, drives the rising magma upward at greater speeds. Thus, the magma interacts more vigorously with the meltwater or meltwater/tephra mixtures into which it is injected, initiating molten fuel–coolant interactions and violently explosive hydromagmatic behaviour. Some of this explosivity might also be initiated high in the bedrock conduit where the magma intersects a groundwater aquifer but the characteristic absence of bedrock-derived clasts is probably an indication that the explosions mainly take place high in the tuff cone/tindar itself (Kokelaar, 1983; Smellie, 2001).

Following the upward transition to explosive activity, the eruptive style and explosive intensity for mafic magmas is probably controlled mainly by the mixing mass ratio of water: magma in the system (Wohletz and Sheridan, 1983). Murtagh and White (2013) noted that variations in magma:water ratios may not be critical for eruptions in a vent flooded with water as the amount of available water is effectively unlimited. However, even then some control of magma:water ratios might be effected by the variable proportions of solids and water in the water-saturated unconsolidated tephras (which will depend on grain size and porosity) with which the magma interacts, and the rise rate and mass flux of magma being discharged at any time (e.g. Hooten and Ort, 2002). At decreasing depths below ice or water (i.e. diminishing ambient pressures), the volatile pressure in the magma begins to overcome the hydrostatic pressure, thus permitting bubbles to nucleate. As the vesicularity increases, it reaches a stage at which the magma becomes fragmented. If the magma reaches a vesicularity of above 70% vesicles by volume in the conduit, it will fragment. The depth at which it occurs is known as the vesicle fragmentation depth (VFD). The VFD varies according to magma composition and the volume of dissolved volatiles. For example, tholeiitic basalts with water contents up to c. 0.5% will evolve to explosive fragmentation at c. 100–200 m; for alkali basalts, with dissolved water contents of c. 1–1.5%, the same transition can occur at c. 400–600 m (Kokelaar, 1986; Edwards et al., 2015a). However, when magmas come into contact with water, additional processes come into play, particularly steam explosivity, i.e. the explosive expansion of water to steam, which itself can disrupt a magma ahead of the VFD, creating additional hot magma surfaces and causing the transition to self-sustaining molten fuel–coolant explosivity. Explosive expansion of water to steam can happen at any depth until the critical pressure for water is approached (c. 2.2 km for pure water), at which the rate of steam expansion is much more limited and probably not explosive (Kokelaar, 1986, 1992; White et al., 2003). The net result is that there are so many variables, including magma rise rate, volume flux and volatile content, ambient pressure, timing of water interaction, magma:water mixing ratios, etc., that the nature of the resulting erupting mixture and its explosivity can be highly variable.

Activity in Surtseyan eruptions (i.e. the stage of shallow-water hydromagmatic explosivity from a flooded vent during emergence) occurs in two main forms: intermittent tephra jets (also called rooster tail jets); and continuous uprush (Thorarinsson et al., 1964;

Kokelaar, 1983, 1986; Moore, 1985). The two styles of eruption may be related to irregular pulsatory degassing involving different volumes of magma and coupling versus decoupling of gas and rising magma. Each eruptive style is driven by variably energetic hydromagmatic explosivity and results in copious volumes of tephra. The tephra-jet explosions transform into concentrated subaqueous density flows or wet subaerial pyroclastic density currents that flow down the volcano flanks depending on whether the eruption column is wholly subaqueous or partly or wholly subaerial, respectively. Continuous uprush activity occurs during more sustained eruptions. It generates hyperconcentrated and concentrated density flows in both subaqueous and subaerial settings. Activity of these types causes rapid vertical aggradation of the tuff cone or tindar ridge (White, 2000; Smellie, 2001, 2013). For example, at Surtsey (although marine erupted), the early explosive tuff cone phase represented just 11% of the entire eruptive period (c. 150 days versus 1300 days) yet the edifice built up to sea level in that time (Thordarson and Sigmarsson, 2009). In addition, abundant tephra contained in any subaerial column will fall onto the surrounding meltwater lake, rapidly overloading the water column and creating a continuous series of downward-cascading vertical gravity flows that then travel as subaqueous density currents (including turbidites) on the submerged portion of the tuff cone flanks (cf. Manville and Wilson, 2004). An unusual feature of many tindar and glaciovolcanic tuff cone successions is a paucity of fine ash, which is abundantly created during explosive hydromagmatic eruptions. By comparison with marine tuff cones, the fine ash is assumed to have somehow bypassed the subaqueous edifice (White, 2000; Smellie, 2001). However, it is harder to argue for glaciovolcanic examples given their confinement within an ice cavity unless the fine ash is efficiently winnowed (in air or by currents in water) or else remains mainly in suspension and is discharged subglacially along with escaping meltwater (see also Section 9.4.3).

Large-scale gravitational wasting is represented by sector collapses. They can significantly modify the flanks of a subglacial edifice and are probably caused mainly by the progressive melt-back of the surrounding ice walls, which removes the buttressing support and leaves the flanks gravitationally oversteepened. In an environment characterised by frequent eruption-related tremors, the flanks are highly susceptible to collapse. Such collapses contribute large volumes of unstructured coarse sediment to the ice cavity, which presumably piles up at lower elevations and becomes banked up against the surrounding ice, thus setting up the tephra pile for further collapses after further melt-back. Evidence for slope collapses is common and conspicuous in glaciovolcanic tuff cones and tindars (cf. Fig. 9.25). They have been called slump-scar surfaces (Smellie, 2000). As the accommodation space in the cavity fills up, bedding in the tephra pile changes up-section from slope parallel (close to the angle of repose) to more flat-lying higher up, reflecting filling-up of the cavity. It may even be back-tilted (Smellie, 2000; Fig. 10.10). Back-tilting and flat-lying beds usually occur at mid to high elevations in tindars and glaciovolcanic tuff cones. Similar features are absent in marine-erupted tuff cones, in which the bedding is slope parallel. This difference has been used to distinguish between confined (i.e. glaciovolcanic) and unconfined (e.g. marine or lacustrine) settings (Smellie et al., 2006a; see Chapter 13). Its occurrence has been attributed to ice-constrained deposition

Fig. 10.10 Lapilli tuffs overlying pillow lava in a tindar at Kalfstindar, Iceland (from Smellie, 2000). Note how the bedding dip decreases upward and becomes back-tilted towards the source (left) at the top of the section. This is diagnostic of impoundment by the ice walls of an englacial vault, against which the beds were banked. The rock face is c. 35 m high.

within an englacial chimney following the collapse of the overlying ice (Bennett et al., 2009). The relations were inferred to reflect a change in the geometry of the subglacial cavity from a wider base to narrower and more closely confined higher up, i.e. an ink bottle shape.

Following emergence, it is typical for the eruptive style to lose its vigour as magma in the vent becomes isolated from the ambient water and the activity shifts to weak Hawaiian fire fountaining, Strombolian bubble bursts and lava effusion. A lava shield is constructed with a broad central cone surrounded by a lava apron that is narrow or wider depending on an interplay between periods of steady, low-discharge (low-profile shields, broad lava apron) and more prolonged periods of relatively high magma and fountain-fed surface discharge (higher profile shields, narrow lava apron; Rossi, 1996; Thordarson and Sigmarsson, 2009; cf. Fig. 8.3). Subaerial lava effusion defines the tuya stage, when lava flows into and advances across the encircling meltwater lake and creates a lava-fed delta. It is also a stage of relatively rapid progradation (Smellie, 2007, 2009, 2013). This is true for the emplacement of pāhoehoe lava-fed deltas, with their moderate effusion rates (generally <5–10 m^3 s^{-1}; Walker, 1973a; Rowland and Walker, 1990; Griffiths and Fink, 1992). Ramalho et al. (2013) have suggested that, due to the significantly different emplacement mechanisms of 'a'ā lava-fed deltas, with their higher effusion rates (>10 m^3 s^{-1}) and thicker flow units with the ability to remain coherent for greater distances underwater (also Tucker and Scott, 2009; Smellie et al., 2013a), 'a'ā lava-fed deltas are more aggradational compared with pāhoehoe deltas (Ramalho et al., 2013; Fig. 10.11). However, 'a'ā lava-fed deltas are progradational overall.

Subaqueous–subaerial transitions are important and conspicuous features of the lava-fed deltas in tuyas. Passage zones (*sensu* Jones, 1969b) are the best known type (see Section 13.4.1)

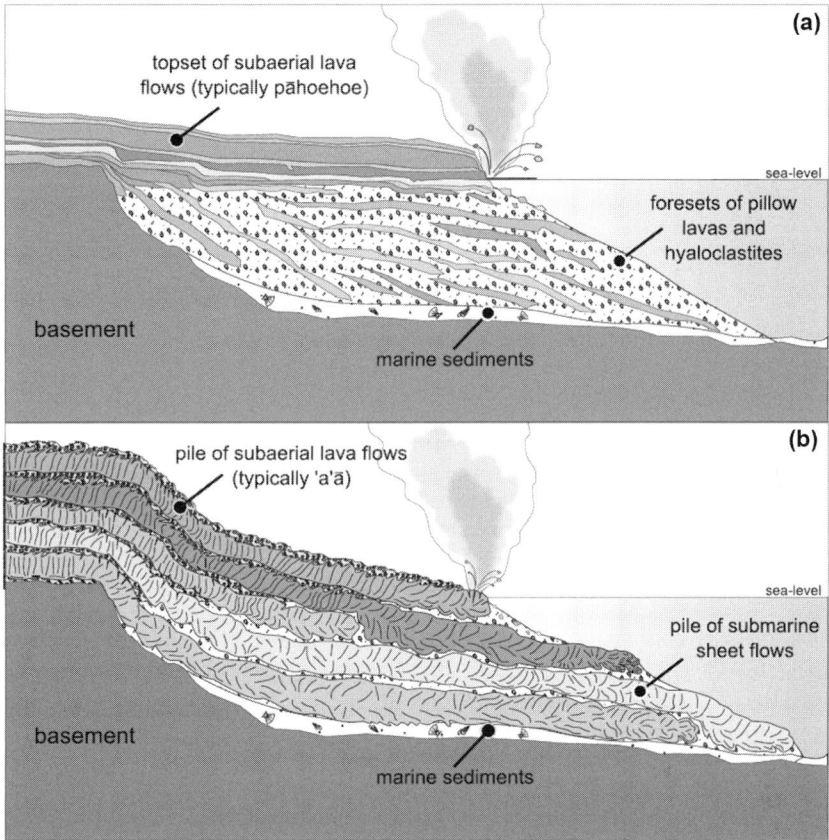

Fig. 10.11 Sketches illustrating some major differences in lithofacies and architecture between pāhoehoe and 'a'ā lava-fed deltas (marine-emplaced in these examples; from Ramalho et al., 2013). The generally greater effusion rates of 'a'ā compared with pāhoehoe result in many 'a'ā lava sheets continuing to flow underwater, whereas pāhoehoe typically shatters into breccia at the delta brinkpoint. The glaciovolcanic equivalents of 'a'ā lava-fed deltas described so far have a greater proportion of breccia and the lava masses are much more irregular (Smellie et al., 2011a,b, 2013a).

and they differ in appearance and lithofacies between lava-fed deltas sourced in pāhoehoe and 'a'ā (cf. Figs 10.12, 10.13). Pyroclastic passage zones may exist theoretically in tindars and glaciovolcanic tuff cones (Russell et al., 2013) but they will be hard to identify because of the great difficulty distinguishing between tephra deposited above and below water (i.e. they are effectively invisible except for relatively ambiguous criteria such as possible better preservation of armoured lapilli above the water line, fossil strandlines, lack of fine-ash coatings on larger ash/lapilli, or symmetric ripples indicative of shallow water waves). Moreover, because of the relatively short time in which tindars and tuff cones are formed compared with lava-fed deltas, it is unlikely that any subaqueous–subaerial transitions would be capable of showing the complexity of elevation changes that can occur very conspicuously in long-lived and laterally extensive lava-fed deltas (cf. Smellie, 2006). Many lava-fed deltas show no evidence for lake

Fig. 10.12 Passage zone in a basaltic pāhoehoe lava-fed delta, James Ross Island, Antarctica. Although a relatively coarse feature, the demarcation between the flat-lying subaerial pāhoehoe lava cap and the underlying tuff breccia foreset beds is quite well defined. The rock face is c. 75 m high. Compare with Fig. 10.13. (Image: Ian Skilling.)

instability, attesting to a stable lake surface (Skilling, 2009). Most monogenetic tuyas are probably active for only a short period of time (months or perhaps a few years; e.g. effusive phases at Surtsey each lasted only about a year; Thordarson and Sigmarsson, 2009), compared with more voluminous, longer-duration eruptions in some stratovolcanoes (possibly decades), which enhances the likelihood of instability being developed in the latter (Smellie, 2006). In tuyas, the transition is typically preserved as a lava-fed delta passage zone. The passage zone represents the location (also known as the brinkpoint; Skilling, 2002) at which subaerial lava enters the coeval meltwater lake, quenches, becomes fractured and mechanically disintegrates into breccia. It is a very prominent feature, essentially a fossil water level. Several distinctive types of transitions have been defined, including rising, falling and horizontal passage zones (Smellie, 2006). They have different consequences for interpreting the eruptive setting, glacier hydraulics and glacial thermal regime (see Section 13.4.1).

10.5 Pillow sheets: fact or fiction?

Mafic pillow sheets are a poorly known and ill-defined glaciovolcanic sequence and landform type that currently has been identified only in central Iceland (Snorrason and Vilmundardóttir, 2000; Vilmundardóttir et al., 2000). They are described as outcrops composed mainly of pillow lava and lesser hyaloclastite breccia that occur as widespread (tens of kilometres) yet thin (tens of metres) sheets. There are no published detailed descriptions of the constituent lithofacies and architecture and they have also been described as some type of 'móberg sheet' (Jakobsson and Gudmundsson, 2008). The status and mode of formation of pillow sheets are thus uncertain. Unpublished studies by one of us (JLS) suggest that there are substantial

Fig. 10.13 View of a passage zone in an 'a'ā lava-fed delta, Minna Hook, Mt Discovery, Antarctica. The passage zone is a very coarse feature that extends many metres. Note also that the subaerial capping lavas of the delta are markedly thinner and more evenly planar compared with those within the underlying passage zone, which are much thicker, have very irregular shapes and are enclosed in variably palagonite-altered breccia. The sharp top of the delta is glacially eroded (white line) and overlain by stratified sediments and a much thinner 'a'ā lava-fed delta with a very thin subaqueous section. Compare with Fig. 10.12. (A black and white version of this figure will appear in some formats. For the colour version, please refer to the plate section.)

similarities with some sheet-like sequences and some others may simply be pillow mounds. Rare possible examples have been identified on satellite imagery of Mars (Scanlon et al., 2014). A brief provisional assessment of examples in Iceland is provided here.

Although not mapped in detail, the examined outcrops are extensive, ranging up to 6 km wide and potentially 10 km long and larger examples may exist. They consist mainly of sheet lava c. 30–100 m thick, yielding very low aspect ratios (i.e. thickness divided by width or length). The lava shows prominent blocky and (mainly) prismatic jointing, including a thin basal colonnade and much thicker entablature. The lava surfaces have prominent glassy chills and they contain rare steep joints showing multiple glassy chills and tiny normal joints (cf. column on column structures of Sporli and Rowland, 2006); pseudopillow joints are also present. Pillows occur at the margins, including on top of the lava; they are apparently only locally developed and may be volumetrically unrepresentative of pillow sheets as a whole. Minor rust-coloured fine gravelly hyaloclastite is present in interpillow spaces. Lava in one outcrop was observed to overlie 8 m of crudely stratified muddy grey diamictite with facetted and striated abraded clasts; a striated surface within the diamictite deposit was also seen. Another sheet lava is overlain by planar and trough cross-stratified glassy sand and fine gravel. All three outcrops examined are the youngest volcanic units present and their margins are prominent steep flow fronts with perpendicular (i.e. subhorizontal) prismatic joints, some with small developments of lava pillows.

There is abundant evidence for water chilling (e.g. glassy surfaces, blocky and entablature jointing, lava pillows, hyaloclastite, pseudopillow joints) and the steep lava margins with subhorizontal fine-scale columns suggests that they were ice-confined. The presence of local pillows at those margins suggests that the pillows were formed by breakouts as the adjacent ice barrier melted back and created a water-filled space (similar to steep ice-chilled surfaces and pillow lava dribbles at Hlödufell; Skilling, 2009; see also Hungerford et al., 2014). Although pillows are often present, they seem to be restricted to marginal locations and are thus volumetrically minor; the name 'pillow sheet' may be a misnomer. The lavas overlie diamictites with a strong glacial influence and they are locally overlain by fluvial epiclastic sediments potentially deposited by meltwater during ice decay. The very young age of the sequences suggests that they were emplaced during the Last Glacial Maximum, which probably attained thicknesses exceeding 1000 m in central Iceland (Bourgeois et al., 1998; Licciardi et al., 2007). In many respects, apart from a conspicuous absence of thick overlying lapilli tuffs, the lavas examined resemble the basal lavas or sills found in sheet-like sequences (cf. Smellie, 2008; Section 10.6). Such subglacial sills were predicted theoretically by Wilson and Head (2002) and in some circumstances they need not be accompanied by thick lapilli tuffs (the 'truncated sequences' of Smellie, 2008; Section 10.6.2). Finally, at least some so-called pillow sheets mapped in the Jökulheimar area by Vilmundardóttir et al. (2000; see also Jakobsson and Gudmundsson, 2008) appear to be simple pillow mounds (pillow volcanoes).

10.6 Sheet-like sequences

Mafic sheet-like sequences are common in some large polygenetic glaciovolcanic centres (e.g. Eyjafjallajökull and Öraefajökull, Iceland; Fig. 8.8; Carsewell, 1983; Loughlin, 2002, and unpublished information of JLS) but they are generally only rarely described (e.g. Smellie et al., 1993, 2011a). Two types have been identified, with inferred different implications for coeval ice thicknesses (Smellie, 2008). More recent studies suggest that there are unresolved problems with the original interpretations and there may be insufficient differences to justify distinguishing two types (unpublished information of JLS, and

Section 10.6.3). Sheet-like sequences are distinctive products of effusive eruptions, with subglacial emplacement. Unlike other glaciovolcanic sequence or landform types, individual sheet-like sequences are not volcanic edifices. They are outflow sequences, i.e. they are deposited away from the coeval volcanic edifices. Their sources are poorly known but, based on the characteristics of the lithofacies, probably comprise environmentally undiagnostic pyroclastic mounds, either tuff cones or tindar ridges (Mt Pinafore types), or pillow mounds (Dalsheidi types; but see Section 10.6.3; Smellie, 2008, 2009, 2013).

10.6.1 Sheet-like sequences of Mt Pinafore type

These are the most commonly described sheet-like sequences (Carsewell, 1983; Smellie et al., 1993; Smellie and Skilling, 1994; Loughlin, 2002). Outcrops are thin (few to several tens of metres) and ribbon-like or lobate. They may also be sinuous and anastomosing and some have steep margins. The morphologies have been described as esker-like (Smellie and Skilling, 1994; Loughlin, 2002; but see Sections 8.2.5 and 8.2.6). Although their full lateral extent is unknown, most probably extend a few kilometres at least, depending on the duration of the eruption and the volume of magma discharged. Volumes are small, typically $\ll 1$ km^3, but very few data on volumes have been published (Smellie, 2008, 2013). Many are valley confined. The sequences were defined using examples at Mt Pinafore, Antarctica, which were regarded as sequence holotypes (Smellie et al., 1993).

The sequences are bounded by sharp glacially modified erosional surfaces that are usually overlain by massive or crudely bedded polymict diamictite interpreted as types of till (lodgement and flow tills). A 'standard sequence' of the overlying volcanic lithofacies consists of the following, from base up:

(1) Planar and cross stratified sand- and fine gravel-grade lapilli tuffs and tuffs showing channel structures and formed of variably palagonitised sideromelane with blocky–angular shapes and variable vesicularity. The deposits are monomict and are interpreted to be tephra formed during explosive hydromagmatic subglacial eruptions. The tephra were then flushed away subglacially by abundant meltwater generated by the eruptions. The amplitude of associated dune bedforms suggests periods of relatively deep water (c. 2 m) despite the subglacial location, and transport and deposition probably in tunnels similar to eskers (Smellie and Skilling, 1994). The characteristics of some beds suggest that they accumulated from more laterally extensive mass flows, including hyperconcentrated density flows (Loughlin, 2002). Epiclastic terminology was used in the original descriptions in deference to the evidence for fluvial transport and deposition but, since the deposits are composed of syn-eruptively redeposited monomict tephra (*sensu* McPhie et al., 1993), a primary volcanic terminology is preferable and is used here (White and Houghton, 2006). The abundant evidence for transport of the lapilli tuffs by traction currents was an important factor in distinguishing the formation of sheet-like sequences from the pooled-water conditions in which lava-fed deltas formed (Smellie, 2000).

(2) The tephra deposits are overlain by fine-grained and aphanitic sheet lava with glassy chilled margins and blocky jointing or prominent entablature overlying a thin layer of colonnade columnar jointing. The lava may be locally pillowed or have platy jointing that mimics the lava shape. Ice-block meltout cavities are rarely present in the lava.

(3) The lava is encased in further sand- and fine gravel-grade lapilli tuff that is usually massive or shows weak planar stratification. It is monomict, formed of finely crystalline clasts and sideromelane described as redeposited hyaloclastite formed by mechanical breakage at the chilled lava margins. That explanation is probably too simplistic and was based on inadequate textural examination of thin sections. Greater similarity in clast characteristics to the underlying lapilli tuffs is now evident in many deposits, although these upper deposits are thicker than the basal planar and cross-stratified lapilli tuffs. Deposition mainly from hyperconcentrated density flows is indicated.

(4) Finally, a few sequences are capped by sheet lavas with oxidised autobreccias emplaced in subaerial conditions. However, they are rarely observed and the precise relationships with the underlying water-chilled and water-deposited lithofacies are uncertain (e.g. Smellie et al., 1993; Loughlin, 2002).

The absence of internal unconformities and significant compositional variations suggests that these sequences are a cogenetic association of lithofacies generated in a single eruption. From the shapes of the basal unconformities, many are valley confined and some contain lava that was inflated due to topographical confinement. A 'thin ice' or alpine-type setting was inferred because of the abundant evidence for flowing water, which implies emplacement in a permeable glacial cover in contrast to the ponded water setting required by lava-fed deltas (Smellie et al., 1993; Smellie, 2000, 2001). A permeable cover is one comprising some combination of snow, firn and fractured ice. Measured depths to the firn–ice transition are up to c. 150 m and most fractures (crevasses) pinch out at <30 m depth (see Chapter 4), yielding a theoretical maximum cumulative thickness for the permeable layer of <200 m. Deposition was probably in tunnels but transport in sheets is not precluded. Under thin ice, it is likely that the tunnel roof will rapidly disintegrate. Thus, later lava effusion within the same eruption may be subaerial, which may explain the rare occurrences of subaerial lava described.

10.6.2 Sheet-like sequences of Dalsheidi type

Dalsheidi-type sheet-like sequences have only been described in Iceland. Although the first described occurrence was that at Dalsheidi, after which the sequences are named, the best and most numerous examples occur in the Sida–Fljotsfverfi district; other examples also exist at Eyjafjallajökull (Walker and Blake, 1966; Loughlin, 2002; Smellie, 2008). The Sida–Fljotsfverfi examples were originally interpreted as products of subglacial fissure eruptions emplaced in a submarine setting below an ice shelf (Bergh and Sigvaldason, 1991). Compared with sheet-like sequences of Mt Pinafore type, most Dalsheidi-type sequences are thicker and much more extensive. Thicknesses vary between 15 and 220 m, with along-strike extents of 12–30 km and 6–14 km down-dip. They have blanket-like (Sida–Fljotsfverfi) or lobate (Eyjafjallajökull) geometries, with individual erupted volumes potentially up to 31.4 km^3 (average: c. 13 km^3; Bergh, 1985). Each Dalsheidi-type sequence is a product of a single eruption, generally bounded by sharp laterally extensive erosional surfaces that are uneven on a scale of a few tens of metres vertically. The overlying lithofacies thicken into palaeovalleys although some examples have planar upper surfaces. The surfaces rarely show glacial smoothing, striations and brecciation. The Dalsheidi-type sequences are locally separated by unfossiliferous polymict diamictites or conglomerates 2–50 m thick that were used by Bergh (1985) as a means of stratigraphical correlation. The diamictites contain facetted and rarer striated clasts and locally display anastomosing parting joints or are injected by underlying lapilli tuff along fractures (Smellie 2008, Fig. 5). The diamictites are probably glacial deposits (types of till). A 'standard sequence' of volcanic lithofacies overlying the diamictites consists of the following, from base up:

(1) A basal sheet lava generally 10–35 m thick (mean c. 22 m) overlies the diamict and conglomerate. It is laterally extensive and can typically be traced tens to hundreds of metres (Eyjafjallajökull) and even several kilometres (Sida–Fljotsfverfi). The lava in the outcrop at Dalsheidi is the thickest known so far. It reaches 100 m but is locally absent. At its base the Dalsheidi lava contains discontinuous patches of diamictite incorporated within it and it has also sunk locally into diamict and tuff. The lava is poorly to non-vesicular. It is associated with aphanitic to glassy fines-poor angular lava breccia described by Walker and Blake (1966) as hyaloclastite. The breccia is also mingled with dispersed abraded pebbles, cobbles and mudstone derived from underlying diamict.

Dalsheidi-type lavas generally show spectacular entablature and much thinner to absent colonnade columnar jointing (Fig. 9.7c). Hackly and blocky jointing is common around the margins. A distinctive feature of the upper surfaces of the lavas is the presence of multiple irregular flame-like apophyses up to 20 m high that both intrude and contribute spalled fragments to the surrounding fragmental deposits (Fig. 10.14). Some apophyses are 15 m wide and 200 m long. A rare feature of the upper lava surface is the presence of steep lenses a few decimetres wide and a few metres deep filled by coarse hyaloclastite and minor small lava pillows. The lenses are surrounded by hackly jointed lava and are truncated at the lava top. They were interpreted as former folded lava tops that underwent erosion by fast-moving sediment-charged density flows (Smellie, 2008). Ice-block meltout cavities up to a few metres in diameter are also present at the base of the lava (also Walker and Blake, 1966). The elevation of the basal lava may rise >60 m over a distance of 4 km away from source (Bergh, 1985).

(2) Voluminous thick sheet-like deposits of volcaniclastic rocks overlie and are spectacularly intruded by the basal lava. They have been described as fine-grained hyaloclastite because of abundant poorly to non-vesicular sideromelane. The deposits are 10–120 m thick (average c. 40 m) and they typically form at least two-thirds of each Dalsheidi-type sequence. They may be coarser to base and fine up (Walker and Blake, 1966) but they are

Fig. 10.14 Relationships between an interface sill and overlying faintly stratified lapilli tuffs in a sheet-like sequence. The sill upper surface (seen at lower right) varies from a dyke-like apophysis with well-defined margins to brecciated (blocky peperite) and pillowed. Eyjafjallajökull, Iceland. The rock face is c. 20 m high.

often massive and of relatively uniform grain size (fine gravel, clasts ≤ 1–2 cm). Bedding is poorly developed, discontinuous and planar; amalgamation may be common.

(3) Many Dalsheidi-type sequences are capped by a much thinner upper unit of tuff showing fine planar or rare ripple cross laminations. According to Bergh and Sigvaldason (1991), the sideromelane in the tuffs is highly vesiculated in contrast to vesicle-poor or non-vesicular sideromelane in underlying deposits.

The mode of formation of Dalsheidi-type sheet-like sequences proposed by Smellie (2008) is as follows: Voluminous eruption was envisaged from fissures under ice probably over 1000 m thick in order to suppress degassing in the non-vesicular basal lava and sideromelane in the overlying volcaniclastic deposits. The distinction in vesicularities was important in this interpretation of the deposits. The basal lava was envisaged as a sill intruding and inflating along the ice–bedrock junction, a process predicted by Wilson and Head (2002). It was called an interface sill by Smellie (2008). Cooling by ice and any accompanying meltwater was responsible for the spectacular entablature jointing. It was suggested that eruptions could cease at that point, creating truncated sheet-like sequences of Dalsheidi type (see pillow sheets, Section 10.5). However, in most cases, meltwater escaped from the englacial cavity in voluminous jökulhlaups in which collapsing coeval pillow lava mounds and ridges were envisaged providing the source for the abundant fragmental detritus that was deposited subglacially from concentrated and hyperconcentrated density flows. Some erosion of the underlying sill may have taken place at that time. The final beds were transported mainly under waning flow conditions. They contain abundant vesicular sideromelane that was explained as a consequence of thinning ice and reduced overburden pressures above the source vent permitting juvenile gases to exsolve and the eruption to become explosive. Injection of lava apophyses from the basal sill into the overlying hyaloclastites was ascribed to the effect on pressure of Bernoulli's principle, i.e. the corresponding pressure reduction caused by the fast-moving sediment flow acting on the basal sill facilitated upward injection of lava from the still-molten core as stringers and sheets. Most were broken up and carried away in the sediment flows and may have been responsible for scarce fluidal bomb-like clasts seen, but the final injections of lava were preserved as the sediment flows froze in place.

10.6.3 Weaknesses in the existing models for emplacement of sheet-like sequences

Interpretation of Mt Pinafore-type sheet-like sequences as products of eruptions under thin ice relies on accepting that the associated basal waterlain sediments were emplaced within a permeable glacial cover composed of snow, firn and fractured ice, for which an approximate maximum thickness of 150–200 m can be estimated (Smellie et al., 1993; Smellie, 2000, 2001). That estimated thickness is a crude theoretical one as the sequences do not display subaqueous–subaerial transitions from which a more accurate estimate can be deduced (see Chapter 13). At the time the hypothesis was formulated (Smellie et al., 1993), orthodox glaciological theory indicated that ice was essentially impermeable and basal discharge would not occur, hence any meltwater must be impounded in an englacial cavity melted by the eruption unless released in a jökulhlaup (Björnsson, 1988; Chapter 4). It is now known that englacial cavities created during subglacial eruptions in unfractured thicker ice are essentially 'leaky' because of the presence of pre-existing subglacial Nye–Röthlisberger channels (Fig. 4.5). The channels are in hydraulic continuity with basal meltwater beyond any cavity ice 'seal' and they rapidly enlarge due to thermal erosion by the volcanically warmed meltwater, thus increasing the basal discharge (Gudmundsson et al., 2004; Smellie, 2006). Thus, evidence for flowing water can occur even in glaciovolcanic sequences formed within much thicker ice, including under tuyas (e.g. Smellie, 2001; Skilling, 2009).

There are other intrinsic weaknesses in the interpretation of Mt Pinafore-type sheet-like sequences. The evidence for pervasive water chilling of the associated lava is consistent with drenching by water, which is unlikely under thin ice with efficient basal water discharge – i.e. how does the lava get submerged? Moreover, thin ice will probably melt through completely during an eruption and the final lithofacies will be emplaced under essentially dry subaerial conditions. In reality, although rare examples of associated coeval subaerial lithofacies have been described (Smellie et al., 1993; Loughlin, 2002) none are wholly convincing; i.e. they may have formed under ice-poor interglacial conditions and are not sheet-like sequences (see also Loughlin, 2002; Fig. 8.8).

Important potential weaknesses in the interpretation of Dalsheidi-type sequences include (1) a requirement for the basal lava/sill and associated volcaniclastic deposits ('hyaloclastite') to represent undegassed lava, with volatile release suppressed until a late stage by very thick overlying ice; (2) derivation of the volcaniclastic deposits from pillow mounds or ridges for which only minimal evidence has been demonstrated; (3) early emplacement of the basal lava as an interface sill intruded ahead of any fragmental deposits; (4) evidence for contemporaneous erosion of the basal lava by sediment-charged density flows during jökulhlaups; and (5) injection of still-molten cores of the interface sills as multiple irregular apophyses into the overlying fragmental deposits predicated by the Bernoulli principle, i.e. that pressures above the sill were lowered during emplacement of the sediment flows and thus facilitated lava breakouts.

10.6.4 A unifying model for the formation of sheet-like sequences

Important new observations include the following: The basal sills show many prominent features diagnostic of peperites, i.e. formed when magma intrudes or lava flows over wet sediments (e.g. Busby-Spera and White, 1987; Skilling et al., 2002). For example, the presence of abundant mini pillows ≤10 cm in diameter, with botryoidal glassy surfaces, are globular peperite whilst the much more dominant breccias, which show a progressive transition from jointed sill interior, through blocky jointed and jigsaw-fit fractures, and out into breccias with lapilli tuff matrix (Fig. 10.14), are blocky peperite. Together with the conspicuous intrusive lava apophyses, they indicate unambiguously that mafic magma intruded the thick deposits of lapilli tuff. Conversely, the evidence for rare erosion of the sill surface described by Smellie (2008) is ambiguous and may simply indicate local spalling of the glassy sill surface caused by interaction with the overlying wet lapilli tuff during emplacement. The abundance of sideromelane in the lapilli tuffs indicates explosive formation during hydromagmatic eruptions from the earliest stages (as in Mt Pinafore-types) and not just in the closing stages of the Dalsheidi-type eruptions. It is important to note that *both* types of sheet-like sequences display these features. The numerous different types of lithofacies variants described by Loughlin (2002) simply display the range of features to be expected in *any* sheet-like sequence. The overwhelming similarities in lithofacies and architecture are consistent with comparable eruptive conditions for all sheet-like sequences. The distinction into Mt Pinafore and Dalsheidi types (Smellie, 2008) is therefore probably unrealistic and there is probably only a single generic sheet-like sequence type. However, the gross disparities in volumes and areal extent probably suggest that most Pinafore types were erupted from small central vents (tuff cones) or short-lived and relatively short fissures whilst the Dalsheidi types were erupted from longer-lived and more extensive fissures.

Eruptions of vesiculating magma are envisaged occurring in an englacial cavity under an ice sheet of unknown thickness. Estimates of pressure conditions over the vent using volatile compositions suggest that the magma was significantly degassed in H_2O, S and CO_2 and the ice may have been quite thin (Banik et al., 2014). An englacial cavity was thus created that was largely filled by meltwater. The abundance of blocky sideromelane with variable vesicularities indicates that the eruptions were explosive and hydromagmatic, constructing a tuff cone (mound) or tindar, although the occurrence of a pillow mound below the tephra (and therefore relatively thicker ice) is not precluded and would provide a ready explanation for the presence of (admittedly uncommon) pillows and pillow lava megaclasts within the lapilli tuffs (Bergh, 1985; Smellie, 2008). At some stage following construction of the tephra pile, the hydraulic pressure at the base of the cavity floated the surrounding ice resulting in a jökulhlaup, causing substantial collapse of the unconsolidated edifice and resulting in the transport and subglacial emplacement of tephra sheets by concentrated density flows. It is inferred that this is when mafic magma then intruded the remnant tephra pile and was emplaced subglacially as one or more sill(s) intruding the lapilli tuff sheets. The magma interacted with the wet tephra to yield the spectacular peperite textures and apophyses. In the model by Banik et al. (2014), an unspecified but potentially large proportion of the lapilli tuff–lava pile was emplaced subaerially, well beyond the coeval glacier front. As evidence, they cited relationships for similar but poorly characterised deposits known as the Kriki hyaloclastite flow sourced in a subglacial fissure crossing Katla volcano (Larsen, 2000), whose very young age (tenth century?) implied subaerial emplacement of the outcrop.

10.7 Large polygenetic glaciovolcanic centres

Large polygenetic glaciovolcanic centres constructed from mafic magmas are common and widespread in Antarctica, where they typically evolved to trachytic or phonolitic compositions (LeMasurier and Rex, 1982; LeMasurier and Thomson, 1990; LeMasurier et al., 1994; LeMasurier, 2002; Wilch and McIntosh, 2002; Smellie et al., 2006c, 2008, 2011a,b, 2013a,b). Other mafic and intermediate-composition examples are much rarer elsewhere, with a few in Iceland and British Columbia (Brown et al., 1991; Edwards and Russell, 2000; Graettinger et al., 2013a–c).

The Antarctic examples are glaciovolcanic shields with ages mainly younger than 10 Ma. Some evolved into stratovolcanoes with the eruption of more felsic products. Particularly prominent examples occur at James Ross Island, Mt Murphy, Crary Mountains, Minna Bluff, Cape Washington and the numerous large centres between Coulman Island and Cape Adare (see Fig. 2.1; Section 2.2). They are dominated by mafic glaciovolcanic deltas, fed by both pāhoehoe (mainly Antarctic Peninsula) and ʻaʻā lava (LeMasurier et al., 1994; Skilling, 2002; Smellie, 2006; Smellie et al., 2011a, 2014) and they include the first described examples of ʻaʻā lava-fed deltas (Smellie et al., 2013a). Some of the individual pāhoehoe deltas are extremely voluminous, with erupted volumes on James Ross Island typically several tens of cubic kilometres and a few probably exceeding 100 km^3. They correspond to glaciovolcanic flood lavas (Smellie et al., 2013b).

Polygenetic volcanic centres in British Columbia that have glaciovolcanic histories include broad plateaux of coalesced basaltic shield and stratovolcanoes at two localities (Level Mountain, Mt Edziza; Wood and Kienle, 1993; see Chapter 2). They were both long-lived and range in age from Neogene to Holocene (<15 Ma; Edwards and Russell, 2000). Level Mountain is the largest, comprising more than 860 km^3 of flat-lying mafic and felsic lavas and felsic domes. Mount Edziza (below) is the most comprehensively described. It is a stratovolcano founded on mafic volcanic rocks but which evolved to felsic products and has a complicated history of glaciovolcanic and subaerial eruptions (Souther, 1992). Another polygenetic mafic centre that has at least a small glaciovolcanic component is located at Fort Selkirk in central Yukon Territory; it is intermediate in size between the large shields and the much more common monogenetic centres and Quaternary in age (1.08 ± 0.05 Ma).

The Mt Edziza volcanic complex, which is within the southern segment of the Neogene to Holocene Northern Cordilleran Volcanic Province in northwestern British Columbia and southern Yukon (Edwards and Russell, 2000), is a large compound volcanic construct (Souther, 1992). The complex has had a long history of eruptions extending from c. 10 Ma to a few thousand years ago (post-glacial; see Section 2.4.5). The successive eruptions have formed a composite mainly basaltic volcanic plateau surmounted by four large steep-sided evolved centres that lie along a north–south axis. The activity occurred through several episodes of regional and local glaciation. However, although there are abundant indications of associated ice, in the form of striated eroded lava surfaces, tills and glaciofluvial beds, evidence for glaciovolcanism is restricted to just five formations, all younger than c. 3–4 Ma, out of 13 formations present. The oldest rocks showing evidence for associated ice (but not glaciovolcanism) are c. 4–5 Ma. One of the important points made by Souther (1992) is that evidence for water chilling and fragmentation of lava, and even rising passage zones, need not imply a glaciovolcanic setting and the effects of the local topography must be carefully assessed in any interpretation; he described several examples where river drainages blocked by newly erupted lava flows created new lakes whose surfaces rose until the water overspilled. Examples of the latter were not only basaltic but also included rhyolite. Souther (1992) also described possibly the first example of a mafic lava-fed delta advancing though a draping ice cover and creating a slope-parallel dipping passage zone (Souther, 1992, Fig. 122; cf. Smellie et al., 2011b, Fig. 4) and he also noted the likelihood that any coeval ash that fell on surrounding ice would be advected away. In addition to mafic tindars erupted through ice, subglacial piles of mafic pillow lava and pillow breccia were described, ice-impounded mafic lavas with quenched margins showing horizontal or radiating small columns, and mafic lava overriding a glacier margin and subsequently collapsing after the ice melted. More recent studies of the Edziza volcanic complex have focused on a subglacial pillow ridge or pillow-dominated tindar (Pillow Ridge; Edwards et al., 2009b), and a subglacial tuff ridge with associated lavas emplaced in subglacial meltwater channels (Tennena volcanic centre; Hungerford et al., 2014). Pillow Ridge is described in Section 10.3. The Tennena cone outcrop is a subglacial ridge 200 m high and 1200 m long associated with fissure-fed pillow lava and lesser massive lava sheets and pillow mounds. Like Pillow Ridge, the cone has no subaerial volcanic lithofacies. The outcrops were formed under a regional Cordilleran ice cover 500–1400 m thick (from analyses of volatiles) with initial eruption of tuff breccia, followed at a later stage by lapilli tuffs formed explosively when the ice cover melted and vent pressures were reduced. The changes between pillowed and massive lava were attributed to different flux rates (higher for massive lava flows). Evidence for confinement of lava in ice tunnels includes unusual vertical lava pillows with distinctive corrugated surfaces that draped steep (impounded) lava flow fronts, and the sinuous shapes of the pillow lava ridges (akin to Röthlisberger tunnels; see Chapter 4). The resumption of pillow lava effusion after explosive lapilli tuff activity may be due to meltwater depths in the englacial vault increasing somewhat, or to partial degassing of the magma.

In Iceland, several large glaciovolcanic polygenetic centres are present but poorly exposed within the present-day ice caps of Vatnajökull, Langjökull and Hofsjökull, including the well-known active centres of Grímsvötn and Bárdarbunga. The best known older centre is Askja. A novel initial interpretation that Askja was constructed by subglacial eruptions of hyaloclastite along ice-confined fractures and that the caldera structure is a constructive feature rather than collapse-related (Brown et al., 1991) is currently being reinvestigated (Graettinger et al., 2013a,b; McGarvie et al., 2013; Skilling et al., 2013). Despite its size, the Askja volcano appears to have been unusually short-lived (<75 k.y. McGarvie et al., 2013). By contrast, Eyjafjallajökull is a large polygenetic glaciovolcanic centre dominated by sheet-like sequences (Loughlin, 2002) and the same may be true for mafic rocks that dominate Katla and Öraefajökull, although they are both much less well studied (Carsewell, 1983; Walker, 2011; unpublished information of JLS). The mafic glaciovolcanic sheet-like sequences of the Sida–Fljotsfverfi district (Section 10.6.2) may have formed several coalesced low-profile shield volcanoes erupted from fissures (Bergh, 1985), but the original volcano morphologies are not well known.

11

Intermediate-composition glaciovolcanic sequences

11.1 Introduction

Intermediate-composition sequences include basaltic andesite and andesite in calc-alkaline systems, and mugearite, benmoreiite, tephriphonolite and phonolite in alkaline systems. In general the compositional range has physical properties (e.g. viscosity, glass transition temperatures) that vary enough from mafic systems to produce deposits that are distinctive in the field (Chapter 1; Fig. 5.6). For example, Walker (1992) showed that andesitic lava compositions could produce recognisable pillow lavas, but that the pillows had distinctly larger sizes than the much more common basaltic pillow lavas. The deposits can also be more ambiguous to interpret as glaciovolcanic for at least two reasons. Firstly, because they have higher viscosities, intermediate-composition lava flows are inherently thicker than mafic ones. Thus, while a 50 m-thick mafic lava flow might be considered to be 'anomalously thick', with the overthickening interpreted as due to impoundment by ice, intermediate-composition lava flows in subaerial environments can be over 100 m thick without any impoundment. Secondly, whereas prominent vitric surfaces on mafic lavas are characteristic indicators of 'enhanced' cooling caused by eruption in water-rich environments, more evolved magmas more readily form thick volcanic glass surfaces even in dry, subaerial settings. These issues become most extreme for felsic compositions, as discussed in Chapter 12.

Sequence types discussed below include ice-impounded lava flows, lava domes, tuyas, other miscellaneous sequences, and sequences found at long-lived stratovolcanoes that have had both glaciovolcanic and non-glaciovolcanic eruptions. While most of the descriptions for these deposits have been obtained from a limited number of areas (e.g. British Columbia and Iceland), increasingly new mapping and geochronology at strato-volcanoes in the western USA, Alaska, South America and Kamchatka will probably increase the number of known intermediate-composition glaciovolcanic deposits significantly, as well as improving our understanding and interpretation of these deposits. Intermediate-composition glaciovolcanism currently remains a major area in need of focused field-based studies.

11.2 Ice-impounded lavas

The pioneering work on ice-impounded lavas by Mathews (1951, 1952a) foreshadowed and in many respects provided the template for all future studies of these distinctive glaciovolcanic sequence types. They are also referred to as ice-confined, ice-dammed or ridge-forming, and several spectacular examples are known. The described examples are overwhelmingly andesitic (also see Chapters 10 and 12). Most of the intermediate-composition outcrops are in British Columbia (Kerr, 1948; Mathews, 1952a,b; Bye et al., 2000; Edwards et al., 2002; Kelman et al., 2002; Kelman, 2005) but well-described examples occur in the Cascade Range, USA (Lescinsky and Sisson, 1998; Lescinsky and Fink, 2000; Lodge and Lescinsky, 2009a). In general, the lavas show no signs of chilling by ice on their upper surfaces but they show anomalous features downslope or along their flanks where they came to rest against ice. Distinctive characteristics attributed to ice impoundment include (1) precipitous, locally concave marginal rock faces up to 500 m high subject to major modification by landsliding after the supporting ice has retreated by melting; (2) unusually thick lavas, up to 270 m in a single lava flow, inferred to have been caused by ponding; (3) occurrence of 'lava dribbles' interpreted as small lava breakouts emplaced between the chilled sheer lava face and the retreating ice wall; and (4) distinctive patterns of cooling fractures preserved on steep rock faces caused by rapid chilling against a former ice barrier (Mathews, 1952a; Lescinsky and Fink, 2000; Edwards and Russell, 2002; Kelman, 2005; LaMoreaux, 2008; Lodge and Lescinsky, 2009a).

11.2.1 British Columbia

The earliest described examples are those described by Kerr (1948). Kerr (1948) worked in the Iskut area of western British Columbia from 1926 to 1929, and provided the first written descriptions of Hoodoo Mountain volcano. He noted that on the western sides of the mountain, lava flows thickened substantially between the higher and lower outcrops, and he suggested that 'such damming can only have been caused by the presence of glacial ice'.

Kelman (2005) has described intermediate-composition ice-impounded glaciovolcanic sequences in the Mt Cayley Volcanic Field of the Garibaldi segment of the Cascade volcanic arc. They include Tricouni Peak, Ring Mountain, Pali Dome, Cauldron Dome and Slag Hill (Fig. 11.1). All but one are andesitic. The exception is at Tricouni Peak (southwest outcrop), which is a basaltic andesite. The outcrops are typically complex piles of overlapping lavas with small erupted volumes ($0.05–0.7$ km^3). No evidence for explosive eruptions has been found and the lavas are poorly vesicular (generally <3% vesicles). Many of the lavas are entirely subaerial but each outcrop includes one or more lavas displaying some combination of steep, finely jointed margins and fanning or subhorizontal polygonal columns. The lava groundmasses contain only a small amount of glass (up to 20%, typically much less) and a few outcrops (Pali Dome, Slag Hill) contain minor hyaloclastite.

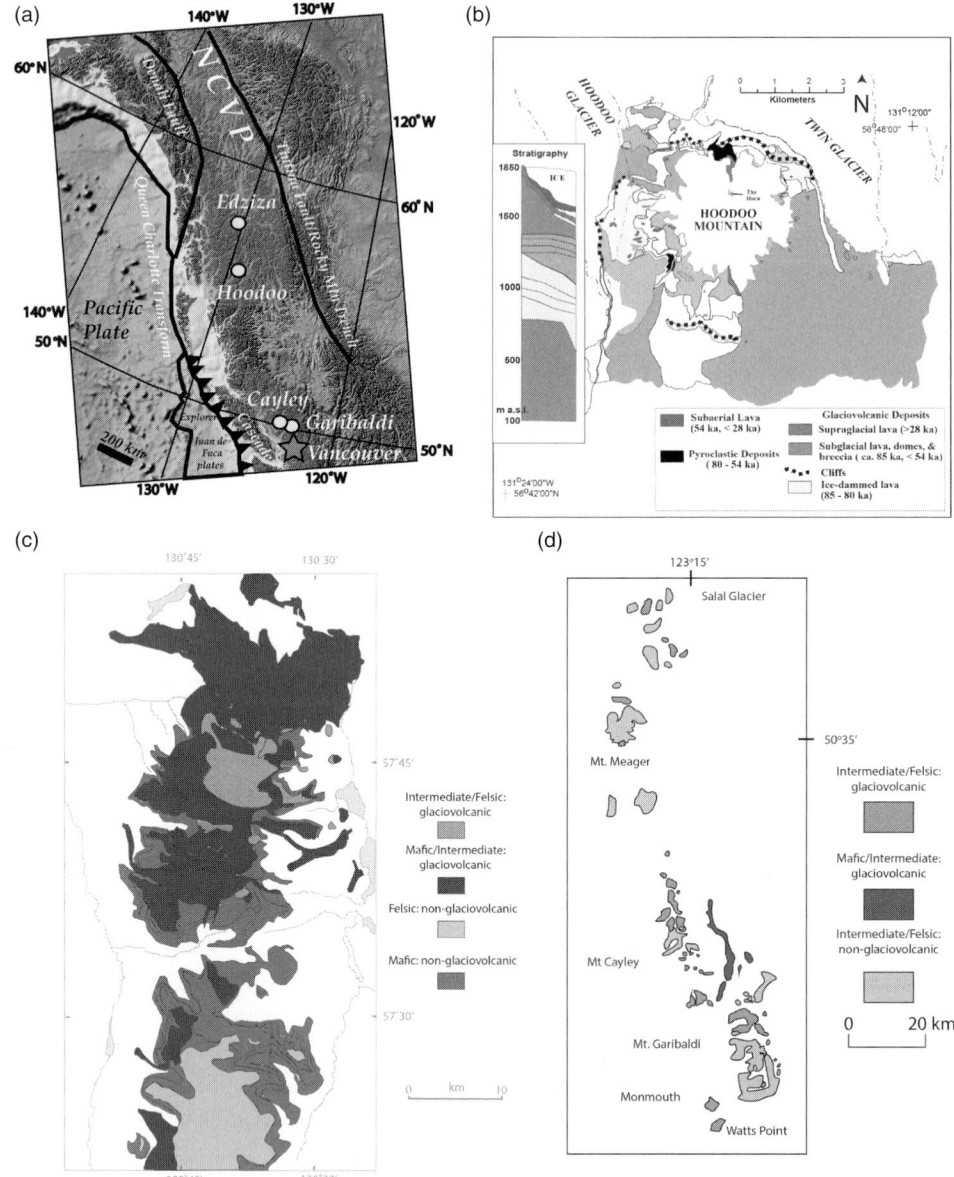

Fig. 11.1 Maps showing the main areas of intermediate glaciovolcanism in British Columbia. (a) Map of western Canada showing locations of detailed maps (b)–(d). (b) Schematic geological map of Hoodoo Mountain volcano showing glaciovolcanic deposits (after Edwards et al., 2000). (c) Generalised geological map of the Mt Edziza volcanic complex showing glaciovolcanic and non-glaciovolcanic deposits (after Souther, 1992). (d) Schematic map showing glaciovolcanic and non-glaciovolcanic deposits in the Garibaldi segment of the Cascade volcanic arc (after Kelman, 2005). (A black and white version of this figure will appear in some formats. For the colour version, please refer to the plate section.)

No evidence for coeval deposition of epiclastic sediments has been reported (cf. Llangorse Mountain ice-impounded basanite and nephelinite lavas, Section 10.2). The glaciovolcanic occurrences were interpreted as products of eruptions below alpine ice (i.e. thin ice). The ice cover rapidly melted through and was followed by essentially subaerial emplacement, with lava flowing downslope until its advance was blocked by much thicker valley-filling ice. The Slag Hill outcrop probably flowed at least 4 km within an ice-confining channel about 1 km wide. By contrast, that southwest of Tricouni Peak is probably a perched lava. It onlapped pre-volcanic basement on one side under subaerial conditions but formed a prominent 200 m-high subvertical cliff at the opposite margin overlooking a valley that would have contained a former glacier. The limited evidence for interaction with water in all the outcrops (i.e. lack of pillows, paucity of groundmass glass in lavas, and minor hyaloclastite) suggests that water was not ponded at the vents but flowed freely away down the steep bedrock slopes.

Ice-dammed phonolite and trachyte lavas at Hoodoo Mountain, British Columbia, form a prominent set of cliffs encircling the base of the volcano (Kerr, 1948; Edwards et al., 2002; Edwards and Russell, 2002; Fig. 11.2). The cliffs are 50–200 m high and are formed of massive, aphanitic flow-banded lavas showing pervasive jointing on a variety of scales. Cliff-face samples are nearly glassy and the joints are mainly small-diameter columns orientated perpendicular to the vertical cliff or forming radiating patterns. The individual lavas are often very thick and may occupy entire cliff sections, but multiple lavas occur elsewhere. They apparently began as relatively thin lava flows (<30 m) and thickened substantially downslope despite low and essentially unchanging

Fig. 11.2 Aerial view showing lava cliffs more than 200 m in height on the southwestern flank of Hoodoo Mountain volcano, British Columbia. The cliffs were most likely formed when lava was dammed by valley-filling ice.

bedrock gradients. The relationships suggest that the lava flows inflated due to impounding by ice that surrounded the lower flanks of the volcano.

11.2.2 Cascades

Numerous (more than 23) ice-bounded andesitic and dacitic lavas a few tens of ka in age occur at Mt Rainier, a stratovolcano in the Cascade volcanic arc, western USA (Lescinsky and Sisson, 1998). Both compositional types show similar features. Many of the lavas form steep-sided ridges up to 25 km long separated by valleys that radiate from the summit. The lower-elevation lavas are unusually thick, up to 450 m, as opposed to c. 20 m on the higher steeper volcano slopes (25° gradients compared with <8° lower down). The lavas' occurrence on topographical highs rather than in depressions might be attributed to post-eruptive river erosion resulting in an inverted topography (Fiske et al., 1963). By contrast, Lescinsky and Sisson (1998) reinterpreted the relationships to suggest that the lavas were confined to their ridge-crest locations by impounding ice. The lava bases contain thin autobreccias and broad (30–40 cm diameter) polygonal joints oriented normal to the lava margins, whereas lava interiors have platy joints spaced at 1–5 cm intervals and parallel to the lava bases. The lava sides are little-eroded and their subvertical slopes are original. They overlook deep U-shaped valleys interpreted as having existed prior to the lavas being emplaced. The lava sides and front have thick glassy zones and subhorizontal columns 8–20 cm in diameter. The lava tops are rarely preserved but locally reveal autobreccia and lava spines consistent with subaerial conditions. Several features were listed as indicative of ice-contact, including (1) thick glassy flow margins indicating rapid quenching, probably by meltwater; (2) steep lava sides suggesting an adjacent near-vertical cooling surface, presumably an ice wall; and (3) an abrupt transition between the glassy margins to crystalline interior, thought to be caused by molten lava flowing within a sheath-like quenched margin similar to lava tubes. The formation of these features was examined in greater detail by Lescinsky and Fink (2000). They showed that there is an inward progression in fracture types that is related to the cooling history of the lava, varying from irregular, closely spaced and hackly at lava margins, becoming more regular and widely spaced, sheet-like and polygonal towards lava interiors (Fig. 9.8). They also drew generic cross sections showing the internal structure (fracture distribution) of several types of ice-contact lavas (Fig. 11.3). Lavas were envisaged erupting subaerially high up on the stratovolcano and flowing down until they encountered valley-filling ice. The lava snouts were deflected sideways to flow along the gully separating the glacier margin from the valley wall. If a ridge separating two valleys was covered by ice (e.g. at lower elevations or at the confluence of two valleys where ice thickens greatly), the lava may melt the thin ice cover on the ridge crest and thus become channelled, a process aided by meltwater running ahead of the advancing lava. Its margins are buttressed by the enclosing ice walls that are melted more slowly by the lava because the off-crest ice is thicker and takes

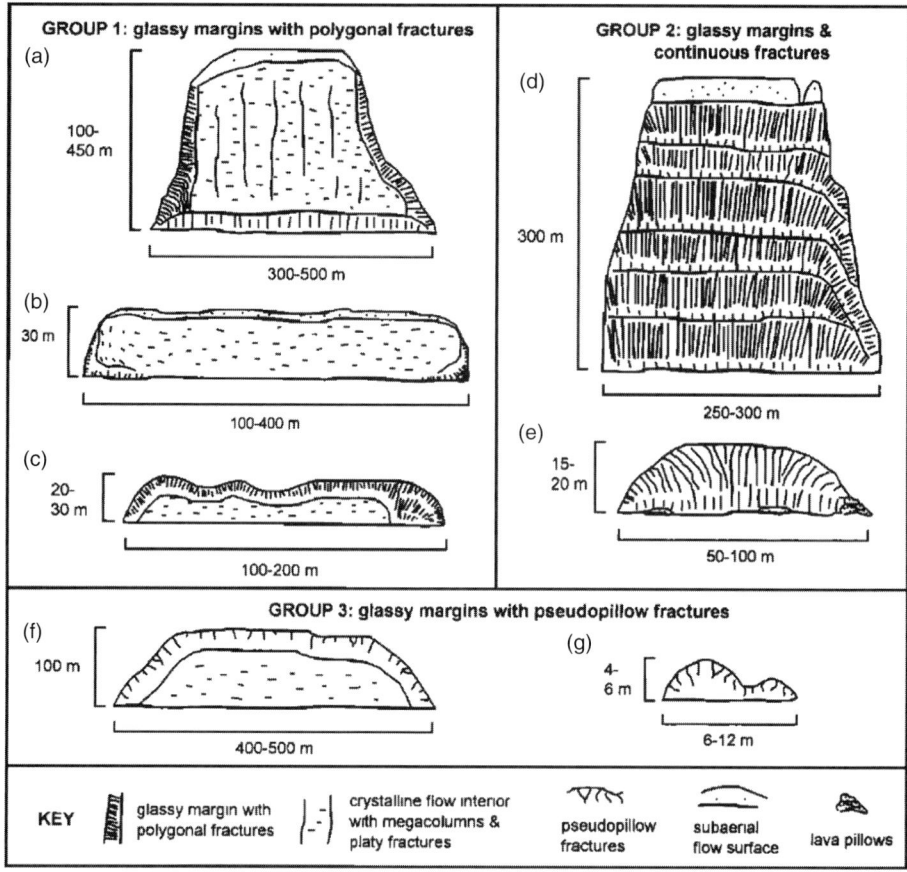

Fig. 11.3 Sketches of the joint distributions of various types of ice-dammed lavas (from Lescinsky and Fink, 2000): (a) ridge-forming lava; (b) smaller lava; (c) polygonally jointed subglacial dome; (d) lava-dominated tuya (felsic); (e) esker-like lava; (f) subglacial dome with pseudopillow fractures; and (g) pillow lava lobe.

longer to melt. The adjacent ice also may get undercut and form steep faces by collapsing. Following climate-related ice decay, the lavas are left as ridge-forming or perched features (Fig. 11.4). Using these features, Lescinsky and Sissons (1998) and Lescinsky and Fink (2000) were able to reconstruct ice configuration and volumes at Mt Rainier (Washington) and South Sister (Oregon) for the last glacial and to predict the likely flow path and hazards associated with a hypothetical eruption at Mt Rainier.

11.3 Domes

Subglacial eruptions of intermediate-composition magmas also produce lava domes. Evidence for subglacial emplacement includes (1) the presence of prominent glassy

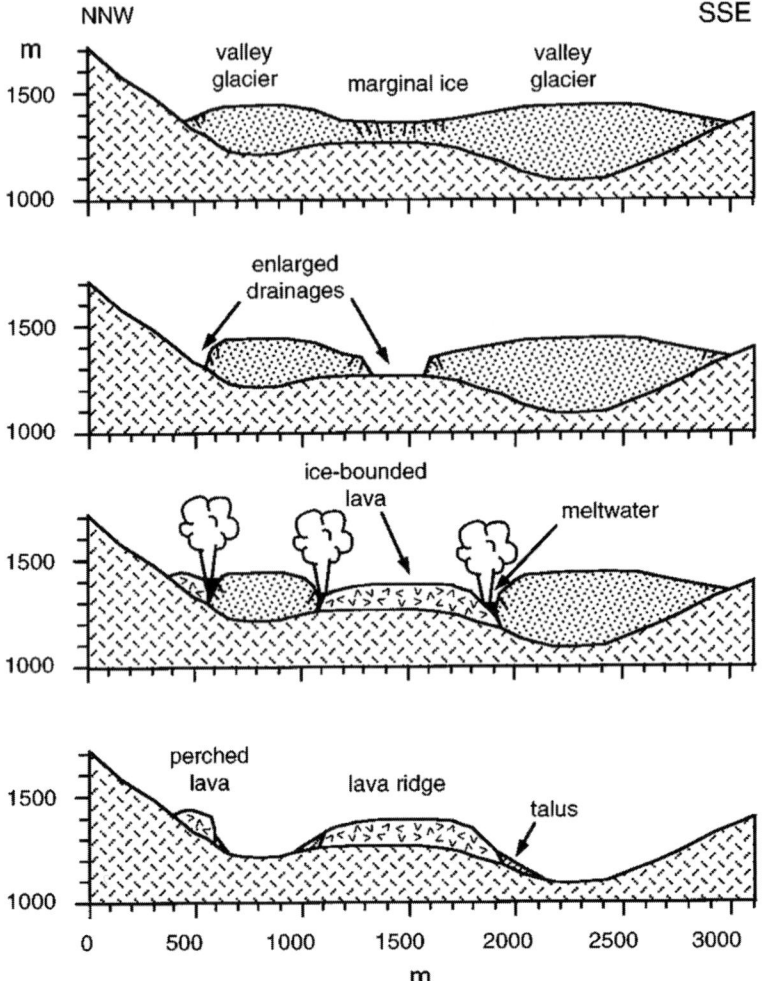

Fig. 11.4 Cartoons illustrating the formation of ridge-forming lavas (from Lescinsky and Sisson, 1998). Once the buttressing glaciers have melted, the lavas remain as prominent ridge crests.

margins with distinctive patterns and orientations of cooling joints; (2) steep-sided morphology attributed to ice-confinement (e.g. Lescinsky and Fink, 2000; Kelman, 2005), similar to criteria used for ice-impounded lavas (Section 11.2); and (3) a lack of flat upper surfaces. Dome morphologies are interpreted as reflecting the shapes of the ice cavities within which they were emplaced. Thus, domes are distinguished from ice-impounded lavas and lava-dominated tuyas by being entirely subglacial (Kelman, 2005). All intermediate-composition examples currently known are in British Columbia (Bye et al., 2000; Edwards et al., 2002; Edwards and Russell, 2002; Kelman et al., 2002; Kelman, 2005).

Fig. 11.5 View of one of the Ember Ridge subglacial domes. The outcrop extends to the columnar jointed rocks on the skyline in the background. The prominent central rock is about 150 m high and probably hosted the vent. It is a dome-like mass of lava with some irregular lava protrusions whose radial columnar joint orientations suggest it is comparatively little-eroded. Dashed line shows original dome surface inferred from column orientations (image: Paul Adams).

11.3.1 Mt Cayley Volcanic Field

Nine probably coeval masses crop out in the Mt Cayley Volcanic Field erupted from eight vents at Ember Ridge. The total volume of erupted rock is 0.13 km^3 and the individual masses vary from 0.0007 to 0.05 km^3. The Ember Ridge outcrops are jagged to bulbous, steep-sided piles of lava with irregular to fanning arrangements of small columns and closely spaced planar jointing and flow banding perpendicular to columns that together reflect dome-like shapes (Fig. 11.5). Heights are typically 140–300 m and diameters are 180–1100 m. The lava masses are composed of both single and multiple cooling units and erosion has probably been minimal. Vesicularity is generally low (<1%) and groundmass glass is minor. Hyaloclastite is also uncommon except at one outcrop where it occurs as pods up to 10 m across separated by screens of lava 2 m wide with well-developed columnar joints perpendicular to their margins. The scarcity of hyaloclastite and glass suggests that water generally did not accumulate over the vents. Their pristine shapes suggest that they were not subsequently overridden by ice and, thus, that the higher masses were emplaced before lower masses within a regime of decaying and lowering ice surfaces. Thermodynamic limitations of intermediate magmas and efficient basal meltwater drainage enhanced by a suitably steep basal topography were cited by Kelman (2005) and Kelman et al. (2002) as likely reasons for the absence of water interaction at subglacial intermediate-composition domes in British Columbia. The thermodynamic considerations indicate that the conversion of ice to water is relatively inefficient and insufficient space is melted to accommodate magma emplacement regardless of magma composition (e.g. heat transfer efficiency may be as low as 45% even in mafic subglacial eruptions; see Chapter 6). Thus, meltwater-filled cavities will rapidly become overpressured leading to flotation of the overlying ice and the sudden release of meltwater in a jökulhlaup and a 'dry' cavity (Höskuldsson and Sparks, 1997; Höskuldsson et al., 2006). Ice thicknesses coeval with the eruptions were calculated to be <650 m (and mostly < c. 400 m) for subglacial domes in British Columbia (Kelman, 2005). The absence of associated tephras associated with intermediate-composition subglacial domes is comparable with the emplacement of small subaerial silicic domes, in which tephra typically makes up <10 vol % of the erupted material (Heiken and Wohletz, 1987). Such a small amount of tephra may be prone to being flushed away in the early stages in the eruption in areas with steep topography, leaving no visible record.

11.3.2 Hoodoo Mountain

At Hoodoo Mountain, bulbous subglacial domes consist of highly jointed non-vesicular phonolite–trachyte lava encased in poorly vesicular monomict lava breccia (Fig. 11.6).

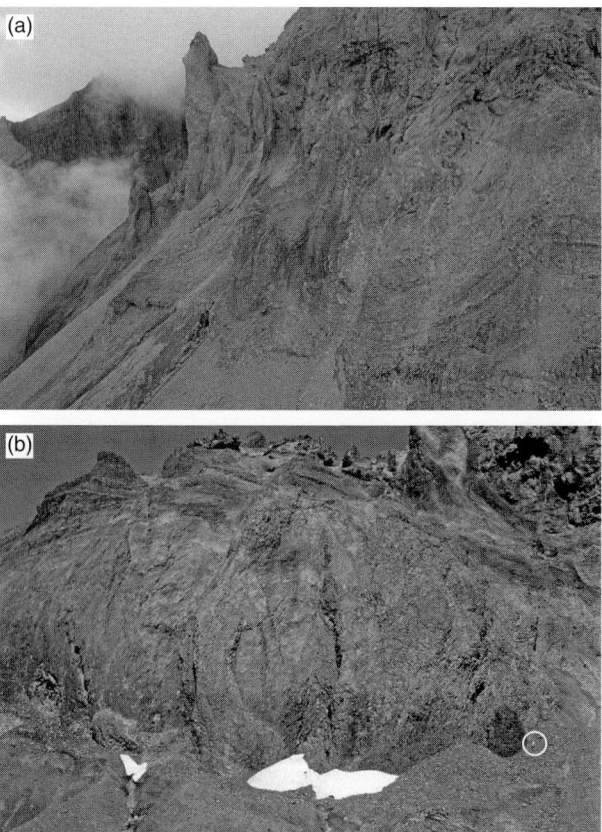

Fig. 11.6 Photographs of lithofacies associated with domes on the upper south flank of Hoodoo Mountain volcano. (a) View showing dome- and spire-intrusions encased in tuff breccia. (b) Closer view showing pervasive jointing in dome. Person (ringed) for scale.

They are found at the base and top of the volcano (Edwards et al., 2002; Edwards and Russell, 2002). The latter occurrences form a discontinuous ring of cliffs that partially surround the volcano at elevations between c. 1300 and 1700 m below the summit ice cap. The breccias are locally crudely bedded, poorly sorted and comprise a mixture of lapilli and blocks in an ash matrix that is reddish-orange coloured due to variable degrees of incipient clay(?) alteration. They correspond to autobreccias formed during dome extrusion and quenching and are inferred to be palagonitised hyaloclastite. Cooling joints are pervasive and comprise small-diameter columns with irregular orientations that are perpendicular to the uneven contacts with breccia. Emplacement took place under relatively thick ice that is thought to have suppressed vesiculation and was over 500 m thick.

11.4 Tuyas

Only four intermediate-composition tuyas have been described, all in British Columbia: at Slag Hill, Ring Mountain and Little Ring Mountain, in the Mt Cayley Volcanic Field, and The Table in the Mt Garibaldi Volcanic Field (Mathews, 1951; Kelman et al., 2002; Kelman, 2005). All are lava-dominated and no intermediate-composition tephra-dominated equivalents are currently known. The Table is the best known and was the first of these landform and sequence types to be described. It is also the most prominent. Blade-like (elliptical) in plan view, it has a basal diameter of only c. 250–300 m and it rises in sheer cliffs over 250 m high to an elevation exceeding 2000 m a.s.l. (Fig. 11.7). It also has a conspicuous nearly horizontal upper surface, the diagnostic feature of tuyas irrespective of composition and internal lithofacies architecture (see Chapter 8).

All of the intermediate-composition tuyas are formed of stacked piles of lava (i.e. they are lava-dominated tuyas, *sensu* Smellie, 2007, 2009, 2013; called flow-dominated by Kelman, 2005) and they mostly form near-vertical-sided cylindrical outcrops 400 m to 2 km in diameter and 400–900 m high with flat summits. The lavas are andesitic (including The Table, originally called a dacite), and individual lavas are several tens of metres thick (down to a few metres at The Table) with well-developed columnar joints showing a variety of attitudes (vertical, horizontal and radiating) and rare planar, hackly or irregular joints. Vesicularity is low (c. 2−<10%). The uppermost lavas are more coarsely jointed and separated by layers of oxidised autobreccia signifying subaerial emplacement of the final erupted units. None of the centres preserve evidence for explosive eruptions. Uniquely, the large Ring Mountain centre subsequently emitted a single thick lava that shows ice-contact features at high elevations but not lower down, suggesting that the associated ice was stagnant and that flow probably took place in pre-existing drainage channels. At The Table, several lavas change orientation from horizontal to steeply dipping at the southwestern periphery, with well-defined cooling surfaces on three sides suggesting that they flowed down a bergschrund. No fragmental lithofacies have been described apart from a layer of

Fig. 11.7 Lava dominated tuya. The Table, British Columbia (from Hickson, 2000). The feature rises over 250 m and is composed of multiple flat-lying andesite lavas 30–80 m thick, some of which also adhere to the vertical sides where they are presumed to have flowed down a narrow gap (bergschrund) between the growing volcanic edifice and encircling coeval ice (Mathews, 1951).

till beneath The Table. The elliptical shape of The Table was ascribed to migration of the cavity melted in the ice by the eruption due to coeval advection of that ice. The suggestion is supported by the restriction of 'bergschrund lavas' to the downstream (southwestern) end of the outcrop, where a gap would have developed.

11.5 Other intermediate-composition glaciovolcanic sequences

Other types of intermediate-composition monogenetic glaciovolcanic sequences are extremely rare. They are currently only known to occur as three monogenetic centres at Kerlingarfjöll, central Iceland. The Kerlingarfjöll examples are particularly interesting in that they are dominated by large amounts of vitroclastic material and show unambiguous evidence for water accumulation over the vents, both of which contrast with intermediate-composition glaciovolcanic centres in British Columbia and Antarctica (Kelman et al., 2002; Kelman, 2005; Stevenson et al., 2009; unpublished information of JLS).

Kerlingarfjöll is an active central volcano complex, a volcanic massif composed of clusters of small intermediate and felsic centres spread over an area of 200 km^2 and constructed on a plateau of subglacially erupted basalt lavas and other basaltic products (Grönvold, 1972). It has been ice covered for most of its volcanic history. The complex contains Iceland's second largest volume of rhyolite but also includes products of andesite and dacite eruptions. Two clusters of edifices are present. The western cluster comprises three major rhyolite centres with ages of c. 345–125 ka (Flude et al., 2010) that overlie andesite–dacite sequences at three sites: Tindur, Haraldur and the informally named 'Campsite Gully' (Stevenson et al., 2009). The sequences at each of the three latter sites are products of small-volume eruptions along fissures that produced the elongated outcrop shapes. Two discrete subglacially erupted deposits of andesite are present, together with one that is dacitic. Although felsic, the dacitic example is also described here with the andesitic occurrences because of overlapping characteristics.

The largest andesite deposit has an estimated volume of 0.15–0.45 km^3 and crops out at Campsite Gully and Bringur. It is dominated by two major vitroclastic lithofacies comprising massive lapilli tuff with intrusions overlain by stratified lapilli tuff (Fig. 11.8). A subordinate

Fig. 11.8 Irregular andesite intrusions in massive lapilli tuffs that become crudely stratified upward. The intrusions have locally disrupted bedding in the lapilli tuffs. Kerlingarfjöll, Iceland.

lithofacies of fluidal-clast-bearing lapilli tuff is also present. The second andesite deposit forms the basal pedestal of Tindur, and is an andesite sequence composed of two major lithofacies (contorted pillow-fragment breccia (Fig. 9.19a) and stratified lava breccia), 200 m thick in aggregate and with a volume of 0.06–0.18 km^3. The massive andesite lapilli tuff is indurated and up to 100 m thick, composed of vesicular (20%) blue-black andesite glass fragments and minor sand-size grains; lenses of microvesicular yellow pumice are also present in the upper parts. The lapilli tuff is intruded by numerous (30–70% by volume) rounded, pillow-like, columnar-jointed lava lobes up to 2 m across with prominent glassy rims up to 6 cm thick, and the outcrop is dominated by a large irregular intrusion 200 m wide by 100 m tall with glassy margins and an irregularly jointed interior that has locally altered adjacent lapilli tuff. By contrast, intrusions are absent in the stratified lapilli tuff, which dominates the stratigraphically higher outcrop at Bringur. Stratification is crudely developed in beds 20–50 cm thick that include cross stratification, reverse grading and pumiceous lenses. Clasts resemble those in the under-lying massive lapilli tuff lithofacies and are only incipiently vesicular apart from within lenses of yellow pumice. The fluidal clast-bearing lapilli tuff is also vitroclastic. It forms cliffs 10 m high and is massive, distinguished by coarsely vesicular flattened clasts 10–20 cm long dispersed through the deposit. Clasts in the Tindur deposit are blocky poorly vesicular pillow fragments.

The dacite sequence at Kerlingarfjöll crops out at Haraldur and is stratigraphically above the Bringur andesites. It is a crudely bedded vitroclastic lapilli tuff that lies conformably on pillow lava and columnar-jointed intrusions with slope-parallel orientations. The vitroclasts are angular, blue-black and poorly vesicular (c. 30% vesicles), dispersed in a yellowish sand-grade matrix.

In many characteristics the Kerlingarfjöll sequences are the andesite–dacite equivalents of the basaltic tindars that are widespread in Iceland (Fig. 11.9; Jones, 1969b, 1970; Werner and Schmincke, 1999; Schopka et al., 2006). The Haraldur dacitic deposit commenced with pillow lava effusion, including slope-parallel megapillows, and cogenetic intrusions. Intrusive lobes have also been described as products of some subaqueous felsic and basaltic eruptions (Yamagishi and Dimroth, 1985; Skilling, 1994). Similarly, the contorted pillow-fragment andesite breccia at Tindur probably formed by the spalling and mechanical disintegration of lava pillows and intrusive lobes, aided by collapses of the unstable pillow pile. The overlying stratified lava breccias at Tindur were also formed by disintegration of pillow lava and repeated subaqueous avalanching on steep slopes. Pillow dimensions in the Tindur and Bringur outcrops are larger (0.5–1.5 m) than in basalt pillow piles in Iceland (<0.4 m; Fridleifsson et al., 1982). By contrast, the blocky form and poor vesicularity of vitroclasts in the lapilli tuffs in the other outcrops suggest that eruptive activity was hydromagmatic and the increased occurrence of pumiceous vitroclasts at higher stratigra-phical elevations suggests that subglacial cavity pressures may have diminished with time in each, perhaps due to thinner overlying ice or eruption of more gas-rich magma, resulting in greater vesiculation of the erupting melt. The large contorted clasts in the locally developed fluidal clast-bearing lapilli tuff in the Campsite Gully deposit were formed by

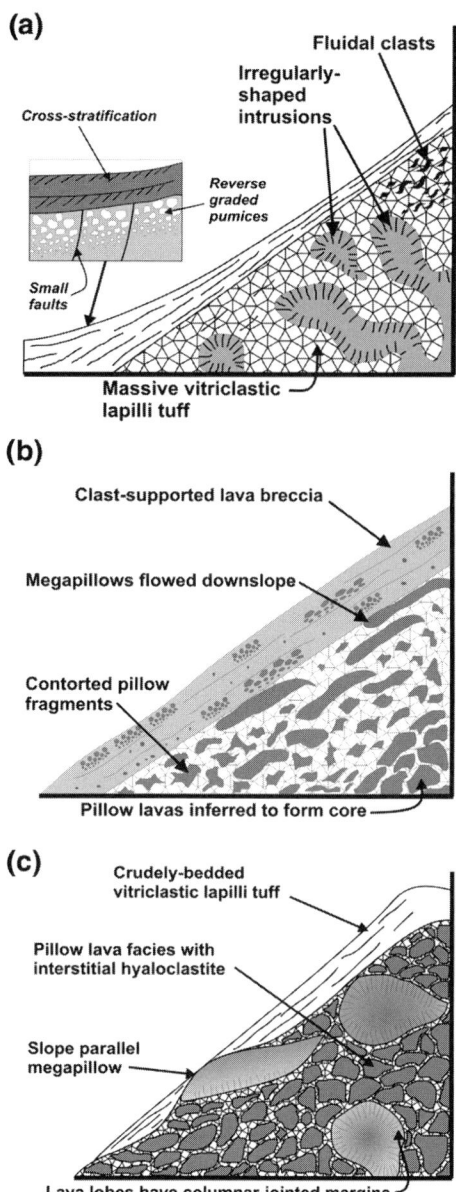

Fig. 11.9 Schematic cross sections across three intermediate-composition glaciovolcanic outcrops at Kerlingarfjöll, Iceland, showing the variable upward transition from pillow lava cores into stratified breccias and lapilli tuffs, analogous to pillow volcano and tindar stages of mafic glaciovolcanoes, respectively (from Stevenson et al., 2009).

tearing apart of fluidal magma during subaqueous fire fountaining, representing high discharge conditions in vent-proximal locations (cf. Simpson and McPhie, 2001). They accumulated in a matrix of blockier fragments formed by the quench-related disintegration of the fluidal clasts in water. The presence of large water-cooled intrusions is perhaps linked to feeder dykes nearby and their emplacement may have been responsible for the absence of stratification in associated massive lapilli tuffs (e.g. in the Campsite Gully deposit). Similar intrusions are present in the products of subaqueous felsic eruptions (e.g. Yamagishi and Dimroth, 1985). Ice thicknesses coeval with the eruptions at Kerlingarfjöll are poorly known but may have been between 550 and 2600 m (Stevenson et al., 2009).

Despite the evidence for the former presence of abundant meltwater in the subglacial cavity during eruptions, there is no indication for the emplacement of the lava-fed deltas that are so characteristic of mafic subglacial eruptions. Stevenson et al. (2009) argued that meltwater retention may be the critical factor determining whether the magmas interacted with water. In other words, with inefficient drainage, sufficient meltwater might be retained and enable explosive hydromagmatic eruptions to occur. At Kerlingarfjöll, inefficient basal meltwater drainage is probably promoted by the relatively subdued pre-eruption bedrock relief, which is much less pronounced (<100 m) compared with the much steeper relief beneath the British Columbia intermediate-composition centres described by Kelman et al. (2002) and Kelman (2005) and enhanced the likelihood of meltwater interaction (Stevenson et al., 2009).

11.6 Large polygenetic glaciovolcanic centres

Glaciovolcanic intermediate-composition lavas are also present in some polygenetic stratovolcanoes but descriptions are still uncommon (e.g. Minna Bluff, Antarctica: Wright-Grassham, 1987, and unpublished information of JLS; Hoodoo Mountain, British Columbia: Edwards et al., 2000; Crater Lake, Oregon: Bacon and Lanphere, 2006; and Okmok, Alaska: Larsen et al., 2007). Those at Mt Rainier (USA Cascades) are described above and are andesites and dacites showing well-developed ice-impounded features (Section 11.2.2; Lescinsky and Sisson, 1998; Lescinsky and Fink, 2000). Other examples include mugearite, benmoreite and tephriphonolite lavas in large glaciovolcanic centres in Victoria Land, Antarctica (Smellie et al., 2011a). The majority of the Victoria Land examples occur as 'a'ā lava-fed deltas and are the type examples for such volcanic land-forms (Smellie et al., 2013a, 2014). However, the Victoria Land deltas are mainly mafic and the intermediate-composition lava-fed deltas are indistinguishable in lithofacies and architecture from mafic examples. They are described in detail in Chapter 10.

11.6.1 Minna Hook

Multiple volcanic units with mainly phonolite compositions are well exposed at Minna Hook, Antarctica, in a cliff face 12 km long and 700–900 m high constructed mainly

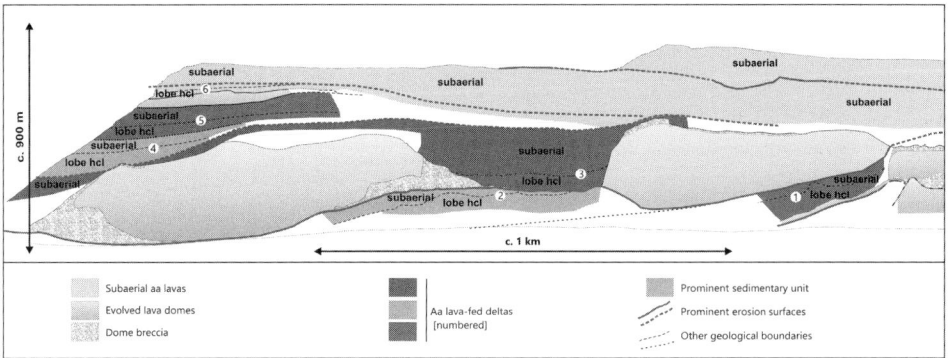

Fig. 11.10 Sketch map of Late Miocene volcanic rocks at Minna Hook, Mt Discovery, Antarctica, showing inferred and observed relationships between felsic domes or coulées (probably glaciovolcanic), glaciovolcanic 'a'ā lava-fed deltas and subaerial capping 'a'ā lavas. The erosional surfaces shown are probably all glacial (JLS, unpublished).

from products of mafic 'a'ā lava-fed deltas (unpublished information of JLS; Fig. 11.10). The units vary in thickness from c. 50 to 200 m and are c. 180–1000 m in length. They form prominent convex-up outcrops with steep flanks and relatively high aspect ratios formed of mainly non-vesicular greenish-grey-coloured rock. Some are formed of several thick superimposed lavas similar to exogenous domes. The units contain prominent platy cooling joints parallel to the lava surfaces that are closely spaced in a metre-thick marginal zone. The joints become more widely spaced above and are accompanied by similarly widely spaced subvertical joints that give a crudely columnar appearance to outcrops. Basal surfaces of domes are either glassy or aphanitic, with jigsaw-fit jointing in a narrow border c. 25 cm thick that passes up into closely spaced platy joints, then subvertical columns. The domes either directly overlie erosional surfaces cut in mafic lavas or they rest on local diamict or a few metres of stratified volcaniclastic sandstone and conglomerate showing channelling and low-angle cross stratification (Fig. 11.11). Massive to crudely stratified lithic or glassy breccia intervenes between the sediments and the massive dome rock. Lithic breccia is most common and glassy breccia is rare. The latter also contains isolated lobes ≤0.5 m in diameter, formed of closely jointed lava that breaks up marginally into glassy breccia. The breccias also form prominent yellow or pale green flanking deposits that wedge out along the basal unconformity. The breccias are monomict and clast supported, formed of angular fragments of the associated domes and minor silt or very fine sand-grade matrix. In a few places, the breccias are underlain by ≤1 m of crudely stratified yellow lapilli tuff full of pumice and fine ash. The pumice is probably cogenetic with dome emplacement (i.e. syn-eruptive). The mode of formation of the subglacial domes at Minna Hook has not been investigated in detail yet and not all the domes may be glaciovolcanic. However, it appears that many were emplaced in association with wet-based ice. Initial interpretation suggests that meltwater generated by the eruption reworked pre-existing

Fig. 11.11 Stratified gravelly sediments sandwiched between overlying pale-coloured subglacial(?) phonolite dome breccia and underlying subaerial 'a'ā lavas of a phonotephrite 'a'ā lava-fed delta. The basal lavas have been sharply truncated by erosive ice. Minna Hook, Mt Discovery, Antarctica.

subglacial sediments, eruption-generated tephra and dome-derived clasts ahead of lava effusion. The lava advanced over cogenetic autobreccias that either show no clear evidence for interaction with water (lithic breccia) or clear water chilling (hyaloclastite breccia).

The Minna Hook units are difficult to categorise unambiguously. Although in many respects (e.g. high aspect ratios; convex-up profiles) they strongly resemble subglacial domes, several are encased in lithic autobreccia showing no obvious signs of water cooling and they generally have the prominent thick columnar-jointed obsidian margins of sub-glacial domes (Section 11.3, above). If they were emplaced entirely subglacially (required to be a subglacial dome; Chapter 8), at least some units were not always in direct contact with ice. In addition, some upper surfaces may have been exposed subaerially and most overlie fluvial sediments and diamictons. These are characteristics of ice-impounded or sheet-like sequences. The ambiguous characteristics illustrate how one category of glacio-volcanic sequence type, e.g. subglacial domes, may grade into another, e.g. sheet-like sequences, depending on local circumstances. In the present context, the effusion of degassed viscous lava as a dome emplaced under thin ice might give rise to the features observed. This sequence also has similarities with dyke-sourced thin rhyolite lavas emplaced in thin ice at Hrafntinnuhryggur, northern Iceland, described by Tuffen and Castro (2009; see Section 12.6.1).

11.6.2 Hoodoo Mountain

Hoodoo Mountain is a well-exposed, Quaternary, polygenetic alkaline volcano in British Columbia, in which eruptive activity spanned the last 100 ka (Edwards et al., 2000, 2002; Edwards and Russell, 2002; see Sections 11.2.1 and 11.3.2). It is part of the Stikine subprovince of the Northern Cordilleran Volcanic Province and has erupted only phonolite and trachyte lavas. The edifice rises almost 2000 m in height, and has a present-day estimated volume of 17 km^3. Early observations from Hoodoo noted its step-like topography, and inferred that the steep lower cliffs probably formed when lavas flowing down from vents near the summit were dammed by surrounding glaciers (Kerr, 1948). This gives Hoodoo a topographical profile that is unique among stratovolcanoes, and it may be a rare example of a polygenetic volcano so dominated by the coeval ice-rich environment that it is morphologically distinctive (see Section 8.2.1). Flanks that have been covered by Holocene lavas on the southeastern side have typical slopes of 15–20°, but the summit is broad and domed, with a diameter of 3 km. In addition to subaerial lavas and tephra, probably erupted during interstadials, a wide range of glaciovolcanic sequence types was emplaced. The latter include ice-dammed lavas, subglacial domes and dyke-like fissure sequences erupted into relatively thin ice.They are described in Section 11.3.2. Six stages of volcano development have been identified and given distinct lithostratigraphical designations, making it one of the best-known glaciovolcanic stratovolcanoes: (1) subglacial eruptions commencing at c. 85 ka (Unit $Qvap_1$) with thick aphanitic lavas and domes intercalated with monomict tuff-breccia followed by (2) ice-dammed lavas erupted at c. 80 ka ($Qvap_2$); (3) subaerial pyroclastic eruptions and minor lavas between 80 and 54 ka (Qvpy and $Qvap_3$); (4) another series of subaerial lava flows at c. 54 ka ($Qvap_4$); (5) subglacial eruptions beneath thick and subsequently thinner ice between 54 and 30 ka ($Qvap_{5,6}$) that produced lavas and domes encased in monomict tuff-breccia; and (6) post-glacial subaerial lavas erupted at c. 9 ka (Qvpp) on the northwestern, southwestern and southeastern flanks. Glacial diamicts interpreted as tills also have been found underlying and overlying $Qvap_1$, bounding vitric tuff breccia/lapilli tuff associated with $Qvap_6$, and underlying Qvpp.

12

Felsic glaciovolcanic sequences

12.1 Introduction

Felsic glaciovolcanic sequences have mainly dacitic–rhyolitic compositions and are generally uncommon. The greatest number occurs in Iceland, but others may be present in British Columbia and a few in Antarctica. Most are tuyas (both lava- and tephra-dominated), but a few examples of tindars, ice-impounded lavas, domes and sheet-like sequences have also been described and there are rare outcrops that defy categorisation in a simple classification of sequence type. There are no polygenetic volcanoes formed wholly or dominated by felsic glaciovolcanic sequences.

12.2 Ice-impounded lavas

Ice-impounded felsic lavas are uncommon and, unlike other occurrences with mafic or intermediate compositions, published details are generally sparse. Examples have been described from British Columbia (Clinker Mountain) and south Iceland (Öraefajökull).

12.2.1 British Columbia

The earliest described examples are those at Clinker Mountain, British Columbia; they include the iconic ice-impounded feature known as The Barrier (Mathews, 1951, 1952a). Clinker Mountain is a steep unglaciated Holocene cone rising 600 m above the surrounding terrain, with two satellite cones. It erupted at a time when the shrinking Cordilleran Ice Sheet had receded from peaks and ridges above c. 1200 m a.s.l. but while ice still filled the valleys. Two dacite lava flows issued from one of the satellite cones. One lava, in Culliton Creek, increases in thickness from c. 70 m thick near its source to over 300 m at its steep termination. The terminal face is distinctly concave in plan view due to two prong-like subsidiary lava tongues. The other lava, in Rubble Creek, has several broad lobes and a convex terminus up to 100 m high. The Barrier is a spectacular cliff up to 250 m high at the margin of one of those lobes (Fig. 12.1). Much of its vertical face shows prominent colour banding and a variety of joint patterns varying from columnar to tabular (sheet-like? cf. Lescinsky and Fink, 2000). However, the uppermost 70 m comprises highly fractured red

Fig. 12.1 View of The Barrier, British Columbia, a massive wall of dacite lava 250 m high on the western flank of Garibaldi Mountain volcano. The unusual thickness of lava forming the cliff face was caused by impoundment of the lava against former valley-filling ice.

dacite and dacite rubble similar to the subaerial surface of the lava seen upstream. It suffers frequent rock falls and has been the source of several recent major landslides. These effects have modified the rock face, which is estimated to have originally risen c. 400 m above the floor of Rubble Creek valley. One of the other lobes of this lava still retains an original cliff face of black glassy dacite c. 300 m high unmodified by significant collapses. It also has well-developed columnar jointing and the columns commonly have a gentle dip indicating cooling against a subvertical ice surface. Another steep lava lobe face shows small sheets and driblets of lava indicating minor breakouts of lava emplaced as the adjacent ice barrier melted back. Similar breakouts were observed on the steep faces of The Table, a lava-dominated andesite tuya in British Columbia, situated not far from Culliton Creek (Mathews, 1951). The concave lava terminus is presumed to form by moulding against an original convex ice front (glacier snout) and is emphasised by collapse of the lava following removal of the support during ice recession.

12.2.2 Iceland

Ice-confined felsic lavas occur on Öraefajökull (Stevenson et al., 2006). Öraefajökull, situated on the southeast coast of Iceland, is that country's largest active stratovolcano. It has a basal diameter of 22 km, rises to 2119 m and has a broad ice cap containing a summit caldera 5 km across. Formed mainly of basaltic subglacial and subaerial lavas, it also has several small trachydacite and rhyolite outcrops (Thorarinsson, 1958; Prestvik, 1979, 1985; Stevenson et al., 2006). Ice-confined lavas have been described at (1) Vatnafjall, capping the ridge on the east flank of Kvíarjökull glacier, and (2) at Hvalvördugil about 10 km WSW of Kviarjökull (Stevenson et al., 2006; McGarvie, 2009; Walkes, 2011).

At Vatnafjall, the lowest and most laterally extensive is a rhyolite lava ('Unit D' of Stevenson et al., 2006; probably ice-impounded) formed of two flow units 75–150 m thick possibly separated by a thin tephra layer, and has a volume of c. 0.2 km^3. Vertical columnar joints up to 2 m wide in the centre of the outcrop narrow to 10–30 cm in a basal zone 2–10 m thick, together with pervasive flow banding that is horizontal to base but variably orientated on the ridge crest. The base also contains several hollows up to 1 m high surrounded by glassy rhyolite with hackly fractures and locally developed breccia (cf. Lescinsky and Fink, 2000). The lava lacks a basal autobreccia and is chilled and glassy due either to interaction with cold wet ground or possibly snow. The hollows are interpreted as meltout cavities representing former ice blocks engulfed by the flowing lava. The elongate NW–SE orientation of the outcrop probably reflects confinement by flanking ice but an origin by erosion cannot be excluded.

The eroded upper surface of rhyolite lava Unit D is overlain unconformably by a trachydacite lava ('Unit E') that increases in thickness from 25 to 75 m southeast along the ridge crest. The volume of the latter is c. 0.02 km^3. Its base along its eastern margin comprises pink or red-coloured clast-supported breccia formed of lapilli-size dark glassy lava. The coloration is associated with the presence of minor very fine matrix. The top of the lava locally shows circular crater-like features up to 20 m across and 3–4 m deep with flat breccia-covered floors and raised rims. Elsewhere the lava surface is covered in a layer of angular lava blocks 2–4 m thick. On the west side overlooking Kviarjökull glacier, the lava surface is glassy and shows prominent fanning and subhorizontal columnar joints and rests on a basal yellow deposit of small lava blocks, pumice and fine ash that is locally penetrated by glassy granules (peperite) from the overlying lava and vice versa (Stevenson et al., 2006, Fig. 8). Agglutinate composed of fluidal glassy trachydacite clasts capped by pumice lapilli and blocks are present on the lava top above the basal zone of interaction. The Unit E lava also forms a 70 m-high pillar-like lava tongue with fanning columnar joints and a glassy base that overlies the steep valley side below the main trachydacite outcrop at its northwest end. Stevenson et al. (2006) suggested that the laterally extensive red-stained glassy basal breccias formed at least in part from disarticulation of basal columnar joints associated with steam penetration along the joints, possibly associated with interaction with glacial melt-water or snow. This is also consistent with the evidence for local peperite. The crater-like depressions were attributed to possible steam blisters associated with water penetrating the lava surface. Explosions driven by basally derived steam may also have penetrated the full flow thickness and created the local agglutinate in a manner similar to the formation of rootless cones (Fagents et al., 2002; Section 9.3.5), which implies that the lava had a subaerial surface. The absence of a flow-front breccia might be due to the flow abutting ice, with the clasts carried away during subsequent melting of that ice. The marginal fanning column orientations and local overthickening suggest that the lava was confined by ice and flowed down the steep underlying topography under that ice in at least one lava lobe. The lava may be sourced in a small dome-like outcrop measuring 200 m in diameter

Fig. 12.2 View of a ridge-confined massive and columnar jointed rhyolite lava at Rotarfjall near Hrútsfjall, Öraefajökull, Iceland. About 400 m of relief is shown. The lava has apparently fed an offshoot that flowed down the steep slope in the foreground and was probably confined in an ice tunnel. The white lines are a photo-interpretation of the lava boundaries.

and 100 m high at the upslope end of the ridge. It was suggested that the Kviarjökull valley may have contained ice c. 500 m thick at the time of the eruption of Unit E and that the trachydacite lava was ice-impounded (ridge-forming, *sensu* Lescinsky and Sisson, 1998).

A thick ice-confined (ridge-forming, *sensu* Lescinsky and Sisson, 1998) rhyolite lava also crops out on a ridge crest on the south flank of Öraefajökull, at Rotarfjall north of Hvalvördugil about 10 km WSW of Kviarjökull (McGarvie, 2009; Fig. 12.2). It is connected to a series of lava lobes with well-developed small (5–10 cm diameter) glassy columnar joints normal to the lobe margins, suggesting that ice filled the adjacent valley, the lobes were emplaced subglacially and the main lava mass was ice bounded. A second much larger rhyolite lava 220 m thick extends northeast from Hrutsfjall, just northwest of Hvalvördugil. It has a quench-fragmented base that rests on tillite (McGarvie, 2009). The lava is columnar throughout and the variable column orientations suggest that it is compound. From its situation at an elevation far above any likely marine influence, and the evidence for quenching, the lava may also be ice-impounded (i.e. ridge-forming).

12.3 Domes

Descriptions of felsic subglacial domes are uncommon and currently restricted to an unusual subglacial rhyolite edifice at Prestahnúkur, in Iceland's Western Rift Zone (McGarvie et al., 2007), which might have formed as a subglacial exogenous extrusion

Fig. 12.3 View of Prestanúkur, Iceland, a rhyolitic edifice with enigmatic morphological affinities (felsic glaciovolcanic dome or tuya (from McGarvie et al., 2007). The letters refer to features described by McGarvie et al.

(Fig. 12.3), and Hoodoo Mountain and Mt Edziza in British Columbia. The Hoodoo Mountain occurrences (phonolites–trachytes) are described in Section 11.3.2.

Prestahnúkur is an unusual outcrop dominantly formed of effusive rocks and McGarvie et al. (2007) compared it with felsic tuyas in Iceland. It thus might be a lava-dominated felsic tuya (cf. Smellie, 2013). However, because it was emplaced entirely subglacially, with no evidence for subaerial conditions, it might also be regarded as an exogenous felsic subglacial dome. The Prestahnúkur outcrop is 570 m high, has a basal diameter of 2 km and a volume of 0.6 km^3. It is dominated by rhyolite lavas and breccias are subordinate. The lavas form three tiers, the uppermost of which varies from slope-draping (dips up to 40°) to flat-lying. The flat plateau-like upper surface strongly influenced the interpretation of Prestahnúkur as a felsic tuya. The three tiers are composed of lavas 5–60 m thick. The basal tier consists of a single thick lava whilst the upper two tiers comprise single and multiple sheet lavas, strongly flow banded and with abundant lithophysae and spherulites. Lava bases are characterised by sandy matrix-supported glassy breccia typically 2–5 m thick (up to 12 m) with domains of jigsaw-fit lava. They are interpreted as the products of quench fragmentation where lava flowed across a wet substrate accompanied by local ash-generating explosive interaction. By contrast, the poorly seen lava tops consist of pale- to reddish-coloured pumice breccia up to 30 m thick (generally <10 m) that overlie glassy lava with numerous small-diameter columns that also forms steep ramp structures within the pumice breccia. Although similar in some characteristics to subaerial rhyolite autobreccias, the Prestahnúkur examples contain smaller blocks and are much thicker (cf. Anderson et al., 1998). They resemble the hyaloclastite carapaces of felsic sills (cf. Hunns and McPhie, 1999; Orth and McPhie, 2003), formed due to sustained superficial wetting, quenching and mechanical disintegration. Some of the lavas have steep glassy margins whose orientation is reflected in steep flow banding in the subjacent rhyolite. These are interpreted as original ice-contact features, i.e. quenching against a steep ice-retaining surface. The lavas were interpreted as subglacial sills emplaced successively at a bedrock–ice interface.

Four groups of fragmental lithofacies are also present at Prestahnúkur but are volumetrically subordinate to lavas. The thickest deposit is lobe-bearing breccia, which dominates the basal parts of the edifice and constitutes more than 30% of its total volume. It is composed of ash-matrixed, clast-supported vesicular glassy and microcrystalline rhyolite lava blocks often prismatically and jigsaw-fit jointed together, with heavily fractured but otherwise coherent lava lobes with irregular columns and hackly and pseudopillow fractures that grade into surrounding breccia. The breccia domains are interpreted as quenched hyaloclastite carapaces of the associated lava lobes and the whole deposit is situated in an analogous structural position to the basal pillow mounds found below many basaltic tuyas, with probably a broadly similar mode of formation (see Chapter 10). Coarse polymict breccias occur at several stratigraphical positions and are composed of mixtures of obsidian, flow-banded and microvesicular rhyolite, variably with an ash matrix or clast-supported. They are interpreted as deposits of pyroclastic density currents (i.e. block and ash flow deposits) or locally developed hyaloclastite carapace breccias, possibly glacially reworked. Well-bedded to diffusely cross-stratified, volcaniclastic, monomict and polymict sandstones and gravels, sometimes showing bedding dipping back into the edifice, occur in isolated exposures. They may be products of debris flows and, if so, attest to the former presence of syn-eruptive standing water at times. Finally, beds of pumice-ash breccia up to 5 m thick are also present. They have been interpreted as products of rare magmatic eruptions, which would imply that dry conditions also sometimes prevailed in the vents. Two possible vents consisting of steep-dipping flow-banded rhyolite enclosing a core of microvesicular rhyolite have been recognised in different locations, suggesting that activity at the Prestahnúkur edifice might have been polygenetic. However, the entire edifice appears to be similar in composition, with only minor variability, suggesting construction from a single batch or related batches of high-silica (76–78% SiO_2) rhyolite probably over a relatively short period, estimated at c. 2–19 years assuming average mass discharge rates of c. 1–10 $m^3 s^{-1}$. It is the most silica-rich rhyolite in Iceland, which may make the magma particularly viscous (cf. Shaw, 1972) and might help to explain the relatively high aspect ratio of the edifice, which is more akin to felsic tuyas than to subglacial domes (Smellie, 2013).

McGarvie et al. (2007) envisaged a three-stage model for construction of the Prestahnúkur edifice involving (1) subglacial effusion with extensive meltwater interaction causing quenching and fragmentation and generation of a large volume of felsic hyaloclastite; a combination of an essentially flat bedrock surface and inefficient meltwater drainage were cited as likely reasons why meltwater was present during the early stages of the eruption. This contrasts with eruption of intermediate-composition subglacial domes in British Columbia emplaced on a steep bedrock topography, which lack such evidence (Kelman et al., 2002; Kelman, 2005). Effusion of volatile-poor magma at Prestahnúkur was also suggested (and supported by FTIR analyses of dissolved H_2O) as a reason for the

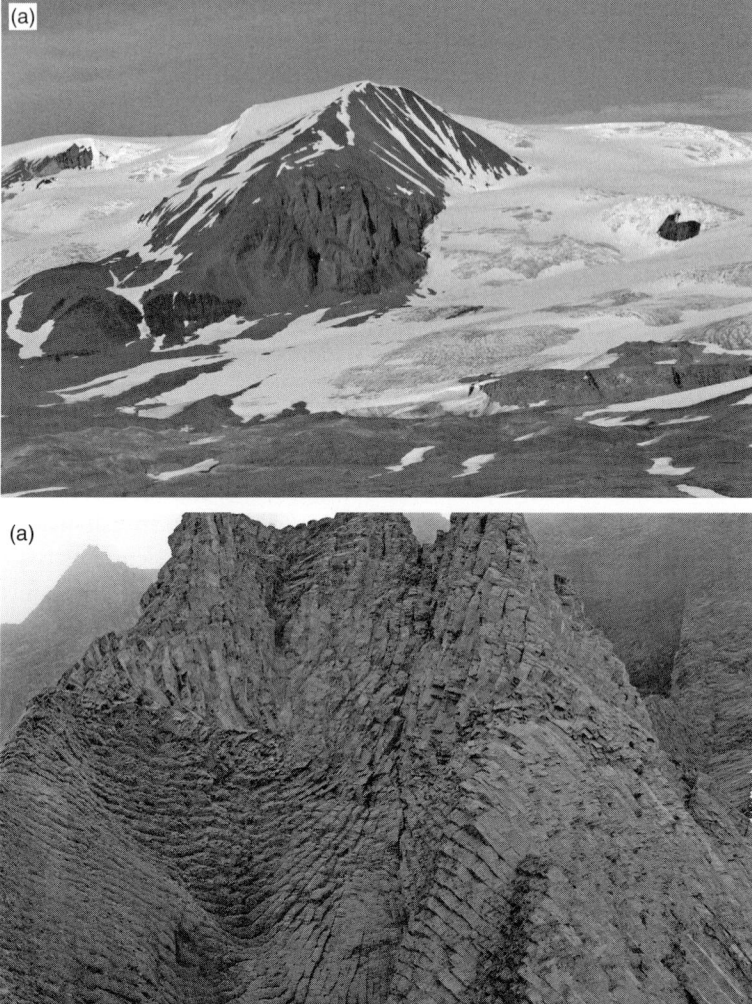

Fig. 12.4   Views of Triangle Dome on the western flank of Mt Edziza, northwestern British Columbia, Canada. (a) View of the southwestern flank of the dome; the dome is 250 m high. (b) Close-up view of jointing at the base of the dome. Joints are c. 1–2 m apart.

scarcity of evidence for explosive hydromagmatic activity, compared with the abundant evidence for explosivity in the tephra-rich pedestals of felsic tuyas in Iceland (see Section 12.4, below); (2) effusion of the thick basal felsic lava tier, accompanied by limited basal interaction with a wet substrate and surface water flooding to yield the generation of hyaloclastite along surfaces; and (3) effusion of the middle and upper lava tiers with generation of associated hyaloclastite and steeply ramped lava termini due to ice ponding, together with local explosive activity, edifice collapses and meltwater reworking to generate associated breccias.

With an age of 84 ± 24 ka (by ^{40}Ar/^{39}Ar), emplacement of the Prestahnúkur edifice coincided in timing with the transition between the Eemian interglacial and Weichselian glacial, corresponding mainly to the lower Weichselian (i.e. MIS 4–5d; probably 5a-5d). At that time, ice thicknesses over Prestahnúkur were estimated to be at least 700 m (assuming ice sag of c. 150 m over the vent). McGarvie et al. (2007) argued that the evidence for substantial ice thicknesses at Prestahnúkur broadly supports published models suggesting the rapid build-up of terrestrial ice even during short-lived cold periods, such as the series of rapid cold–warm fluctuations that characterised the Weichselian period (Raymo et al., 1998), although the large 1-sigma error on the Ar age precludes a more definitive association.

The Mt Edziza volcanic complex (Souther, 1992; Fig. 11.1) has at least one lava dome that has been interpreted as glaciovolcanic (Souther, 1992; LaMoreaux, 2008). Triangle Dome, which is exposed on the northwestern flank of the complex, is a pervasively jointed trachyte dome that is elongated in an east–west direction (Fig. 12.4). Most of the outer carapace has been removed by subsequent glaciation, exposing a cross section through the dome's interior. The three-dimensional orientation of the cooling joints is consistent with emplacement in an ice-confined cavity, and the pervasive nature of the jointing is consistent with cooling enhanced by the presence of meltwater.

12.4 Tuyas

All evolved (i.e. felsic and intermediate-composition) tuyas are divided into two distinctive types: lava-dominated (cf. flow-dominated: Kelman, 2005) and tephra-dominated (Smellie, 2007, 2009, 2013). The only published examples of lava-dominated tuyas are for intermediate-composition magmas (see Chapter 11). No felsic equivalents are presently known, possibly apart from the unusual effusive sequence at Prestahnúkur (McGarvie et al., 2007), which may be a subglacial dome because it lacks evidence for subaerial capping lavas (Section 12.3). Thus, all the published descriptions of felsic tuyas are for tephra-dominated examples.

Tephra-dominated felsic tuyas are prominent and common in Iceland, which is the only region in which they have been described (Tuffen et al., 2002a; McGarvie, 2009; Stevenson et al., 2011). They each follow a similar twofold lithosome architecture comprising a core of ash-rich tephra overlain by a much thinner caprock of flat-lying lavas. There is no evidence that any felsic tuya ever gave rise to lava-fed deltas, and that is a major difference from tuyas of mafic composition (see Chapter 10). The difference is principally thought to be due to the significantly lower thermal energy available for melting during felsic subglacial eruptions, which is capable of melting much less space than during mafic eruptions, thus permitting the incoming magma and any meltwater to lift the overlying ice and allowing the meltwater to escape (see Chapter 6). The published examples of tephra-dominated tuyas crop out in the Torfajökull and Kerlingarfjöll central volcanoes (Tuffen et al., 2002a; McGarvie, 2009; Stevenson et al., 2011). All are rhyolitic in composition.

12.4.1 Torfajökull

The Torfajökull volcano is the largest felsic complex in Iceland, measuring c. 18 km by 12 km across. It has erupted over 250 km^3 of peralkaline rhyolite over a period of c. 1 m.y. (McGarvie et al., 1990) and included subglacial eruptions on several occasions. The subglacially erupted centres mostly crop out as a prominent ring of flat-topped volcanoes surrounding the hydrothermally altered interior. They rise to 370–550 m above the surrounding terrain with summits at c. 924–1235 m a.s.l., and have similar compositions, suggesting that they might have been emplaced during a single eruptive period, probably corresponding to the last glacial (McGarvie, 1984). Of these, the group of four separate flat-topped subglacial centres at Raudufossafjöll, situated at the western margin of the complex, is the most voluminous (c. 6 km^3). The southeastern centre is the best exposed in the group, with good sections over 100 m high revealed by multiple sector collapse scars, and it has been intensively examined by Tuffen et al. (2002a). The capping lavas are best exposed, with much of the underlying tephra pile extensively obscured by modern scree deposits. This is typical of tephra-dominated felsic tuyas in Iceland and it has significantly limited our knowledge of the early tephra-forming history of these edifices (but see Kerlingarfjöll, Section 12.4.2).

The southeast centre at Raudufossafjöll is a flat-topped ridge whose summit platform (c. 1200 m a.s.l.) is 350–450 m above the surrounding terrain (Fig. 12.5). It extends 1.5 km and is 35–250 m wide. It has a prominent gently inclined platform of associated volcanic rocks on the southeastern flank measuring up to 500 m wide and at an elevation of c. 900 m a.s.l. (Fig. 12.6). The core lithosome is formed dominantly of a very poorly exposed rhyolitic ash lithofacies together with volumetrically minor volcaniclastic sediments. The ash is essentially unconsolidated and varies from grey ash in basal sections containing poorly vesicular (<20% vesicles) blocky shards and c. 5% angular fragments of obsidian, to similar ash in higher locations containing c. 40% angular pumice up to 10 cm in diameter and highly sheared ribbons of obsidian up to 1 m long with pumiceous margins. The core ash deposit is envisaged as a product of explosive hydromagmatic eruptions, with glacial meltwater the likeliest source of the water involved. The high proportion of fine ash present may indicate that it was largely unable to escape the vent area because of confinement within a large glacial vault or cavity. The tephra deposits, poorly sorted overall, may be products of low-temperature and relatively low-concentration pyroclastic density currents ('surges'), with the subordinate proportion of obsidian clasts reflecting a minor component entrained in the surges derived from localised spatter-forming eruptive phases. The volcaniclastic sediments are exposed in a single 15 m-thick section in the Blautakvísl stream gully draping the northeast flank of the edifice. They consist of a complex succession of discontinuous beds of monomict and polymict sandstone and breccia dominated by clasts of rhyolite and obsidian and variably matrix- and clast-supported. The characteristics of the volcaniclastic lithofacies suggest that they were deposited from gravitational collapses of coeval lava domes and as syn-eruptive volcanic detritus reworked by flowing (melt)water and as debris flows.

Fig. 12.5 Felsic tuya, Raudufossafjöll, Iceland. The tuya is tephra-dominated and rises about 400 m above the surrounding plain. The flat top is formed of two flat-lying subaerial rhyolite lavas that overlie a tephra core (now extensively covered by scree). Cf. Fig. 12.6.

Several lavas have been distinguished in the Raudufossafjöll tuya and are associated with both the capping lava lithosome and the low platform on the southeastern flank. The summit platform consists of two laterally extensive horizontal microcrystalline rhyolite lavas, each extending c. 1.5 km and up to 100 m thick. A large proportion of the lavas has been removed by post-glacial sector collapses and is now preserved as lobate debris-avalanche deposits on the western side of the edifice (Fig. 12.6). The capping lavas are non-vesicular and show prominent flow banding, bands of pale pumiceous obsidian and local poorly developed columnar joints. The upper surfaces of both lavas are glassy and vesicular. They appear to have been subaerially emplaced (i.e. there is no hyaloclastite, perlitisation or distinctive fractures caused by rapid cooling), with minor evidence for ice impoundment (i.e. upturned marginal flow banding and high aspect ratios of the lavas). At one locality, a steep downslope-dipping columnar lava mass attached to the margin of the lower lava (lava A) was interpreted as a possible lava dribble into a coeval bergschrund (cf. interpretation of similar features in the intermediate-composition lava-dominated tuya at The Table; Mathews, 1951; see Section 11.4; also Tuffen and Castro, 2009).

Lavas are also present at lower elevations around the periphery of the main tuya edifice and illustrate additional magmatic events associated with tuya construction. They include (1) a group of ten microcrystalline lava masses (lava C; Fig. 12.6) with pervasive columnar joints with horizontal orientations indicating cooling against a near-vertical ice surface; (2) several lava outcrops on the eastern platform (lava D) and on the southern flank (lava E), the largest more than 1 km long, 250 m wide and 80 m thick, composed of rhyolite and mainly columnar jointed obsidian capped by a pumiceous and obsidian orthobreccia; and (3) a final lava (lava F), which consists of a dome-like obsidian mass surrounded by

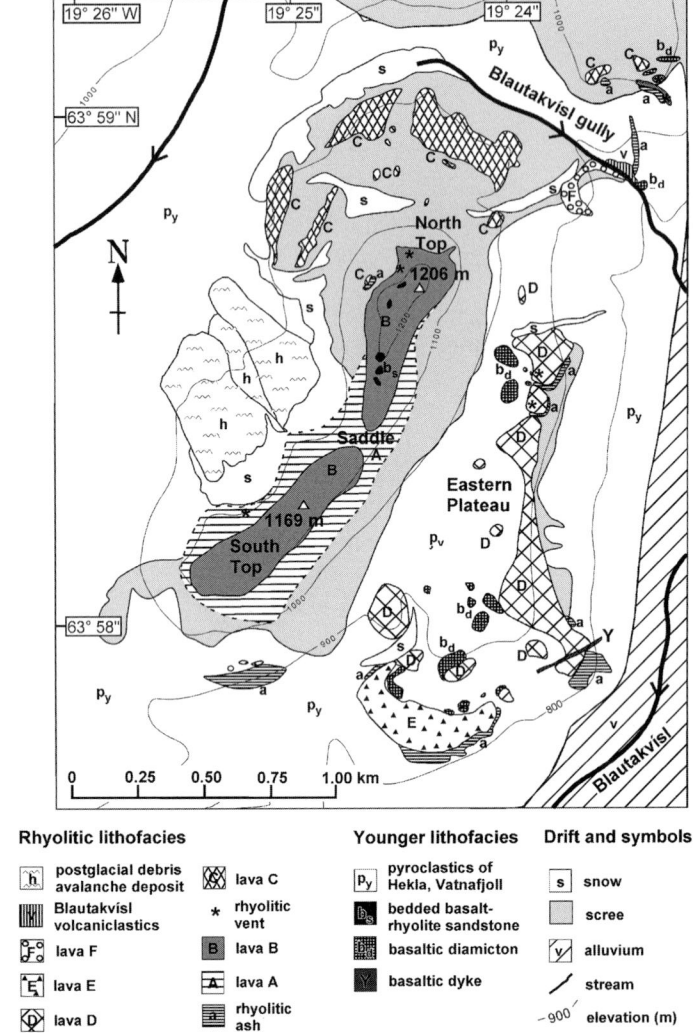

Fig. 12.6 Outcrop map of a tephra-dominated felsic tuya Raudufossafjöll, Iceland (from Tuffen et al., 2002a).

obsidian breccia that extends downslope into a crudely bedded sheet. Lavas D and E lavas rest on ashy matrix-supported obsidian and pumice breccia containing 'rafts' of bedded ash up to 2 m across. The contact between massive lava and the basal breccia is marked by a zone of peperite that grades up into jigsaw breccia. Lava E is also associated with minor cross-laminated rhyolitic sandstone. The lithofacies and relationships for lavas D and E were interpreted to be a consequence of subaerial rhyolite lavas that flowed over a wet substrate of poorly consolidated breccia on the low eastern flank of the tuya and were impounded by ice surrounding the tuya; they were also associated with minor

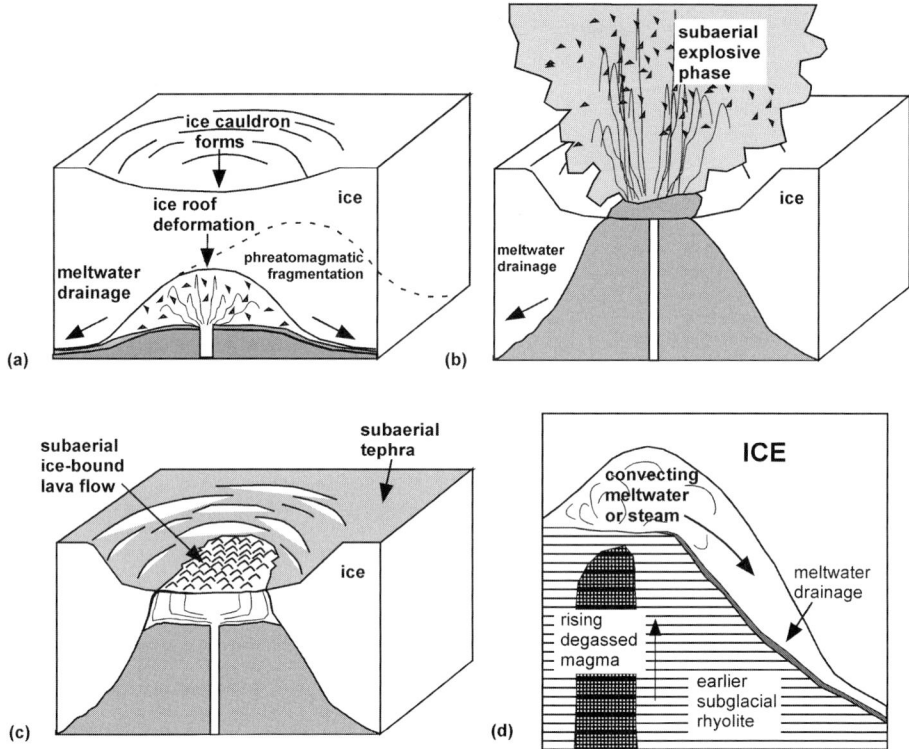

Fig. 12.7 Cartoons illustrating the possible evolution of the Raudufossafjöll felsic tuya (from Tuffen et al., 2002a).

sedimentation by flowing meltwater. Lava F is interpreted as a small subglacially emplaced dome and associated carapace breccia; the latter underwent gravitational collapse to form the small downslope-directed sheet.

Construction of the Raudufossafjöll tuya (Fig. 12.7) was envisaged commencing as hydromagmatic explosivity in a subglacial cavity, but the apparent absence of bedding and associated subaqueous gravity-flow deposits in the resulting ash pile suggests that any meltwater present was not pooled. Meltwater was envisaged draining continuously, as in the 'leaky vault' scenario of Smellie (2000; see also Smellie, 2006), and near-atmospheric pressure may have been maintained at the vent if it was connected hydraulically with the glacier margins. If the cavity became completely filled by rhyolite products, cavity pressures may have increased to near-hydrostatic (i.e. glaciostatic) and the eruption could have transformed to an intrusive phase, a scenario further explored by Tuffen et al. (2002b). It is unclear if the overlying ice was penetrated during the explosive phase, as speculatively depicted in Fig. 12.7. However, the presence of subaerial capping lavas indicates that the eruption must have ultimately pierced the overlying ice and became subaerial. The lavas were emplaced within substantial ice cauldrons. Later activity was by effusion of

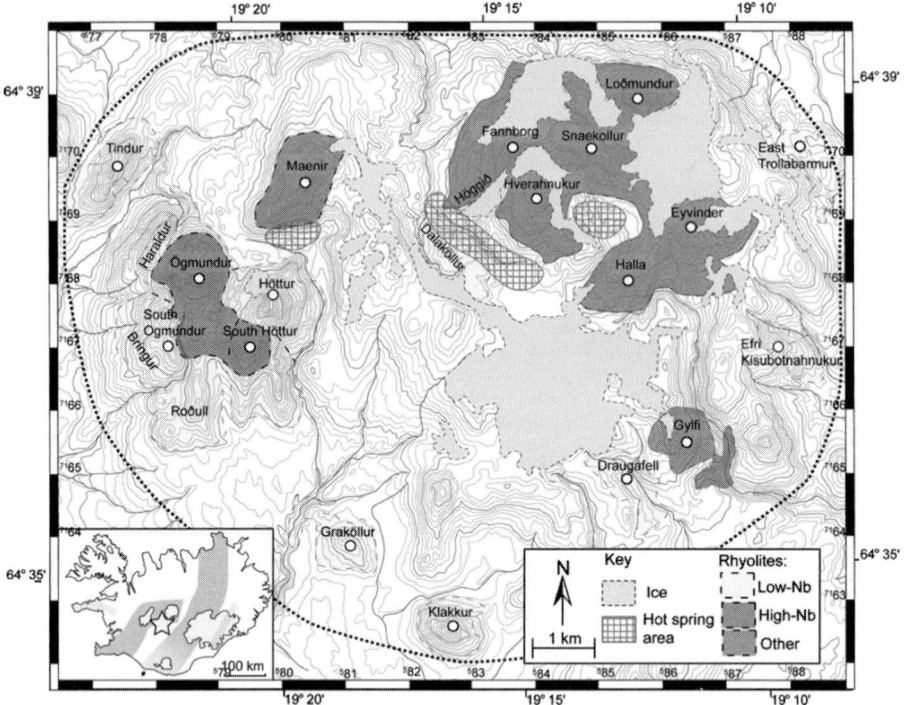

Fig. 12.8 Geological map showing the distribution of rhyolite volcanic centres in the Kerlingarfjöll volcanic complex, central Iceland (from Flude et al., 2010). The pecked line represents the approximate limit of the volcanic complex. (A black and white version of this figure will appear in some formats. For the colour version, please refer to the plate section.)

essentially degassed lavas from a series of peripheral vents situated low on the flanks of the tuya (lavas C–F) where they interacted with surrounding ice and probably melted local cavities in that ice.

12.4.2 Kerlingarfjöll

The Kerlingarfjöll volcano is described in Section 11.5. Like Torfajökull, it contains multiple subglacially erupted volcanic centres of rhyolite composition and has the second-largest volume of rhyolite in Iceland (Fig. 12.8). Three centres (Ögmundur, Höttur and a smaller un-named centre informally named South Ögmundur) were studied because their tephra cores are better exposed than in other felsic tuyas in Iceland (Stevenson et al., 2011). The three edifices rise 300 m above the local topography to elevations of c. 1200–1300 m a.s.l. They vary in age between 345 and 125 ka (Flude et al., 2010). The study by Stevenson et al. (2011) included the first quantified grain size and vesicularity determinations for a felsic tuya. Ice thicknesses coeval with eruptions are unknown but were at least 300 m.

The tephra core lithofacies are dominated by varieties of tuff and lapilli tuff, variably massive and stratified and ash-rich or ash-poor, together with bedded polymict tuffaceous breccia. Because of their unconsolidated condition, the pyroclastic lithofacies probably should be referred to as ash and lapilli ash. Despite the greater exposure of the fragmental deposits compared with other tephra-dominated felsic tuyas, it is seldom possible to observe relationships between the major lithofacies, although from the elevations of exposures it is possible to infer their relative positions within an upward evolving sequence. The earliest-formed deposits (lithofacies A, lithic-rich massive tuff of Stevenson et al., 2011) are massive and fine ash rich, comprising abundant blocky poorly to non-vesicular ash shards and minor pumice together with accessory obsidian fragments, matrix vesicles and ash aggregates. They were probably deposited relatively rapidly from wet concentrated pyroclastic density currents (*sensu* Branney and Kokelaar, 2002) during explosive hydromagmatic eruptions. The succeeding lithofacies, of which four were described (B–E), are broadly similar but are characterised by spinose pumiceous lapilli and cuspate bubble-wall shards, sometimes accompanied by bombs and subordinate obsidian. However, the pumice vesicularities are variable. Whereas some pumices are highly vesicular consistent with magmatic fragmentation, most samples have much lower mean vesicularities. Fragmentation was thus not purely magmatic but must have involved some interaction with meltwater, although the magma was already vesicular when it encountered that water. Deposition was mainly from high-concentration pyroclastic density currents. The presence of fine ash and matrix vesicles is also consistent with a minor influence of water in the eruptions. Lithofacies D, which crops out in the upper reaches of the edifices and was probably the final pyroclastic lithofacies to be erupted, is distinguished by steep bedding dips and a wide variety of clast types (including coarse bombs) and sizes. It may have formed in a vent proximal position in which the different clasts formed simultaneously but during different explosive events and were subsequently juxtaposed. Lithofacies E, described as coarse obsidian-rich massive lapilli tuff, is distinctive. It has a restricted distribution and comprises coarse tube pumice, up to 30% obsidian and little fine ash. It occurs as domains within lithofacies D, especially close to intrusive lava lobes. The obsidians are unusual. They are up to 3 m long and have vesiculated margins and non-vesicular interiors. Lithofacies E is thought to have been formed under unusual eruptive conditions close to the explosive–effusive transition. The preferred hypothesis presented by Stevenson et al. (2011) is that lithofacies E deposits were emplaced within lithofacies D as 'pyroclastic intrusions', with the combined deposits fluidised by upward-streaming gases such as occurs in vents. Emplacement of the lithofacies was envisaged under essentially non-explosive conditions.

Lithofacies F, bedded polymict tuffaceous breccia, is a resedimented deposit less than 50 m thick. It has a known distribution restricted to two gully exposures between Haraldur and Ögmundur and between South Ögmundur and Ögmundur, and comprises alternating obsidian-rich (with yellow ash matrix) and obsidian-poor (with grey ash matrix) beds of grey and white pumice, some hydrothermally altered, and lesser flow-banded rhyolite up to

50 cm in diameter. Beds are up to 50 cm thick and show normal and less common reverse grading. The lithofacies appears to have been derived by erosion of the associated unconsolidated pyroclastic deposits with the sediment transported as concentrated mass flows (i.e. sedimentary density currents) similar to the tephras but epiclastic. Deposition was therefore subaqueous and was either broadly contemporaneous with, or occurred very soon after, eruptions ceased.

An idealised lithofacies progression was suggested for each tuya, comprising initiation as violently explosive hydromagmatic eruptions in the presence of abundant water (lithofacies A). The water was probably mainly meltwater but the association with abundant bedrock-derived obsidians suggests the additional involvement of shallow groundwater. The resulting fine-grained tephra were then deposited from moist (<100 °C) granular pyroclastic density currents. The deposition of abundant fine ash was facilitated by moisture-bound ash aggregates and those, together with the presence of matrix vesicles, signify that ponded water was probably not present even at this early stage. Subsequent tephra-forming eruptions involved a greater role for magmatic volatiles in driving fragmentation, and may have been particularly important for lithofacies C with its lesser proportion of fine ash, but access of water at the vent was still significant in order to generate the fine ash present and the bimodal grain size distributions. The importance of water declined up through the sequences at each centre, but deposition of all the tephra lithofacies continued to be from moist pyroclastic density currents. Lithofacies D, the coarsest lithofacies, was probably the final tephra unit to be erupted at each edifice. Its likely vent-proximal deposition probably caused the poor sorting and lesser clast breakage (i.e. preservation of many large clasts, as coarse pumiceous bombs and obsidians), suggesting that water played a more minor role in fragmentation. This is consistent with a relatively dry vent and decline in explosivity. The steep inward-dipping bedding in lithofacies D suggests the presence of an oversteepened tephra cone prone to inward collapse and that the tephra was unable to build outward significantly owing to confinement by the surrounding ice. Finally, lithofacies E was interpreted to be as a result of intrusion of pyroclastic tephra during fluidisation caused by streaming gases during a declining rate of mass discharge, which also elutriated much of any fine ash present. This final explosive phase probably corresponded to an eruptive mode transitional to the effusive phase that took over to form the overlying lava capping units.

The tephra piles also contain numerous irregular intrusions 1–100 m in diameter. Many are unusual in that they have fragmented margins gradational with the host tephras. The margins are peperitic and form zones up to 10 m wide. They are particularly common in lithofacies A. Other intrusions are dominant in the overlying lithofacies and have either compact obsidian margins and cores of poorly vesicular rhyolite or are composed of obsidian with dense cores and outwardly vesiculating margins; columnar joints perpendicular to intrusion margins are ubiquitous. Interaction between these higher intrusions and the host tephras is much less extensive. From their irregular shapes, the host tephras were probably unconsolidated at the time of emplacement of all three intrusion types, and those with peperitic margins are consistent with the host tephras being wet. Conversely, based on a lack of evidence for interaction, the compact-margined intrusions were presumably emplaced in much drier tephras. Those with outwardly increasing vesiculation, the textural opposite of bombs which are more vesicular in their cores, are found within the last-erupted, driest lithofacies D. Texturally, they resemble the obsidian bodies found within lithofacies E but, as intrusions, they are presumed to be connected at depth to feeder dykes. Similar characteristics, but on a much larger scale, have been described in submarine pumiceous peperites by Hunns and McPhie (1999) and Gifkins et al. (2002) and obsidian sheets in a subglacial outcrop elsewhere at Dalakvísl, part of the Torfajökull volcano, by Tuffen et al. (2008). The latter interpreted the textures in the obsidian sheets as collapsed foam domains in small vesicular intrusions compressed and welded by the weight of the overlying tephra pile. Alternatively, because the Kerlingarfjöll occurrences are not flattened like those at Dalakvísl, Stevenson et al. (2011) speculated that they may be intrusions whose obsidian margins vesiculated by shearing during emplacement. Whatever the explanation, it is clear that similar features characterise many glaciovolcanic rhyolite edifices.

Finally, each of the Kerlingarfjöll felsic tuyas examined by Stevenson et al. (2011) is capped by flat-lying rhyolite lavas individually up to 75 m thick, with columnar joints and flow banding. The caprock at Ögmundur is c. 200 m thick but it is unclear if it is composed of multiple lavas. That on Höttur has an aspect ratio of 0.25 (75 m thick; 200–300 m wide). Basal contacts between lava and tephra vary from compact, as described for the intrusions, to jigsaw-fractured obsidian breccia with an ash matrix up to 10 m thick. Upper surfaces are poorly seen below a blockfield of rhyolite fragments but are locally vesicular and almost pumiceous. Lava caps at Tindur and Haraldur are less well developed. All the capping lavas are presumed to have been emplaced subaerially on the basis of the presence of broadly pumiceous rhyolite flow tops and an absence of abundant obsidian. The compact basal contact zones lack evidence for water interaction, and it was suggested that the tephra piles were essentially dry during lava effusion, whereas the basal obsidian breccia seen on Ögmundur is an indication of emplacement on wet tephra. The high aspect ratios suggest ice impoundment within a chimney melted through the surrounding ice.

In their summary model for felsic tephra-dominated tuya eruptions (Fig. 12.9), Stevenson et al. (2011) noted major differences from mafic tuyas, including the absence of felsic lava-fed deltas at Kerlingarfjöll and that the transition from explosive to effusive eruption is not due to the vent drying out but is probably controlled by the degassing of the felsic magma, comparable in subaerial felsic eruptions to the transition to dome effusion (Jaupart, 2000). Thus, in felsic subglacial eruptions, the role of meltwater is comparatively minor (it probably escapes continuously, without accumulating) and, instead of lava-fed deltas forming in a meltwater lake, the felsic magma is emplaced through moist unconsolidated tephra, mainly forming intrusions but also creating a subaerial lava caprock. The predominant role of the surrounding ice is of confinement, which promotes vertical aggradation over lateral progradation and ensuring high aspect ratios in the edifices (also Smellie, 2007, 2009, 2013).

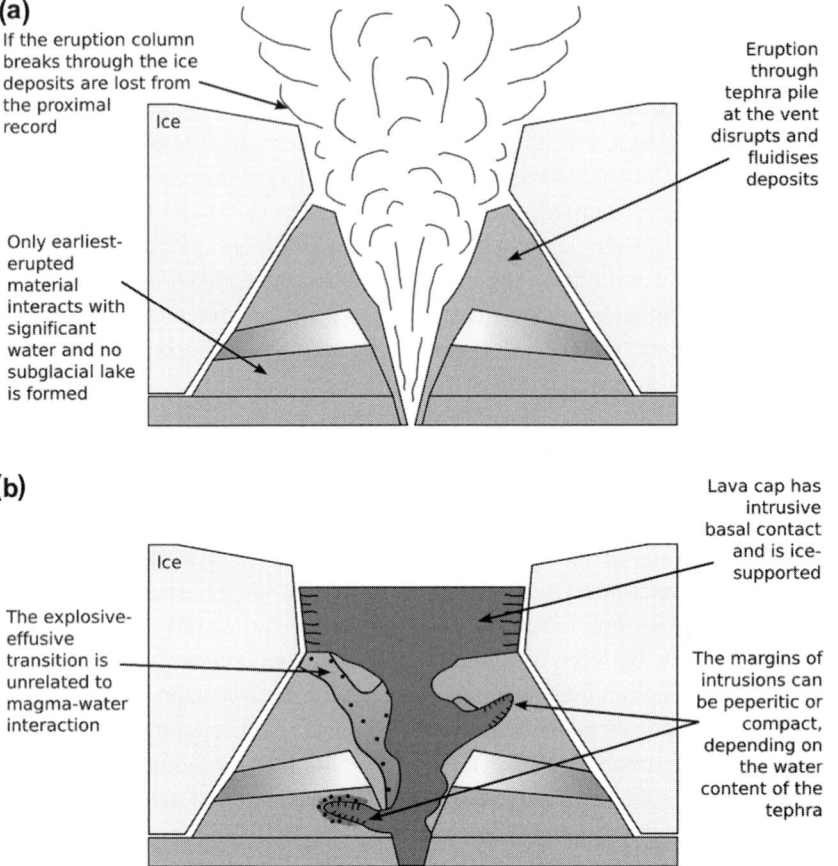

Fig. 12.9 Schematic diagram illustrating the construction of a tephra-dominated felsic tuya at Kerlingarfjöll, central Iceland (from Stevenson et al., 2011). (a) Early explosive eruptive phase generating the tephra core of the tuya; (b) later effusive eruptive phase, which formed the subaerial lava cap. Melt-back of the enclosing ice walls generated collapses of the oversteepened flanks.

12.5 Tindars

Felsic tindars are present in Iceland but are not currently known anywhere else. For example, they occur as a series of probably fissure-erupted centres at Kerlingarfjöll. However, because they lack a protective cap of resistant lavas and are formed mainly of weakly consolidated or unconsolidated tephra, they are extensively covered by thick scree and good exposures are rare. There are no published descriptions of felsic tindars. The tephra piles appear to contain numerous high-level intrusions but little is known about felsic tindars, their internal structure and construction. They are presumably formed of volcaniclastic piles and coeval intrusions similar to the cores of tephra-dominated felsic tuyas (Section 12.4). These features strongly resemble the assemblage of lithofacies present

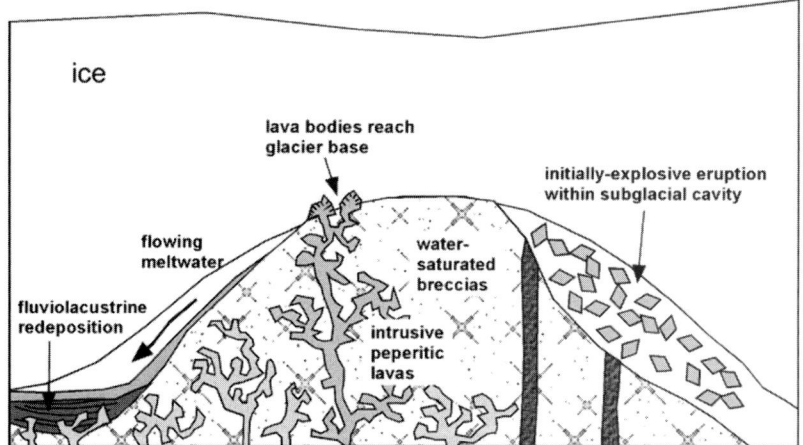

Fig. 12.10 Diagram showing the formation and eruptive behaviour of Dalakvisl outcrop, central Iceland (from Tuffen et al., 2008). The massive breccia pile (see Fig. 12.11) is assumed to be water saturated. It is intruded by anastomosing lava masses forming peperite, some of which reach the glacier base and become columnar jointed on cooling in contact with ice. The cavity above the breccias may be filled by air or steam and any meltwater flowed down the breccia pile flanks to form lacustrine and fluvial deposits, whilst at other times explosive activity may occur.

in a small glaciovolcanic outcrop at Dalakvísl, Iceland, and it is suggested that it may be the eroded roots of a small-volume felsic tindar (see also description of an outcrop at Bláhnúkur; Section 12.7).

The Dalakvísl outcrop is also situated in the Torfajökull volcanic complex, a few kilometres east of the Raudufossafjöll felsic tuyas (Section 12.4.1). Although compositionally similar to the Raudufossafjöll tuyas and therefore probably related petrogenetically and in time, it is morphologically very different, lacking a flat top and having a much smaller volume (<0.2 km³; Tuffen et al., 2008). Four fragmental and three lava lithofacies have been described.

The three lava lithofacies comprise sheets and dome-like masses of microcrystalline rhyolite with perlitised columnar-jointed obsidian margins and localised domains of perlitic obsidian breccia (their lithofacies *pl* and *cj*), and irregular vesicular peperitic obsidian lava lacking columnar joints (lithofacies *iv*). Lithofacies *pl* and *cj* are thought to represent lava bodies that intruded into and across unconsolidated wet breccias (lithofacies *mb* and *os*; below), accompanied by autobrecciation and gravitational instability, thus contributing to the fragmental pile. They may have locally chilled against ice in small subglacial cavities similar to lava lobes exposed at Bláhnúkur (cf. Tuffen et al., 2001; Section 12.7). By contrast, lithofacies *cj* lavas were simply intruded into waterlogged ash and breccia (lithofacies *cba* and *wba*, below) and they did not reach the glacier base (Fig. 12.10).

Two of the fragmental lithofacies are breccias, described as massive obsidian and pumice breccia (designated *mb*) and massive breccia with obsidian sheets and vesicular bombs (lithofacies *os*). Together, they dominate the Dalakvísl outcrop, with the *mb* lithofacies

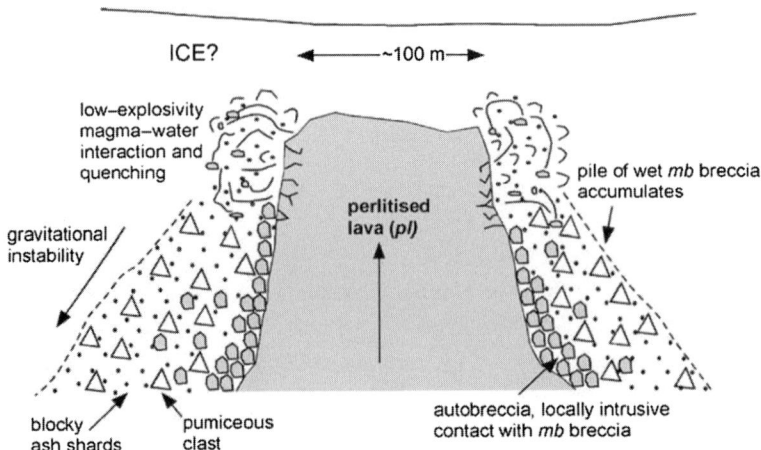

Fig. 12.11 Model showing the formation of the massive breccia lithofacies at Dalakvisl, Iceland (from Tuffen et al., 2008). The rising felsic magma generates a pile of wet breccia by low to moderately explosive magma–water interaction. Further intrusive phases subsequently rise through the breccia and form autobreccia at their margins.

most widespread and the *os* lithofacies confined to a prominent central ridge, described as esker-like in reference to its slightly sinuous form and subglacial setting. The two lithofacies are similarly composed of breccias with ash matrices and perlitised obsidian clasts. The ash is blocky and has low vesicularity in *mb* but is significantly more vesicular in *os*. Lithofacies *os* also contains highly vesicular bombs and obsidian sheets distributed along the ridge axis and down its flanks. The *mb* breccia is envisaged as a composite deposit formed from the products of low to moderate hydromagmatic explosivity (ash) and hyaloclastite (breccia), the latter created when rising magma (lithofacies *pl*) intruded wet fragmental deposits (initially ash (tephra), later tephra and hyaloclastite), chilling and forming marginal autobreccia that mingled with the enclosing deposit, thus adding to it with every fresh injection (Fig. 12.11). Lithofacies *os* probably formed during explosive hydromagmatic eruptions in a vent-proximal location, associated with collapsed-foam domains (obsidian sheets). Explosivity may have diminished with time and the eruption became dominated by hyaloclastite rather than ash. It was speculated that the original subglacial cavity might have become filled, causing the eruption to become more intrusive (Tuffen et al., 2007; Fig. 12.10). The two additional ash-rich lithofacies (crudely bedded ash (*cba*) and well-bedded ash and massive breccia (*wba*)) have a much more restricted distribution. They are dominated by vesicular and blocky, low-vesicularity shards that in lithofacies *cba* are well sorted. Those in lithofacies wba are interlaminated coarse and fine ash with interbeds of matrix-supported breccia. The ash lithofacies were probably fed by debris created during gravitational collapses of the flanks of the growing edifice. They were deposited from subaqueous density flows and traction currents in a low-energy lacustrine (lithofacies *cba*) and higher-energy fluvial and lacustrine (*wba*) settings. The local accumulation of meltwater implied by this interpretation is unusual for felsic subglacial eruptions but is known to occur in some glaciovolcanic eruptions of intermediate-composition magmas (Stevenson et al., 2009) and is characteristic of mafic glaciovolcanic eruptions, when they may also be accompanied by lacustrine sediments (Bennett et al., 2006, 2009). At Dalakvísl, inefficient escape of meltwater was probably promoted by the pre-eruption topography (a stream valley), with eruption-generated detritus blocking the drainage pathways. The inferred palaeo-ice thickness is more than 150 m based on the thickness and relief of lithofacies *mb* but may have been much greater (c. 300–750 m) based on dissolved water contents.

A conceptual model for the formation of the Dalakvísl outcrop was presented by Tuffen et al. (2008) involving eruption of rhyolite magma into an ice cavity. The eruption was initially explosive then, as the cavity became filled and the vent blocked, the magma was forced to invade the poorly consolidated tephra, causing arrested fragmentation and a transition to intrusive emplacement. However, the sequence of events described above is what would be inferred on a-priori grounds for the early stages of eruption of felsic tindars, based on interpretations of the tephra cores of felsic tuyas (cf. Tuffen et al., 2002a; Stevenson et al., 2011; Section 12.4). It is therefore suggested that Dalakvísl might represent the early-formed roots of a felsic tindar extensively exhumed by glacial erosion. In this view, the different lithofacies can be characterised as follows: (1) vent region, a short fissure (300 m long) that gave rise to explosively erupted tephra (lithofacies *os*); (2) over-lying tephra together with endogenic breccias formed *in situ* by multiple intrusions, with the erupting magma increasingly degassed with time (lithofacies *mb*); (3) intrusions (lithofacies *pl, cj* and *iv*); and (4) flanking sediments related to subglacial meltwater drainage (lithofacies *cba* and *wba*). The outcrop characteristics are consistent with eruption of a small-volume felsic magma that degassed and rapidly exhausted itself. If the estimated palaeo-ice thickness of c. 700 m is correct, it is unlikely that the edifice ever vented to the atmosphere and it was probably entirely confined to a subglacial cavity (Tuffen et al., 2008).

12.6 Sheet-like sequences

Felsic sheet-like sequences have been reported from only three localities: (1) Hrafntinnuh-ryggur, Iceland (Tuffen and Castro, 2009); (2) Mandible Cirque, Antarctica (Smellie et al., 2011a); and (3) Kerlingarfjöll, Iceland (Stevenson et al., 2011).

12.6.1 Hrafntinnuhryggur

Tuffen and Castro (2009) described the products of a fissure eruption that resulted in the emplacement of a 2.5 km-long rhyolite dyke at Hrafntinnuhryggur, northern Iceland, during the last glacial period. The dyke was emplaced within the caldera of the Krafla central volcano and the eruption created a ridge that today rises at least 80 m above the surrounding land. Two main phases of rhyolitic volcanism occurred at Krafla between 100 ka and 24 ka, each producing c. 1 km^3 of magma and forming a rhyolitic dome, 2.5 km^3 of welded airfall tuff and three subglacial rhyolite ridges composed of lava lobes and hyaloclastite. A third smaller phase created the Hrafntinnuhryggur outcrop. The Hrafntinnuhryggur magma intruded the easternmost of three parallelc fissures that follow a regional tectonic trend within the caldera.

The Hrafntinnuhryggur outcrop is a ridge composed of up to 12 small-volume lava flow and dome-like bodies with columnar-jointed margins and volumetrically minor perlitic basal hyaloclastite and pumiceous upper surfaces. The high vesicularity of the carapace pumices contrasts with basal hyaloclastites, which are vesicle-free. The western ridge flank is extensively hydrothermally altered and, at the southern end, the feeder dyke is exposed, flaring upward to feed one of the overlying lava bodies. The individual lava masses are near-horizontal and sheet-like with dimensions of c. 20–50 m thick, 20–200 m wide and 70–750 m in length. They are characterised by steep flow banding parallel to the ridge. Four textural zones were identified in the effusive lavas (Fig. 12.12). They comprise (1) an outer zone 1–2 m thick of black obsidian, which is locally perlitised, extensively fractured (decimetre-spaced columnar and platy joints) and occurs only in a basal position in the outcrop, beneath textural zone 4, and is associated with hyaloclastite; (2) mm- to cm-scale flow-banded, multicoloured (blue, grey, green, brown, black) obsidian showing small-scale faults, forming a zone up to 1 m thick and with pervasive microscale textures indicative of ductile shearing, vesiculation, bubble collapse, fragmentation and annealing indicating deformation spanning the brittle–ductile transition and attributed to high strain rates associated with cracking of the outer rhyolite carapace and sudden vaporisation of external water; (3) a zone of lithophysae-rich multicoloured banded obsidian enveloped by zone 2, which is locally developed, varies in thickness and is partly vesiculated, with lithophysae up to 7 cm in diameter typically making up 5–10 vol % of the rock; and (4) an inner spherulitic, massive to banded, poorly vesicular obsidian zone that dominates the lava outcrops; the spherulites are up to 10 cm across and may form up to 100% of the rock. The range of textures were interpreted to be a consequence of variations in cooling rate controlled by proximity to the lava margin, with zone 1 fractured obsidian representing a quenched outer zone. The steep orientation of the flow banding is consistent with limited viscous gravitational spreading laterally of small lava flows and dome-like bodies that budded from the feeder dyke. Radiating and low-angle columnar jointing resembling entablature characterises the sides of the lavas at mid to high elevations on the ridge, consistent with quenching of the margins against confining ice. By comparison, joints in the feeder dyke are hackly. Lateral field relationships suggest that the lava did not quench basally away from the main ridge and the attitudes of associated polygonal columns suggest that the lava flowed down the dipping bedrock surface within a well-drained cavity ('bergschrund') that had opened up between the bedrock and overlying ice. Associated fragmental lithofacies are volumetrically minor and consist of massive pumiceous tuff over 5 m thick possibly formed during explosive magmatic fragmentation. Jigsaw-fit perlitised hyaloclastite breccia may be related to localised quenching of the lava flow margins.

Tuffen and Castro (2009) interpreted the Hrafntinnuhryggur outcrop as the products of a small-volume rhyolite fissure eruption beneath thin ice, i.e. a cover estimated as only

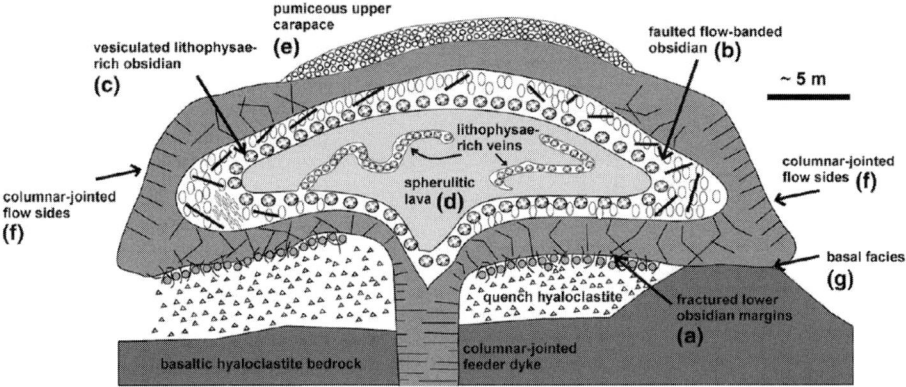

Fig. 12.12 Schematic cross section showing the main textural zones of felsic lava bodies at Hrafntinnuhryggur, northern Iceland (from Tuffen and Castro, 2009). The columnar-jointed outer zone has a pumiceous upper carapace and a fractured base resting on hyaloclastite. The lava interior is composed of vesiculated and spherulitic zones, which dominate the outcrop (not drawn to scale).

35–55 m thick and composed largely of firn. The setting thus corresponds broadly to that of a sheet-like sequence (cf. Smellie et al., 1993; Smellie and Skilling, 1994). The eruption breached the glacial cover and comprised essentially subaerial lava effusion with only minor explosive activity. The firn/ice cover was inferred to be highly permeable and the eruption site well drained. It probably would not have generated a significant jökulhlaup (probably <70 m^3 s^{-1}). Calculations suggested that the eruption lasted between 2 and 20 months. The Hrafntinnuhryggur outcrop was the first described example of a rhyolite eruption in thin ice. It is an example of the types of glaciovolcanic lithofacies that will be found in rhyolitic sheet-like sequences in vent-proximal locations. Other examples (below) include more far-travelled felsic lava and additional associated mostly epiclastic lithofacies.

12.6.2 Mandible Cirque

A spectacular succession at least 900 m thick comprising multiple rhyolite and trachyte domes or coulées (collectively called rhyolite below, for ease of description; Fig. 12.13) occurs at Mandible Cirque (Daniell Peninsula, Antarctica; Smellie et al., 2011a). Because of access difficulties, only the basal and uppermost few tens of metres of section have been examined and the lavas at higher elevations are not glaciovolcanic. The basal section is dominated by flow folded lavas individually c. 50–200 m in length and a few tens of metres thick encased in autobreccia. Those examined are separated by sharp erosional surfaces. The most prominent has a relief of over 200 m, is traceable laterally for a few kilometres and shows prominent 'U-shaped' profiles (Fig. 12.13). Crudely stratified polymict volcaniclastic diamictite 8 m thick overlies the erosion surface, followed by c. 8 m of well-stratified rhyolite lava-derived sandstone and granule conglomerate with

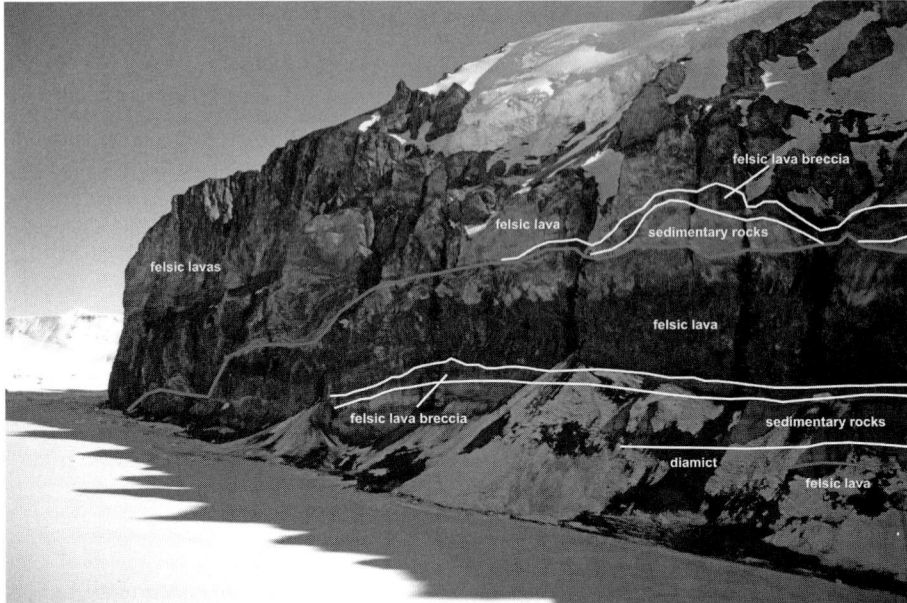

Fig. 12.13 Sequence of rhyolite lava domes at Mandible Cirque, Victoria Land, Antarctica. The basal lava domes were subglacially emplaced and are separated by sharp uneven surfaces (red lines), at least some of which are glacial, overlain by diamict (including tillite) and stratified fluvial sediments. The rock face is c. 400 m high. (A black and white version of this figure will appear in some formats. For the colour version, please refer to the plate section.)

compaction-flattened pumices. The stratification is on a centimetre scale and sandstone beds prograde locally over internal surfaces in the deposit. The sandstones and conglomerates are overlain by originally glassy lava autobreccia, now perlitised and with poikilitic textures caused by pervasive devitrification and hydration under warm temperatures in a wet environment. Stringer-like masses of the lava breccia have injected down into the sandstones and conglomerates and the overlying lava also contains tall flame-like masses of sandy-gravelly sediment up to 8 m high resembling spiracles that are surrounded by steep close-spaced platy joints in the enclosing lava that also mimic an internal flow foliation, often tightly folded. The breccia changes progressively upward, via jigsaw-fractured lava, into massive lava a few tens of metres thick that dominates each cogenetic eruptive unit. The lavas are aphanitic, becoming finely crystalline upward where they are topped by a typically brownish-coloured rhyolite carapace breccia. The smaller rhyolite lavas (<20 m thick) may be wholly encased in colourful massive glassy autobreccia composed of dark green and black obsidian set in a bright yellow-green clay-altered matrix.

Smellie et al. (2011a,b) interpreted the sequence of events involved in the formation of each eruptive sequence as follows (Fig. 12.14): (1) wet-based ice eroded the prominent unconformities and deposited the overlying polymict diamictites as redeposited

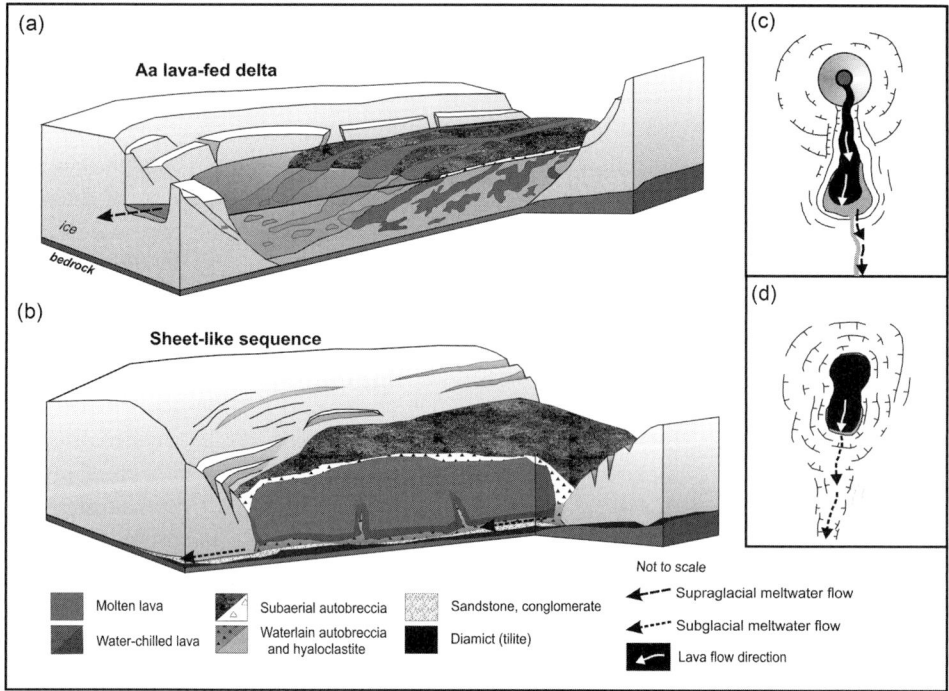

Fig. 12.14 Schematic perspective views depicting the major constituent lithofacies and inferred mode of emplacement of (a) 'a'ā lava-fed deltas and (b) felsic sheet-like sequences based on glaciovolcanic outcrops in northern Victoria Land, Antarctica (from Smellie et al., 2011b). (A black and white version of this figure will appear in some formats. For the colour version, please refer to the plate section.)

tills (flow tills); (2) meltwater generated by volcanic heat was flushed below the wet-based ice, probably as sheets and in tunnels eroded up into the ice, and transported and deposited mainly glassy tephra as fluvial sediments; (3) lava effusion resulted in the creation of basal hyaloclastite breccia by interaction with the underlying coeval water-saturated soft sediments; basal breccia was also injected downwards and sediment injected upwards due to trapped basal water flashing to steam (cf. similar relationships described in trachydacite lava by Stevenson et al., 2006, their unit E); and (4) the thicker lavas melted through the entire thickness of overlying ice, as shown by the upward increase in crystallinity and the presence of brownish subaerial carapace breccias; thinner lavas were emplaced wholly subglacially, probably in broad low tunnels, as demonstrated by the presence of hyaloclastite breccias on top of those lavas as well as below. There is no evidence for ponded water conditions in the sequences examined and the meltwater generated was able to escape subglacially. This succession of events was then repeated with each eruption after more wet-based ice re-formed on top of the new ground surface provided by the earlier felsic lava.

12.6.3 Tindur

Stevenson et al. (2011) described rhyolite lava and associated cogenetic breccia at Tindur, Kerlingarfjöll (Iceland), as a likely sheet-like sequence involving felsic lava. It is a lava lobe attached to rhyolite lava forming the ridge crest and consists of flow-banded rhyolite lava sheathed in gravel-grade monomict obsidian breccia. The lava lobe thickens and widens where it reaches flatter ground and it has small-diameter (10–15 cm wide) columnar joints perpendicular to its margins. The breccia is formed of pebble- and granule-size (<4 cm) clasts of poorly vesicular black obsidian and has a coarse sand-grade matrix. Because of the absence of fine ash and pumice, the breccia was interpreted as felsic hyaloclastite, formed by mechanical granulation and localised steam explosions as the lava flowed downslope from the ridge crest. It was probably confined within a subglacial tunnel. Analogies were drawn with mafic sheet-like sequences described by Smellie et al. (1993) and Loughlin (2002; see also Smellie, 2008), and a downward-thickening columnar-jointed lava lobe attached to the flanks of an ice-impounded (ridge-forming) trachydacite lava at Öraefajökull (Stevenson et al., 2006). The latter lacks a basal hyaloclastite breccia, thought to be due to the very steep underlying bedrock slopes able to efficiently drain away any meltwater. Both Icelandic occurrences also show broad similarities with mafic esker-like sequences, also envisaged forming in ice tunnels (Section 8.2.6).

12.7 Other felsic glaciovolcanic sequences

An unusual felsic glaciovolcanic sequence crops out at Bláhnúkur, Iceland. It is difficult to categorise. The sequence was interpreted by Tuffen et al. (2001) as the product of effusive rhyolite subglacial volcanism (see also Furnes et al., 1980; Tuffen et al., 2002b). In that context, it is therefore possible that it is a distinctive category of felsic subglacial dome. However, the abundance of fragmental rocks present is different from any other subglacial dome described so far and the outcrop is difficult to place in any glaciovolcanic category. There are substantial similarities in lithofacies with the glaciovolcanic outcrop at Dalakvísl, which may be the roots of a small-volume felsic tindar (Tuffen et al., 2008; Section 12.5). Only a vent region with explosively generated tephra is missing from the Bláhnúkur outcrop, and that may simply be due to later erosion or a lack of suitable exposure.

Bláhnúkur is a small and roughly pyramidal hill that rises 350 m. The outcrop occurs within the Torfajökull volcanic complex, the largest rhyolite centre in Iceland and Bláhnúkur may be the youngest centre. Bláhnúkur may correspond to a complex exogenous dome or series of small syn-eruptive domes, all perhaps sharing a common dyke source, emplaced on a pre-existing steep-flanked rhyolite hill. Interpretation of the edifice suggests that it underwent repeated gravitational collapses to generate the abundant breccias present. The outcrop measures c. 1.4 km in diameter, comprising a core of older altered rhyolite and minor basaltic hyaloclastite draped by a veneer of younger subglacially erupted rhyolite c. 50 m thick

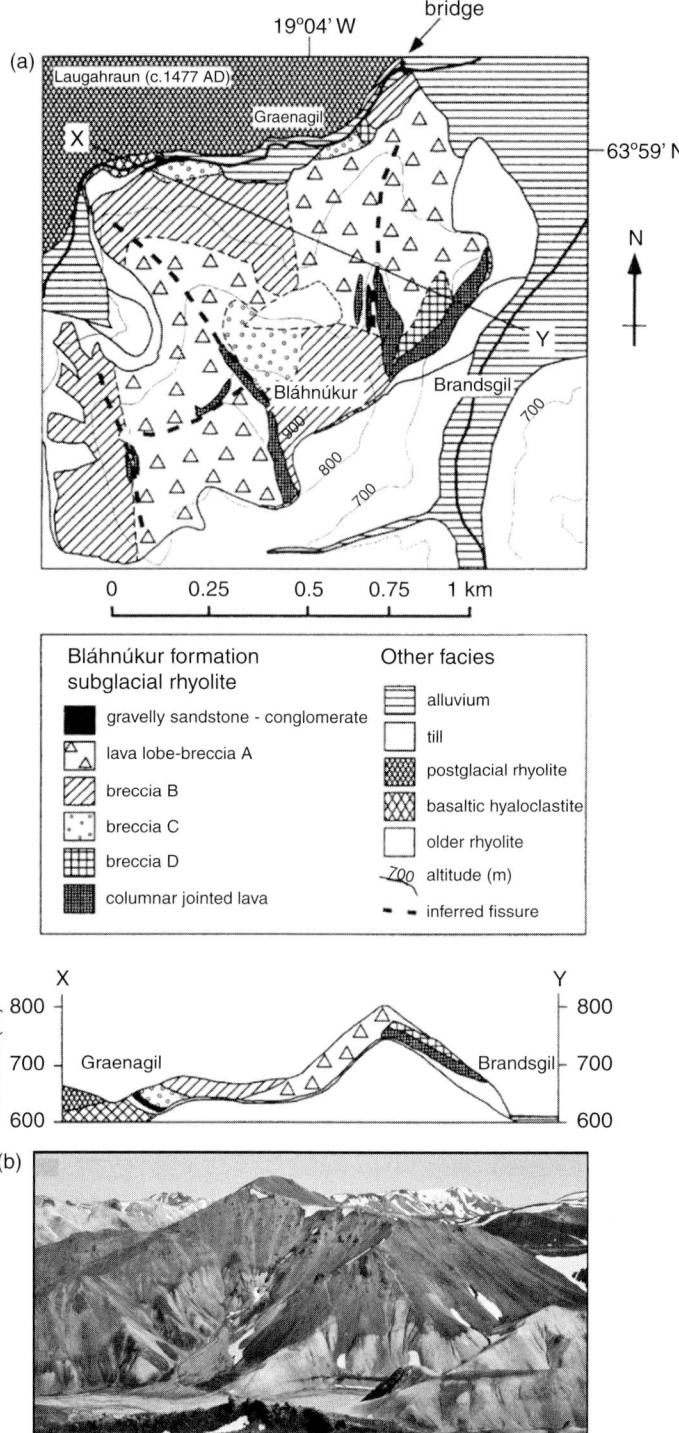

Fig. 12.15 (a) Outcrop map and cross section of Bláhnúkur, central Iceland (from Tuffen et al., 2001). (b) View of the Bláhnúkur outcrop (from Owen et al., 2012). The grey draping outcrop consists of hyaloclastite breccia intruded by numerous felsic lava lobes.

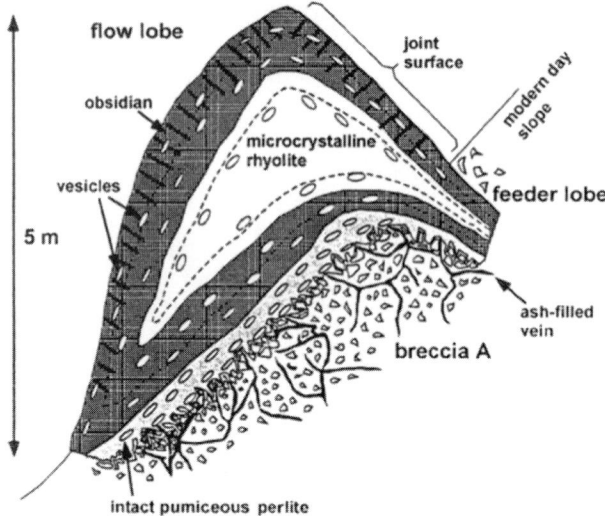

Fig. 12.16 Sketch showing the formation of felsic lava lobes at the base of a glacier, based on outcrops at Bláhnúkur (from Tuffen et al., 2001). The lava lobes are encased in pumiceous perlitised breccia in contact with earlier-formed felsic breccia and were emplaced into the overlying glacier ice by melting. Contrast with Fig. 12.17.

(Fig. 12.15). The subglacial rhyolite seems to have been erupted from a series of at least four variably orientated fissures. It overlies up to 20 m of diamicton and eruption-derived channelised gravelly sandstone and conglomerate dominated by clasts of obsidian and altered rhyolite set in obsidian sand and showing evidence for deposition from flowing water. Most of the volcanic lithofacies are types of volcanic breccias.

Lava lobe breccia (breccia A) comprises irregular and cylindrical lobes and sheets 5–20 m across composed of marginal obsidian and internal microcrystalline rhyolite set in pale grey breccia with a blocky ash matrix. Vesicularity is poor except at the base of lobes where it increases to c. 30% and gives the rocks a pumiceous appearance. The lobes also have fanning columnar-jointed upper surfaces whereas their bases grade into breccia (Fig. 12.16). Breccia A was interpreted to represent emplacement of multiple rhyolite lava masses in well-drained essentially dry vaults melted in the overlying ice, which constrained the lava and (together with limited gravity flow downslope) influenced the final lobe shapes. By contrast, the lobe bases chilled and fragmented by mechanical deformation and steam explosions while flowing over a water-saturated substrate. An alternative explanation is that the lobes were emplaced as intrusions entirely within wet breccia, with later erosion removing the surrounding fragmental rocks to expose the more coherent lobe cores (Fig. 12.17; Stevenson et al., 2011).

The other breccias include (1) a crudely bedded, ash-matrixed, pumiceous perlitised obsidian deposit with elongate to lenticular, vesicular obsidian ribbons and amoeboid masses (breccia B) interpreted as formed by the syn-emplacement gravitational collapse of still-hot lobe breccia A deposits, i.e. a hot avalanche deposit; (2) a chaotic sequence of

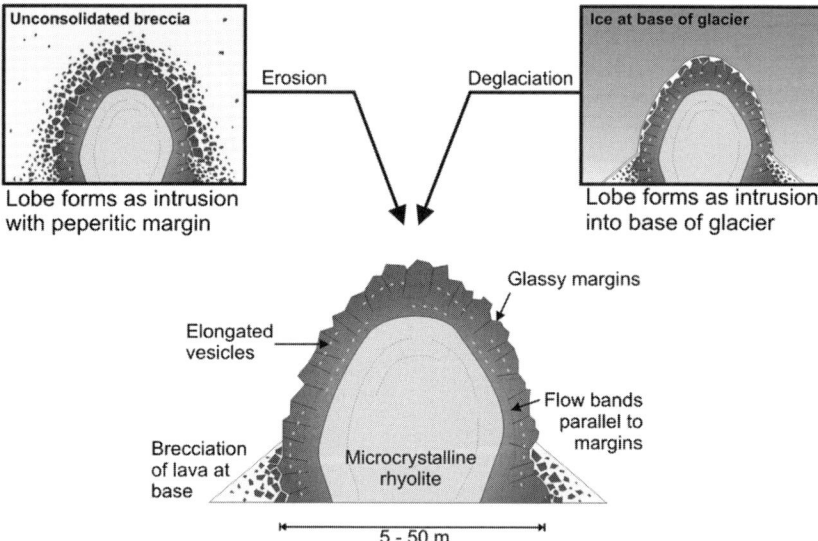

Fig. 12.17 Two scenarios showing alternative inferred modes of formation of felsic lava lobes with glassy and columnar-jointed margins based on outcrops at Kerlingarfjöll, central Iceland (from Stevenson et al., 2011). In one (upper right), the lobe intrudes the base of a glacier (cf. Fig. 12.16), whereas in the other (upper left) it intrudes unconsolidated breccia. The resulting lithofacies may be very similar in both cases.

fragmented rhyolite lava lobes and lenses of breccia cut by numerous small faults (breccia C) also formed by gravitational collapse of lava lobe breccia A but at cooler temperatures; and (3) breccia D, which comprises c. 20 m of faintly bedded, polymict obsidian–microcrystalline rhyolite breccia with an ash matrix and intruded by irregular masses of obsidian with fine marginal polygonal columns. Breccia D probably also formed by gravitational collapse, perhaps as a cohesive (matrix-supported) mass flow rooted in cooled lava lobe breccia A. Columnar jointed rhyolite lavas are also present up to 20 m thick and 400 m long, with obsidian margins and microcrystalline interiors. The lava tops are also columnar jointed and they may have flowed and chilled within subglacial tunnels; like the rhyolite lava lobes (breccia A), the lack of evidence for magma–water interaction suggests that the tunnels were essentially dry.

Calculations by Tuffen et al. (2001) indicated that the advance rate of rhyolite lava with a chilled carapace (much more viscous than unchilled rhyolite magma) was much greater than the rate of ice melt-back due to conductive heat transfer from magma to ice (see Chapter 6). The rhyolite lavas would therefore be impounded by the surrounding ice, consistent with interpretations of the cooling joint patterns in the lobes and lavas present, which were thought to mimic the shape of ice cavities melted at the glacier base during eruption. The cavities themselves may have been largely dry except at their bases where the presence of perlitic obsidian breccia with a blocky ash matrix indicates that they retained at least some meltwater with which the rhyolite interacted (i.e. chilled and fragmented). It was

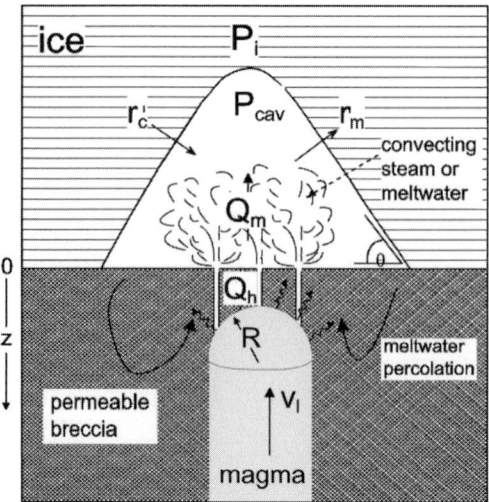

Fig. 12.18 Cartoon illustrating a model for melting an ice cavity during the ascent of felsic magma (from Tuffen et al., 2002b). The rising magma loses heat by conduction and steam convects that heat in vapour-escape pipes to the glacier base, where cavities are melted ahead of the arrival of the magma. See Tuffen et al. (2002b) for further description and explanation of symbols.

postulated that some of the cavity melting might be caused by convecting steam (or possibly convecting meltwater; Tuffen et al., 2002b; Fig. 12.18), possibly linked to lava lobe advance. Initial cavity dimensions were quite small, typically measuring 2–5 m high, and increased to 20 m high by 20 m wide later in the eruption, with steep walls. After emplacement of each hot rhyolite mass ceased, it was speculated that further heat loss from the lavas allowed the subglacial cavities to continue melting back, thus removing support from the steep lava flanks, which then underwent gravitational collapse. In addition, with subsequent ice retreat, parts of lobe breccia A were destabilised and collapsed down the steep gradients of the underlying older rhyolite hill. Thus different breccias were created and were either still hot or cooled depending on when the collapses occurred. The Bláhnúkur outcrop was probably emplaced beneath ice over 400 m thick (the edifice is 350 m high and was emplaced entirely subglacially). It was envisaged that syn-eruptive meltwater was able to drain away continuously from the numerous small rhyolite vents at Bláhnúkur, similar to intermediate-composition subglacial domes in British Columbia, which would have reduced the extent of any magma:water interaction.

12.8 Large polygenetic glaciovolcanic centres

Although eruptions of felsic magma occur in many stratovolcanoes with glaciovolcanic histories, the felsic sequences are subordinate to the much more voluminous products of mafic and intermediate-composition magmas (e.g. Öraefajökull, Iceland: Prestvik, 1985;

Section 12.2; Mt Edziza, British Columbia: Souther, 1992). The most evolved formation at the Edziza complex is the Spectrum Formation, which has a significant rhyolite component but which has been dated to be 3.1 Ma, suggesting that most of the emplacement history was pre-glacial (Souther, 1992). Polygenetic stratovolcanoes composed mainly of felsic glacio-volcanic products are unknown.

13

Glaciovolcanic sequences as palaeoenvironmental proxies

13.1 Introduction

The use of glaciovolcanic sequences as indicators (proxies) for palaeoenvironments is still in its infancy. However, it is a novel and increasingly important methodology whose strength lies in its ability to determine, often quantitatively, a wide range of palaeo-ice sheet characteristics, including ice extent, ice thickness and subglacial hydrology. At many locations these parameters cannot be obtained in any other way. Glaciovolcanic sequences are thus able to provide a more holistic view of past ice sheets than any other method currently existing and the use of glaciovolcanic investigations is set to increase. In this chapter we describe the rationales that underpin the palaeoenvironmental interpretations of glaciovolcanic sequences and show, with selected published examples, how they can be used successfully to determine multiple critical parameters of past ice sheets.

13.2 Ancient ice

In general, ice melts during interglacials and is not preserved in the rock record. Yet identifying its presence and principal characteristics in Earth's past and on other water-bearing planetary bodies, such as Mars, is critically important for understanding planetary climate evolution, for constructing accurate global water inventories and, on Earth, demonstrating how ice sheets contributed to eustasy (Smellie and Chapman, 2002). The sole exception of ancient preserved ice known to us may be an occurrence at least 8.1 m.y. old (Late Miocene) that occurs under exceptional circumstances in the Dry Valleys region of East Antarctica (Sugden et al., 1995). The ice is buried beneath about 4 m of glacial debris (till) in Beacon Valley, Victoria Land, about 5 km from the periphery of the Antarctic Ice Sheet at elevations of 1300–1450 m a.s.l. The current mean annual temperature at the site is −30 to −35 °C, with precipitation <10 mm water equivalent per year and the presence of the buried ice suggests that the local climate has not changed significantly since that ice was formed. The ice extends over an area of c. 4 km^2 and may be at least 15 m thick. Contraction-crack polygons occur on the surface of the till and extend as V-shaped troughs 2–3 m deep that penetrate the underlying ice. Well-preserved volcanic ash occurs in several of the polygons. The ash shows no signs of reworking and is probably in situ (ash-fall tephra). Its age was determined by laser-fusion ^{40}Ar/^{39}Ar dating as 8.1–7.9 Ma, implying that the till on which the ash was deposited must be older than 8.1 Ma. Structural analysis of fabrics in the buried ice indicates that it was derived from an expanded Taylor Valley outlet glacier (i.e. Antarctic Ice Sheet) rather than local valley glaciers. Moreover, δ^{18}O values of the fossil ice cluster around −33‰, significantly higher than modern Antarctic ice locally at greater elevation but comparable with modern ice at the head of Beacon Valley. The evidence was interpreted to indicate that the ice was a remnant of an ancient outlet glacier formed in Middle Miocene time under cold climatic conditions and a polar glacier thermal regime comparable with those of today. Objections to the interpretation were raised by Hindmarsh et al. (1998), who argued on theoretical grounds that any estimated sublimation rates should be sufficient to destroy the Miocene ice, but the field relationships and an ancient age seem unequivocal (Schäfer et al., 2000; Marchant et al., 2002). However, apart from this unique example, ice is either not preserved once a glacial succumbs to the global warmth of interglacials, or else it is very much reduced in geographical extent.

13.3 Sedimentary evidence for glacial conditions and basal thermal regime

Landforms become significantly modified or are not preserved after multiple successive glaciations and thus can often only be used to assign glacial characteristics qualitatively to a particular glacial period, especially prior to the Last Glacial Maximum (e.g. Bennett and Glasser, 2009). Thus, sedimentary rocks are overwhelmingly used to infer the presence of past ice and to derive a limited range of characteristics, principally recording the local presence of ice and, particularly, its thermal regime. Many books describe the sedimentary characteristics used to infer a glacial regime (e.g. Hambrey, 1994; Benn and Evans, 1998; Hambrey and Glasser, 2012). Important sedimentary features indicative of an association with overlying glacial ice include the presence of poorly sorted clast-bearing deposits (diamict) in which the clasts show faceting, bullet shapes and striations, together with evidence for basal glacial deposition (e.g. shear-induced clast trains and/or low-angle shear surfaces). Stratified diamicts may also be indicative of former ice but form as melted-out, remobilised basal debris. The surfaces of the underlying bedrock may also show striations, polishing and surface moulding. However, although probably indicative in most cases, the association of striated, glacially modified surfaces and diamicts containing striated clasts is not always diagnostic of a glacial origin (Eyles et al., 1983; Geirsdóttir, 1991).

The sedimentary and landscape evidence for ancient ice is determined by the basal thermal regime of the associated ice. The most practical division is into wet-based and cold-based ice. The highly distinctive features just described are characteristic of wet-based ice (i.e. temperate or warm ice, and sub-polar or polythermal ice; see Section 4.3). To recap: wet-based ice, also descriptively called thawed-bed ice, has a basal film of meltwater, moves mainly by sliding and is highly erosive. By contrast, cold-based (polar) ice is below freezing throughout, is frozen to its bed (hence it is also called frozen-bed ice) and acts largely to protect bedrock surfaces, including even quite fine details such as loose clasts resting on that surface. Such basal erosion as occurs beneath polar ice creates geographically restricted coarse features that should be easily distinguished from features of wet-based ice in most cases (cf. Atkins et al., 2002; Lloyd Davies et al., 2009). Temperate ice is at the pressure melting point throughout except in a thin surface layer that is below the freezing point in winter. Sub-polar ice differs from temperate ice. Unlike temperate ice, it is below the melting point everywhere except in a thin wet basal layer and thus has a thawed bed and movement is mainly by sliding. However, although debris entrainment and deposition can be substantial (by riding up internal shear surfaces where wet-based interior ice meets cold-based marginal ice), meltwater and fluvial activity are much less well developed than for temperate ice (Hambrey and Glasser, 2012). Finally, a category known as polythermal ice also exists. Although polythermal and sub-polar are often used interchangeably, another usage of polythermal is as a descriptor for an ice sheet with a basal thermal regime that is a coarse temperature *mosaic* of frozen-bed and thawed-bed patches (Kleman and Glasser, 2007; Smellie et al., 2014). In this book we will refer to the three basal regime categories predominantly as warm ice, cold ice and polythermal ice, with warm and polythermal ice collectively referred to as wet-based when required.

The distinction between the different types of basal thermal regime in ice sheets is important for determining environmental evolution, itself linked to climate change. Ice sheets are effectively pinned by their areas of frozen bed and any unstable behaviour is initiated in those areas of thawed bed. Thus, determining the pattern of frozen and thawed basal ice is of primary importance because it determines the relative stability of an ice sheet and, thereby, its potential contribution to global sea levels. The problem is a key one not just for environmental geologists but also for glaciologists. For example, the Neogene evolution of the basal thermal regime of the East Antarctic Ice Sheet has been the subject of active debate amongst geologists for over 30 years and is still contended (Sugden et al., 1993; Sugden, 1996; Wilson, 1995; Lewis et al., 2007). The debate is important because it concerns nothing less than the evolution of polar ice sheets and the global climate system, which has serious implications for a world experiencing greenhouse warming.

13.4 Glaciovolcanic evidence for ancient ice

The presence of interbedded glacial sediments and glacially modified surfaces within a volcanic sequence are good indicators that eruptions may have taken place in a glacial setting. However, such evidence can be ambiguous; it simply indicates that ice was present before eruptions took place. Additional signs, such as evidence that the glacial sedimentary rocks were still soft, wet and therefore probably contemporaneous, are required to diminish

Fig. 13.1 Pale-coloured diamict (tillite) overlain by irregular lava masses and breccia at the base of a glaciovolcanic basaltic lava-fed delta. The volcanic rocks have sunk into the diamict, which must have been soft when the delta was emplaced and therefore probably contemporaneous. About 15 m of vertical section is shown. James Ross Island, Antarctica.

the uncertainty. That evidence includes features such as (1) load structures and flame-like protrusions of glacial sediment that penetrate up several metres into overlying volcanic rocks (Fig. 13.1); (the development of lava pillows and/or peperite at contacts between lava and sediments; and (3) the presence of lonestones dropped by floating ice, e.g. in lake and tuff cone sequences formed of volcanic (ash) turbidites (Smellie et al., 2011a,b) or in hyaloclastite breccias foreset deposits of marine-emplaced lava-fed deltas (although the delta itself in this example is not glaciovolcanic). Compared with their sedimentary equivalents, the evidence for a glacial setting contained by glaciovolcanic successions is much less well known, even though glaciovolcanic sequences are currently the most informative rock records of former ice sheets. This is particularly true for periods prior to the Last Glacial Maximum, since successive wet-based ice sheets routinely remove much of the evidence of older glacial deposits because of the thinness of the associated terrestrial sediments (often just metres thick). Even the much thicker shallow marine record of continental shelves fed by glaciers can be radically modified by younger events. For example, almost the entire Middle and Late Miocene period, a period of about 10 m.y. (c. 15–5 Ma), is missing from the Ross Sea continental shelf (Smellie et al., 2011b). The period is an important one climatically. It includes part of the Middle Miocene Climatic Optimum (c. 17.0–14.5 Ma), a complex stage of fluctuating $\delta^{18}O$ records suggesting a varied Antarctic ice volume, and the Middle Miocene Climatic Transition (MMCT; c. 14.5–13.0 Ma), commonly regarded as marking the onset of climatic cooling and ice sheet expansion when the EAIS may have acquired the cold polar (frozen-bed) glacial regime that it displays today (Lewis et al., 2007). There are also no onshore sedimentary outcrops of appropriate age in the region but glaciovolcanic rocks fill the missing record almost exactly and have been used to reconstruct the glacial conditions (Smellie et al., 2011b, 2014). Geomorphological signs of past ice, such as trim lines indicative of past ice thicknesses, are also either destroyed or significantly modified and are typically evidence of only the most recent ice sheet or in the case of trim lines the highest ice stand. They are difficult to date. Conversely, glaciovolcanic sequences are also proxies for past ice. They are generally tens to hundreds of metres thick and contain numerous interbedded lavas that are hard to erode and are usually easily dated isotopically. Moreover, heat emitted by the slow cooling of the edifices effectively shields them from erosion by ice for many years and permits the initially unconsolidated volcaniclastic piles to lithify by authigenesis and thus rapidly become more resistant (Jarosh et al., 2008). The sequences typically contain unambiguous evidence for past ice thickness and they often overlie and thus preferentially protect glacial sediments associated with the coeval ice. The association of volcanic rocks with those sedimentary rocks is an indirect way of inferring a glacial eruptive setting for the volcanism, assuming that contemporaneity between the volcanics and sediments can be proven. That is not always an easy task. However, glaciovolcanic sequences themselves have an increasingly important part to play in recognising the presence of former ice and for deducing critical parameters of that ice, such as its age, thickness, thermal regime and surface elevation. This is a greater number of parameters than any other proxy can provide.

13.4.1 *Subaqueous–subaerial transitions*

Identification of a glacial eruptive environment from characteristics of glaciovolcanic rocks relies primarily on recognition of subaqueous and subaerial lithofacies, with the latter deposited on top of the former after the enclosing ice is completely melted through. Identification of subaerial lithofacies is relatively straightforward. The surface separating coeval subaqueous and subaerial volcanic lithofacies is known as a *subaqueous–subaerial transition*. It is recognised by distinctive differences in the lithofacies that are formed.

Subaerial lava, i.e. air cooled, is crystalline and it contains widely spaced cooling fractures, including large polygonal columns formed by slow conductive cooling (Chapters 6 and 9; Fig. 9.7; e.g. Long and Wood, 1986). In 'a'ā lava, subaerial auto-breccia is commonly oxidised to a distinctive maroon-red coloration and pedogenesis creates orange-red soils in the breccias between eruptions, commonly containing plant roots and fossil leaves. In pāhoehoe lava, the glassy chilled margins of lava tubes and toes are formed by comparatively slow chilling in air and ought to be thinner (<1 cm) than in water-cooled lava (1–2 cm or even thicker). The water in which subaqueous lithofacies form is derived by melting of the associated ice. It can often be inferred to have been ponded by former ice if the volcanic rocks are relatively young (Neogene) and there is no preserved topography that might have acted as a barrier capable of impounding water in a (non-glacial, pluvial) lake.

Several types of subaqueous–subaerial transition are known. They have been described principally in association with pāhoehoe-fed lava deltas, although many glaciovolcanic sequences can contain such transitions. Smellie (2006) described and defined five different types of subaqueous–subaerial transitions, which differ mainly in their geometries. The study followed from a remarkably prescient pioneering study by Nelson (1975) based on 1960s field work on James Ross Island, Antarctica. Nelson described and categorised volcanic features in lava-fed deltas that were thought at that time to be marine-emplaced but are now known to be glaciovolcanic (Skilling, 2002; Smellie et al., 2008); recognition of this type of transition was in part the basis for the early recognition of and models for tuya formation in British Columbia by Mathews (1947). Further analysis of subaqueous–subaerial transitions in marine-emplaced lava-fed deltas was subsequently published by Jones and Nelson (1970). It is clear that identical subaqueous–subaerial transitions can occur in both glacial and non-glacial settings, and only some are diagnostic of a glacial environment. It is important to establish initially that the transition is not simply due to erosion, i.e. by demonstrating that the surface is depositional and coeval with the volcanic edifice in which it occurs. In cases where the surface is erosional, it is simply an unconformity and has no environmental significance for the volcanic structure. The five types of subaqueous–subaerial transition defined by Smellie (2006) and Russell et al. (2013) are as follows and are particularly characteristic of mafic pāhoehoe lava-fed deltas (Fig. 13.2):

Cone surface overlap structure: In this type of transition, subaerial lava of a lava-fed delta caprock overlies bedded tephra comprising the subaqueous dipping flank of a coeval tuff cone (produced during the preceding tindar or subaqueous tuff cone stage). It may be difficult to determine unambiguously that the tephra beds were deposited subaqueously but that distinction is important since a surface composed of subaerial lavas abutting subaerial tephra beds is not a subaqueous–subaerial transition. Diagnostic characteristics for tephras deposited in the two settings have not been described (but see Section 9.4.4). In a true transition, lavas terminate against subaqueous tephra and beds in the latter are typically parallel to the transition surface. The surface is thus uneroded, although coeval flank collapse of the subaqueous pyroclastic cone can result in truncation of the tephra beds and is a possibility that needs to be objectively assessed before it is discounted. The surface itself is sharply defined. If subaqueous deposition of the tephra beds can be determined, which is not always possible, this type of transition is probably diagnostic of a glacial setting since it implies that an important episode of falling water level has occurred in the coeval lake, which is more likely if that lake is glacially impounded meltwater.

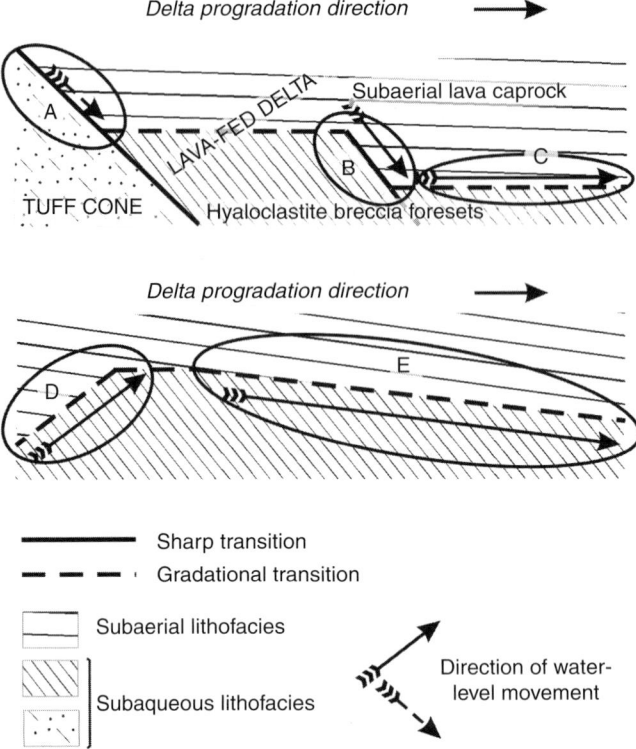

Fig. 13.2 Terminology of lava-fed delta subaqueous–subaerial transitions (from Smellie, 2006). A: cone surface overlap structure; B: delta front overlap structure; C: horizontal passage zone; D: rising passage zone; E: falling passage zone.

Delta front overlap structure: This surface consists of subhorizontal subaerial lava of a lava-fed delta caprock abutting homoclinal, steep-dipping, subaqueously deposited delta foreset breccia beds (Fig. 13.3). Despite the onlapping arrangement of lithofacies, all parts of this transition comprise a single lava-fed delta. As in cone surface overlap structures, beds in the underlying breccias dip parallel to the transition (i.e. it is unlikely to be an erosional surface) signifying likely contemporaneity with the adjacent lavas. The vertical extent of this structure is limited by two bounding horizontal passage zone surfaces, i.e. the higher-elevation passage zone represents an earlier time of higher water level and the lower passage zone corresponds to a new younger water level, a process anticipated in early descriptions by Mathews (1947). The difference in elevation between the two passage zones is commonly several tens of metres, far greater than could be caused by a change in sea level during eruption. This type of transition in a single lava-fed delta is diagnostic of a glacial setting.

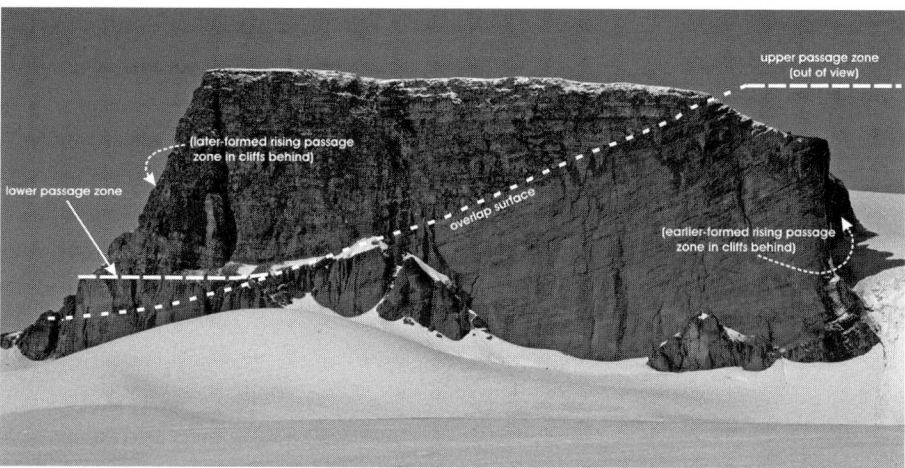

Fig. 13.3 View of a spectacular delta front overlap structure, caused by a drop in the elevation of the coeval meltwater lake (from Smellie, 2006). The rock face is over 200 m high. James Ross Island, Antarctica.

Horizontal passage zone: A passage zone is a fossil water level comprising a subhorizontal planar surface separating overlying subhorizontal subaerial lava from subaqueous steeply dipping homoclinal beds of subaqueously deposited breccia. Together the two lithosomes define a classic Gilbert-type of delta (i.e. topsets overlying foresets) but composed entirely of primary volcanic materials (Fig. 9.14). Passage zones occur in all mafic and some intermediate-composition lava-fed deltas and form wherever subaerial lava enters and progrades across a body of water (Nelson, 1975). Felsic lavas do not seem to form lava-fed deltas (Section 12.4.1). A passage zone simply records a palaeo-shoreline signifying the presence of ponded water (sea or lake). It is the most commonly described type of subaqueous–subaerial transition described. Some features may be diagnostic of a marine shoreline, e.g. recording tidal effects (Jones and Nelson, 1970; Furnes and Fridleifsson, 1974; Ramalho et al., 2013). The passage zone itself is a coarse feature. It is best defined for pāhoehoe-fed deltas in which lava lobes either terminate at the delta brinkpoint (i.e. the position on the delta front at which the molten lava enters water and disintegrates into breccia) or else flow a short distance down the delta front, typically just a few metres but occasionally 10 or 20 m, before freezing (Moore et al., 1973; Tribble, 1991). Thus, the passage zone in a pāhoehoe-fed delta comprises an uneven zone usually a few metres thick consisting of intermixed lava and breccia, giving way downward into comparatively well-developed steep beds of breccia (Fig. 10.12). Conversely, passage zones in 'a'ā lava-fed deltas are much less well defined as 'a'ā lava is generally thicker than pāhoehoe lobes and travels much further down the delta front before disintegrating (Mitchell et al., 2008; Stevenson et al., 2012; Smellie et al., 2011a, 2013a). The 'a'ā lava also advects subaerially oxidised autobreccia clinkers into the subaqueous setting, thus further masking the essential characteristics of the subaerial and subaqueous lithosomes. Thus the transition is typically developed over 5–10 m and examples 25 m thick are known (Smellie et al., 2011a,b, 2013a; see Chapter 10). This type of transition by itself is not diagnostic of a glacial setting.

Rising passage zone: Non-horizontal passage zones are diagnostic of a glacial eruptive setting. A rising passage zone (originally called a terminal slope overlap structure by Nelson, 1975) is caused by increasing elevation of the associated meltwater lake ahead of the advancing lava-fed delta due to an increasing accumulation of meltwater. In a temperate glacier situation (i.e. with basal drainage), the lake surface elevation is a balance between the volume of meltwater escaping subglacially and that overflowing and escaping supraglacially (Smellie, 2001, 2006). Thus, should the basal drainage be impeded, e.g. by flank collapses of the volcanic pile blocking subglacial tunnels, meltwater will accumulate ~~and overflow~~ causing a rise in water level until pinned by the elevation of the lowest overflow point. A distinctive feature of passage zones associated with a rise in meltwater is that the meltwater floods over and onto stalled parts of the lava-fed delta surface. When the active part of the delta then returns to flow over those flooded areas it then forms a horizontal unit of massive hyaloclastic breccia overlain by subaerial lava. These breccia–lava couplets may occur several times in succession and they are distinctive. Although characteristic of a rising water level, the individual passage zones in these lava–breccia alternations are typically subhorizontal but they unequivocally indicate meltwater accumulation relative to delta advance and are diagnostic of a glacial situation, even when a rising passage zone geometry cannot be discerned in the outcrop.

Falling passage zone: With efficient basal drainage and a slowly advancing lava-fed delta, the rate of meltwater generation due to heat from the delta relative to meltwater escape will fall and cause the lake level to lower. This should result in a passage zone that dips down in the direction of lava-fed delta advance. It should also be diagnostic of a glacial situation. However, taken to its logical conclusion, all the ponded water ultimately will escape ahead of the delta and the delta itself will cease and transform into simple subaerial effusion. It may be significant that no examples forming in this way have been described. In most cases, a falling water level is recorded as a single step, resulting in a delta front overlap structure (as described above). Thus, the formation of falling passage zones due to a slow progressive fall in water level may be theoretical only or at least very rare. However, a second type of falling passage zone occurs when a lava-fed delta advances through a glacial cover that simply drapes (rather than drowns) the slopes of a volcano. The upper surface of the glacial cover thus mimics the slope of the volcano flank and the passage zone developed in the lava-fed delta dips similarly (Fig. 13.4). Thus, the orientation of the passage zone is governed by the morphology of the surface of the glacial cover (or rather, the firn–ice transition within that glacial cover; see Chapter 4), a process anticipated remarkably presciently by Souther (1992, Fig. 122), rather than controlled by

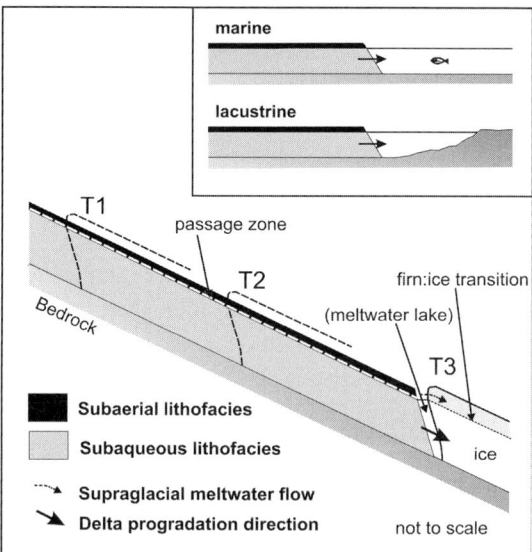

Fig. 13.4 Sketch showing how falling passage zones form in a lava-fed delta advancing down a sloping bedrock draped by a thin ice sheet (from Smellie et al., 2011b). The passage zone orientation mirrors the slope of the coeval ice sheet and bedrock surface. In an absence of tectonic effects (tilting), the presence of a falling passage zone is diagnostic of a glaciovolcanic setting.

a falling meltwater surface caused by the hydraulics of water escape from a meltwater vault. Such features are common in the rock record and are extensively present in many large volcanic centres in Victoria Land, Antarctica (Smellie et al., 2011a,b). Mapped in three dimensions, the passage zones of multiple lava-fed deltas will dip radially away from the main erupting vent (summit crater; Smellie et al., 2011b, Fig. 8). They are diagnostic of a glacial setting.

Pyroclastic passage zone: These were described by Russell et al. (2013) based on the concept that tuff cones constructed in ice-confined glacial lakes should show contrasts in depositional conditions (subaerial, subaqueous) according to their position relative to the coeval lake surface and that they may even show variable geometries caused by fluctuations in elevation of the lake, as occur in passage zones in lava-fed deltas. The recognition of pyroclastic passage zones is based on a range of criteria (listed in Section 9.4.4). However, the criteria are comparatively ill-defined features of pyroclastic density current deposits. They are much more ambiguous for establishing precisely (i.e. within tens of metres vertically) the location of the passage zone surface and no photograph of a well-defined pyroclastic passage zone has yet been published, which contrasts with the very visible passage zones associated with lava-fed deltas. Moreover, it is unclear if the duration of eruptive activity at glaciovolcanic tuff cones (typically weeks or months) is long enough to permit the fluctuations in lake elevations required to register in the tuff cones (even at longer-lived vents, the vent tends to get enclosed by subaerial lavas and dries out, thus terminating the tuff cone activity). By contrast, the emplacement duration of lava-fed deltas may occur over years to decades in voluminous eruptions. It may be significant that the prominent fluctuations in passage zone elevations in single lava-fed deltas described by Smellie (2006), for flood basalts on James Ross Island (Antarctica), have only rarely been observed in smaller-volume monogenetic tuyas, probably due to the shorter timescale in which the latter eruptions occurred. Thus, the recognition of pyroclastic passage zones will be very difficult.

Most of the transitions described above were observed in the very well-exposed large-volume (tens of km^3) far-travelled (tens of km) pāhoehoe lava-fed basaltic deltas that constructed the Mio-Pliocene Mt Haddington volcano on James Ross Island, Antarctica (Skilling, 2002; Smellie, 2006; Smellie et al., 2008). That they have not been described in any of the very common tuyas in Iceland and rarely in British Columbia, also mainly pāhoehoe lava-fed, may be because of the greater glacial dissection and much larger scale, especially lateral extent (tens of km), of the Mt Haddington deltas and the prolonged history of glaciations represented on James Ross Island extending back to Late Miocene times at least. The examples in Iceland and British Columbia are much smaller (typically 2–5 km in diameter) and are inferred to be short-lived monogenetic tuyas. However, while Icelandic examples are not much dissected, some examples in British Columbia show passage zone elevation variations attributable to fluctuating water levels (Mathews, 1947; Edwards et al., 2011; Ryane et al., 2011; Russell et al., 2013).

From this short discussion of passage zones, it is evident that glacial meltwater lakes associated with lava-fed delta progradation are dynamic features and lake surface elevation will migrate up and down several times depending on the duration of the eruption and the glacier hydraulics (see Chapter 4). Passage zones fed by pāhoehoe are a particularly clear and unambiguous structure that can be used to make environmental interpretations. Descriptions of similar transitions in 'a'ā lava-fed deltas are much less often described and the transition is a much coarser zone. The most commonly observed transition in 'a'ā lava-fed deltas is a falling passage zone that dips parallel to the underlying volcano flank. Other types of transitions analogous to those in pāhoehoe-fed lava deltas probably exist in 'a'ā-fed deltas but they will likely be cruder in construction and thus less clearly defined.

Subaqueous–subaerial transitions in intermediate and felsic magmas also occur but are typically even coarser features and it is consequently harder to determine their position with great accuracy. This is probably due principally to the higher viscosities that cause the associated subaerial lava to retain coherence underwater rather than disintegrating into glassy breccia. Examples are known in which felsic sheet lava (coulées) a few tens of metres thick overlies thinner (few dm to 1 or 2 m) coeval hyaloclastic breccia (Fig. 9.10b). The glass commonly has a sugary texture in hand specimens caused by secondary perlitisation and growth of micropoikilitic textures due to hydration of the glass in a warm environment (i.e. heat from the overlying lava; Smellie et al., 2011a).

The lava contains relatively closely spaced irregular cooling joints and may be associated with intrusive flames of glassy breccia penetrating up several metres into the lava from its base. The cooling joints become more widely spaced higher in the lava, the lava becomes more crystalline, and it may be capped by lithic autobreccia indicative of slower cooling and likely subaerial conditions at its upper levels if the lava is thick enough to melt completely through the surrounding ice (Smellie et al., 2011a; Fig. 12.14). Distinctive patterns of cooling joints in intermediate lavas caused by water chilling in ice-contact locations have also been described by Lescinsky and Fink (2000; see Section 9.3.7). In felsic tuyas, although subaerial capping lavas are easily identified (Tuffen et al., 2002a), it is often very difficult to locate a subaerial–subaqueous transition. This is because the glacial vaults associated with felsic tuyas are generally thought to be drained due to the thermodynamics of felsic lava–ice interaction (less heat available; Höskuldsson and Sparks, 1997; see Chapters 4 and 6) and therefore are merely moist and not water filled (Stevenson et al., 2011). Thus, a transition *sensu stricto* is probably not present in most cases, although any basal tephras are typically phreatomagmatic and attest to the presence of meltwater (Tuffen et al., 2002a; Stevenson et al., 2011). However, it is possible to infer the thickness of associated ice (see below).

13.4.2 Other volcanogenic features diagnostic of eruptions in association with former ice

Ice-block meltout cavities are uncommon but distinctive features of glaciovolcanic sequences. They are diagnostic of a glacial setting as they cannot occur in sequences emplaced in seawater nor in pluvial lakes since ice floats and thus cannot easily be trapped. The features consist of broadly equant open cavities a few dm in diameter to rarely more than 2 m, showing thick glassy surfaces (glass c. 1–5 mm thick) and occasional small (<10 cm) lava breakouts on walls that also show polygonal cracking (Smellie, 2008; Skilling, 2009; Fig. 9.9). The cavities may occur in clusters spaced at over 1 m or occur in isolation. Their mode of formation is not well understood but they are currently envisaged forming where lava engulfs displaced ice blocks trapped between the lava and an overlying ice wall with the lava chilling against the ice blocks (Skilling, 2009, Fig. 6; Fig. 10.5).

Overthickened lavas ponded against or otherwise confined by former ice are common features particularly of felsic glaciovolcanic sequences and may be the principal evidence in those sequences for a glacial setting (cf. Figs 11.2, 12.1; Kerr, 1948; Mathews, 1952a; Lescinsky and Fink, 2000; Tuffen et al., 2001, 2002a; Edwards et al., 2002; Edwards and Russell, 2002; Stevenson et al., 2006; Harder and Russell, 2007). A lava upper surface that is subhorizontal despite emplacement on a dipping bedrock is additional confirmation of ponding by a barrier. Overthickening may be very obvious. For example, unconfined subaerially emplaced basanite lavas have aspect ratios (lava thickness:lava width) of c. 10^{-4}

but ice-impounded basanite may have an aspect ratio as high as 0.3 (Harder and Russell, 2007). In other cases, lavas just a few tens of metres thick close to their source become substantially thicker downslope, reaching thicknesses of a few hundred metres at their termination (Edwards et al., 2002). The subvertical lava margins are also associated with glassy surfaces and well-developed patterns of distinctive cooling fractures, particularly subhorizontal polygonal columns, which signify rapid cooling against a near-vertical ice wall.

The presence of **pillow lavas** is an indication of effusion of mafic and intermediate lava underwater, whether seawater, freshwater or glacially confined meltwater. Variations in size are related to magmatic composition. Because of rheological differences, intermediate composition lava pillows are commonly larger than those with mafic compositions (Walker 1992). Lava pillows are only diagnostic of emplacement under water. However, the presence of additional features such as lava pillows moulded against a surface now no longer preserved and that can reasonably be presumed to be former ice are diagnostic (Skilling, 2009). Described as **ice-contact lava confinement surfaces**, they consist of distinctive steep (75–85°) pillow lava surfaces that are flat or billowy and glassy (Fig. 13.5). The surfaces are interpreted as the steep glassy chilled margins of pillow lava where it directly rested against an ice wall. They are extremely rare features in pillow mounds probably because the subvertical surfaces are gravitationally unstable and will usually collapse when the adjacent ice melts back and removes its support. Alternative interpretations involving confinement against earlier-deposited tephra or hyaloclastite were rejected by the absence of peperite and other features indicative of pre-existing rocks, although complete erosion of those rocks remains a possibility (but only a slight possibility). Moreover, the presence of distinctive downward-pointing lava pillow 'breakouts' on the steep moulded surface and subvertical pillows (Fig. 13.5; Skilling, 2009) are important in this context as they presumably indicate that a narrow gap of variable thickness developed by heating along the lava–ice contact and into which new pillow 'drips' were emplaced and cooled, although such features can also form rarely in open water (non-glacial) situations (Batiza and White, 2000; Hungerford et al., 2014). Such features are unlikely to occur at a former contact confined by other rocks now removed by erosion and they are probably diagnostic of an ice-contact.

Lava pillows may often closely resemble subaerial pāhoehoe lava toes and their distinction is clearly important environmentally (Walker, 1992). Because chilling in water is much more rapid than in air (because of the very high heat capacity and thermal conductivity of water compared to air; White et al., 2003), water-chilled margins should be significantly thicker than their air-chilled equivalents. Moreover, because of buoyancy effects (buoyancy contrasts are greater between lava and air than lava and water), lava pillows forming a pillow pile in water are typically more rounded than those formed in air (cf. Figs. 9.2, 9.3) and many examples can probably be distinguished on that basis (see Section 13.4.3).

Fig. 13.5 Views of steeply plunging lava pillows with smooth ice-moulded surfaces caused by confinement by ice. Both views taken at Hlödufell, Iceland.

13.4.3 Effects of lava buoyancy underwater and environmental application

The observation that subaerial 'a'ā lavas commonly thicken significantly when they pass across the passage zone in lava-fed deltas (Smellie et al., 2013a) is probably a consequence of (1) buoyancy associated with submergence and (2) possible Rayleigh–Taylor instability whereby denser magma subsides into water-saturated breccias, with a lower bulk density, such as those occurring at the unbuttressed snouts of lava-fed deltas. The latter effect is probably responsible for the highly bulbous flow noses, chaotic lobe shapes and deformed stratification steepened well beyond the angle of repose seen at some lobe snouts (Figs. 9.16, 10.13, 12.14).

Buoyancy effects on lava thicknesses and the palaeoenvironmental significance are investigated here empirically for the first time. If we equate the lava fluxes above and below water (i.e. assuming continuity across the waterline) and assume that viscosity remains constant, buoyancy will affect the thickness of a subaerial lava continuing to flow underwater, as follows:

$$h_w/h_a = \sqrt[3]{\{\rho/(\rho - \rho_w)\}}\{\sin \alpha / \sin \alpha_w\} \tag{13.1}$$

where h_w and h_a are the thickness of lava in water and air, respectively, and their ratio depicts the relative change in flow thickness of a lava flowing into water; ρ and ρ_w are density of magma (related to vesiculation %) and water, respectively; α and α_w are the subaerial and subaqueous slope angles (adapted from Mitchell et al., 2008). Equation (13.1) is empirical and ignores yield stresses, natural variations in bedrock slope, and assumes equilibrium conditions, but it still illustrates a first-order range of behaviour for flows entering water (Fig. 13.6).

For the simplest case where there is no change in the underlying slope across the waterline, lava thicknesses below water should increase for any range of vesicularities (bold curves in Fig. 13.6). The increase in lava thickness is due to buoyancy because of the reduced density contrast between lava in water compared with air. However, the flow fronts will also tend to inflate because the delivery of magma is often faster than can be distributed by flow advance or spreading, probably due to increased effective viscosity caused by the thicker cooled margins underwater. The greatest buoyancy-related thickening is associated with the most highly vesiculated (least dense) lavas. Since 'a'ā lavas degas and densify with distance travelled (Lipman and Banks, 1987), the greatest inflation of subaqueous lava will be associated with deltas proximal to the vent. A consequence of the greater degree of inflation in highly vesicular lavas will be to create steeper subaqueous delta fronts, which might also enhance the likelihood of gravitational collapses, effects that will be less likely for lavas with fewer vesicles. Lavas continuing to flow underwater will also become thinner at greater depths as vesicles are suppressed and they densify (Moore et al., 1985; Mitchell et al., 2008; Smellie et al., 2013a). These predictions apply to lavas and their deltas irrespective of whether they are fed by 'a'ā or pāhoehoe, although most pāhoehoe flow lobes (thinner than their 'a'ā equivalents) will disintegrate at the delta front.

Figure 13.6 also investigates what happens to lava thickness when there is a change in slope across the waterline, as happens in lava-fed deltas. In Fig. 13.6a, lava flowing on a subaerial slope of 10° passes across the waterline onto steeper subaqueous delta front slopes of up to 25°. In all examples, the lava becomes thicker but the thickening only affects the most vesicular (least dense) lavas for the steepest subaqueous slope (25°) depicted. Since many volcanoes have slopes ≥ 10°, this result probably applies only to those glaciovolcanic eruptions in which a relatively thin ice cover simply *drapes* the volcano flanks, rather than drowning it (see Fig. 13.4). Conversely, for glaciovolcanic shields, with slopes of c. 5°, most lava flows will become thinner below the waterline (Fig. 13.6b). Neither of these scenarios is realistic for non-glacial eruptions since a lava delta will rapidly

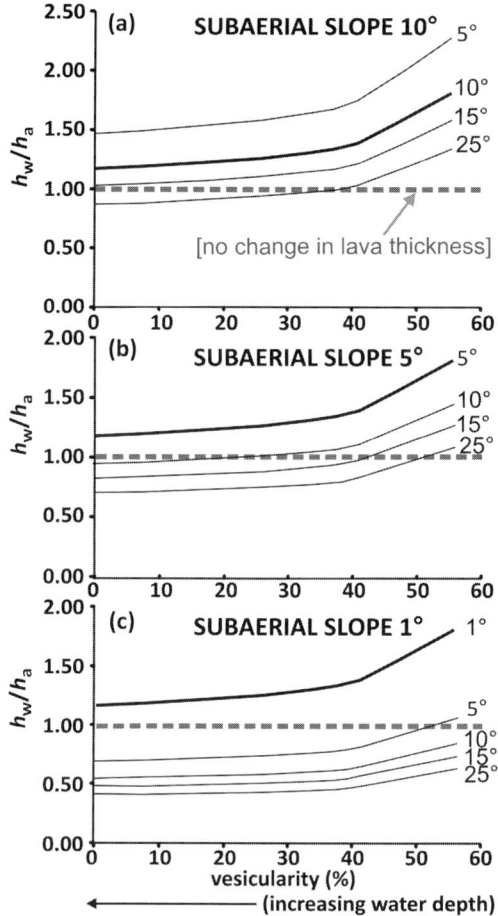

Fig. 13.6 Diagrams illustrating effects of buoyancy on lava thickness for various subaerial and subaqueous slopes as the lava passes below a waterline. Bold lines are for examples with no slope change above and below water. See text for further description.

form a subhorizontal subaerial upper surface as it advances into the (horizontal) sea or lake, nor are they applicable to glaciovolcanic settings in which the landscape is drowned by a very thick ice sheet, which will also have an almost flat surface. These types of eruptions are investigated in Fig. 13.6c, in which a subaerial slope of just 1° is assumed. The subaqueous lava becomes *thinner* in every case except those deltas with very low angle (≪5°) subaqueous slopes when the lavas become slightly thicker. This result implies that what happens to lava thicknesses as the lavas pass below a waterline in lava-fed deltas might be a useful environmental discriminant; i.e. subaqueous lavas in glaciovolcanic deltas will generally become inflated compared with their subaerial feeders, whereas in non-glaciovolcanic lava-fed deltas and glaciovolcanic deltas affected by a much thicker (topography-drowning) ice sheet, the subaqueous lavas will generally become thinner. However, the additional ingestion and expansion of water (as vesicles) trapped by overriding lava in any subaqueous setting may mask these environmental-specific differences.

13.4.4 *Authigenic mineral compositions and eruptive palaeoenvironments*

The compositions of authigenic minerals formed in the subaqueous lithofacies during the immediately post-eruptive period have been investigated by several authors (e.g. Furnes, 1978; Staudigel and Hart, 1983; Jercinovic, 1988; Ellerman, 1992; Stroncik and Schmincke, 2001, 2002; Johnson and Smellie, 2007; Antibus et al., 2014; see Chapter 7). Despite these studies, using authigenic minerals as palaeoenvironmental proxies has met with limited success except for the study by Johnson and Smellie (2007). The principal aim is to find a geochemical signature that can be used to distinguish between authigenic palagonite and zeolites generated by low-temperature (<100 °C) marine and freshwater alteration of basaltic glass. It is an attractive idea since obtaining rock samples is seldom very difficult and it raises the possibility that simply analysing the samples to determine the eruptive environment might supersede the current reliance on recognising a broad range of diagnostic field criteria (above) that may be absent, hard to deduce or ambiguous in poorly exposed or heavily eroded terrains. Moreover, lithofacies formed during submarine and subglacial eruptions can look very similar and a clear compositional difference could be crucial in their distinction especially when fossils are absent. Palagonite, phillipsite and chabazite are also formed in both seawater and freshwater (meltwater) so they are potentially a good choice for environmental discriminant studies (e.g. Smellie, 2008, Fig. 10).

Sample selection is important for all authigenic mineral studies. Only samples with relict fresh glass and filled pore spaces should be chosen. This is intended to ensure, so far as possible, that the system was closed at least locally and thus more likely to retain an original pore-water signature by minimising any influence by subsequent meteoric groundwater or marine incursions (Johnson and Smellie, 2007; Fig. 13.7). In addition, whole-rock lava compositions should be broadly similar to avoid variations between datasets from different lava units caused by magmatic variations.

However, despite a history of investigation going back several decades, the nature of elemental mobility during palagonitisation is still elusive and unquantified with respect to eruptive environment (Pauly et al., 2011; see Chapter 7). The published information on compositional trends is uncertain and not diagnostic, and there is an important dependence of palagonite composition on original sideromelane composition. It is also unclear (and has not been tested rigorously) whether palagonite is stable after formation or whether it will readily change over time and in the presence of later fluids (i.e. during burial and diagenetic alteration unrelated to eruptive setting). Thus, the composition of palagonite currently does not appear to be a promising proxy for identifying eruptive palaeoenvironments.

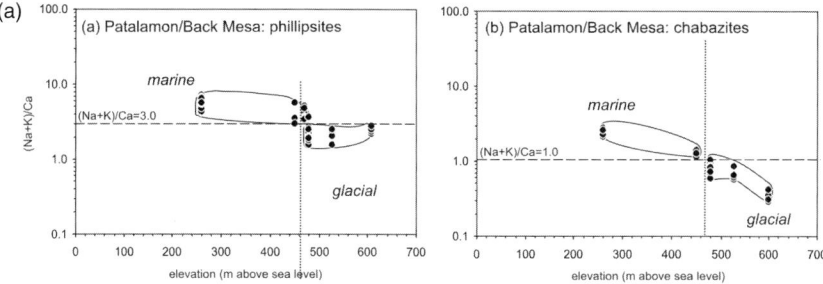

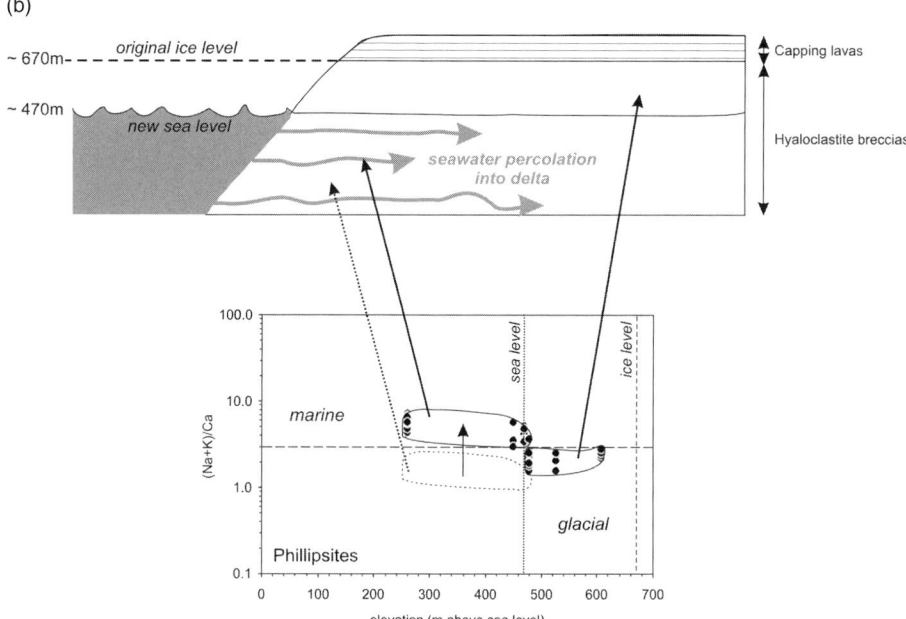

Fig. 13.7 (a) Diagram showing sample elevation (metres above present-day sea level) versus (Na + K)/ Ca for phillipsites and chabazites in a single lava-fed delta on James Ross Island. The samples from lower elevations show compositions of both zeolite species that are characteristic of marine alteration whereas compositions of the higher-elevation samples are more akin to freshwater alteration. (b) Sketches illustrating a possible explanation for the varied environmental conditions implied by the contrasting zeolite compositions shown in (a). It is inferred that the delta was initially emplaced in a glacial setting, when the zeolite compositions at all elevations would have reflected freshwater (glacial) alteration (pecked line). The lower part of the delta was then affected by a marine incursion, probably during an interglacial period, and the affected zeolites took on a marine compositional signature. (Adapted from Johnson and Smellie, 2007.)

Microbial activity has also been implicated in palagonitisation, and hence may have a use in environmental discrimination, but its effect remains uncertain (Thorseth et al., 1992; Alt and Mata, 2000; Stroncik and Schmincke, 2001; Furnes et al., 2002). Microbial activity is revealed by the presence of texturally distinctive endolithic microborings around the margins of clasts. They are regarded as a unique form of trace fossils, for which a putative ichnotaxonomy has been proposed (McLoughlin et al., 2009). Such biotic alteration of basalt glass is significantly different texturally from that formed abiotically (see below; Fig. 13.9). Although microbial traces are well known in submarine-altered basalts, they are rarely present in subglacially erupted basalts that were subsequently affected by much later marine incursions. It has thus been suggested that the microbial activity may be predicated by marine conditions (Cousins et al., 2009). The impact of biotic alteration and its use as a potential environmental discriminant is not well established although at least it is texturally distinctive and thus readily recognised.

Zeolites

Greater success has been achieved in deducing eruptive palaeoenvironments using the compositions of authigenic zeolites. In a study by Johnson and Smellie (2007), phillipsite and chabazite in basaltic hyaloclastite samples obtained from six different lava-fed deltas on James Ross Island (Antarctica) with known eruptive environments (marine and glacial) were analysed for major oxides by electron microprobe. The results were reported as ratios of anhydrous cations. They found that phillipsites from known freshwater settings have generally lower Si/Al (1.65–2.17) and lower (Na + K)/Ca (0.57–1.96) than phillipsites from known marine alteration environments. Fewer chabazite data were obtained, reflecting the less common occurrence of chabazite in their samples, but the results suggested that Si/Al ratios (1.86–2.30) were similar for samples from both freshwater and marine environments, whereas (Na + K)/Ca ratios are statistically different but have a broad overlap (0.17–1.91 (freshwater) and 0.55–24.06 (marine)). (Na + K)/Ca ratios of 3.0 and 1.0 for were suggested for phillipsite and chabazite, respectively, to discriminate between marine and freshwater environments (Fig. 13.8). The two ratios were chosen visually as best-fit

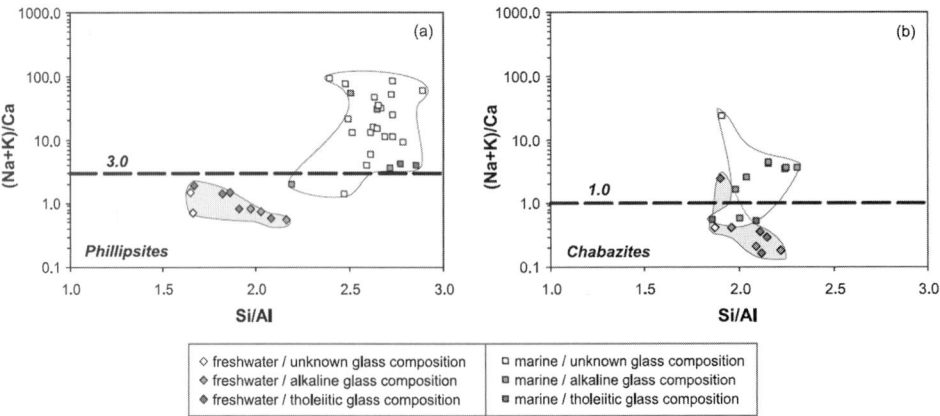

Fig. 13.8 Si/Al versus (Na + K)/Ca for (a) phillipsites and (b) chabazites from known alteration environments and a variety of original glass compositions (from Johnson and Smellie, 2007). The bold dashed lines represent the best-fit discriminant values for distinguishing between marine- and freshwater-altered samples. In reality, the discrimination between the two fields is likely to be gradational, as acknowledged by Johnson and Smellie (2007).

discriminants but Johnson and Smellie (2007) suggested that, in reality, the two environmental fields are likely to be gradational. The occurrence of marine zeolites with higher (Na + K)/Ca ratios than in freshwater-altered samples is unsurprising since seawater contains much more Na than K or Ca. Other combinations of major oxide indices yielded large overlapping datasets that are of no apparent practical value in discriminating alteration environment. Conversely, a study of authigenic zeolites in glaciovolcanic 'a'ā lava-fed deltas by Antibus et al. (2014) revealed variable but high ('marine') (Na + K)/Ca ratios that were explained as a consequence of alteration in a closed (rather than open) hydrological system. However, since both studies were of zeolites in subaqueous breccias in lava-fed deltas, intuitively the alteration conditions ought to have been broadly similar, so the reasons for the contrasting results are not well established.

An unexpected result of the original study was that samples from individual lava-fed deltas on James Ross Island showed ambiguous and often conflicting compositional affinities despite being formed in well-known environmental settings (glacial and marine). The problem was resolved when samples from each lava-fed delta were plotted against height, when a clear separation into 'marine' and 'freshwater' values became evident, with freshwater values (i.e. representing glacial emplacement consistent with the environmental information from the lithofacies) restricted to higher elevations indicating that the delta underwent a much later episode of alteration involving infiltration by seawater (Fig. 13.7). It is clear that, despite the rigorous sample selection following the criteria outlined earlier, the zeolite compositions were not fully stable and were affected by much younger water penetrating several kilometres into the volcanic piles. Thus, breccia piles probably never become completely sealed to fluids, despite authigenic minerals crystallising in the pore spaces and reducing the permeability. In the James Ross Island examples, the results indicate that the glacially emplaced lava-fed deltas were affected by younger seawater incursions, with the seawater probably migrating along fractures and grain boundaries. Although later published studies using these discriminant values are few (Smellie, 2008; Nawrocki et al., 2011), they appear to verify the results of Johnson and Smellie (2007), although further studies are needed. In addition, an independent investigation into the incidence of endolithic microborings in the James Ross Island lava-fed deltas caused by microbes (Fig. 13.9) revealed that endolithic microborings are distinctly more abundant in samples that had undergone marine alteration, either during primary marine emplacement of a delta or its later permeation by seawater. This is consistent with infiltration of breccia initially emplaced and altered in freshwater resulting in few or no microborings, and 'contamination' by later seawater resulting in an increased incidence of microborings (Cousins et al., 2009). The study by Johnson and Smellie (2007) emphasises how the results of compositional investigations need to be treated with great care. Ideally, multiple samples should be taken from the units being investigated and their relative elevations recorded.

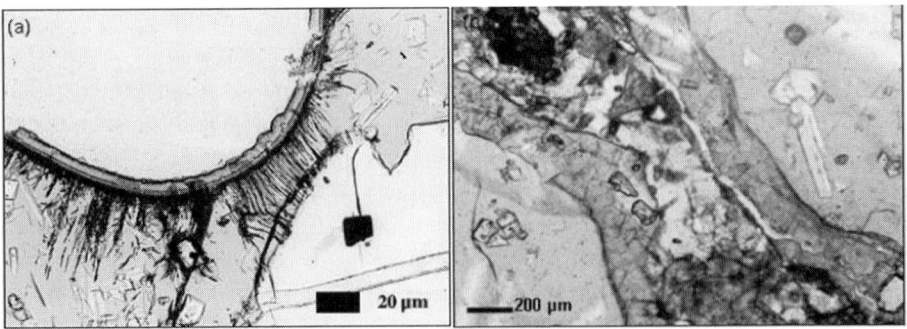

Fig. 13.9 Photomicrographs showing the textural differences between (a) biotic and (b) abiotic alteration of mafic sideromelane (from Cousins et al., 2009).

13.5 Calculating ice thicknesses and surface elevation
from glaciovolcanic sequences

Ice thickness and surface elevation are some of the most valuable and unique criteria that can be obtained by interpreting glaciovolcanic sequences. Non-volcanic geomorphological indicators of past ice thickness are often relatively subtle and frequently impersistent. For example, trim lines are a clear, usually paler line on the side of a valley formed by a glacier representing the most recent highest extent of that glacier. They are associated with glacier decay but they typically do not survive ice overriding during successive glaciations and cannot easily be dated. By contrast, ice thicknesses can be quite precisely measured using unambiguous features in glaciovolcanic successions. The calculations rely on identifying robustly the subaqueous–subaerial transitions described above. The transitions record the contemporaneous water level coeval with an eruption. That corresponds approximately to the thickness of impermeable unfractured ice.

There are two principal methods for calculating ice thicknesses: (1) measuring the concentration of dissolved magmatic volatiles (especially H_2O, CO_2) in volcanic glasses and applying their pressure-dependent solubilities to estimate palaeo-pressures at the eruption site (Dixon et al., 2002; Tuffen et al., 2010; see Chapters 5–7); and (2) measuring the thickness of water-chilled lithofacies in individual monogenetic (i.e. single-episode) eruption sequences.

13.5.1 Calculating ice thicknesses from dissolved volatile
contents in lavas

The quenched glassy rinds found on lavas erupted subglacially preserve information about eruptive volatile contents and the quantification of magmatic water, CO_2 and other volatile concentrations (e.g. S, Cl) can be used to deduce coeval ice thicknesses (Dixon et al., 2002; see Sections 5.2.3 and 7.4). The dissolved volatile contents are particularly useful for glaciovolcanic successions that lack a subaerial caprock and for which the measured

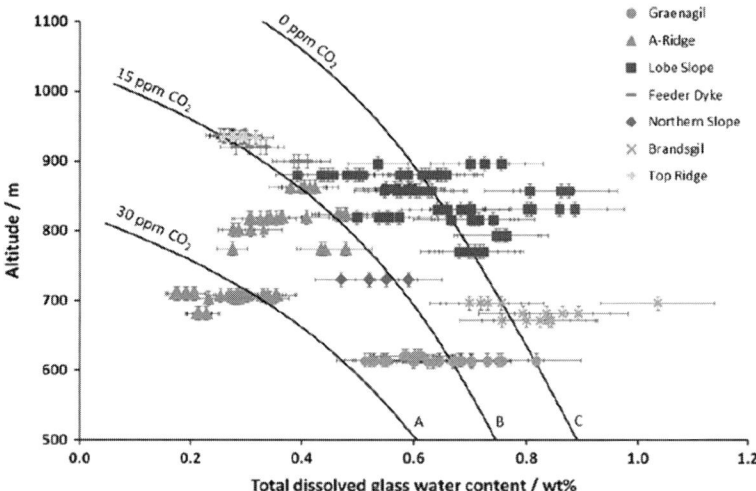

Fig. 13.10 Diagram showing total dissolved H_2O in glass samples from Bláhnúkur, Iceland, as a function of sample elevation (from Owen et al., 2012). The curved lines are solubility–pressure curves (SPC) marking the expected water contents for a coeval ice surface elevation (hence ice thickness) fixed at 1200 m and eruptive temperature of 800 °C. SPCs A, B and C show the effect of 30, 15 and 0 ppm CO_2, respectively. Note the wide spread of data points for samples from each locality, which makes estimating original eruptive pressures (and palaeo-ice thicknesses) very uncertain (see also Fig. 7.3).

lithofacies thicknesses can only yield a minimum thickness estimate for coeval ice (e.g. Walker and Blake, 1966; Dixon et al., 2002; Tuffen et al., 2001). The methodology determines dissolved water and CO_2 concentrations in glasses to calculate vapour saturation pressures and the composition of the vapour phase in equilibrium with the melt at that pressure. However, the errors in the ice thicknesses calculated by this method are typically greater than using the a-priori lithofacies method (reviewed by Tuffen et al., 2010; also Owen et al., 2012; Fig. 13.10; see Section 7.4). Despite this, the method is still valuable and, in some cases, may be the best or only method available for deriving ice thicknesses. Indeed, the method has been widely and enthusiastically adopted by many workers and papers that contain volatile-based calculations of former ice thickness include Dixon et al. (2002), Schopka et al. (2006), Höskuldsson et al. (2006), Tuffen et al. (2008), Edwards et al. (2009b), Stevenson et al. (2009), Tuffen and Castro (2009), Owen et al. (2012) and Hungerford et al. (2014), for glaciovolcanic centres in British Columbia and in Iceland.

Reconstructed ice thicknesses also rely on an assumption that the palaeo-ice surface was essentially horizontal. This important assumption breaks down if an ice cauldron develops (as was well documented during the 1996 Gjálp eruption (mafic); Gudmundsson et al., 1997, 2002, 2004). A similar uncertainty related to cauldron development affects ice thicknesses deduced from lithofacies transitions, but to a lesser degree (see Section 13.5.2).

In summary, the method of using dissolved volatiles to deduce palaeo-ice thicknesses has undergone extensive and rapid development and is still the best method to use for glaciovolcanic edifices lacking a subaerial section and for which the lithofacies can only yield a minimum estimate. However, it is bedevilled by the very large number of uncertainties that have to be incorporated into the calculations and subsequently justified (see Owen et al., 2012, Table 5), with decisions often made unavoidably on a very weak basis. Moreover, the lack of precise CO_2 determinations means that the calculated ice thickness estimates are also extremely broad. Thus, studies using dissolved volatiles to determine eruption emplacement pressures have resulted in very crude estimates (i.e. a wide range of potential pressures). This is a severe limitation on their practical use in palaeoenvironmental investigations. However, one study using volatile contents, by Banik et al. (2014), determined that subglacial lavas in southern Iceland were effectively degassed and erupted under relatively thin ice. Their data provided a strong refutation of a previously published lithofacies-based model for emplacement of the lavas that had argued, on a-priori grounds, for eruption of un degassed lava beneath very thick ice (c. 1000 m; Smellie 2008; see Section 10.6.4).

13.5.2 Calculating ice thicknesses from measuring the thickness of water-chilled lithofacies

Unlike the dissolved-volatiles method, the lithofacies method uses simple measurements of rock unit thicknesses made in the field. Far fewer assumptions are required but a familiarity with lithofacies is needed. As long as the eruption penetrated the overlying ice, estimates of ice thickness can be deduced with considerable confidence (Smellie and Skilling, 1994; Smellie, et al., 2011b). The following explanation (and Section 13.5.3) is adapted from Smellie et al. (2011b, Supplementary Information).

Ice thicknesses are calculated by measuring the thickness of subaqueous volcanogenic lithofacies. That thickness is determined by the elevation of the transition from subaqueous to subaerial conditions. In lava-fed deltas, the transition corresponds to the passage zone, which is a conspicuous feature unlikely to be confused with any other structure (Jones, 1969b; Smellie, 2000, 2006; Skilling, 2002; Smellie et al., 2013a). Passage zones mark the position of the surface of the coeval meltwater lake created by melting the surrounding ice during the eruption. Relating the thickness of subaqueous lithofacies to ice thickness has the following rationale: Since unfractured glacial ice is effectively impermeable on the time-scale of most volcanic eruptions (Smellie, 2009), it is therefore able to impound the resulting meltwater; the meltwater then interacts with the (hot) volcanic rocks, causing rapid chilling of lava and the creation of new lithofacies, particularly hyaloclastite breccia formed *in situ* and lava pillows; because the maximum thickness of subaqueous lithofacies formed cannot exceed the thickness of the surrounding impermeable ice, the thickness of subaqueous lithofacies is therefore a proxy for the former (impermeable) ice thickness. Although it is a crude measure, it can be refined (Section 13.5.3).

13.5.3 *Refining ice thickness calculations using glaciovolcanic sequences*

In most cases, the maximum elevation that a passage zone can attain does *not* correspond precisely to the surrounding glacier surface. Away from the lower-elevation ablation zone, ice sheets and glaciers are layered, with a basal dense impermeable ice mass overlain by a permeable layer composed of any fractured ice (e.g. crevasses), firn and snow (Fig. 4.1). Within the ablation zone, snow and firn are absent (melted away) due to normal climatic effects (ablation rates exceed those of accumulation) and ice is present right up to the surface apart from a thin transient surface layer of winter snow. In polar ice, depths of snow and firn can exceed 120 m although thicknesses of 70 m or so are more common (Paterson, 1994, Table 2.2). Another 30 m, or even more, of fractured ice might be added (ice fractures (crevasses) typically pinch out at depths of 20–30 m; Glen, 1954). Thus, simply measuring the thickness of subaqueous lithofacies is only a measure of the minimum ice (*sensu stricto*) thickness, and does not take into account the permeable layers above. For polar ice situations, therefore, as much as 150 m or more may need to be added to more accurately reflect the true surface elevation of the ice sheet. The problem is less important for temperate ice since the firn–ice transition depth is much less. It is commonly only a few tens of metres and may be closer to 10 m in some cases. Therefore, together with the presence of a fractured layer (≤30 m), the relevant thickness measured from the subaqueous lithofacies present may require the addition of a few to several tens of metres for sequences formed by eruptions in temperate ice. An empirical thickness of 100 m has been used for the thickness of a former permeable capping layer of snow, firn and fractured ice for glacio-volcanic sequences erupted under both wet-based (Antarctic Peninsula) and cold-based (northern Victoria Land) ice (Smellie et al., 2008, 2009, 2011b). It is a relatively large thickness probably most suitable for eruptions under cold-based ice, as occurred in northern Victoria Land, in which the thicknesses of firn and snow are potentially greatest. For the studies of glaciovolcanic sequences in the Antarctic Peninsula, which were erupted within wet-based ice, it is also probably appropriate since the thermal regime of the coeval ice was probably sub-polar (i.e. ice that was below the pressure melting point except in a thin basal layer, hence principally cold stiff ice). However, glaciovolcanic sequences erupted in association with temperate ice, such as for most Icelandic sequences, should probably use a much thinner estimate, perhaps a maximum of 50 m. For eruptions within the much greater area of a glacier corresponding to the accumulation zone, the meltwater surface in an eruption-related englacial cavity will rise until it reaches the lowest elevation of the permeable surface layer, at which point it will overflow (Smellie, 2001, 2006). However, as has been demonstrated, meltwater hydraulics in a glaciovolcanic vault are a dynamic balance between basal escape and supraglacial overflow at a spillway (Smellie, 2006). Thus, only the maximum elevation reached reflects the coeval ice thickness. All lesser thicknesses simply record periods when basal meltwater escape was dominant and over-flowing was not possible.

Erosion caused by escaping supraglacial meltwater can also potentially affect ice thickness estimates. It has been observed in non-volcanic situations in which ice dams are breached (Xia and Woo, 1992; Gore and Pickard, 1998; Raymond and Nolan, 2002). The escaping meltwater either carves a tunnel during infiltration along the firn–ice transition before downcutting and faster flow as water escapes the system, thus creating a distinctive 'keyhole'-shaped channel (Fig. 4.2), or else it will simply carve a prominent ice gorge at spillways where any firn or snow is very thin (see Chapter 4 for further discussion). The maximum elevation of meltwater that can be attained in an overflowing englacial lake is determined by these supraglacial spillways (Smellie, 2006). The measured thicknesses of water-chilled lithofacies might thus not represent true thicknesses of the enclosing ice if the ice surface at the spillway has been downcut by overflowing meltwater. With time, spillway elevations diminish due to thermal erosion by the escaping volcanically heated meltwater. The meltwater will also carve a supraglacial channel ahead of an advancing delta. The net effect is to lower the apparent thickness of the coeval ice, as reflected by the thickness of subaqueous lithofacies formed. The thicknesses of breccia formed at an early stage in lava-fed delta progradation, i.e. closer to source with less time elapsed for down-cutting, might therefore more closely reflect the true thickness of the ice at that location than later-formed more distal delta sections due to the progressive down-cutting by the warm meltwater. However, rates of surface downcutting are likely to be much slower in cold ice compared with warmer (i.e. temperate) ice due to the requirement to raise the temperature of cold ice to its melting point; temperate ice is already at its melting point except possibly in a thin surface layer (Paterson, 1994; Smellie, 2009). As the overflowing meltwater migrates downslope, it will become cooler and the rate of downcutting will diminish as sensible heat is progressively lost and the water ultimately refreezes. Thus the depth of the channel cut by the meltwater and its overall gradient measured from the spillway will also diminish outward as will the flow velocity, and the system may ultimately reach an approximately steady state with a channel base situated not far below the elevation recorded by the advancing delta passage zone. Therefore, although measured thicknesses of breccia foresets in a delta can be somewhat less than the thickness of the coeval impermeable ice, the differences are probably small in such a self-regulating system. These effects will be more pronounced for more steeply sloping ice sheet surfaces, since the meltwater will escape continuously as the lava-fed delta advances downslope and thus it will continue to erode, whereas erosive effects are probably quite small for gently sloping to subhorizontal ice sheets since the water will quickly 'back up' in the absence of a significant ice surface gradient.

An additional influence on meltwater lake elevation, alluded to in the Section 13.5.2, is ice sheet surface sagging towards the glaciovolcanic vault. During an eruption, surface sagging has been observed amounting to 100 m or more in areas 2–3 km wide on either side of the erupting vent (Gudmundsson et al., 2002, 2004). Intuitively, this might also lower the apparent ice thickness based on measured thickness of the subaqueous lithofacies formed. This will be true for eruptions in a sloping ice sheet since the water will continue to escape from the system, as was documented for the 1996 Gjálp eruption in Iceland (Fig. 13.11; Gudmundsson et al., 2002). By contrast, for eruptions in a subhorizontal ice sheet, sagging will normally not alter the elevation of a meltwater

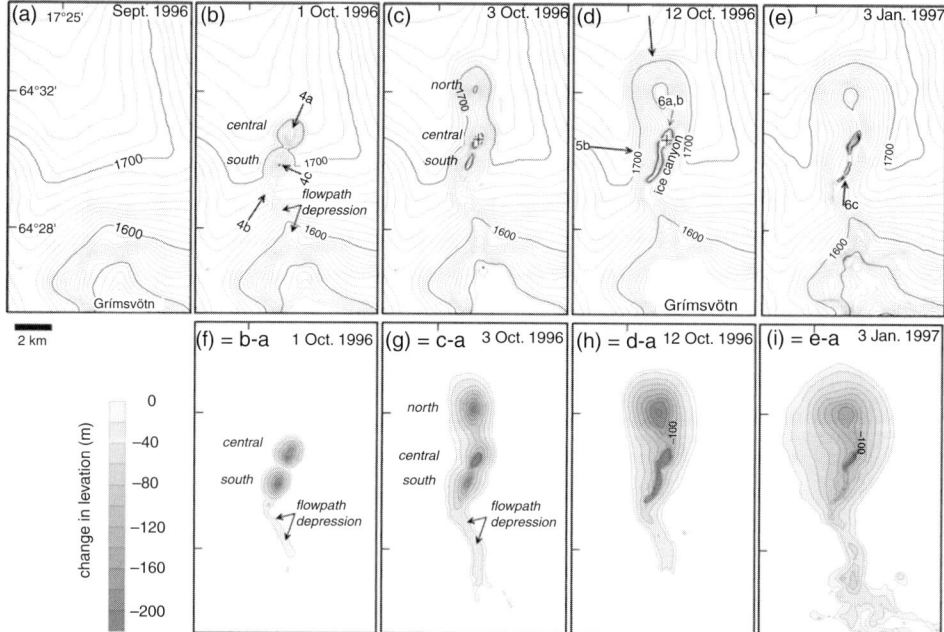

Fig. 13.11 Maps showing the development of sagging in the ice surface surrounding the 1996 eruption of Gjálp, Vatnajökull, Iceland (from Gudmundsson et al., 2004). Contour intervals are 10 m (a)–(e) and 20 m (f)–(i). The location of the subaerial crater (when the tuff cone emerged from the meltwater pond) is indicated by a cross in (c) and (d). The maximum amount of subsidence achieved during the eruption was c. 140–160 m.

lake significantly since the thickness of impermeable ice is unaffected outside of the zone affected by sagging and the meltwater will simply be ponded at the elevation that existed prior to the eruption (Fig. 13.12b). Thus, simply measuring the thickness of subaqueous lithofacies in lava-fed deltas is a relatively crude proxy for ice sheet thickness unless it is used carefully.

For glaciovolcanic sequences that do not form in sustained lakes, sagging can have an important influence on estimates of ice sheet thicknesses. In eruptions producing many mafic and felsic sheet-like sequences, meltwater escapes continuously along a sloping bedrock surface because the meltwater is never actually ponded. This is because, for these sequence types, the surrounding 'ice' is wet-based (thus 'leaky') and has a high proportion of snow, firn and/or fractured ice relative to impermeable ice. Because of the different ice sheet structure and mode of emplacement of these 'thin ice' sequences (i.e. sheet-like sequences; see Section 10.6) compared with deltas, the elevation of any subaqueous–subaerial transition is determined simply by the thickness of the glacial cover above the volcanic rocks, which needs to be fully melted through for the uppermost volcanic lithofacies to reflect dry conditions (Fig. 13.12a). However, not all sheet-like sequences melt completely through their glacial cover.

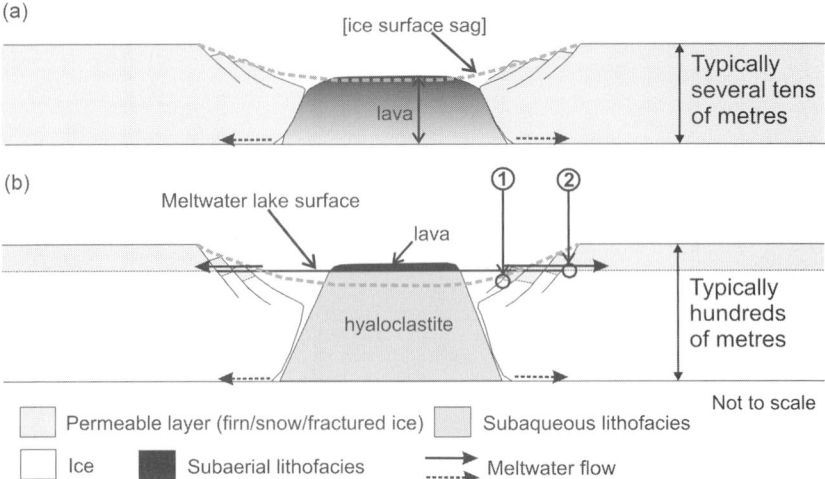

Fig. 13.12 Sketches illustrating the effects of ice surface sagging on the thickness of subaqueous lithofacies developed during (a) glaciovolcanic sheet-like sequences and (b) lava-fed deltas (from Smellie et al., 2011b). The situations depicted are empirical and assume a horizontal bedrock surface. In such situations, there is no impact on subaqueous lithofacies thickness for lava-fed deltas but it is potentially more important for sheet-like sequences (although probably limited mainly to a few tens of metres).

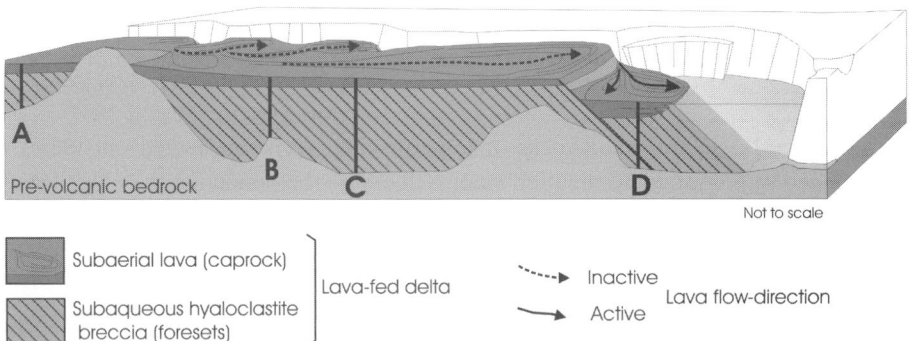

Fig. 13.13 Cartoon cross section through a glacially emplaced lava-fed delta showing how delta thicknesses vary with uneven bedrock topography. For any calculations of coeval ice sheet thickness, the maximum measured thickness of subaqueous lithofacies (section C in the figure) should be used (from Smellie et al., 2008).

It is also important to know whether the measured section of water-chilled lithofacies is the maximum developed thickness for a selected sequence. An uneven bedrock morphology can significantly influence inferred ice thicknesses (Fig. 13.13). This problem can be intractable for sequences only exposed at a single locality (and therefore yield just a single thickness estimate that has to be assumed is representative) and in strike sections (since a draping ice sheet can thicken towards lower elevations so a thickness estimate based on a strike section is only appropriate at that elevation). Thus, ice sheet estimates based on these sections should be regarded as minima. To reduce these effects, maximum measured thicknesses from multiple widely distributed exposures of water-chilled lithofacies should be used in calculating ice sheet thicknesses whenever possible.

In summary, quantifying the effects of ice sheet sag and estimating thicknesses of a permeable glacial capping layer are the least precisely defined variables in ice thickness calculations based on lithofacies. However, the effect of those errors reduces rapidly for sequences erupted under substantial ice thicknesses (hundreds of metres). Since the capping permeable layer has a fixed limit relative to its thickness, its thickness is relatively less important for eruptions beneath thicker ice (Smellie, 2009). Despite these errors, estimating palaeo-ice thicknesses from glaciovolcanic sequences is the only method currently available for which ice thicknesses can be routinely derived and the estimates are remarkably accurate. The principal caveat is that it is only effective for those glaciovolcanic sequences that have subaerial capping lithofacies.

13.6 Effect of volcanic heat on basal thermal regime of ice

Ice sheets exist in different glacial states and are classified empirically by their basal thermal regime (see Section 4.3). The basal regime is a complicated function of several variables, including precipitation, ice thickness, ambient temperature and geothermal heat flux, which vary spatially and temporally. All these factors are climate-related apart from geothermal heat flux. The basal regime determines the relative stability of an ice mass. As ice sheets are effectively pinned by areas of frozen bed and unstable behaviour is initiated by thawed-bed areas, determining the basal regime of ice is thus of primary importance and forms a major part of glaciovolcanic palaeoenvironmental investigations.

Volcanoes give off heat. Volcanic heat is simply a form of geothermal heat, albeit often enhanced compared to regional geothermal heat and it is especially raised around the vents during eruptions. Since the basal thermal regime is strongly influenced by geothermal heat, it is important to question whether the basal thermal regime of palaeo-ice determined from glaciovolcanic sequences will truly reflect that determined by climate or whether it becomes overprinted by the local volcano-related heat flow. The problem probably does not affect small monogenetic glaciovolcanic centres as the timescale for their eruptions is small and they will accurately preserve a snapshot of the local climate-related basal regime. It is, however, potentially important for the long-lived, much larger polygenetic volcanoes with their long-lasting cooling magma storage zones that increase local geothermal gradients over longer timescales.

Heat flow or heat flux (measured in $W\,m^{-2}$) in a volcano relies on several factors, including: the regional heat flow through the crust (related to the presence and abundance of radioactive elements); heat from the cooling magma chamber; hydrothermal activity; and the permeability of the volcanic rocks. The highest temperature geothermal fields on a volcano occur in hydrothermal systems above heat sources corresponding to hot intrusions at depth associated with active volcanism. The heat is transferred via fracture zones (i.e. they are fracture-related groundwater upflow zones).

Fassett and Head (2006, 2007) used a transient conductive heat flow model to examine the magnitude and pattern of heat flow at the bedrock surface below 'snowpack' (i.e. snow or ice) resulting from cooling of a magmatic intrusion (magma chamber) beneath two volcanoes on Mars. They found that most of the heat transport is vertical and it is concentrated near the summit, a relationship that holds for a variety of magma chamber geometries and is enhanced by any convection present (Gulick, 1993, 1998). The results suggest that peak heat flow is sufficient to induce basal melting of the snowpack on both volcanoes for periods of c. 50 ka. Their model showed the distribution pattern of raised bedrock surface temperatures as a broad annular zone of increased diffused heat flow extending roughly a further caldera diameter from the margins of the summit caldera (Fig. 13.14). Since calderas probably closely approximate the diameter of the underlying magma chamber, the results confirm the hypothesis that most of the heat travels vertically, with comparatively limited lateral spreading.

Conversely, Gulick (1993) modelled the heat effects associated with a hydrothermal system and showed that it rises in a narrow zone surrounding the summit (Fig. 13.15). Her focus on groundwater flow (unlike Fassett and Head, 2006, 2007) is probably more realistic for most terrestrial glaciovolcanic situations. Hydrothermally driven groundwater is cycled beneath volcanoes in a vigorous circulatory system that replenishes itself by continually drawing in colder denser groundwater from the surrounding aquifer. Any warm groundwater discharging beneath ice at the surface will flow and cool along equipotential gradients (see Section 4.5.2), re-entering and recharging the system (together with melted overlying snow and ice) along fractures and permeable lava autobreccias. Any large-scale anisotropy in permeability will significantly influence the fluid flow directions in a volcano. A typical glaciovolcanic architecture comprises alternating eruptive units dominated by thick lava-fed deltas, which dominate the volcano flanks and dip radially out away from the summit region. They surround a central vent complex composed of pyroclastic cones that, if glaciovolcanic, correspond to tindars or tuff cones. Lapilli ash-dominated pyroclastic cones are relatively impermeable due to the fine grain size and authigenic mineralisation (Jakobsson and Moore, 1986; Jarosh et al., 2008) and any groundwater flow will be focused in fractures. Heat transfer will mainly be by conduction (as modelled by Fassett and Head, 2006, 2007). Moreover, the coarse breccia foreset beds that dominate each lava-fed delta are also likely to be aquicludes or aquitards (Johnson and Smellie, 2007; see Section 13.4.4). By contrast, the subaerial capping lavas in the lava-fed deltas are excellent aquifers, via pervasive cooling fractures in the massive lava and via highly permeable coarse autobreccias. Together with deep-penetrating caldera margin

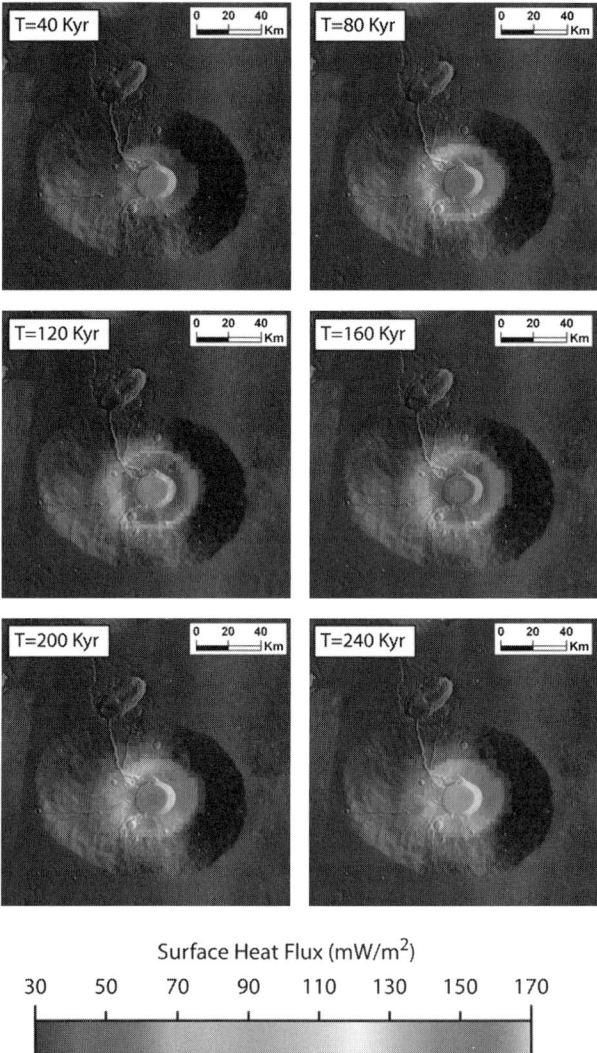

Fig. 13.14 Results of a conductive heating model on heat flow at the surface of a Mars volcano (from Fassett and Head, 2007). Note that the maximum lateral extent of the raised isotherms is about one caldera diameter on either side of the summit caldera and that most of the volcano surface area is unaffected. (A black and white version of this figure will appear in some formats. For the colour version, please refer to the plate section.)

faults, the lavas will form the major stratabound routes preferentially exploited by hydrothermally driven groundwater flow, both on the way up to the surface and then returning as recharge (Flovenz and Saemundsson, 1993). Because of the radially outward-dipping strata, the heat flow becomes preferentially focused on the summit region.

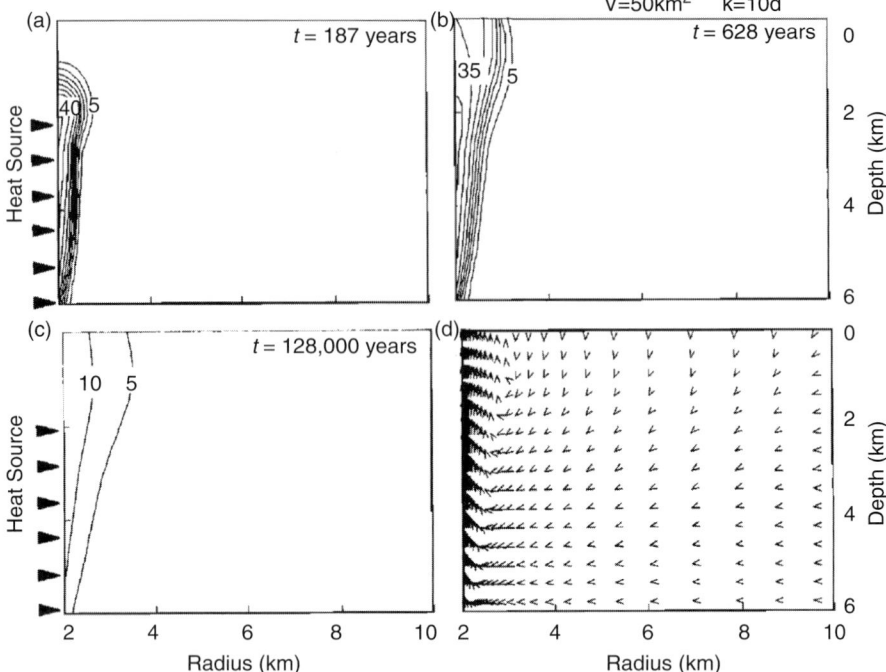

Fig. 13.15 (a)–(c) Temperature contours (5° intervals) for a rising hydrothermal plume at various times after emplacement of a 50 km³ (i.e. 2 km radius) cylindrical magmatic intrusion; permeability in the country rock is assumed to be isotropic (from Gulick, 1998). (d) Groundwater flow field for the system after 628 years. Note the influx of cold waters replenished by the environment. Although this is an empirical system (based on the Mars environment), it illustrates clearly how the heated groundwater is focused in a narrow zone above the heat source, similar to predictions of the model by Fassett and Head (2007; Fig. 13.15). Even greater focusing is predicted for a volcano with outward-dipping beds acting as alternating aquicludes and aquifers.

Gulick (1993, 1998) demonstrated that hydrothermal convection focuses the upflow in narrow zones close to the chamber margins and the area of heated ground at the surface is geographically restricted to approximately one magma chamber diameter beyond the chamber margin, a result similar to that for conduction (cf. Figs. 13.14, 13.15). She showed how any impermeable layers impede and divert the buoyant groundwater flow. Convergence of higher temperatures in the summit area will be imposed by outward-dipping impermeable layers (not modelled by Gulick), an effect that is also reinforced by inflowing cold groundwater, which helps to keep the volcano flanks cool, and summit focusing of heat is enhanced for more shallowly emplaced magma chambers.

Thus, the basal thermal regime of an ice cover on a volcano with a cooling magma chamber is likely to be changed only in a relatively well-defined area restricted principally to the summit. This is shown in modern active glaciovolcanoes by the presence of ice cauldrons formed over the subglacial hydrothermal discharge points, which are situated principally within the calderas (Gudmundsson et al., 2007). Thus, for most practical purposes the basal regime is probably unchanged over much of the edifice outside of a summit zone perhaps as much as three caldera diameters wide. This suggestion is supported by the occurrence of glaciovolcanic sequences erupted in association with cold-based ice widely preserved in several large volcanoes in Antarctica (Wilch and McIntosh, 2002; Smellie et al., 2011a,b).

The preceding discussion relates essentially to mafic glaciovolcanic centres. By comparison, felsic centres are less well-structured internally. Felsic lavas typically are thicker (few tens of metres) and fewer than in mafic lava sequences and they tend to have a very low permeability (see Gulick, 1993, p. 72). They will presumably be aquitards at best. They are generally associated with minor autoclastic hyaloclastite breccia and relatively thin (metres) volcaniclastic sediments (Kelman et al., 2002) or else thick piles of altered phreatomagmatic tephra (lapilli tuff; Tuffen et al., 2002a; Stevenson et al., 2011), again essentially aquitards or aquicludes. Any hydrothermal circulation set up in the volcanic pile will be dominated by localised fracture flow, but much of the heat escape will be by conduction. The impact on heating the volcano surface will probably more closely resemble the conditions modelled by Fasset and Head (2006, 2007).

13.7 Strengths and disadvantages of glaciovolcanic palaeoenvironmental investigations

Using glaciovolcanic sequences as palaeoenvironmental proxies has several important advantages:

1. Glaciovolcanic sequences are ice-contact ultra-proximal terrestrial deposits. They form directly in association with ice and record ice conditions at geographically well-fixed locations. Conversely, sedimentary sequences offshore typically record ice stream deposition and retain characteristics that may be averaged over a large hinterland (drainage-basin scale) in which basal regime may vary (Smellie et al., 2014).
2. Glaciovolcanic centres record inland ice and not ice streams/outlet glaciers. Since ice streams and outlet glaciers occupy just 10% of the current Antarctic Ice Sheet, inland ice corresponds to c. 90% of that ice sheet and the glaciovolcanic record is representative of the greater glacially covered area (Smellie et al., 2011b).
3. The lavas in glaciovolcanic sequences are usually very fresh and can be easily dated isotopically. In an absence of interbedded tephras, marine sedimentary sequences are typically relatively coarsely dated by biostratigraphical methods; terrestrial diamicts (tills) are notoriously difficult to date.
4. The volcanic sequences are relatively thick and are often covered by erosion-resistant lavas or rapidly lithified tuffs and breccias. They are therefore typically robust to removal and usually survive multiple episodes of overriding ice.
5. Each eruption of lava creates a new surface on which ice reforms once the lavas have cooled. Those surfaces are then affected by the ice according to its basal thermal regime

and the temporal evolution of the basal regime can be documented by dating the lavas associated with each eruption (Smellie et al., 2014). This is a unique property of glaciovolcanic sequences.

6. Rock outcrops are often logistically simple to examine directly, much cheaper and involve far fewer persons than obtaining marine records by very expensive offshore drilling.

7. Studies of glaciovolcanism can deliver the widest range of critical parameters of ancient ice of any proxy method, including: presence/absence of ice, the age of that ice, ice thickness, ice surface elevation (mainly in a relative sense but, where a region's uplift history is well known, in an absolute sense too), ice structure and basal thermal regime (Smellie, 2000; Smellie et al., 2009, 2011b). Some of these parameters arc obtainable no other way (ice thickness and surface elevation, temporal evolution of basal regime).

There are also some important disadvantages, including:

1. The presence of volcanic rock outcrops is essential. This is a major reason why glaciovolcanic studies will never be used to investigate the history of the Greenland Ice Sheet.

2. The glaciovolcanic record has a low resolution since eruptions typically occur at long intervals. For example, in the Antarctic Peninsula, eruptions took place at intervals of 100–150 ka (Smellie et al., 2008). By contrast, many marine shelfal sequences have a much higher resolution because sediments are often deposited continuously over long periods and thus can often contain a clear glacial–interglacial cyclicity in the patterns of sedimentation (e.g. Naish et al., 2009).

3. Although it is relatively easy to date glaciovolcanic sequences by the ^{40}Ar/^{39}Ar method, the precision of the ages is generally quite low and typical 2-sigma errors are currently c. 40–60 k.y. (Wilch and McIntosh, 2002; Smellie et al., 2008, 2011b). ^{40}Ar/^{39}Ar ages are also not absolute and it may be impossible to place an ^{40}Ar/^{39}Ar age within a specific glacial cycle owing to instrumental errors and uncertainties over the precise ages of the dating standards used. This means that, although many characteristics of past ice can be deduced, it is often difficult to know exactly to what part of a glacial cycle they relate. However, this problem is slowly being reduced as advances in dating techniques bring errors down to well within glacial cycles, particularly for Quaternary felsic sequences when 2-sigma can reduce to ≪10 k.y (e.g. Guillou et al., 2010; Matchan and Phillips, 2014; McIntosh et al., 2014) and with further improvements in reliability of defining the ages of the standards, which are constantly being revised.

4. It is only in exceptional circumstances that glaciovolcanic sequences can be related to conditions of maximum ice thickness and some other measure for maximum ice thickness will usually need to be sought (Smellie et al., 2008; Johnson et al., 2009). This is because of the loading effect an overlying ice sheet exerts on crustal magma chambers and/or mantle plumes, which may bias eruptions towards periods of rapid ice sheet

decay (e.g. Jull and Mackenzie 1996; Maclennan et al., 2002; see Chapter 14). What is reasonable to infer is that glaciovolcanic sequences will often record average ice conditions for any glacial cycle rather than maximum ice conditions (Smellie et al., 2011b).

13.8 Examples of palaeoenvironmental investigations using glaciovolcanic sequences

13.8.1 Antarctic Peninsula

Glaciovolcanic sequences are common and widespread in the Antarctic Peninsula (Fig. 2.1; see Chapter 2). The outcrops mainly comprise several mafic, monogenetic fields with areas up to 2000 km^2 or more (Smellie, 1999). However, a few large polygenetic volcanoes are also present at the northern end of the Antarctic Peninsula and include James Ross, Brabant and Anvers islands (Smellie et al., 2006c, 2008). The James Ross Island area is the largest and best known volcanic field in the region and is dominated by the large Mt Haddington polygenetic volcano, surrounded by multiple satellite and flank centres. It has an area of 7000 km^2 and a minimum volume of 4500 km^3, of which 3000 km^3 of volcanic rock forms the unexposed root of Mt Haddington (Jordan et al., 2009; Smellie et al., 2013b). The volcanic field has an eruptive history extending back to 12 Ma, at least, but the oldest ages were obtained only on clasts in interbedded glacial deposits (Sykes, 1988; Jonkers et al., 2002; Marenssi et al., 2010) and the oldest *in situ* volcanic outcrops are 6.16 Ma in age (Smellie et al., 2008). The Neogene outcrops on James Ross Island are included in the James Ross Island Volcanic Group, in which 24 volcanic formations and 4 glacial sedimentary formations have been distinguished (Nelson, 1975; Pirrie et al., 1997; Jonkers, 1998; Jonkers et al., 2002; Gaździcki et al., 2004; Nelson et al., 2009; Nývlt et al., 2011; Smellie et al., 2013b). The intimate association between the volcanic and sedimentary rocks is important evidence favouring a glacial setting for the eruptions (Fig. 13.1) but the volcanic sequences themselves, overwhelmingly pāhoehoe lava-fed deltas, show diagnostic criteria mainly for glaciovolcanic eruptions (Fig. 13.3; Hambrey and Smellie, 2006; Hambrey et al., 2008; Smellie, 2006).

About 50 eruptive episodes have been identified in the James Ross Island region, with an eruptive frequency of c. 120 k.y., and the volcanic field is probably still active (Smellie et al., 2008). The environmental record is therefore low-resolution, a characteristic of glaciovolcanic investigations. Almost all of the lava-fed deltas were emplaced within an ice sheet. Conversely, most of the associated tuff cones were marine erupted. Some of the tuff cones are fossiliferous but most are not, and a marine setting is suggested for them based on the lateral continuity of correlated beds, which crop out over distances of several kilometres from their eruptive vent, which are far too large for the centres to have been erupted subglacially (Fig. 13.16). Delta thicknesses (thickness of subaqueously deposited breccia foreset units) were used to deduce coeval ice thicknesses. The results show that for

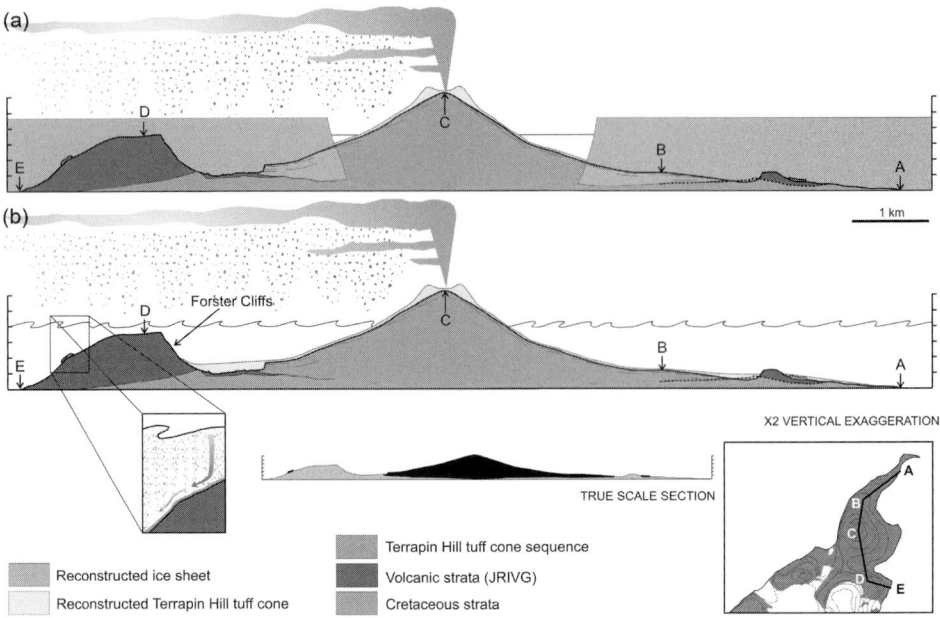

Fig. 13.16 Cartoons illustrating how volcano morphology can help to discriminate between eruptive environments. In (a), an ice sheet is reconstructed around a tuff cone (<660 ka) on James Ross Island, Antarctica. It is unlikely the tuff cone could have melted a hole in any overlying ice wider than the 3 km diameter hole sketched. The ice walls impinge unacceptably on the asymptotic tuff cone flanks. (b) A marine setting is thus favoured, helping to explain distal occurrences of tephra up to 5 km from the vent, some behind a prominent bedrock barrier (D), and the homoclinal dips of the constituent tuff beds, which are parallel to the surface slope all the way to the base of the cone (from Smellie et al., 2006a). (A black and white version of this figure will appear in some formats. For the colour version, please refer to the plate section.)

most of the period between 6.16 Ma and present, the coeval ice was just 200–300 m thick, calculated allowing for c. 100 m of ice surface sag (probably a maximum figure to use). However, the calculated thicknesses also show a progressive increase with time, and statistically greater thicknesses of c. 300–500 m characterise the younger deltas (Fig. 13.17). There is substantial data scatter, however, with a few periods of significantly greater ice thickness, c. 600–750 m. The greatest ice thickness is associated with a satellite tuya known as Dobson Dome, dated as <80 ka; it probably corresponds to ice conditions at, or close to, the Last Glacial Maximum. Other indicators within the sequences suggest that the ice was wet-based, but the absence of evidence for substantial volumes of meltwater associated with the interbedded glacial sedimentary rocks suggest that the ice was probably sub-polar/polythermal (Smellie et al., 2008; Nelson et al., 2009). Recent studies of landscape features and glacial sediments associated with the last glacial suggest that the ice was polythermal, with different basal regimes prevailing in different areas (Reinardy et al., 2009; Davies et al., 2013).

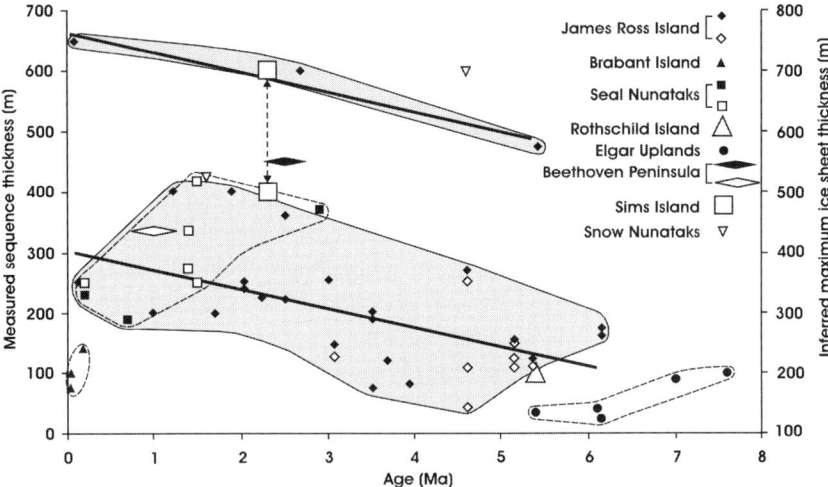

Fig. 13.17 Summary of thicknesses of the Antarctic Peninsula Ice Sheet calculated from glaciovolcanic sequences using lithofacies analysis (from Smellie et al., 2009). This is the largest dataset of
palaeo-ice thicknesses calculated for any ice sheet using any proxy methodology. No other proxy
methodology is capable of calculating pre-late Quaternary ice sheet thicknesses.

The environmental information derived from the James Ross Island Volcanic Group
was combined with comparable information obtained from other glaciovolcanic outcrops throughout the Antarctic Peninsula. Together, they demonstrate that the Antarctic
Peninsula Ice Sheet was never more than c. 750–850 m thick during the past 7.5 m.y.
Moreover, in the James Ross Island region, the evidence was interpreted to suggest that
from its early history the area was covered by a geographically extensive Antarctic
Peninsula Ice Sheet. The growing Mt Haddington volcano was draped by a thin mantling
ice dome that dominated the local ice sheet morphodynamics. In particular, ice flowing
west of Mt Haddington met the regional ice flow from the higher Antarctic Peninsula and
it was diverted north and south in Prince Gustav Channel. Diversion of the Antarctic
Peninsula-based ice was probably aided considerably by the construction of steep
ramparts by the leading edges of successive early lava-fed deltas, and by the subsurface
deformation of the soft Cretaceous bedrock caused by the volcano settling on its
substrate and pushing it out sideways and up (Jordan et al., 2009). The combination of
superimposed thick lava-fed deltas and local Cretaceous bedrock uplift has created
a rampart in Prince Gustav Channel that is over 600 m high today. Finally, despite the
low-resolution record with its time gaps, the lack of evidence for long-lived ice-free
conditions was used to suggest that the Antarctic Peninsula Ice Sheet on land may be
robust to climate warmth at least equivalent to Pliocene conditions, a conclusion that is
supported by a wide array of investigations, including climate modelling, cosmogenic
surface exposure dating and palynological studies (Johnson et al., 2009; Smellie et al.,
2009; Salzmann et al., 2011).

13.8.2 Victoria Land, Antarctica

Volcanic sequences situated on the west flank of the Ross Sea, in northern Victoria Land, are also glaciovolcanic but differ in many characteristics from those in the Antarctic Peninsula (see Chapter 2). The compositional range is broader (mafic to felsic) and they have a wider age range than glaciovolcanic rocks in the Antarctic Peninsula, extending between c. 12 Ma and present. Most ages are between c. 10 and 5 Ma (Late Miocene; Smellie et al., 2011b). The sequences are dominated by lava-fed deltas fed by basalt–mugearite 'a'ā lava rather than pāhoehoe, with rare occurrences of sheet-like sequences (Smellie et al., 2011a). The latter include the first felsic sheet-like sequences to be described. 'A'ā lava-fed deltas are very different in lithofacies and architecture than pāhoehoe deltas. In particular, the structural division into subaerial capping lavas ('topset beds') and subaqueous breccias ('foreset beds'), so prominent in pāhoehoe deltas, is much less well defined (Smellie et al., 2013a). However, they can be used as indicators of past ice in exactly the same way as the much better known pāhoehoe deltas.

The results of the glaciovolcanic investigations in Victoria Land indicate that eruptions took place under ice that was typically a few hundred metres thick (<300 m; Smellie et al., 2011b). For most of the period, however, the ice merely draped the volcanoes and they formed a series of prominent ice domes up to 2 km high along the Ross Sea margin of the Transantarctic Mountains. Unlike in the Antarctic Peninsula, where there is geomorphological evidence for maximum ice thicknesses (Smellie et al., 2008; Johnson et al., 2009), the maximum thickness attained by ice sheets in northern Victoria Land is unknown. However, the presence of a small number of prominent erosional unconformities suggests that, at times, the volcanoes might have been over-ridden by much thicker ice (Smellie et al., 2014). The basal thermal regime was predominantly cold-based but it included episodes of wet-based ice. Extending these studies to Mt Discovery, where glaciovolcanic sequences crop out that are similar in age to those in northern Victoria Land, has demonstrated comparable ice thicknesses along 800 km of the Ross Sea margin of the East Antarctic Ice Sheet. More importantly, the studies have demonstrated for the first time that the East Antarctic Ice Sheet was polythermal overall and probably comprised a geographically and temporally varying coarse temperature mosaic of frozen-bed and thawed-bed patches, similar to the East Antarctic Ice Sheet today (Smellie et al., 2014). Prior to the study by Smellie et al. (2014), the prevailing hypothesis for the evolution of the basal regime suggested a single step-change from wet-based to cold-based. Thus, the results of the glaciovolcanic study indicate that an important shift is required in the prevailing paradigm.

13.8.3 Marie Byrd Land, Antarctica

Lithofacies analysis of three large polygenetic basalt–trachyte volcanoes in eastern Marie Byrd Land, i.e. at Mt Rees and Mt Steere in the Crary Mountains, and Mt Murphy, suggest geographically variable surface elevations and basal thermal regime for the West Antarctic Ice Sheet during the period 9–8 Ma (Wilch and Mcintosh, 2002). Both volcanoes expose basal sections of mainly basaltic glaciovolcanic rocks overlain by thick sequences of subaerial basalt–trachyte lavas. The basal sequences were described as alternating 'wet' and 'dry' volcanic lithofacies with slope-parallel surfaces. LeMasurier (2002) interpreted the sequences at Mt Murphy as lava-fed deltas but did not specify the feeding lava (pāhoehoe or 'a'ā). The evidence, including associated glacial beds, was interpreted to represent products of rapid changes in ice level that varied up to c. 1300 m above present ice level (i.e. a thick topography-drowning ice sheet, although not specified). Conversely, the Mt Steere and Mt Rees sequences, exposed above 2000 m a.s.l., were interpreted as a reflection of the interaction between lavas and a thin draping cover of ice or snow. The sequences in all three volcanoes are reinterpreted by one of us (JLS) as products of 'a'ā lava-fed deltas emplaced in a relatively thin ice cover at both localities (see also Smellie, 2000). The individual deltas at Mt Steere and Mt Rees are complete whereas those at Mt Murphy have locally had the subaerial capping lavas stripped off by glacial erosion, resulting in truncated sequences. The surfaces of the individual sequences at Mt Murphy are also sharp, well-defined and glacially modified and are often associated with glacial diamict (till), whereas such features are entirely absent at Mt Steere and Mt Rees. Thus, the coastal Miocene ice at Mt Murphy was wet-based whereas that 250 km inland and higher at Mt Rees and Mt Steere may have been cold-based (Wilch and McIntosh, 2002). By contrast, three volcanic nunataks located at lower elevations just west of Mt Murphy and with younger ages (6.80–4.70 Ma) are eroded remnants of small polygenetic tuya-like outcrops. In contrast to glaciovolcanic sequences in the main shield, they have multiple horizontal passage zones at elevations of c. 250–600 m a.s.l. that record the fluctuating surface of an ice sheet surface that was up to c. 200 m higher than ice levels today (Fig. 13.18; Smellie, 2001).

Fig. 13.18 Nunatak with a prominent flat top originally interpreted as a tuya (LeMasurier et al., 1994). Despite its flat top, the nunatak is constructed of at least five superimposed lava-fed deltas (numbered labels), each separated by erosional surfaces (white lines). It is an eroded remnant of a polygenetic shield volcano with a main vent situated somewhere to the right of the view (Smellie, unpublished). Hedin Nunatak, Antarctica. The rock face is c. 300 m high.

A factor that complicates interpretations of past ice thicknesses inferred for these localities is that ice surrounding Mt Murphy currently descends from c. 800 m a.s.l. to c. 200 m a.s.l. possibly due in part to a damming effect of the substantial volcanic edifice on ice flow. Wilch and McIntosh (2002) also highlighted additional potential problems caused by the coastal location of Mt Murphy and the unknown impact of raised sea levels, together with the possibility of premature englacial lake drainage due to the proximity of the sea and the unknown effect it may have had on delta passage zone elevation(s). Because of these complicating factors, Wilch and McIntosh (2002) avoided making quantitative interpretations of palaeo-ice levels.

The Miocene–Pliocene (11.4–2.6 Ma) glaciovolcanic record in the Hobbs Coast, Marie Byrd Land, was also investigated by Wilch and McIntosh (2008) and it illustrates some of the difficulties of interpreting monogenetic volcanic centres. The remnants of seven small mainly basaltic centres were described. The outcrops were used to infer past ice surface elevations of the West Antarctic Ice Sheet. However, the Hobbs Coast centres are situated on basement ridges that form uplifted horst blocks separated by troughs representing downfaulted graben. Because the centres occupy positions on interfluves much higher than the glacial troughs, it introduces another topographical complication in the interpretation of past ice thicknesses (cf. Wilch and McIntosh, 2002). The outcrops comprise a lava-fed delta remnant, phreatomagmatic and magmatic pyroclastic cones and subaerial lavas. Despite the volcanic strata overlying glacial diamict and glacially modified basement rock surfaces, there is only limited evidence that the volcanism and glacial setting were contemporaneous. However, the lava-fed delta outcrop was interpreted as glaciovolcanic and may have formed either within ice with a surface slightly higher than present or else that surface has been locally lowered due to downcutting by ice in the adjacent glacier-filled graben. The emergent to subaerial lithofacies in the other outcrops and the lack of evidence for erosion of most by overriding ice were taken to indicate that regional ice levels were never much greater than at present during and following the brief eruptive periods.

13.8.4 Iceland

Despite having the longest historical association with glaciovolcanic studies, Iceland has a remarkable paucity of palaeoenvironmental investigations using the products of glaciovolcanism, in contrast to numerous studies using glacial sedimentary rocks (e.g. Geirsdóttir, 1991; Geirsdóttir and Eiríksson, 1994; Geirsdóttir et al., 2007). Most of the former are investigations of the physical volcanology of small monogenetic centres for which a contemporaneous ice thicknesses is deduced simply to refine the eruptive environment of the volcano being investigated (e.g. Tuffen et al., 2002a; McGarvie et al., 2007), whereas Bourgeois et al. (1998) compiled the summit elevations of multiple Pleistocene tuyas across Iceland to reconstruct the likely surface profile of the Icelandic ice cap at Last Glacial Maximum (after Walker, 1965; also Licciardi et al., 2007). The only palaeoenvironmental study of a long-lived polygenetic glaciovolcanic centre is that by Helgason and Duncan (2001) for the Skaftafell district.

The Skaftafell district occupies an embayment in the southern flank of the Vatnajökull ice cap c. 50 km east of the active Eastern Volcanic Zone in southern Iceland. Outlet glaciers from Vatnajökull have carved the landscape into several prominent ridges. The well-exposed outcrops comprise alternating subaerially erupted lavas and subglacially erupted breccias, glaciofluvial sediments and pillow lava. Together with beds of glacial diamict (till), the sequences provide a detailed window into the glacial–interglacial history of Iceland. The strata were interpreted in terms of at least 16 glacial and interglacial intervals erupted over the past 5 m.y. The sequence is unusually well dated for Iceland, with a range of K–Ar isotopic ages supported by numerous palaeomagnetic determinations. Multiple logged sections from 70 cliff profiles (Helgason, 2007) were used to reconstruct two composite sections. The geological units are divided into numerous formations that can be traced laterally for up to c. 6 km, whereas individual beds are often only traceable for a few hundred metres. The Skaftafell section is composed of 30 geological formations (i.e. eruptive units). It is 2.8 km thick, whereas the Hafrafell section comprises 30 formations and is 1.9 km thick (Fig. 13.19). During glacial periods, the volcanism constructed discontinuous pillow lava ridges, breccias and sedimentary rocks including local tills, whereas interglacials are represented by relatively continuous subaerial lavas and thin volcaniclastic sedimentary interbeds. Erosional unconformities separate the formations but stratigraphical markers are rare. Although eight glacials can potentially be traced between the two composite sections, three other glacials in the Hafrafell section are not represented at Skaftafell, whilst five glacials in the latter have no counterparts at Hafrafell. Some substantial gaps are also present, e.g. 3–2.4 Ma at Hafrafell (but it may be present in the Skaftafell section). These observations illustrate the high level of stratigraphical complexity present in some glaciovolcanic sequences. The oldest glacial period is dated as c. 4.7–4.6 Ma. For comparison, studies of volcanic units similar in age to those in the Skaftafell district were made in Borgarfjordur, western Iceland. Despite an absence of erosional hiatuses, only 7 glacial intervals were recorded compared with possibly up to 14 at Skaftafell, illustrating the considerable value of studying glaciovolcanic sequences in proximal locations relative to their source vents and where they are closer to the centre of the local ice sheet.

13.8.5 British Columbia

The ages of extensive glacial deposits in western Canada are not well constrained but are thought to be generally younger than 3 Ma (Duk-Rodkin et al., 2004), and formed mainly after the onset of global glaciations in the latest Pliocene. British Columbia has been repeatedly covered by multiple incarnations of the Cordilleran Ice Sheet (CIS; Clague and Ward, 2011; Duk-Rodkin and Barendregt, 2011). This is an important region for understanding the dynamics of ice-sheet construction and destruction, as it is now largely

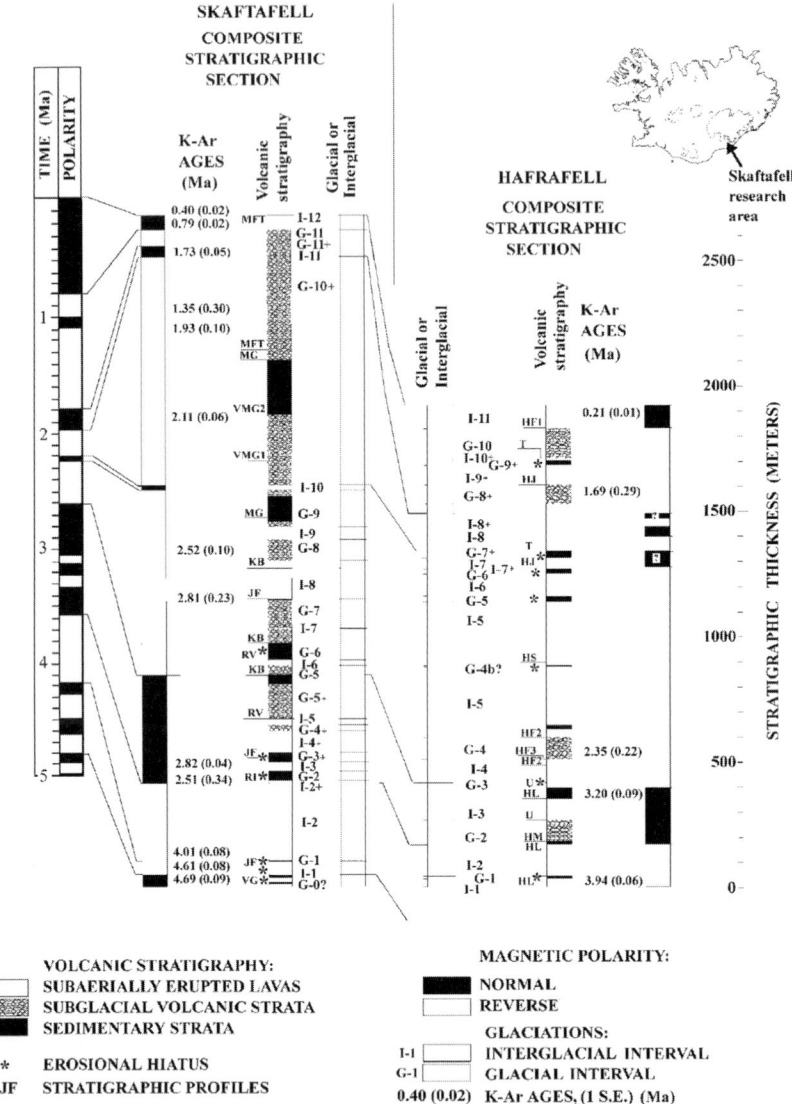

Fig. 13.19 Stratigraphical sections of volcanic successions at Skaftafell and Hafrafell, Iceland, summarising magnetic polarities, lithological characteristics, K–Ar ages and eruptive setting (G: glacial; I: interglacial) (from Helgason and Duncan, 2001). This is the most detailed onshore geological record of glacial/interglacial periods for Iceland.

ice-free; thus bedrock features formed by glacial erosion are visible. Compared to Iceland, this region appears to have experienced less tectonically driven morphological change since 3 Ma, has had less active volcanism to bury and obscure older deposits, and the ice sheet is topographically constrained by the coastal mountain ranges of Alaska and British Columbia

Fig. 13.20 View looking to the south showing the profile and stratigraphy of Hoodoo Mountain volcano, northwestern British Columbia. Vertical relief from the base to summit ice cap is c. 1000 m.

to the west, and the Rocky Mountains to the east (which were likely a major ice divide separating the CIS from the Laurentide Ice Sheet). Three geographical areas have been studied and demonstrate local versus regional approaches for using glaciovolcanism to reconstruct the palaeoenvironments: Hoodoo Mountain (Edwards et al., 2000, 2002; Edwards and Russell, 2002); Mt Edziza (Souther, 1992; Edwards et al., 2009b; Hungerford et al., 2014); and the Kawdy–Tuya region (Edwards et al., 2011; Edwards et al., 2014b).

The Hoodoo volcanic complex (see Chapter 11) is one of the better studied polygenetic glaciovolcanoes outside of Antarctica. It has been active over the last 100 k.y. The complex is located within the Coast Mountains of western British Columbia, which are still glacierised, and Hoodoo Mountain, which is the main edifice in the complex, has an ice cap with an area of c. 10 km^2 and a maximum thickness of c. 200 m (Fig. 13.20; Russell et al., 1998). The complex is bounded along the eastern and western sides by valley glaciers: Twin Glacier to the east and Hoodoo Glacier to the west. Retreat of the two glaciers shows that lavas from previous eruptions now partly underlie both. Given that at least two of the most recent lava flows are Holocene, the volcano could produce future glaciovolcanism. Although Hoodoo Mountain itself appears to mainly comprise intermediate to felsic products (trachyte–phonolite), the complex also includes an adjacent basaltic volcano to the north (informally referred to as 'Little Bear Mountain'; Edwards and Russell, 1994; Edwards et al., 1995) that not only erupted subglacially at 157 ka ± 3 k.y. (Blondes et al., 2007) but has also been geomorphologically modified by subsequent glaciation, and it is directly overlain by at least two glacial tills and by units from Hoodoo Mountain (Edwards et al., 1995). The 2000 m a.s.l., geomorphologically distinct edifice shows a number of ways in which ice thicknesses can be inferred through time; however, given its location within the Coast Mountains that have elevations locally of over 3000 m a.s.l., it is uncertain whether changes in ice extent were local or regional. As is discussed below, the timing constraints are consistent with the local ice fluctuations being generally synchronous with global climate records. The initiation of volcanism began during Marine Isotope Stage (MIS) 6 with eruption at Little Bear Mountain at 157 ka. Because its summit is now c. 100 m above the surrounding ice, that is a minimum estimate for syn-eruption ice surface elevation. Given the presence of glacial erosion and two overlying till units, the evidence is consistent with subsequent re-covering by ice during MIS 4 and 2. The next identified glaciovolcanic deposits (Qvap$_1$) are along the northern and western base of Hoodoo Mountain, and erupted subglacially at 85 ka (MIS 5 to 4 transition). The dense nature of the lavas and clasts in associated monomict tuff breccia suggest a glaciovolcanic origin, and the maximum relative elevations of these deposits with respect to modern ice are consistent with the coeval ice thicknesses 200–400 m greater than today. Ice-confined lavas (Qvap$_2$), emplaced at 80 ka, form cliffs up to 300 m high that overlie Qvap$_1$ and are consistent with similar minimum ice thicknesses. Between 80 and 54 ka, subaerial pyroclastic eruptions (Qvpy) deposited lapilli tuff at elevations up to 1300 m a.s.l.; either these eruptions deposited debris within an ice-confined cavity, or ice elevations had decreased. Although overlying lavas erupted at c. 54 ka (Qvap$_4$) show no obvious signs of ice confinement, their absolute elevations (generally above 1300 m a.s.l.) could mean they erupted above surrounding ice. An obvious return to subglacial eruptions beneath thick ice, also based on lava morphologies and clast densities in associated tuff breccia, occurred between 54 and 30 ka (Qvap$_{5,6}$), coincident with MIS 2. Given the elevations of the deposits, up to 1700 m a.s.l., these may be the best record of maximum ice thicknesses throughout the entire history of the complex, and could indicate regional ice more than 1 km thick. Glacial diamicts underlying the Holocene units that cap the rest of the volcanic stratigraphy (Qvpp) indicate re-covering by ice that has persisted until the present. Recent studies show that the Hoodoo ice cap is still relatively stable compared to the rapidly retreating, lower elevation bounding glaciers (Kargel et al., 2014).

While an order of magnitude larger, the Edziza volcanic complex is similar to Hoodoo in that is it still glacierised by approximately 60 km^2 of ice. Deposits there need precise constraints in order to assess whether they represent local or regional (e.g. Cordilleran-wide) glaciations (e.g. Edwards et al., 2009b). Souther (1992) reconstructed the volcanic and palaeoenvironmental history of the Mt Edziza volcanic complex, where volcanism began at 10–7 Ma and has continued through to the Holocene. Glacial diamicts are present that contain diverse clast lithologies indicative of former regionally extensive glaciations (Souther, 1992; Spooner et al., 1995). In particular, during the emplacement of the Ice Peak Formation c. 1 Ma (Fig. 13.21), Souther (1992) inferred that extensive ice was present during much of the eruptive activity. Subsequent work at Pillow Ridge (Edwards et al., 2009b) and Tennena Cone (Hungerford et al., 2014) showed that these deposits formed under ice much thicker than today (>400 m, at least), but which might also be related to expansion of the existing ice cap. Given the abundance of Holocene activity at Edziza (>10 Holocene pyroclastic cones and lavas) and its sizeable ice cap, it could also produce glaciovolcanic eruptions in the future.

While the previous two examples provide detailed information on changes in local ice conditions, a more significant area for constraints on regional ice sheets is presently being documented by multiple studies to the north of Mt Edziza, in the Kawdy–Tuya region (Edwards et al., 2011; Ryane et al., 2011; Russell et al., 2013, 2014; Edwards et al., 2014b, 2015a; Turnbull et al., 2014; see Chapter 2). This area is at lower elevations (generally <2000 m a.s.l.) than Edziza (>3000 m a.s.l.) and so it is not glacierised today, and volcanism there was the first inferred to have subglacial origins in North America (Mathews, 1947).

Fig. 13.21 Evidence for interactions between volcanism and ice in northern British Columbia. (a) View of Tutsingale Mountain tuya in northern British Columbia with glacial mega-grooves carved across its summit. (b) Limestone glacial erratic on the summit of Tutsingale Mountain. (c) Ice Peak Formation lava on the southwestern flank of Mt Edziza, in British Columbia, directly overlying a glacial diamictite.

The Kawdy–Tuya region has more than 50 sites where essentially monogenetic basaltic volcanic deposits, including vents, have been identified, although few have been studied in detail (Allen et al., 1982; Moore et al., 1995; Dixon et al., 2002; Edwards et al., 2011). A new study using $^{40/39}$Ar geochronometry from 25 of those sites has documented the earliest unequivocal presence of regional ice in North America at 2.8 Ma, as well as up to 15 younger incarnations of the ancestral Cordilleran ice sheet. Intriguingly, although many of these edifices were inferred to have been emplaced during the Last Glacial Maximum, the youngest identified so far (Ash Mountain; Fig. 8.7) erupted at c. 66 ka. Although existing publications only describe an area of c. 7500 km^2, together they make up one of the largest regional studies in British Columbia documenting the presence of 'extinct' ice, and they demonstrate the critical nature of glaciovolcanism as a climate proxy: the glaciovolcanic deposits are the only source of information that documents the former presence of ice in this area.

14

Climate triggers for glaciovolcanism

14.1 Introduction

This chapter reviews the history of ideas attempting to show linkages between volcanism and major changes to the cryosphere (e.g. waxing and waning of valley glaciers and ice sheets). The idea that glaciation and volcanism could be linked dates from at least the early 1900s in Iceland (Pjetursson, 1900; Peacock, 1926), and the 1950s in British Columbia (Mathews, 1958; Grove, 1974). Several ideas were subsequently advanced exploring the ways in which climate changes directly related to glacial cycles could trigger eruptions, including the possible influence of isostasy on melting in source regions (Hardarson and Fitton, 1991; Jull and McKenzie, 1996; Slater et al., 1998; Maclennan et al., 2002); a control by extensional crustal stresses favouring or inhibiting dyke propagation (Nakada and Yokose, 1992; Glazner et al., 1999; Sigvaldason et al., 1992; Jellinek et al., 2004); and influences of local stress fields acting on crustal magma chambers (Gee et al., 1998; Glazner et al., 1999; Edwards et al., 2002; Sigmundsson et al., 2010a). Lowering of global sea level during glaciations may influence eruption rates on volcanic systems close to oceans or seas (e.g. Rampino et al., 1979; Wallman et al., 1988; Patterne and Guichard, 1993; McGuire et al., 1997) and recent studies have raised concerns about future melting of present ice caps reducing the lithostatic (glaciostatic) loading on magma chambers and enhancing the potential for increasing eruption rates during periods of deglaciation (Pagli and Sigmundsson, 2008; Schmidt et al., 2013). Climate–volcanism feedbacks may also exist whereby increased rates of volcanism during rapid deglaciation can have a positive feedback on climate, with the volcanism contributing rapid increases in CO_2, enhancing global warming and hence increased rates of ice melting (Huybers and Langmuir, 2009).

14.2 Historical studies

By the early 1900s, the links between glaciation and volcanism were becoming obvious in Iceland (Pjetursson, 1900; Peacock, 1926). Pjetursson (1900) not only recognised that multiple periods of glaciation had affected central Iceland but also that many of the glacial deposits were interstratified with volcanic deposits. A subsequent field and petrographical study of the 'Glacio-palagonite Formation' by Peacock (1926) was probably the first to suggest that extensive ancient deposits of vitroclastic rocks found throughout Iceland had formed by eruptions beneath and within glaciers. From observations of modern eruptions, Thorarinsson (1953) suggested that the 1938 eruption of Grímsvötn was actually triggered by slight pressure changes caused by a jökulhlaup that *preceded* the volcanic activity. The study by Thorarinsson was remarkably prescient and several later ideas incorporated relatively small changes in pressure to influence volcanic systems (e.g. Grove, 1974; Hardarson and Fitton, 1991; Jull and McKenzie, 1996). In British Columbia, Mathews (1958) briefly speculated about potential causal links between glaciation and volcanism based on his field observations around Mt Garibaldi. Grove (1974) also proposed deglacia-tion as a driving force, suggesting without detailed analysis that isostasy could alter crustal stresses and trigger eruptions. He also may have been the first to suggest that increased volcanism during glacial waning would increase rates of deglaciation, an early example of a positive volcanism–climate feedback loop. Simultaneously, Hughes et al. (1974) described a glacier surge thought to be induced by the 1970 Deception Island eruption, by basal melting, lending qualitative support to Grove's prediction of links between volcanism and deglaciation. Hall (1982) also presented evidence from Marion Island that deglaciation might have induced faulting and associated volcanism there, but later recanted these observations when more fieldwork revealed that the 'faults' were more likely erosional scarps unrelated to volcanism (Hall et al., 2011). He did demonstrate, however, that volcanism on the island occurred during both glacial and interglacial periods.

By the 1960s and 1970s scientists were beginning to explore the potential for many different, apparently small changes in lithospheric stresses to generate volcanic eruptions (e.g. Matthews, 1969; Chappel, 1975; Rampino et al., 1979). Several later workers (e.g. Wallman et al., 1988; Patterne and Guichard, 1993; McGuire et al., 1997) presented evidence that sea level drawdown during glaciations could produce enough change in stress fields around volcanoes and their magmatic systems to induce increased rates of eruptions. Patterne and Guichard (1993) found a 24 000 year cyclicity in eruption rates at some Italian volcanoes in the Campanian area of Italy and speculated that it might signal a link between Milankovich climate cyclicity and volcanism via pressure changes from eustatic variations.

Meanwhile, evidence for links between deglaciation and increases in eruption rates strengthened in Iceland. Several studies documented an apparent dramatic increase in volcanism immediately following the onset of major ice sheet retreat from northern (Slater et al., 1998; Maclennan et al., 2002; Licciardi et al., 2007), eastern (Sigvaldason et al., 1992), and southern Iceland (Jakobsson et al., 1978; Gudmundsson, 1986; Hardarson and Fitton, 1991; Gee et al., 1998; Sinton et al., 2005). While explanations for the causality vary and are still being debated (see below), the link between deglaciation and increased volcanism, at least for the end of the Last Glacial Maximum (c. 13 k.y. BP), is strongest for Iceland. Thordarson and Höskuldsson (2008) have estimated that up to one-third of Holocene eruptions in Iceland happened within two to three thousand years after the end of the Last Glacial Maximum, and data from Greenland ice cores have documented a higher rate of overall volcanic sulphur production from 15 to 8 k.y. BP (Zielinski et al., 1996).

In other parts of the world, problems with the accuracy of geochronology have not yet allowed such detailed comparisons. However, work from California (Glazner et al., 1999; Jellinek et al., 2004) has broadly suggested an anti-correlation between glaciation and volcanism (i.e. less volcanism during glaciations and more during interglacials), as have sparse data from northern British Columbia (Edwards and Russell, 1999; Edwards et al., 2010), studies in France and Germany (Nowell et al., 2006), and studies from Chile (Best, 1992; Gardeweg et al., 1998). One of the most recent analyses, by Huybers and Langmuir (2009), found a strong statistical correlation between increases in eruption rates at high northern latitudes and the broad period covered by decreasing ice sheet thicknesses (17 to 7 ka). Their spatial analysis also showed that the largest increases in eruption rates occurred in areas inferred to have the largest volumes of ice during the Last Glacial Maximum (40 to 20 ka).

14.3 How can waxing and waning of glaciers and ice sheets affect volcanism?

The earliest speculations about causal links between surface loading and volcanism suggested that relatively small changes in crustal loading might be significant enough to trigger or suppress volcanism. This led to ideas about (1) increased rates of volcanic activity in the ocean basins during glaciations, when sea level was lower by as much at 100 m, essentially decreasing lithostatic loading by c. 1 MPa, and (2) concomitant suppression of volcanic eruptions beneath glaciers due to excess loading by ice up to 3 km in thickness, equivalent to c. 27 MPa of increased lithostatic loading (although see Edwards et al. (2002) for alternatives). Likewise, after extensive deglaciation the converse would apply, with increased volcanism under thinning ice sheets, and less frequent volcanism in the ocean basins. The three main arguments supporting these interpretations are that (1) glacio-isostatic adjustments can cause changes in source region melting depths and melt production (e.g. Hardarson and Fitton, 1991; Jull and McKenzie, 1996), (2) diminishing stress unloads extensional fractures and facilitates melt migration to the surface (e.g. Gudmundsson, 1986; Wallmann et al., 1988; Jellinek et al., 2004), and (3) the presence or absence of ice affects magma buoyancy in the lithosphere (e.g. Edwards et al., 2002).

14.3.1 Glacio-isostasy and source region processes

Hardarson and Fitton (1991) first identified geochemical consequences that could be related to source region changes in melting caused by deglaciation. They studied samples from the Snæfellsnes volcanic system in southwestern Iceland, and argued that since Snæfellsnes was outside of the main Iceland rift axis, it should be more sensitive to changes in melting associated with deglaciation. They suggested that glacial loading during waxing stages (ice sheet build-up) should generate asthenospheric outflow from beneath Iceland, potentially suppressing volcanism in off-axis systems, while deglaciation would allow for return flow of asthenospheric mantle, producing increases in melting and rates of magma production. They concluded that significant shifts in trace element ratios (Ce/Y and Zr/Nb) for early glacial and post-glacial basalts compared with late glacial basalts were consistent with increases in mantle source region melt fractions of 0.4–0.6%, which corresponds to melting of a 2 km-thick Icelandic ice cap and consequent isostatic adjustments. They also speculated that melting in the central rift axis would be largely unaffected by isostatic adjustments. However, Jull and McKenzie (1996) showed quantitatively that ice loading and unloading could generate variations in the magma production rate beneath the Icelandic rift system, while assuming that the overall, long-term magma production rate would be constant. Thus, while melting of a 2 km ice sheet over the span of 1 k.y. (approximately 2 m per year) could enhance melting rates by up to 30 times the 'normal' rate, it must be balanced by a prolonged period (likely c. 30 k.y.) for ice sheet growth (thickening) with associated suppressed magma production. Essentially, the increase in melting driven by deglaciation is simply the 'release' of the excess magma whose production had been suppressed during crustal loading by glaciation. The distribution of volcanism produced by deglaciation in Iceland was predicted to also be directly related to the variations in thickness of the ice sheet; the highest rate of suppression or production of magma should occur in locations covered by the thickest ice. They concluded that it was very likely that isostatic rebound in extensional environments would increase rates of magma production, at least temporarily. Subsequent work by Slater et al. (1998), Maclennan et al. (2002) and Sims et al. (2013) reached similar conclusions, specifically linking changes in the major and trace element and isotopic compositions between glacial and early post-glacial volcanism to changes in melt regime caused by rapid glacio-isostasy.

While several workers subsequent to Hardarson and Fitton (1991) identified similar compositional changes between basaltic lavas erupted during full glacial and waning/early post-glacial times, the changes could be attributed to at least two causes. Slater et al. (1998) and Maclennan et al. (2002) argued that they were due to differences in total melt fraction as well as depth of melting caused by glacio-isostasy, while Gee et al. (1998) suggested that the increased rates of volcanism in deglacial periods cleared out crustal magma storage areas, thus permitting chemically depleted, early post-glacial melts to move through the crust rapidly enough to avoid the effects of magma mixing, assimilation and fractionation processes that might otherwise have obscured their distinctive chemical signatures.

14.3.2 Isostasy and extension-driven fracture propagation

Another possible causal link between glaciation and volcanism is the role of glacio-isostasy in the inhibition and generation of crustal fracture systems (Gudmundsson, 1986; Nakada and Yokose, 1992; Kelemen et al., 1997; Glazner et al., 1999; Jellinek et al., 2004; Sigmundsson et al., 2010a; Fig. 14.1). Most published studies have attempted to estimate stresses imposed or released on rocks due to the superimposition of mass from growing glaciers, in order to determine if the stresses are enough to

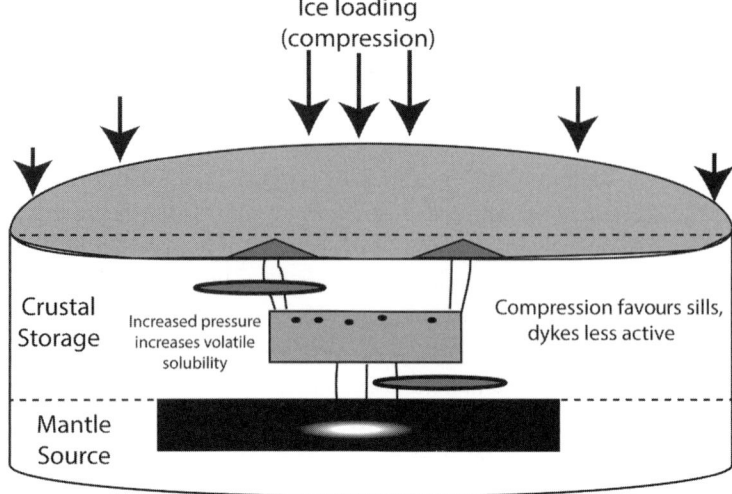

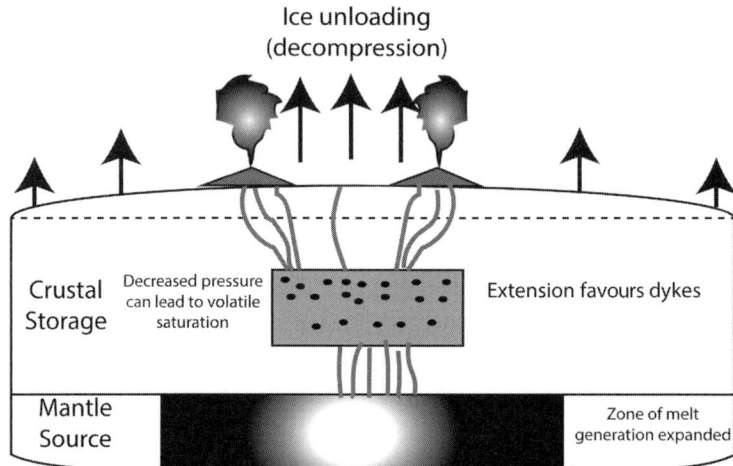

Fig. 14.1 Schematic summary diagram showing linkages and feedbacks between fluctuations in terrestrial ice and styles of volcanic activity.

inhibit or facilitate dyke formation. At least two studies linked variations in ice (and water) loading to tectonic-scale features (Gudmundsson, 1986; Nakada and Yokose, 1992), either associated (Gudmundsson, 1986) or not associated with volcanism (Nakada and Yokose, 1992). Gudmundsson (1986) summarised evidence for increased magmatism immediately following the last disappearance of ice sheets in the Reykjanes Peninsula of southwestern Iceland, and calculated that extensional fault throws of up to 8 m were consistent with an approximate 10 MPa increase in magmastatic driving pressure, which is consistent with the observed post-glacial magma flux. He suggested that the cyclicity of glacial waxing and waning in Iceland throughout the Quaternary was a significant part of the development of Iceland's tectonic and magmatic fissure systems. Nakada and Yokose (1992) examined the combination of stress changes resulted from loading of ice on island arc landmasses during glaciations and the unloading of adjacent ocean crust due to sea level decrease, versus the opposite effects during deglaciation. They estimated that sea level variations of 130 m could generate pressure gradients in 'thin' lithosphere of up to 0.8 MPa km^{-1}, or the equivalent of an excess magma buoyancy of 100 kg m^{-3}, and that deglaciation would generate stresses of 13 MPa for the top and the bottom of the lithosphere. Based on their calculations, they suggested that glaciation produced extensional fractures in the lower lithosphere due to crustal downwarping, but that deglaciation allows the extension to migrate through the mid- and upper lithosphere, generating progressive tectonic- and magma-driven fracturing and increased eruption rates.

Other studies generated further predictions about the role of fractures and magma ascent (Kelemen et al., 1997; Glazner et al., 1999; Jellinek et al., 2004). As part of an analysis of the conditions necessary for melt extraction from mantle source regions, Kelemen et al. (1997) suggested that it might be viable to extract melt from source regions by hydrofracturing using melt production rates for Iceland from Jull and McKenzie (1996). Based on an analysis that found an anti-correlation between glaciation and volcanism in eastern California (i.e. increased volcanism during interglacial periods), Glazner et al. (1999) estimated that loading of c. 300 m of ice on the Sierra Nevada Mountains, or equivalent water depths from pluvial lakes, could generate enough horizontally directed stress in the crust (c. 1 MPa) to arrest dyke formation. Subsequent glacio-isostasy or draining of pluvial lakes would reinvigorate dyking and facilitate renewed volcanism. They also speculated that a 'wetter' upper crust during glaciations could increase hydrothermal circulation of groundwater, generating more efficient removal of heat from the crust, and also impeding dyke propagation in the crust, as well as possible effects of increased lithostatic pressure on volatile saturation and depressed source region solidi. Jellinek et al. (2004) reanalysed the Glazner et al. (1999) dataset and identified a 40 k.y. cyclicity for volcanism in eastern California. They argued against any impact of relatively thin Sierran glaciers on melting in mantle source regions, and they suggested a more specific relationship between deglaciation and volcanism whereby changes in eruption frequency are tied to changes in rates of varying ice volumes. They also identified a lag time between the onset of rapid ice melting and onset of magmatism. A lag of approximately 3 k.y. was suggested for basaltic magmas that required a critical overpressure for dyke formation in crustal storage regions of ≪1 MPa, whereas as much as 11 k.y. might be required for 'silicic' magmas that required a critical overpressure of c. 3 MPa for dyke formation. Based on analysis of magma overpressures at dyke tips, Wilson et al. (2013) further suggested that dykes could easily penetrate up into ice, with many attendant consequences for heat transfer and rates of melting at eruption onset.

The general consensus from all workers seems to be that loading of the lithosphere by ice or water should impede migration of melts towards Earth's surface, while removal of those same loads should facilitate melt migration, depending on the rates of unloading and with the caveat that time lags between unloading and volcanism could be compositionally dependent.

14.3.3 Buoyancy effects of ice loading and unloading

The physics of magma 'buoyancy' has already been discussed (see Section 6.3), and several workers have suggested that changes in ice loading can change the buoyancy potential of magma (e.g. Gee et al., 1998; Wilson et al., 2013). However, Edwards et al. (2002) showed that the potential impacts of ice on magma buoyancy could vary depending on whether the system being investigated was likely to produce a subglacial eruption (i.e. wholly covered by the overlying ice) or a supraglacial eruption (one that penetrates the surface of the ice; Fig. 14.1). They modelled changes to static magma buoyancy for a basaltic magma in the crust undergoing assimilation–fractional crystallisation (AFC), assuming a critical over-pressure of 10 MPa to overcome the strength of crust with a thickness of c. 12 km above a mid-crustal magma storage area. During the AFC process, significant changes in magma density occur. They found that production of lower density melts by AFC processes increased the opportunity for eruption of those magmas. In combination with thinning ice sheets, it would facilitate supraglacial eruptions, while thickening ice sheets could actually facilitate subglacial eruptions of denser magmas, as increasing ice thickness results in a higher load on the magma storage area.

14.4 Likelihood of present-day climate change increasing volcanic activity

Recent work for Iceland has estimated that modern glacial recession is not only con-tributing to a slight volumetric excess of magma production (Schmidt et al., 2013), but also to destabilising crustal magma chambers (Pagli and Sigmundsson, 2008; Sigmundsson et al., 2010a). Schmidt et al. (2013) made estimates for the overall increase in magma production rate beneath Iceland since 1890 and suggested that deglaciation and the resulting glacio-isostasy may have increased overall volumes of erupted material during that time by the equivalent of one Eyjafjallajökull eruption (c. 0.18 km^3 of magma) every 7 years. Depending on melt rise velocities from source areas, they presented five different curves to model the increases in melt volumes supplied to the base of the Icelandic crust: for rise velocities of 500 m per year, they predicted a peak in melt production at c. AD 2100 and a span of increased melt production of almost 300 years. Slower melt rise times would lead to much smaller melt increases spread out over more than 500 years. These predictions, however, are based only on glacial unloading up to 2013. Presumably if the rate of ice mass loss increases in the future, so will rates of melt production and concomitant increases in the volume of erupted material.

Pagli and Sigmundsson (2008) examined increases in eruption frequency beneath Vatnajökull as a result of present-day rates of glacio-isostasy (c. 20 mm per year). Their estimates of changes in crustal stress from the resulting rates of uplift suggested that the outer edges of the ice cap would experience the largest stress increases, having the greatest potential to affect the Bárdarbunga volcanic system. They emphasised that, while changes to mantle melting due to deglaciation were more likely in areas of extensive ice sheets, changes in crustal stress due to deglaciation of smaller ice caps on stratovolcanoes also have the potential to change eruption behaviour. A subsequent study by Sigmundsson et al. (2010a) examined the effects of stress changes on crustal magma storage areas induced by deglaciation, and suggested that even annual seasonal variations due to loading and unloading of snow could affect failure conditions acting on shallow magma storage areas.

14.5 Potential glacial–volcanic climate feedback mechanisms

Grove (1974) appears to have been the first to suggest that an increased frequency of eruptions caused by deglaciation could in turn lead to more rapid loss of ice. Much later, Huybers and Langmuir (2009) presented arguments for the potential of increased volcanism immediately following the destruction of major ice sheets 12–7 k.y. ago to generate a large influx of CO_2 to Earth's atmosphere, causing a greenhouse effect and enhancing further long-term warming and accelerating the loss of ice mass. The return to 'normal' rates of volcanism allows natural carbon sinks to catch up with overproduction of volcanic CO_2, eventually depleting the atmosphere in CO_2 again, and leading to a renewed glaciation. Conversely, Roth and Joos (2012) questioned the predicted increase in volcanically produced CO_2 as atmospheric records indicate that CO_2 *decreased* in concentration during the early Holocene.

In summary, although an intuitive link between volcanism and waxing and waning global glaciations has been postulated since the early 1900s, it is only since the late 1980s that quantitative tests have been devised to understand the details of the relationships involved. Studies in many areas, but especially in Iceland with the greatest number of dated eruptions, seem to show a clear linkage between increases in the rate of volcanism and the end of the Last Glacial Maximum (12–7 ka). Glacio-isostatic unloading can lead to increases in rates of magma production in source regions, as well as the generation of stress forces favouring particular fracture pathways that facilitate magma ascent and eruption (Fig. 14.1). While details of the timing of isostatic adjustments and their correlation with eruption ages needs significantly more work (dependent on improving isotopic dating precision; Section 13.7), the evidence thus far available consistently suggests that glaciation may significantly modulate volcanism.

15

Hazards associated with glaciovolcanic eruptions

15.1 Introduction

Some of the most significant volcanic disasters in history are directly linked with volcano–water interactions. Many of the 36 000 casualties from the massive 1883 eruption (c. >10 km³ DRE) of Krakatoa volcano in Indonesia resulted from tsunami generated by the explosive disintegration of the island volcano (Winchester, 2003). Just over 100 years later, a much smaller eruption (<0.1 km³) from the ice- and snow-capped Nevado del Ruiz volcano in Colombia produced volcanic mudflows that killed over 23 000 people (Pierson et al., 1990). A repeat of the 1877 eruption-induced lahars at ice-clad Cotopaxi volcano in Ecuador today would threaten the safety of more than 100 000 people (Mothes et al., 2004). Ice-draped and fully subglacial volcanoes are unique in that they represent large surface reservoirs of water potentially available at all times of the year to generate floods and phreatomagmatic explosions, whereas lahars and flooding from non-ice-clad volcanoes are dependent on annual climate variations and wet/dry seasonal cycles. The presence of water during eruptions will generally lead to explosions that are more energetic than in 'dry' environments (Mastin et al., 2004), and phreatomagmatic eruptions generally produce larger volumes of fine ash sizes (Liu et al., 2015), which may facilitate longer distance atmospheric transport (Gudmundsson et al., 2008). Like other volcanoes, glaciovolcanoes have summit craters or calderas where meltwater can accumulate. They thus represent a significant flooding hazard. In addition, the presence of ice during effusive eruptions can create oversteepened deposits that, once supporting ice buttresses melt, are more susceptible to gravitational collapse (e.g. Mathews, 1952a; Moore, 1976).

Many regional studies have recognised the inherent dangers of glaciovolcanoes. Gudmundsson et al. (2008) concluded that numerous hazards related to volcanism in Iceland are associated directly with volcano–ice interactions. Many fatal eruptions in South America have occurred at glacierised volcanoes. While Nevado del Ruiz is the largest single historic disaster, eruptions from Cotopaxi in Ecuador (Mothes et al., 2004; Pistolesi et al., 2013) and Villarrica in Chile (Naranjo and Moreno, 2004) have caused more than 1000 casualties since 1877. In North America, 12 of the 16 volcanoes in the 'very high' threat group are snow- or ice-clad (Ewert, 2005). Interactions between snow/ice and juvenile volcanic material have also produced hazards in Italy, Japan, Mexico, New

Zealand and Russia. Major and Newhall (1989) reviewed dozens of historical eruptions from snow/ice-clad volcanoes, and found that explosive eruptions appear to induce the greatest number of flooding/lahar events. They also reviewed several examples of effusive eruptions that produced hazardous flooding, as well as hazards from active geothermal systems not directly associated with eruptions (e.g. many jökulhlaups from Grímsvötn caldera, Iceland).

Volcanic eruptions cause a range of local, regional or global hazards, depending on factors related to properties of the erupting magma (e.g. viscosity, volatile content), or to the eruption environment (e.g. local topography, presence of water/snow/ice). The hazards include (1) lava flows, (2) ash falls, (3) pyroclastic density currents, (4) meltwater floods and associated sediment mass flows, (5) avalanches, (6) volcanic gases, (7) tsunami, and even (8) volcanic lightning. Many texts and videos have described these hazards in detail (e.g. Scarpa and Tilling, 1996; Sigurdsson et al., 2000). In this chapter we focus on how volcanic hazards are compounded or mitigated in the glaciovolcanic environment.

15.2 Lava flows

Because lava flows generally move relatively slowly (typically up to a few kilometres per hour), they are mainly hazardous to permanent infrastructure that is not easily moved. However, there are some spectacular exceptions. For example, eruptions from Nyiragongo volcano (Africa) have produced highly mobile lava flows that travelled tens of kilometres per hour, and even moved quickly enough to overtake large animals (e.g. elephants; Chirico et al., 2009). In addition, some large volcanoes (e.g. Mauna Loa in Hawaii, Etna in Sicily) have sizeable human populations that live on the volcano flanks where volcanic fissures may form. However, because most ice-clad and subglacial volcanoes are geographically remote, damage to infrastructure is often relatively minor. Moreover, most lavas on Earth rarely flow more than 10–20 km. The main hazard from lava flows derived from glacio-volcanic eruptions is their potential to melt snow and ice (causing floods), to interact explosively with meltwater, and to form oversteepened escarpments (lava termini) that are highly fractured and very susceptible to sudden gravity-driven failures (see Section 15.6). Effusive glaciovolcanic eruptions with notable resulting hazards include those from Mt Etna (Romano and Sturiale, 1982), Klyuchevskoy volcano in Kamchatka, Russia (Ozerov et al., 1997), and at least three South American volcanoes (Llaima: Moreno and Fuentealba, 1994; Villarrica: Naranjo and Moreno, 2004; Hudson: Naranjo et al., 1993). Smaller eruptions in Iceland, Alaska and Russia have also provided insights into glaciovolcanic lava flow hazards. Major and Newhall (1989) suggested that, in general, lava flows were unlikely to produce significant meltwater floods, although they noted that some historical floods were associated with casualties. However, recent glaciovolcanic experiments, in which basaltic melts were poured onto ice, showed that significant melt-water can be produced relative to the volume of lava involved (Edwards et al., 2013). In combination with rapid effusion rates, such as those seen during the 2012–13 Tolbachik eruption where lava flows travelled more than 9 km within the first 48 hours of the eruption, voluminous meltwater production by glaciovolcanic lava flows is an underappreciated hazard.

15.2.1 Examples of lava emplaced onto and beneath ice or snow

Only a few volcanoes have historical records of lava flows being emplaced onto or beneath ice. The two oldest reported occurrences of mudflows/lahars potentially generated by effusive eruptions occurred at Mt Etna in 1536 and 1755 (Romano and Sturiale, 1982). While few details are available for either eruption, both events occurred in spring (March, April) when snowpack would have been close to its peak thickness. Eruptions in 2013 also appear to have produced meltwater that drained into uninhabited areas in the Valle del Bove. There are also records of lava-induced lahars associated with eruptions of Villarrica volcano in southern Chile. The lahars are estimated to have killed more than 100 people in the twentieth century (Naranjo and Moreno, 2004) and are the most significant hazards from the volcano. Villarrica is largely covered by glaciers and the volcano has a lava lake in the summit crater that periodically overflows to produce lava flows. Several of Villarrica's eruptions have produced local and damaging lahars, including some that resulted in fatalities (Naranjo and Moreno, 2004). Llaima volcano, north of Villarrica, has also had numerous effusive eruptions that produced damaging lahars, although deaths at Llaima are fewer. Moreno (1994) described the 1994 eruption at Llaima, and inferred that lava flows travelling beneath the glacier were the source of most of the associated lahars. The 1991 eruption at Hudson volcano produced a lava flow that may have generated lahars, which flowed partly beneath and partly above slope ice (Naranjo et al., 1993). Finally, remote Klyuchevskoy volcano on the Kamchatka Peninsula in eastern Russia has also had many effusive eruptions that produced lahars on its steep, snow and ice-covered slopes (Vinogradov and Muravyev, 1985; Ozerov et al., 1997). A 1994 eruption at Klyuchevskoy also produced a secondary hazard when lava flows that collapsed on the steep volcano slopes mixed with snow/ice and interacted explosively (Belosouv et al., 2011; see Section 15.6).

Four effusive eruptions in Iceland and Russia between 2010 and 2013 also provide important information on the potential for lava flows to generate flooding in snow/ice-rich environments: 2010 Fimmvörðuháls, 2010 Eyjafjallajökull, 2013 Tolbachik, and 2013 Veniaminof. The 2010 Fimmvörðuháls eruption in Iceland emplaced a small-volume lava field on top of seasonal snow (Edwards et al., 2012; Fig. 15.1). As snow depths were only 3–5 m, the total amount of meltwater generated was small and, indeed, satellite images and field measurements documented only small water discharges from beneath several lava flows. Similarly, the 2012–13 eruption at Tolbachik volcano, located just south of Kluychevskoy in Kamchatka, produced small, episodic meltwater discharges emanating from beneath lava flows (Edwards et al., 2015b; Fig. 15.2). While the total volume of water produced was small, the discharges were warm (>60 °C) and advected heat from the lava that melted snow further downslope.

Fig. 15.1 IKONOS satellite image showing lava emplaced on top of snow during the 2010 Fimmvörðuháls eruption in Iceland. Small steam clouds are visible along the lava edges (lower right), and dark tracks to the north of the lava are from small lahars (from Edwards et al., 2012). (IKONOS image is courtesy of the GeoEYE Foundation.)

While the major impact of the 2010 eruption at Eyjafjallajökull was due to generation of ash, it also produced a lava flow that travelled both within and beneath Gigjökull glacier (Edwards et al., 2012; Oddsson et al., 2012; Fig. 15.3). The meltwater generated initially from explosive eruptions at the Eyjafjallajökull vent not only travelled over the surface of Gigjökull, it also enlarged pre-existing subglacial drainage pathways. Those passages were later exploited by an intermediate-composition lava flow that emanated from outside the main summit crater. The lava flow slowly advanced down the glacier, initially moving in subglacial tunnels. The tunnel roofs subsequently collapsed to form an open channel in the glacier that accommodated subaerial blocky lava flows. Because meltwater was produced continuously during both the lava flow and explosive phases of the eruption, it is difficult to

Fig. 15.2   Small, ephemeral meltwater stream generated by 'a'ā lava melting underlying snowpack during the 2012–13 Tolbachik eruption, Kamchatka, Russia. Footstep tracks in the snow between the two small streams for scale (from Edwards et al., 2015b).

Fig. 15.3   Aerial view of the northern side of Eyjafjallajökull volcano in Iceland and the lava that flowed within and through Gigjökull glacier during the eruption. Summit craters are obscured by the rising steam. The foreground canyon in the ice is approximately 0.5 km wide. Image taken on 9 July 2010.

assign volumes of meltwater specifically to the lava-producing episode, but observable changes were caused to the Gigjökull glacier.

The Eyjafjallajökull eruption is from one of few ice-bound volcanoes with frequent effusion-dominated eruptions. Veniaminof volcano, Alaska, is a large stratovolcano with

Fig. 15.4 View looking to the southwest of the active intracaldera cone at Veniaminof volcano in southwestern Alaska on 18 August 2013. Ash and lava are visible on and above the cone, while steam in the left foreground rises from areas where lava is in contact with ice or meltwater. The cone is c. 0.5 km high.

an ice-filled caldera (Yount et al., 1985; Welch et al., 2007). It has a small tephra cone that projects above the caldera ice (>400 m thick; Welch et al., 2007). Andesitic lavas are erupted from the cone and flow onto and beneath the surrounding ice. At least three such eruptions have been documented (1983: Yount et al., 1985; 1993–4; and 2013), but only the 1983 eruption generated meltwater (cf. Figs. 1.1b, 3.5 and 15.4). During the 1983 eruption, a lake was formed at the base of the cone within the ice (see Smellie, 2006, Figs. 13 and 14). It eventually drained, but its volume was small and did not significantly increase the normal seasonal meltwater discharge from the caldera (Yount et al., 1985). The Veniaminof eruptions have produced relatively viscous slowly advancing andesitic lava flows. Associated ice melting was correspondingly slow and occurred over several days to a few weeks. While other eruptions at Veniaminof have produced more obvious ice melting, much of the meltwater may remain within the caldera. However, despite being a potential threat from flooding, the remote location of the volcano means that the threat to infrastructure or human populations is small.

15.2.2 Emplacement of lava domes into snow/ice

Many ice-clad volcanoes are high-elevation stratovolcanoes that produce lava domes when they erupt. However, two in particular provide insight on glaciovolcanic hazards: Mt St Helens, southwestern Washington (USA), and Mt Redoubt, south-central Alaska (Figs. 15.5). Both volcanoes erupted multiple times between 1980 and 2011, and volcano–ice interactions were at least partly responsible for the hazards generated at each (e.g. Brantley and Waitt, 1988; Wait et al., 1983; Waitt, 1989; Dorava et al., 1994; Waythomas et al., 2013).

The reawakening of Mt St Helens volcano on 18 May 1980 was a watershed event for knowledge of processes and hazards associated with ice-clad volcanoes. Research on the eruptions and their deposits has produced one of the largest bodies of literature on any modern volcano (>500 publications). From the first fractures in the glaciers on the northern side of the volcano to the end of the lahars as they reached the Pacific Ocean after runouts of more than 100 km, the 1980 eruption provided invaluable insights on the role of ice and surface water in facilitating velocities and travel distances for associated avalanches and lahars (Brantley and Waitt, 1988; Waitt et al., 1983; Waitt, 1989). Very little melting of the glaciers occurred before the 1980 eruption, but the Plinian phase significantly reduced the glacier cover on the volcano. An unknown part of that ice was converted to water that extended the runout distances of many lahars and debris flows. At least one later eruption, in 1982, also produced small lahars from interactions between lava and ice (Waitt et al., 1983; Fig. 15.5a).

After 18 years of quiescence, Mt St Helens awoke again in 2004 and produced a largely effusive eruption. By that time, small cirque glaciers had formed within the crater after the 1980s eruptions, in one of the few documented events of new glacier formation of the twentieth century. Dome formation between 2004 and 2006 at Mt St Helens physically moved the glacier aside and very little melting was observed (Walder et al., 2007). As the dome was emplaced within a much larger crater depression, there is little potential for rock avalanching.

Eruptions from Redoubt volcano, in Alaska, have produced much more significant ice melting, even though associated lava domes are not significantly larger than those formed in the 2004–6 Mt St Helens events (Waythomas et al., 2013). The crater at Redoubt is much smaller than that at Mt St Helens, and it is situated at a much higher elevation, with a dome perched at its outer edge. Growth of the Redoubt lava dome triggers fast-moving lava avalanches and they immediately travel down on top of the Drift River glacier. The rapid mixing of hot dome material with snow and ice has led to very large-volume lahars and debris flows, which are discussed in Section 15.5 (Fig. 15.5b).

15.3 Ash falls

While ash falls are commonly only severe for local residents, entrainment of ash into air currents can cause problems thousands of kilometres from eruption sites, as evidenced by the 2010 eruption of Eyjafjallajökull (e.g. Gudmundsson et al., 2012; Section 3.3.13). During that eruption, ash fall severely affected local farmers on the flanks of the volcano and surrounding flood plains. Light ash fall also reached Iceland's capital, Reykjavik, situated c. 100 km to the west. Airborne ash was dispersed predominantly to the southeast and affected much of Europe, with major economic consequences for all air traffic in the region, which shut down for two weeks. Later stages of the eruption were predominantly magmatic and it is likely the early ash production was phreatomagmatic and characterised by much fine

Fig. 15.5   Views of Mt St Helens in Washington State and Mt Redoubt in Alaska to compare the relative sizes of craters. (a) View of Mt St Helens looking to the south on 19 March 1982 showing a lahar produced by an explosive eruption (image courtesy of the USGS Cascade Volcano Observatory). The valley from which the lahar exits the crater is c. 1.3 km wide. (b) View of Redoubt volcano looking to the south from a tephra sampling station (09RDKLW090) on 21 April 2009 (image courtesy of J. Schaefer, the Alaska Volcano Observatory and the University of Alaska Fairbanks Geophysical Institute). The valley below the steaming summit is c. 1 km wide.

ash due to interactions of the magma with meltwater derived from the thin overlying Eyjafjallajökull ice cap (Gudmundsson et al., 2012a). The tendency for glaciovolcanic eruptions to produce much more fine volcanic ash than 'dry' (magmatic) eruptions, especially for basaltic eruptions, was highlighted by Gudmundsson et al. (2008) in their volcanic hazard assessment for Iceland. This is especially important as basaltic fissure eruptions can be relatively long-lived events, and so can release ash over time periods of months (Gudmundsson et al., 2008). As fine to very fine ash appears to be most problematical for modern jet engines (e.g. Kueppers et al., 2014), glaciovolcanic eruptions that are dominantly phreatomagmatic pose a high risk for aviation.

Ash deposition onto snow or ice can accelerate rates of snow and ice melting in summer months, depending on the thickness of the accumulated ash layer (cf. Cuffey and Patterson, 2010; Wilson et al., 2013), and can lead to contamination of drinking water derived by ice melting. This can create a long-term hazard for eruptions at ice-clad volcanoes where the snowpack/ice is a critical source of water for consumption or agriculture. Heavy ash fall can also temporarily block natural water courses, acting as a temporary dam that, when it fails, can generate lahars (Manville and Cronin, 2007).

15.4 Pyroclastic density currents

Pyroclastic density currents (PDCs) are gravity-driven flows of tephra, heated rock fragments and high temperature (e.g. >100 °C) fluidised gas. Although initially separated into two categories: pyroclastic flows (high particle density) and pyroclastic surges (low particle density), there is a continuum of flow densities (Branney and Kokelaar, 2002). PDCs are amongst the most destructive types of volcanic events as they can travel very quickly (>100 km h^{-1}), and have high temperatures (>700 °C). They can also transport fragments ranging from very fine ash to boulders several metres in diameter, and their momentum enables them to override topographical barriers that would impede other types of flows (e.g. lavas, sediment gravity flows). PDCs likely cause more direct human casualties than other volcanic phenomena (Scarpa and Tilling, 1996). PDCs are generated by three principal processes: volcanic explosions, gravity-driven marginal collapse of eruption columns, and gravitational collapses of oversteepened lava domes/flows (e.g. Belousov et al., 2011).

Pyroclastic density currents may be preferentially associated with glaciovolcanism because of the very common association with phreatomagmatic activity due to interaction with meltwater. The rapid advection of heat carried by dense particles within PDCs will also cause aerially extensive, rapid melting of surrounding snow/ice surfaces at glaciovolcanoes. Thus, production of lahars, which rival PDCs in terms of mortalities, by PDCs is probably very common on glaciovolcanoes (see Section 15.5).

The eruptions of Mt St Helens in the 1980s provided one of the finest opportunities for studying the interactions between PDCs and snow/ice (Brantley and Waitt, 1988; Waitt, 1989). Low-density PDCs (e.g. 'surges') formed at the start of the eruption on 18 May 1980 likely initiated many of the floods/lahars (Section 15.5). Contemporaneous observations suggested that the lahars formed after passing of the surges (Brantley and Waitt, 1988; Waitt, 1989). The lahars travelled down the steep volcanic slopes at velocities estimated to be c. 90 km h^{-1}. However, while the largest lahars formed during the initial phase of the eruption, smaller lahars formed periodically from other processes, for example, from localised impacts of 'ash flows' (Waitt, 1989). Lahars also formed from snow avalanches triggered by volcanic explosions within the crater in 1983 (Waitt et al., 1983).

Because they were so destructive, the initiation of lahars during the 13 November 1985 eruption at Nevado del Ruiz volcano in Colombia has been studied in detail (e.g. Pierson et al., 1990). Although the volcano had shown increasing signs of activity during much of 1985 and hazards from lahars were well known as a result significant fatalities during previous eruptions, the eruptions that triggered the disastrous lahars on 13 November were relatively minor. Four discrete explosions were identified, but only the first two appear to have generated enough mixing between pyroclasts and snow/ice to generate large amounts of water (Pierson et al., 1990). Later analysis showed that the PDCs commenced as pyroclastic flows with significant erosive power, so that they eroded channels up to 100 m wide and 4 m deep within summit glaciers (Pierson et al., 1990). The combination of advected heat to melt ice and transported angular clasts to erode ice quickly makes PDCs especially dangerous at ice-clad volcanoes.

15.5 Meltwater floods and associated mass flows

Floods of meltwater are amongst the most visible and devastating events normally associated with glaciovolcanic eruptions. They even have their own word to describe them: jökulhlaups (Roberts, 2005; see Glossary). Because the meltwater flows subglacially and supraglacially, typically over unconsolidated coeval tephra or relatively poorly consolidated subglacial volcanic and other materials laid down by the glacier, the floods generally bulk up with solids to a variable degree and ultimately transform from water-rich slurries to concentrated flows, hyperconcentrated flows and debris flows (e.g. Carrivick et al., 2004; Major et al., 2005). They are collectively called mass flows (also known as gravity flows) and they contain water as the clast-supporting medium instead of heated gases.

Because they are rich in volcano-derived particulate debris, the flows are generally referred to as lahars rather than floods. Cohesionless lahars are a type of concentrated or hyperconcentrated flow, whereas cohesive flows are known as debris flows (Mulder and Alexander, 2001). Because mass flows require water, they are particularly common hazards at snow- and ice-clad volcanoes. They can be particularly devastating to local populations in remote regions and the damage to infrastructure (roads, bridges, power lines) can be severe and difficult for small populations to sustain. For example, the small eruption of Gjálp in 1996 created a peak meltwater discharge that was approximately four times the mean discharge of the Mississippi River and resulted in widespread damage to infrastructure on the Skeidarársandur flood plain, on the south side of Vatnajökull, southern Iceland. The 1918 glaciovolcanic eruption of Katla is another particularly well documented and graphic example. It had a likely peak discharge of 300 000 $\text{m}^3 \text{s}^{-1}$ (Tomasson, 1996), which exceeds the average annual discharge of the Amazon River. The Katla outburst flood illustrates many of the features associated with large-volume jökulhlaups and is described in detail in Section 3.3.1. Some pre-historical glaciovolcano-sourced jökulhlaups in Iceland were immense floods, and included examples that may have been amongst the largest freshwater discharges in Earth history (e.g. Howard et al., 2012; but see Carrivick et al., 2013). These sudden glaciovolcanic outburst floods are common but they are generally short-lived.

As discussed above, lahar formation can be a direct consequence of lava flows or PDCs travelling across snow or ice, or potentially from the rapid dumping of ash destabilising an unstable snow or ice substrate (Capra et al., 2004). They can also be generated indirectly by longer-term release of volcanic heat into snow- or ice-filled craters (e.g. 1997 Mt Ruapehu: Manville and Cronin, 2007; 2005 Mt Chiginagak: Schaefer et al., 2008) or calderas (e.g. Grímsvötn: Gudmundsson et al., 2008), or even by snow- or ice-rich avalanches (Carrasco-Núñez et al., 1993). Mass flows can have very extensive runout distances, making them a very significant hazard, including those associated with glaciovolcanoes (Gudmundsson et al., 2008). For example, lahars generated by the 1980 glaciovolcanic eruption of Mt St Helens travelled over 100 km; those generated during the much smaller eruption at Nevado del Ruiz in Colombia travelled almost 150 km and caused more than 23 000 deaths; and a prehistoric eruption of Cotopaxi volcano in Ecuador is estimated to have travelled over 300 km from its source (the Chillos Valley Lahar; Mothes et al., 2004). The 1877 Cotopaxi eruption produced lahars on several flanks of the volcano and caused more than 1000 deaths (Mothes et al., 2004). Even in a relatively remote setting such as Alaska, infrastructure far downstream from volcanoes can be damaged by the long reach of these flows (Waythomas et al., 2013). Mass flows rich in clay-size material (>3–5% of fine fraction), referred to as cohesive, can maintain their destructive power longer than cohesionless flows (Capra et al., 2004; Fig. 15.6). Extensive studies of the effects of lahars on glacierised volcanoes have also been undertaken at Mt Rainier, Popocatepetl, Calbuco and Villarrica (e.g. Naranjo and Moreno, 2004; Castruccio et al., 2010; Delgado-Granados et al., 2015).

Snow and ice 'slurry' flows are unique to glaciovolcanic eruptions. They have been observed at several stratovolcanoes, but are best described for Mt Redoubt (Waythomas et al., 2013) and Mt Spurr (Waitt, 1995). The debris in these mass flows is dominated by fragments of ice. The flows appear to be initiated when hot volcanic materials entrain and melt enough ice to create a low-friction, water-rich matrix that enhances runout distances.

15.6 Avalanches

Avalanches at glaciovolcanoes are initiated in several ways and they have the ability to displace enormous rock masses (sometimes known as sector collapses) during volcanic eruptions. Ewert (2005) included the presence of 'permanent snow and ice cover' as one of several criteria used to evaluate the potential for sector collapses at volcanoes. In general, avalanches are essentially dry gravity-driven flows initiated by failure of oversteepened surfaces. During the 1980 Mt St Helens eruption, an earthquake initiated a huge avalanche of rock and ice and uncovered the underlying still-pressurised lava dome. A sudden

Fig. 15.6 Views of distal lahars from the 2009 eruption of Redoubt volcano in Alaska. Views of the Drift River oil terminal on 8 June 2009: (a) looking upstream along the Drift River towards Mt Redoubt showing lahar deposits adjacent to the oil infrastructure; (b) looking downstream towards the Pacific Ocean showing lahar deposits adjacent to the oil infrastructure (both images courtesy of M. Kaufman, the Alaska Volcano Observatory and the University of Alaska Fairbanks/ Geophysical Institute). The oil storage tanks are c. 60 m in diameter.

transition to a Plinian eruption then took place, driven by magmatic gas expansion. During subsequent smaller explosions, ballistic blocks impacting on snow- and ice-covered rock faces within the Mt St Helens crater initiated snow avalanches that transformed into lahars because the snow was melted by heat transferred from hot rock fragments and pyroclasts and by frictional heating.

The ability of thick ice to impound lava flows, leading to substantially increased thicknesses of ponded lava (e.g. Mathews, 1952a; Edwards et al., 2002; Harder and Russell, 2007; Figs. 10.1, 12.1), can also result in avalanches occurring thousands of years after the eruptions ceased. Once the ice has melted, the overthickened lavas are no longer buttressed and can collapse progressively, thus generating significant avalanche deposits (Moore, 1976). Capra (2005) suggested that a broad pattern of massive volcanic collapses can be correlated with major deglaciation events, in which rapid retreat of local supporting ice leads to extensive slope failures. Note that such collapses do not require any particular trigger other than gravity and time.

15.7 Lightning

Volcanic lightning is a localised enigmatic phenomenon that can also form in eruption plumes. Gudmundsson et al. (2008; also Larsen, 2002) reported that at least two people were killed by lightning during a glaciovolcanic eruption at Katla volcano in 1755, and suggested that lightning might be especially common during phreatomagmatic eruptions (probably because of the high volume of small charged particles injected into the atmosphere). Recent experiments confirm that the presence of very fine ash particles, which are abundant during phreatomagmatic activity (Liu et al., 2015), may facilitate or enhance volcanic lightning (Cimarelli et al., 2013).

15.8 Short-term versus long-term climate impacts

All volcanoes can release ash, gases and aerosols into the atmosphere which may block sunlight, reduce insolation and cause short-term global cooling. This is particularly important for eruptions that occur at mid- to low latitudes, where the enhanced widespread dispersal of volcanic ejecta may be on a hemispheric scale. However, any cooling effect will be short term and it can be mitigated or even negated depending on the volume of greenhouse gases, such as CO_2, also released during the eruption. However, aerosols and particulate matter eventually settle out of the atmosphere, whereas greenhouse gases have much longer residence times. Hence, even an eruption that causes significant short-term global cooling may contribute much more to global warming. It has been speculated that enhanced rates of volcanism associated with retreat and rapid thinning of ice sheets during the Last Glacial Maximum in Iceland led to enhanced magma production and higher eruption rates, which in turn promoted long-term global warming via high rates of discharge of volcanogenic CO_2 (see Chapter 14). While at present volcanoes are not a major source of CO_2, an increase in thinning of ice caps in volcanic regions may increase eruption rates, a positive feedback that may significantly enhance global warming.

15.9 Summary

The enhanced potential for the formation and involvement of water can lead to a number of hazards that are particularly characteristic of glaciovolcanic eruptions. Meltwater can be generated by lavas and pyroclastic currents simply by flowing over snow or ice. Turbulent mixing between hot volcanic debris and snow or ice efficiently facilitates the rapid intimate contact required to create the rapid generation of meltwater. On volcanoes with a thick glacial cover or confined within a volcanic structure (e.g. calderas), short-term storage of meltwater during glaciovolcanic eruptions can lead to particularly large floods (e.g. during the 1996 eruption at Gjálp, Iceland; see Preface). While many glaciovolcanic hazards are only important on a local scale, the enhanced production of fine to very fine ash during phreatomagmatic glaciovolcanic activity can have much more widespread effects hazardous to aviation and economically very significant (see Section 3.3.10). The fine ash and moist eruption columns may even facilitate volcanic lightning strikes. Many of these hazards were exhibited during the 2010 eruption of Eyjafjallajökull in Iceland, which highlights the need to better understand how energy is transferred from lava or tephra to snow and ice during glaciovolcanic eruptions.

16

Glaciovolcanism on Mars

16.1 Introduction

Mars is an enigma within Earth's Solar System. In the nineteenth and early twentieth centuries, telescope observations dramatically suggested the presence of linear features identified as canals, which were interpreted as part of an extensive irrigation system built by an intelligent civilisation (e.g. Lowell, 1906). However, the reality proved to be very different, with orbiting satellites and landing craft confirming that the planet is frigid and dry today and seemingly barren of life. Mars has a mean surface temperature of c. 215 K at the equator and it has a mean surface pressure of just 6.1 mbar. Thus, the surface conditions make liquid water unstable, implying that any liquid water that reaches the surface will rapidly freeze. Yet for more than 40 years, widespread geomorphological evidence suggested the former presence of liquid water, even on oceanic scales. More recently, claims have also been made for water melted from snowpack at the surface in geologically recent times (Christensen, 2003). The presence of water is important for it has profound implications for the climatic and atmospheric evolution of the planet. In this debate, the possible existence of subglacially erupted volcanoes has played an important part (e.g. Allen, 1979; Chapman and Tanaka, 2001; Chapman, 2003; Ghatan and Head, 2002). In this chapter, we review the evidence for water on Mars, followed by an appraisal of the theoretical background to Mars' volcanic eruptions and the observational evidence for glaciovolcanism.

16.2 Geological background

Mars is the second smallest planet in the Solar System. It is approximately half the size of Earth, has a surface gravity of 3.71 m s^{-2} and a surface pressure about 150 times lower than on Earth. Mars probably accumulated and differentiated into crust, mantle and core relatively rapidly after the formation of the Solar System, possibly within a few tens of millions of years. However, geological events occurring during the first 1.5 billion years are very poorly known. They were probably dominated by numerous large basin-forming impacts, including the one that created the conspicuous geographical division into two globally dominating terrains known as the southern highlands and northern plains, also called the Mars global dichotomy (Fig. 16.1; Waters et al., 2007). Mars' subsequent history

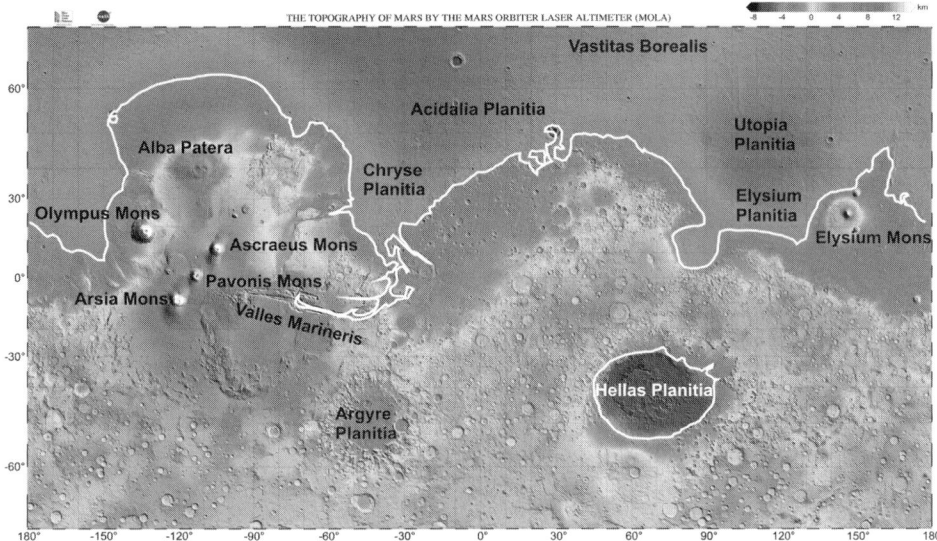

Fig. 16.1 Mars Orbiter laser altimeter (MOLA) map of the topography of Mars showing the pronounced separation into the two discrete physiographical provinces known as the Mars dichotomy. The white line is a depiction of a possible shoreline of a postulated Noachian ocean associated with the northern low-lying plains (shoreline from Clifford and Parker, 2001). (A black and white version of this figure will appear in some formats. For the colour version, please refer to the plate section.)

is better known, and it has been split into three major geological periods known as the Noachian, Hesperian and Amazonian (see Carr and Head, 2010 and references therein; Fig. 16.2).

The base of the Noachian Period is taken as the time of formation of the Hellas impact crater at c. 4.1–3.8 Ga. It ended around 3.7 Ga. The period was characterised by high rates of cratering, valley formation, erosion and construction of much of the Tharsis volcanic bulge. The Tharsis bulge is an enormous volcanic plateau situated in a low-latitude position and is home to the largest volcanoes in the Solar System. It includes the three enormous shield volcanoes of Ascraeus Mons, Arsia Mons and Pavonis Mons (Fig. 16.1). Growth of the Tharsis bulge is probably partly responsible for the tectonic initiation of the huge Valles Marineris canyon system, perhaps by rifting induced by dyke emplacement (McKenzie and Nimmo, 1999). Compared with the rest of Mars history, erosion rates were comparatively high in the Noachian but they are much lower than equivalent rates on Earth. The valley networks are very stream-like in appearance and opinion is undecided on their origin, whether by flowing water or by surface sapping, and what the climatic implications may be (Craddock and Howard, 2002). Warm wet conditions should have been necessary for the stabilisation of water at the surface and may also be responsible for the generation of widespread hydrous weathering products identified, such as phyllosilicates, but a cold hyperarid environment is also commonly assumed (cf. Clifford, 1993; Craddock and Howard, 2002; Hynek and Phillips, 2003; Carr and Head, 2010; Clifford et al., 2010).

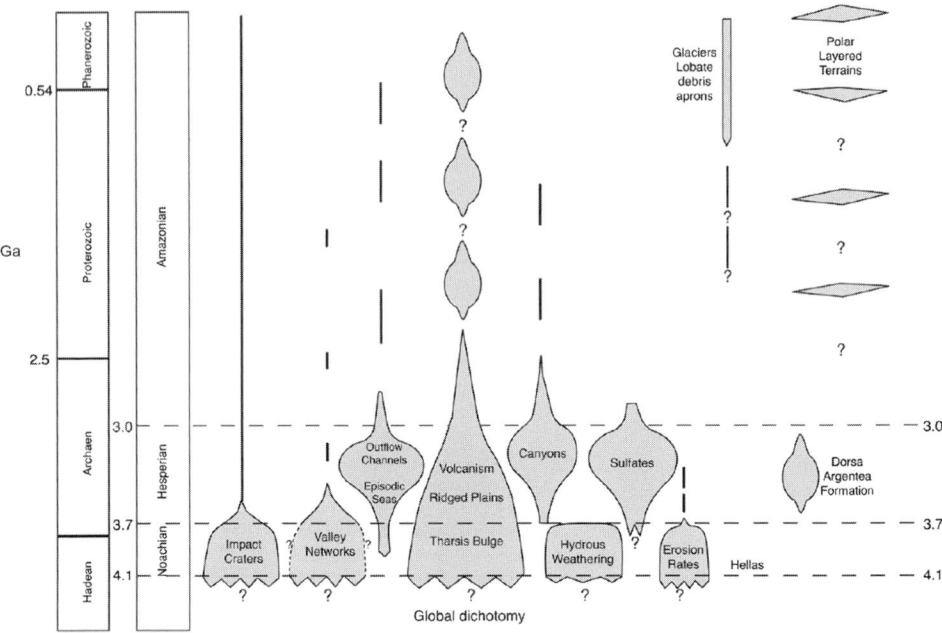

Fig. 16.2 Graphic depiction of geological activity on Mars over time (from Carr and Head, 2010). Many of the events shown were associated with water and may have had associated glaciovolcanism (e.g. outflow channels; Dorsa Argentea Formation).

Episodic warm and wet or dry and cold environments may have affected at least the final stages of the Noachian, perhaps linked to the presence of atmospheric greenhouse gases associated with impacts or volcanism.

The Hesperian Period dates from the end of the Noachian Period of heavy impact bombardment until c. 3 Ga. Its main characteristics are the development of extensive ridged lava plains, possibly episodically formed; formation of large canyons (e.g. Valles Marineris), outflow channels and possibly related lakes or seas; and substantially reduced rates of valley formation, erosion and production of phyllosilicates. Many of the surface changes are probably climate related and the conditions were probably colder and drier generally than in the Noachian, though possibly with reversals. The volcanism that created the lava plains was fed by extensive feeder dykes, but it also includes a number of distinctive low and shield-like central volcanoes known as paterae. Examples of the latter include Alba (now called Alba Mons), Apollinaris and Hadriaca paterae. Although particularly prominent peripheral to Tharsis, and in Hesperia Planum and Syrtius Major Planum, ridged plains also occur in low areas throughout the southern uplands region, including the floor of Hellas.

The remaining two-thirds of Mars geological history <3 Ga is known as the Amazonian Period, when the tempo of geological activity slowed yet again. The generally extremely low rates of erosion and weathering of the Hesperian continued, as did more modest rates of impact cratering and volcanism. For example, average rates of volcanism may have been approximately ten times slower than in the Hesperian, and it was geographically restricted mainly to the Tharsis and Elysium volcanic provinces. However, Olympus Mons, the largest and second tallest volcano in the Solar System, with a height of c. 22 km (almost three times the height of Mt Everest) and an area about three-quarters that of France, was also constructed during the period. The most distinguishing feature of the Amazonian is the extensive evidence for the accumulation and flow of glacial ice (e.g. Head et al., 2005; Baker et al., 2010; Souness and Hubbard, 2012; Souness et al., 2012). In addition to polar ice caps, glacier-like landforms are particularly prominent at mid latitudes and also occur at high elevations at low latitudes, and it is generally accepted that Mars' recent geological past has been dominated by the action of ice-related processes. Although small water floods may have occurred episodically, the main period of flooding that characterised the Hesperian was over.

16.3 A water inventory for Mars

The presence and inventory of water on Mars are amongst the most intriguing and enigmatic aspects of its history, affecting climate, surface weathering conditions, volcanism, the search for exobiology, and human colonisation (Carr, 1996, 2006; Cockell, 2004). Under current cold hyperarid conditions, surface ice is largely confined to the polar regions and only tiny amounts of water are present in the atmosphere. Mars' atmosphere is about 100 times thinner than that on Earth and it is unable to retain much heat energy. Thus, average surface temperatures are only c. 213 K (range: 120–293 K) and liquid water is unstable everywhere but at the poles. Surface ice is also unstable today because summer daytime temperatures are above the frost point even at mid to high latitudes. However, the planet sustains two prominent ice caps that are formed primarily of water ice together with lesser CO_2 ice. Additionally, the ubiquitous presence of subsurface water at mid latitudes and within the tropics is indicated by the widespread distribution of certain morphologically distinctive craters (known as rampart, excess-ejecta, perched and pedestal craters). The characteristic lobate ejecta and (in pedestal craters) steep marginal profiles probably originated from impact into a water- or ice-rich crust and its resulting fluidisation (Kadish and Head, 2011; Schon and Head, 2012). Some pedestal craters strongly resemble tuyas and may also be glaciovolcanic (personal communication, M. Chapman). Moreover, some paterae may have erupted explosively early in their development, producing pyroclastic ejecta most easily explained as a consequence of interaction of rising magma with water or ice in the crust (Crown and Greely, 1993; Wilson and Head, 2007a). Rootless cones signifying hydrovolcanic interactions between lavas and subsurface water ice have also been postulated, including in tropical locations (Fagents et al., 2002; Lanagan et al., 2001; Fuller and Head, 2002). Thus, the potential for a substantial water inventory on Mars is high (e.g. Carr, 1996; Balme et al., 2013). It has been estimated that more than 90% of the inventory probably resides in two distinct subsurface reservoirs: (1) as deep crustal groundwater, and (2) as ice within the cryolithosphere (Carr, 1996). The zonally averaged thickness of the cryolithosphere probably varies from 0–9 km at the equator to c. 10–22 km at the poles (Clifford et al., 2010). Here, we describe evidence from Mars' surface where water or ice are believed to have played an important part.

16.3.1 Integrated valley networks

Much of the Noachian southern highland terrain and some Hesperian features are incised by valley systems, i.e. anastomosing (branching) valleys that together resemble terrestrial drainage systems (Cradock and Howard, 2002; Hynek and Phillips, 2003). The individual valleys are typically a few kilometres wide and they may extend many hundreds to thousands of kilometres in length. Many have rectangular cross sections and are sourced in amphitheatre-like canyon heads, features that suggest an origin principally by groundwater sapping, but others have V-shaped cross sections more consistent with the flow of surface water (Craddock and Howard, 2002). There are problems with both interpretations (cf. Craddock and Howard, 2002; Carr and Head, 2010; Fairen, 2010) but either interpretation is indicative of the former presence of flowing surface water or groundwater that, because of the sub-zero surface temperatures on Mars, is likely to be a brine (Clifford, 1993; Wynn-Williams et al., 2001; Knauth and Burt, 2002; Clifford et al., 2010; Fairen, 2010). Sapping features in the valley networks become more common at the expense of surface runoff characteristics passing from the Noachian to the Hesperian (Carr and Head, 2010). Sedimentary deltas or fans are sometimes present where valleys debouche into topographical lows, such as craters and crater chains, suggesting that the lows were formerly occupied by lakes (Fig. 16.3). Several lake-filled craters, some as large as the Caspian Sea on Earth, have inlet and outlet valleys indicating that they were brim-full and overspilled (Fassett and Head, 2008).

16.3.2 Canyons (chasmata)

The Valles Marineris canyon system is one of the most conspicuous, spectacular and enigmatic features on Mars, with a lateral extent almost a fifth of the circumference of the planet (Fig. 16.1). The series of interconnected and unconnected enclosed troughs appears to have formed late in the Noachian but mainly in the Hesperian probably as a result of volcanotectonic stresses created by the growth of the Tharsis bulge. Dynamic upwarping may have occurred associated with magma rising from a long-lived mantle plume, with resultant movement and possible dyke injection along radial faults causing the formation of east–west striking grabens and horsts (e.g. Lucchitta et al., 1992; Mège and Masson, 1996). Valles Marineris spans c. 4000 km and reaches depths of 10 km, an order of magnitude greater than the Grand Canyon on Earth. Individual canyons range from 50 to 600 km wide (Chapman, 1994). The presence of mounds of layered sediments (which may

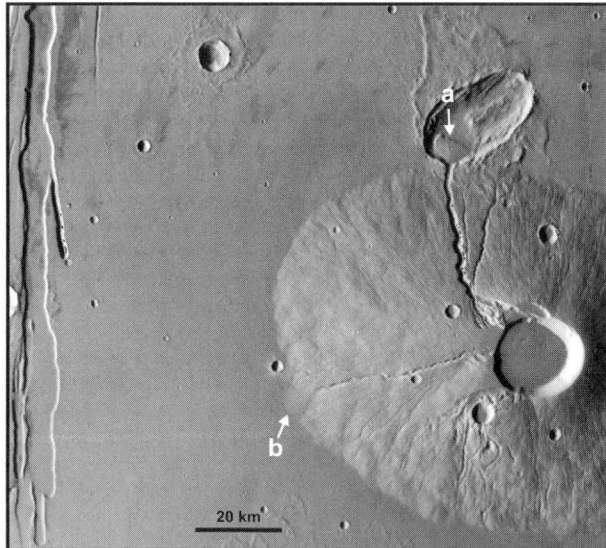

Fig. 16.3 Ceraunius Tholus, a Hesperian-age volcano in the Tharsis region. The volcano shows abundant evidence for water-cut runoff channels, possibly formed by basal melting of a summit ice cap, and marginal depositional sediment fans (two indicated by white arrows). That marked 'a' may have fed a delta, implying that the host crater was once water-filled (see Fassett and Head, 2007).

be volcanic; see Section 16.6.3) and sculpted streamlined landforms led to the hypothesis that the canyons were formerly filled by lakes that were catastrophically released. The water (or ice) filling the canyons may have accumulated due to climatic effects (Section 16.4) or else from repeated injections of groundwater linked to interactions between magmatism, faulting and cryospheric cracking (Section 16.6.5). In the latter scenario, it is likely the water then froze and the Valles Marineris canyons were once filled by ice. This suggestion is consistent with the presence of distinctive topographical features in the canyons known as sackung, including double-crested ridges and uphill-facing scarps formed by stress release on the slopes after the ice has melted, together with evidence for subglacial erosion and glacial landforms (Mège and Bourgeois, 2011; also Lucchita, 1982; Lucchitta et al., 1992). The ice was melted and released as immense jökulhlaups possibly during subglacial eruptions (Chapman et al., 2003). The channel dimensions imply that the deluges of water were as much as 10^7 km^3 in volume (for perspective, the 1918 Katla eruption produced c. 8 km^3 of water). Although large, they are an order of magnitude less than the oceanic volumes implied by some authors (cf. Parker et al., 1993; Carr and Head, 2003).

16.3.3 Outflow channels

Outflow channels are also major features of the Mars landscape, originating at Valles Marineris troughs, fossae (graben) and areas of chaotic terrain formed by ground collapse

(Section 16.3.4). They extend downslope to the low-lying northern plains, broadening and ultimately vanishing at around 40° N latitude. The channels are sinuous to straight, generally flat-floored and steep-walled, with distinctive streamlined islands, flow marks such as deep scours and grooves (striations) hundreds of metres long and several tens of metres deep, and other features characteristic of a flood origin (Baker, 1982; Baker et al., 1992). The channels have a wide range of likely ages but are mainly early Hesperian. Megascale flooding by liquid water is almost universally regarded as the major agent of erosion responsible for creating the channels (e.g. Carr, 1996) although an origin involving ice streams has also been proposed (Lucchitta, 1982, 2001). Similar to the formation of features on Earth such as the Channelled Scablands, Mars' channels may be linked to some combination of a massive staged release of groundwater or draining of surface lakes (i.e. jökulhlaups), probably linked to the geological and hydrological development of Mars' canyons (e.g. de Hon and Pani, 1993; Chapman et al., 2003). The floods may have been 100 times larger than the largest terrestrial floods.

16.3.4 Oceans and lakes

Whether oceans existed on the northern plains of Mars is highly contentious (cf. Baker et al., 1991; Parker et al., 1993; Head et al., 1999; Clifford and Parker, 2001; Kreslavsky and Head, 2002; Carr and Head, 2003, 2010; Fairen et al., 2003; Fig. 16.1). Their presence has been linked to large-scale catastrophic outflow from the major canyons and fossae. The plains are connected to outflow channels that are themselves linked to canyons via an intervening landscape type known as chaotic terrain. Chaotic terrain is a jumbled mess of mesas, buttes and hills cut across by valleys. The juxtaposed blocks individually measure tens of kilometres across and are separated by depressions hundreds of metres deep. However, the evidence for the largest ocean-scale bodies of standing water is equivocal. Shorelines have been variably proposed and refuted. The plains are draped by a widespread geological unit known as the Vastitas Borealis Formation whose boundaries approximately coincide with one of the two proposed major shorelines and that is believed to comprise sediments carried by large floods (Clifford and Parker, 2001; Kreslavsky and Head, 2002). Numerous features mapped on the northern plains, including some that may be glaciovolcanic, suggest that the water rapidly turned to a stagnant ice sheet that possibly exceeded 1000 m in thickness, which then sublimed (Allen, 1979; Chapman, 1994; Kargel et al., 1995; Kreslavsky and Head, 2002). Indeed, younger lava flows of the Late Hesperian Syrtis Major Formation interacted with the Vastitas Borealis Formation, becoming broken up and disrupted as underlying ice melted in the Vastitas Borealis Formation and providing confirmation that the latter was formerly volatile rich (Ivanov and Head, 2003; also Kreslavsky and Head, 2002). Some ice may even remain locally (Murray et al., 2005). Much smaller-scale water-ice lakes have also recently been imaged on Mars and demonstrate unequivocally the presence of surface water (Fig. 16.4).

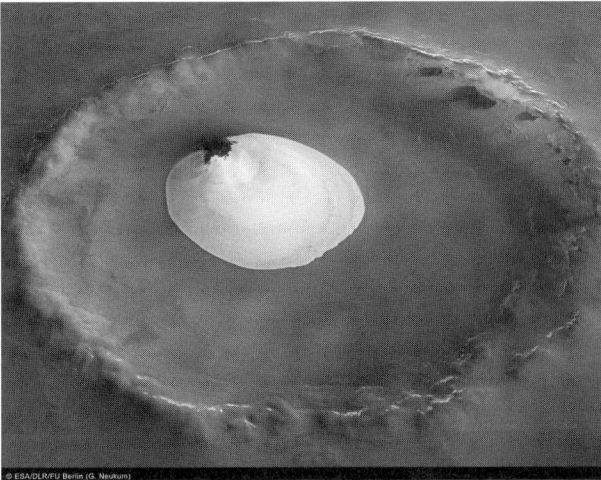

Fig. 16.4 Mars Express high-resolution image of a frozen lake within a crater near the north pole on Mars, proving the presence of liquid surface water (now ice) on the planet. The lake is c. 10 km in diameter. (Image: ESA/DLR/FU Berlin; G. Neukum.)

16.3.5 Polar caps and a circumpolar ice sheet

Mars sustains two polar caps (ice sheets) composed of water ice and lesser CO_2 ice. The northern cap on Mars has a maximum seasonal diameter of c. 1000 km and a thickness of c. 3 km. Its surface has an average age of c. 10^5 years and the deposits are divided into an upper unit of finely interlayered ice and dust up to 2 km thick resting upon a basal massive ice unit c. 1 km thick. By contrast, the south polar cap has a diameter of c. 350 km and a thickness of c. 3 km. Its surface is significantly older than the northern cap, with an age of c. 10^7 years, and it is also thinly layered. Both caps have volumes of c. 1.6×10^6 km^3. The reason for the disparity in ages is unknown (Carr and Head, 2010). The layering in both caps probably preserves a detailed record of geologically recent climatic changes on Mars and may be caused by modulations in orbital and rotational motions of the planet affecting the relative accumulation of dust and ice (e.g. higher incidence of dust storms at higher obliquities; Head et al., 2003).

There is also substantial evidence for a formerly much more extensive south polar cap or ice sheet of Hesperian age on Mars (e.g. Baker et al., 1991). It includes an extensive layered volatile-rich geological deposit known as the Dorsa Argentea Formation and probably some associated units previously mapped as Hesperian–Noachian (undivided) in age (Head and Pratt, 2001). The unit covers a surface area that may be as large as 2.94×10^6 km^2, which is more than twice that of the present south polar cap (Fig. 16.5). Important features that suggest the former presence of ice in the Dorsa Argentea Formation include a northwest-displaced asymmetry of the outcrop mimicking that of the south polar cap, which it partially encircles; numerous sinuous esker-like ridges;

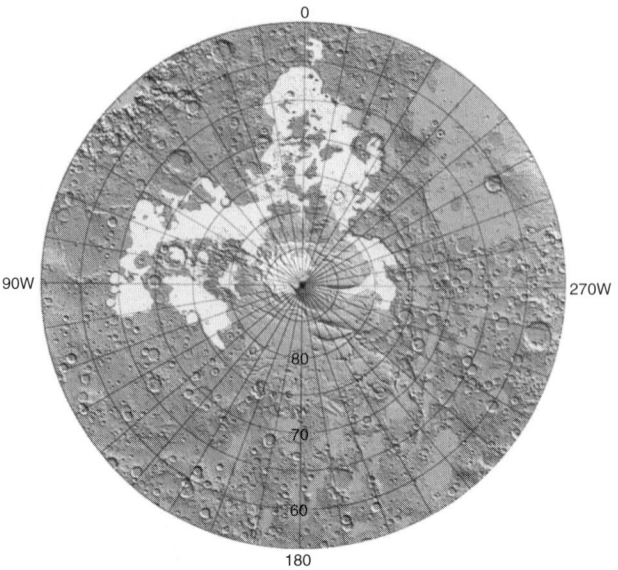

Fig. 16.5 Geological map of Mars' southern hemisphere showing the present ice cap (white), Amazonian polar layered terrain (grey), Hesperian–Noachian undivided (purple), and the Hesperian Dorsa Argentea Formation (yellow) (from Head and Pratt, 2001). The outcrop of the Dorsa Argentea Formation, a likely glacial deposit that contains numerous putative glaciovolcanic edifices (Fig. 16.10), extends far beyond the present polar ice cap and suggests the presence of a much more widespread ice sheet in Hesperian time. (A black and white version of this figure will appear in some formats. For the colour version, please refer to the plate section.)

pedestal craters; wide steep-sided pits with flat interiors (including Angusti and Sisyphi cavi) probably formed by basal melting of a volatile-rich deposit (Ghatan et al., 2003); and distinctive landforms interpreted as subglacially erupted volcanoes (Ghatan and Head, 2002; Section 16.6.2). The depths of the pits and heights of the glaciovolcanic edifices suggest that the original ice sheet was up to 1500 m thick and a maximum thickness of c. 2500 m is possible. A significant volume of ice probably remains in the deposit.

16.3.6 Tropical mountain ice

The possibility that Olympus Mons, the largest volcano on Mars, erupted in association with a tropical mountain ice cap 2–3 km thick in Hesperian–Amazonian times (mainly 3.67–2.54 Ga; Isherwood et al., 2013) was suggested by Hodges and Moore (1979) in order to explain unusual features of the volcano, specifically its encircling 6 km-high cliffs and geographically very extensive aureole deposits. Despite the enormous differences in scale and its polygenetic origin, the formation of the cliffs of Olympus Mons was compared with the steep margins of Icelandic tuyas, whilst the aureoles were regarded as possible tindar ('móberg') deposits created during widespread subglacial eruptions that failed to penetrate the thick ice cover. However, other origins have also been proposed (e.g. the volcano flanks were unstable due to loading, gravity spreading and landsliding following construction of the volcano on a weak substrate; Comer et al., 1985; Tanaka, 1985; McGovern and Solomon, 1993). Helgason (1999) modified the model significantly whilst retaining a subglacial origin. He also suggested that depressions associated with possible jökulhlaup fans on nearby Arsia Mons represented melting and flooding effects above vents covered by ice several kilometres thick and suggested that the ice cover may still exist (see also Scanlon et al., 2014). Cold-based mountain ice caps of late Amazonian age have also been postulated on other Tharsis volcanoes based on glacial geomorphological criteria, including arcuate ridges interpreted as drop moraines, knobbly terrain interpreted as sublimation till, possible rock glaciers, some of which may still be cored by glacier ice (Head and Marchant, 2003; Shean et al., 2005, 2007; Fastook et al., 2008), and the possible recognition of glaciovolcanic features (Section 16.6.4). The landforms are consistently found on the northwest flanks of the Tharsis volcanoes where glacier ice spread out as piedmont-like fans extending several hundred kilometres onto the surrounding low-lying landscape (Fig. 16.6). The studies were supported by Mars atmospheric general circulation models, which showed that

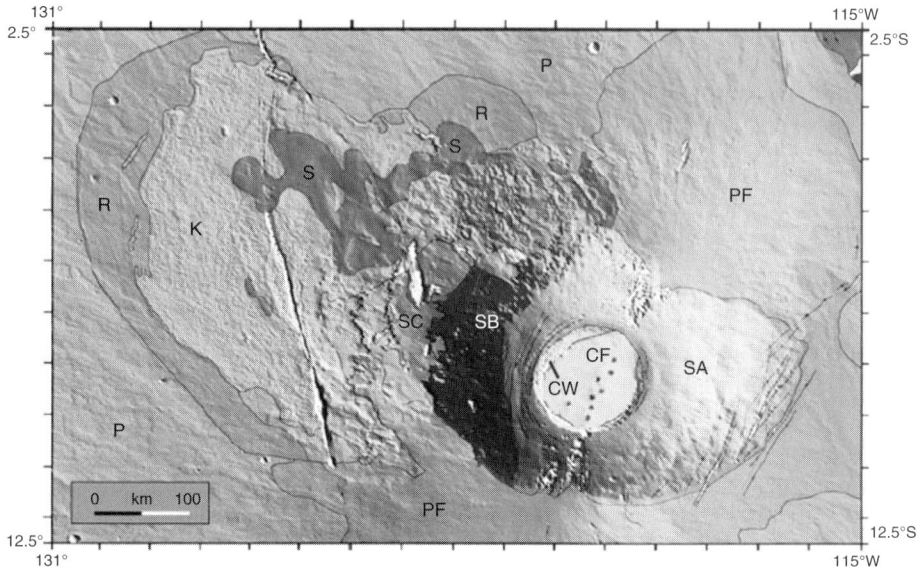

Fig. 16.6 Geological sketch map of Arsia Mons showing geological units believed to be glacial, including ridged (R), knobby (K) and smooth (S) deposits. They form a prominent fan-shaped deposit and are interpreted as drop moraines, sublimation till and rock glacier deposits, respectively (from Head and Marchant, 2003). (A black and white version of this figure will appear in some formats. For the colour version, please refer to the plate section.)

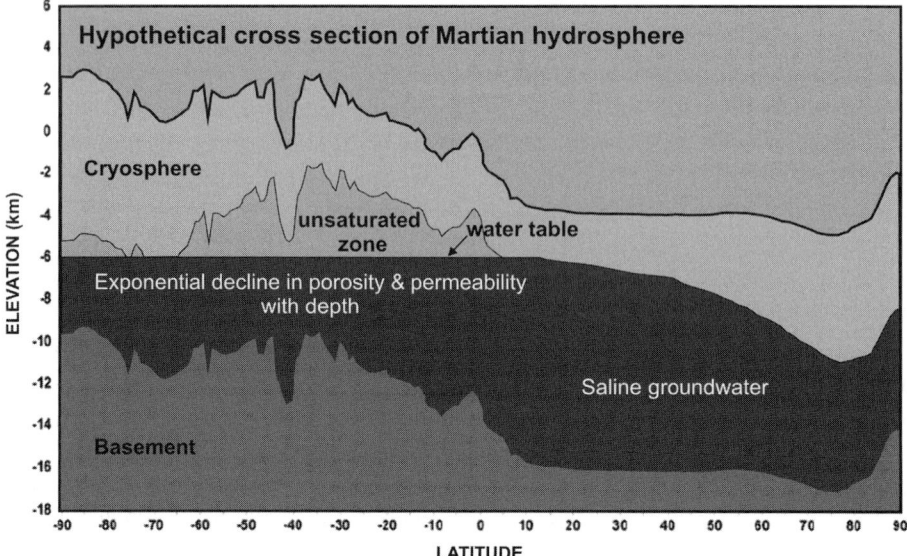

Fig. 16.7 Hydrological conditions on Mars illustrated by a hypothetical pole-to-pole cross section of the present-day crust and inferred relationships between surface topography, ground ice and groundwater (modified after Clifford and Parker, 2001). The basement shown is where porosity and permeability vanish.

tropical mountain glaciers could form during periods of high Mars obliquity, with asymmetrical snow accumulation favoured on the flanks rather than the summits.

16.4 The hydrological cycle on Mars and glacial–interglacial cyclicity

Clifford (1993) demonstrated that water on Mars probably shuttles between the atmosphere and a deep crustal reservoir of groundwater via cold sinks (ice caps) at both poles (Fig. 16.7). The model is a natural consequence if the outgassed inventory of water exceeds the pore volume of the cryolithosphere by at least a small percentage. Under the cold arid climate that prevails today, the thermal instability of ground ice at the low to mid latitudes causes net atmospheric transport from the warm equatorial regions to the cold poles, where the vapour precipitates as snow and forms ice caps. Basal melting takes place below the polar caps when they reach an appropriate thickness (at a depth related to the snow accumulation rate and local geothermal gradient; Clifford et al., 2010). A hydraulic head is thus created in the crust by the raised elevation of the basal melting zone relative to that at lower latitudes. The meltwater thus released is driven away from the polar regions via the sub-cryolithospheric hydrosphere (Fig. 16.7), where the presence of even a small geothermal gradient forces the water to discharge vertically as vapour to higher elevations in the crust. The water then refreezes and ultimately sublimes, passing back into the atmosphere and commencing the next cycle.

On a much longer (geological) timescale, the hydrological cycle is also driven by insolation-related climate change linked to adjustments in orbital parameters, which can reverse the hydrological cycle described by Clifford (1993). Unlike Earth, Mars undergoes radical variations in obliquity (i.e. the angle between Mars' pole of rotation and the ecliptic plane). From numerical modelling, we know that the present obliquity of Mars (25°) is anomalously small. Much higher obliquities, perhaps approaching 47°, occurred in the past (Touma and Wisdom, 1993; Laskar et al., 2004). At the higher obliquities (i.e. exceeding 30°), the increased insolation acting on the polar caps results in loss of water by sublimation. The water vapour is redistributed atmospherically towards the equator where, at the lower prevailing temperatures there, it precipitates. Prior to c. 5 Ma, Mars obliquity regularly exceeded 45°. At those obliquities, the polar sublimation is very high and climate models also envisage that equatorial ice would persist (Jakosky and Carr, 1985; Head et al., 2003; Schon and Head, 2012; see also Murray et al., 2005). By these processes, the models predict that ice thicknesses of tens of metres will be removed from the poles. The planet's ice mass becomes redistributed across its surface until lower obliquities (<30°) again prevail, ice once again sublimes from the lower latitude deposits and the cycle reverses. The result is a glacial–interglacial repetition over geological time and supporting evidence for its existence has been presented for ice ages in the mid latitudes and mountain glaciers in the tropics (e.g. Head and Pratt, 2001; Ghatan and Head, 2002; Ghatan et al., 2003; Head et al., 2003, 2005; Head and Marchant, 2003; Shean et al., 2005, 2007; Fassett and Head, 2006, 2007; Fastook et al., 2008). Mars' glacials were probably characterised by widespread ice sheets that extended from the poles down to mid latitudes (where they may only have been metres thick). Although this sounds similar to the glacial cyclicity on Earth, there are major differences between the two planets. On Earth, glacial periods are characterised by colder temperatures and lower polar insolation. The converse is true on Mars, i.e. polar insolation is higher and glacials are *warmer* climates.

16.5 Theoretical aspects of Mars' glaciovolcanism

Because of the inferred widespread presence of water and ice on Mars, it will have had a critically important impact on the style of eruptions there, as well as on the volcanic products and processes involved, just as on Earth (cf. Smellie, 2000, 2009, 2013; Head and Wilson, 2002, 2007; Wilson and Head, 2007; Wilson et al., 2013). However, there are also fundamental differences between eruptions on the two planets, as a consequence of the different planetary properties. Mention has already been briefly made (Section 16.2) of contrasting planetary properties such as gravity (one-third that on Earth) and surface pressure (150 times lower than on Earth). These and other properties (e.g. surface temperature, magma composition) can cause important variations in eruptive style.

16.5.1 Eruption styles

Eruption styles are modulated by the Mars environment, of which the most important controls are gravity and the low density of the Mars atmosphere (Wilson and Head, 1983, 1994, 2007a). Because of the lower gravity, vertical migrations of magma will be slower and magma volumes will be larger and ascend to shallower levels than on Earth, although magma reservoirs are expected to be deeper on Mars by a factor of about four. Dyke widths will also be greater, by a factor of about two, and that will enhance effusion rates by a factor of five. The lower gravity and atmospheric pressures will also ensure that disruption of rising vesiculating magma on Mars will occur at systematically greater depths than on Earth. Lava heat-loss processes are comparable, but cooling-limited lava flows will be considerably longer and characterised by higher effusion rates because of the lower Mars gravity. This permits the construction of very much larger effusive volcanoes, typically with diameters of c. 500–700 km.

Another consequence of Mars' low gravity and atmospheric pressure is that during pyroclastic eruptions the largest clasts will fall closer to the vent, but the finer ones will be dispersed far more widely. Thus, even for Strombolian eruptions, the cones formed will be somewhat broader and lower compared with Earth, and surrounded by a widely dispersed mantle of very fine ash. Craters will also be broader by a factor of up to at least five. Pyroclast dispersal is enhanced because eruption columns will be substantially taller due to Mars' atmospheric pressure and temperature structure. Columns will rise about five times higher than on Earth for the same eruption rate. Moreover, because it is likely that column height:deposit width relationships will be the same on both planets, the fall deposits on Mars will be dispersed to substantially greater distances, with outcrop widths ranging from several tens to a few hundred kilometres. Finally, because of enhanced magma fragmentation on Mars, pyroclastic grain sizes will be markedly finer than on Earth; at least one order of magnitude finer for Hawaiian eruptions and about a hundred times finer for Plinian eruptions. Another consequence of the enhanced fragmentation is that basaltic Plinian eruptions, rare on Earth, may be relatively common on Mars, as will eruptions producing pyroclastic density currents, since eruption cloud instability and collapse will be favoured. The fountains feeding the pyroclastic density currents will also be more than twice as high on Mars and the distances travelled by the currents will be much greater, perhaps by a factor of three.

16.5.2 Impact of Mars environment on glaciovolcanism

Because of the combination of widespread volcanism and the ubiquitous presence of ice on Mars, particularly an unusually thick cryolithosphere, glaciovolcanism is likely to be similar to Earth but modulated by the different Mars environment (Head and Wilson, 2002, 2007). A wide array of glaciovolcanic eruption styles is possible and is illustrated in Fig. 16.8). Enhanced thermal fluxes surrounding magma chambers significantly alter the hydrology in volcanoes, heating groundwater and melting ice in the cryolithosphere to

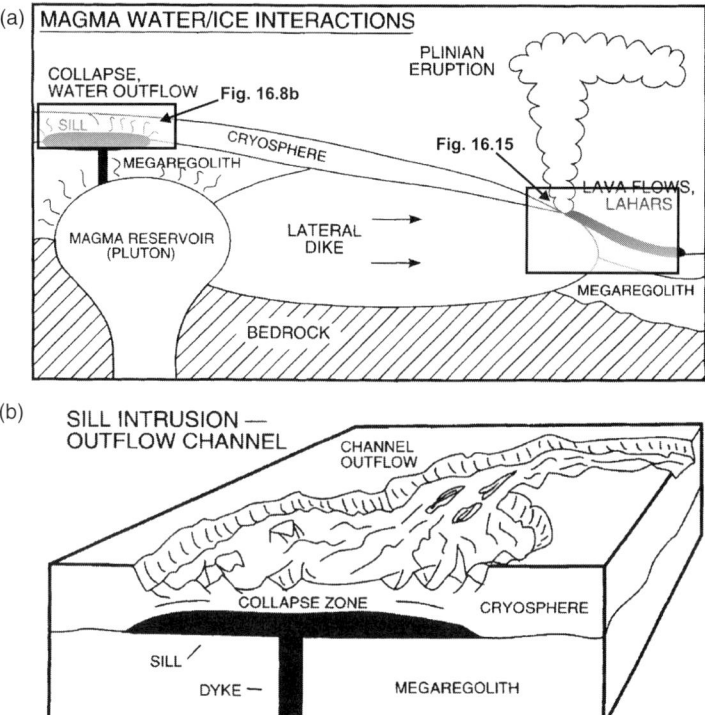

Fig. 16.8 Examples of Mars environments with magma–ice interactions (from Head and Wilson, 2002). Because Mars' cryosphere is underground (i.e. 'cryolithosphere') and it currently lacks surface ice sheets apart from polar caps, no equivalents of Earth's glaciovolcanic landforms (tuyas, tindars, domes, etc.) can form today. The scene depicted in (b) is dependent on the presence of a thick cryolithosphere intruded by a sill, leading to surface collapse, creation of chaotic terrain and release of voluminous meltwater as channel outflows; it also cannot occur on Earth.

create a hydrothermal circulation. This may initiate or enhance the basal melting of an overlying ice cap and lead to the formation of valley networks on the summits of volcanoes (e.g. Gulick, 1998; Fassett and Head, 2006; see Section 13.6). Sills and dykes will also melt pore-space ice in the cryolithosphere or any volatile-rich surface layer (Squyres et al., 1987; McKenzie and Nimmo, 1999; Ghatan and Head, 2002; Ghatan et al., 2003). As a result, the overlying landscape will subside, leaving a steep-sided flat-floored depression or chaotic terrain whose depth reflects the thickness of the volatile-rich layer. A thickness of meltwater much greater than the thickness of the sill will be created which, if the local permafrost seal is breached, will trigger eruption of meltwater deluges onto Mars' surface (e.g. Squyres et al., 1987; Head and Wilson, 2002, 2003; cf. McKenzie and Nimmo, 1999). Dykes intruding faults will also cause melting of ice in the cryolithosphere, melting the ice out to several dyke widths but with heat especially focused at the top of the dyke (Head and Wilson, 2002, 2003). In addition to generating meltwater, violent mechanical and thermal mixing associated with overpressures of a few hundred MPa may lead to fuel–coolant-type interaction and phreatomagmatic eruptions, and ejection of laterally widespread (many tens of km) mud-like deposits (Wilson and Mouginis-Mark, 2003; Russell and Head, 2003). Reactions such as these for sills and dykes injected into ice-rich permafrost should differ from intrusion into a layer of surface ice (i.e. ice sheets or caps). With surface ice, eruptions should result in the construction of pillow volcanoes, tindars and tuyas as they do on Earth (e.g. Chapman, 2003; Ghatan and Head, 2002). However, under Mars' lower gravity and much colder temperatures, and the likelihood of more voluminous magmas, the resulting edifices will be substantially taller and wider compared with their terrestrial equivalents (Head and Wilson, 2002; Smellie, 2009, 2013; Fig. 16.9). Finally, lavas flowing across ice or a volatile-rich layer will cause substrate melting and the lavas will collapse into rubble (Ivanov and Head, 2003; Scanlon et al., 2014). It may also result in the explosive release of water vapour, forming rootless cones (pseudocraters), but with cone and crater diameters much larger than on Earth (cones up to 1500 m across; crater diameters 200–1000 m; e.g. Fagents et al., 2002; Lanagan et al., 2001; Fuller and Head, 2002; Hamilton et al., 2010, 2011).

16.6 Observations of Mars' glaciovolcanism

It was recognised from the earliest days of Mars exploration that the discovery of subglacially erupted volcanic centres on Mars would provide prima facie evidence for the former presence of surface ice, at a time when the existence of water on Mars was very controversial (Allen, 1979; Hodges and Moore, 1979). It was also known that the heights of those centres provide information on the thicknesses of past ice sheets. The morphology and morphometry of glaciovolcanic edifices, especially those of tuyas with their more or less flat tops, summit craters and steep flanks, and to a lesser extent the features of tindars, are very distinctive and generally not mimicked by other landforms (Mathews, 1947; Jones, 1969b; Smellie, 2009, 2013). The discovery, therefore, of numerous landforms on Mars that resemble glaciovolcanic constructs has had an important impact on the development of ideas regarding the water inventory of the planet.

16.6.1 Northern plains

Mars' northern plains may have been the site of accumulation of either vast oceans or large lakes linked to enormous successive meltwater discharges from the canyon systems and fossae (Sections 16.3.2–16.3.4). Because of the very low ambient temperatures, any lakes or oceans would have frozen rapidly and ultimately sublimed. Volcanic edifices have been observed that appear to confirm the former presence of those frozen oceans or lakes. Large-scale surveys have revealed the presence of numerous mesa-like features that resemble tuyas (Allen, 1979; Hodges and Moore, 1979; Parker et al., 1993; Martínez-Alonso et al., 2011). They occur dispersed across the northern plains, near the boundary with the southern

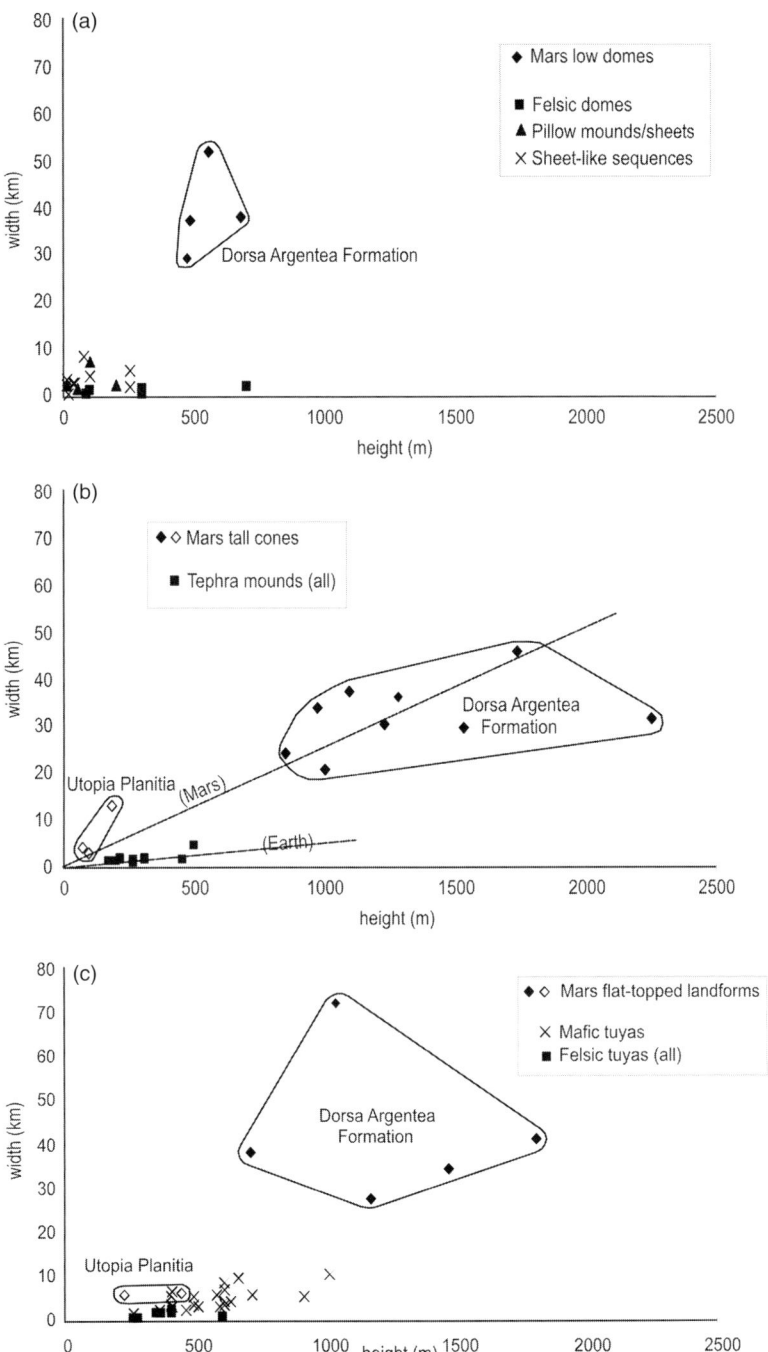

Fig. 16.9 Diagrams illustrating aspect ratios (height vs width) for morphologically distinctive Mars landforms, which may be glaciovolcanic, compared with terrestrial glaciovolcanic landforms. In B and C, note how the Mars examples are generally much higher and wider than their terrestrial equivalents (from Smellie, 2009). The Dorsa Argentea Formation low domes (in A) are probably not glaciovolcanic.

highlands, and concentrated in the prominent embayments known as Chryse and Acidalia planitiae. Some of the mesas are capped by summit ridges, domes, cones, lava flows and craters suggesting a volcanic origin. The mesas are generally 1.5–8 km in width (average: 2400 and 3900 m wide and long, respectively), 100–500 m tall and their steep flanks appear draped in talus. Their summit hills rise between 100 and 300 m above the remarkably flat mesa tops. Linear rocky outcrops also poke through the flanking talus slopes, suggesting layering.

Numerous ridges were also identified in the early studies and assigned, on less diagnostic criteria, as 'móberg' ridges. The thickness of the former ice cover was estimated as c. 100–1200 m. Additionally, a series of tindar-like features downslope of graben-like features known as the Elysium Fossae were identified on the flanks of Elysium Mons where it progresses down into Utopia Planitia (Chapman, 1994, 2003). They comprise dyke-fed volcanic ridges that show no evidence for interaction with ice at their upslope ends in the fossae, where they were probably responsible for subaerial lava effusion. The ridges transform downslope at similar elevations into rough textured linear features resembling tindars (i.e. composed of explosively erupted materials due to volcano–ice interaction). The relationships were interpreted as an indication that the ridges intersected extensive thick surface ice at the lower elevation. Chapman (1994) also suggested that fossae formation was also due to explosive volcano–subsurface ice interactions from fissures, producing chains of maars (pits) that subsided to create the fossae. Alternatively, Russell and Head (2003) interpreted the fossae as the graben-like surface manifestation of dyke emplacement radial to Elysium Mons (cf. McKenzie and Nimmo, 1999). The fossae, which are 1–22 km wide and have depths of 100–1100 m, were also the sources for geographically extensive lobes at lower elevations in Utopia Planitia. The lobes are on average about 50–70 m thick and have a total volume estimated as 14 000–19 000 km^3. They are interpreted as lahar deposits generated either by large jökulhlaups during eruptions in lakes melted in the ice sheet and in ice-marginal lakes situated within the fossae (Chapman, 1994, 2003; see also Christiansen, 1989), or during the discharge of large volumes of groundwater (and entrained materials) overpressured by the capping cryolithosphere and released catastrophically by dyke-induced fracturing that disrupted the cryolithosphere (Russell and Head, 2003). With a likely Early Amazonian eruption age, the lahars represent one of the youngest large outflow events known on Mars. The volume of water was probably enhanced by melted subsurface ice created during dyke intrusion. Conversely, phreatomagmatic explosivity related to sill emplacement in the shallow subsurface above a dyke was probably involved in the formation of Hrad Vallis, a series of fossae on the north flank of Elysium Mons (Wilson and Mouginis-Mark, 2003). The Hrad Vallis fossae differ from others in the area, and in Mars generally, in having constructional ramparts due to the inferred pyroclastic activity.

Martínez-Alonso et al. (2011) described features in Chryse and Acidalia planitiae interpreted as glaciovolcanic. In addition to mesas regarded as tuyas, they also identified other possible glaciovolcanic features, including: buttes 200–1200 m across with stepped or ribbed surfaces that are variably lobate in plan and interpreted as subglacial mounds;

laterally extensive sinuous and sheet-like outcrops formed of tightly packed, rounded 'blocks' individually up to 10 m across, some with an internal zoning structure, interpreted as ice-confined pillow mounds (pillow sheets?; see Section 10.5); and large sinuous ridges 50–600 m wide that can be followed more than 17 km. The sinuous ridges have flat level tops and steep flanks and are interpreted as lavas that were emplaced subglacially in bedrock valleys and that have undergone topographical inversion (cf. Walker and Blake, 1966). Thinner sinuous ridges that drape both topographical highs and lows may be esker-like lavas emplaced on and melted into coeval ice, whereas hollows 260–1900 m in diameter with steep walls, flat floors and lacking raised rims were interpreted as kettle-holes formed in some cases by volcano–ice interaction after far-travelled lavas flowed across ice-containing deposits, melting their substrate.

16.6.2 Southern circumpolar region

The Dorsa Argentea Formation is a middle Hesperian-aged volatile-rich unit with a distribution confined to within c. 9000 km of the south pole. Together with geological units mapped as Noachian–Hesperian (undivided) that crop out within the boundary of the Dorsa Argentea Formation, they are believed to represent the residue of a former south polar ice sheet at least twice as extensive as the present south polar cap. Based on the depths of steep-sided, flat-bottomed subsidence pits developed within the formation and heights of associated glaciovolcanic edifices, the ice sheet may have had an original thickness of c. 1500–2500 m (Head and Pratt, 2001; Ghatan and Head, 2002; Ghatan et al., 2003; see Section 16.3.5). The Dorsa Argentea Formation is also associated with 21 mountain features that morphological and morphometrical analysis suggests are volcanic, and most may be glaciovolcanic (Fig. 16.10; Ghatan and Head, 2002). Several of the mountains occur outside of the Dorsa Argentea Formation outcrop (Fig. 16.5). Basal diameters of the mountains range between 20 and 70 km (mean: 40.3 km) and summit elevations are variably 500–2200 m above their bases. Individual volumes vary between 75 and 1964 km^3 (average: 700 km^3). The mountains are divided into three main morphological types: (1) low domes, (2) flat-topped (with or without a summit crater), and (3) cone-shaped (also with or without a summit crater; Fig. 16.10). Despite their low slopes (<10°, often <5°; lowest for the domes), the gradients tend to be steeper than for other volcanic (non-glaciovolcanic) constructs on Mars. This leads to broad summits on the features and very low aspect ratios (i.e. height to width; cf. Smellie, 2009). The craters also tend to be shallow (c. 200 m) and large, often dominating the mountain summits, and four examples may be calderas. Many of the mountains are also surrounded by roughly circular shallow depressions c. 50–100 m deep (exceptionally 250–300 m) lacking raised rims. They extend out to about one mountain-diameter and grade gently up to the level of the surrounding plain. Sinuous channels carved into the Dorsa Argentea Formation also extend away from some of the mountains.

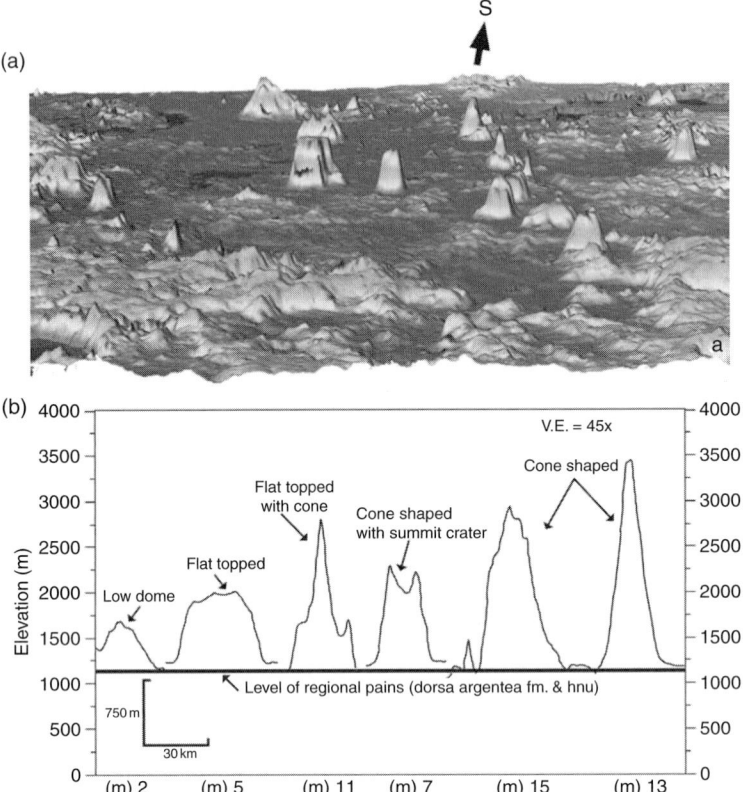

Fig. 16.10 (a) Perspective view of putative glaciovolcanic mountains in the Dorsa Argentea Formation near the south polar cap. Vertical exaggeration 70× (from Ghatan and Head, 2002). (b) Series of MOLA topographical profiles of the mountains shown in (a), illustrating the distinctive morphologies for each. All the features except low domes are thought to be glaciovolcanic.

Because the mountains are closely associated with the Dorsa Argentea Formation, a glacial setting and glaciovolcanic origin are permissive. The distinctive encircling depressions hint at a possible interaction of the volcanoes with the surrounding Dorsa Argentea Formation. They are believed to be subsidence pits caused by base-up melting of the Dorsa Argentea Formation triggered by enhanced geothermal heat associated with the individual volcanoes. Basal melting may be indicated by the sinuous (meltwater?) channels and esker-like ridges also seen in the Dorsa Argentea Formation. However, the interpretation of basal melting relies on an assumption that the depressions are underlain by shallow hypabyssal intrusions (whose cooling provided the heat). The heat requirement based on the dimensions of the depressions necessitates that the volume ratio of subsurface to surface magma is about 4:1 (Ghatan et al., 2003).

The cone-shaped mountains, which are also the tallest constructs, fall within the Dorsa Argentea Formation outcrop close to its centre, where the former ice sheet would have been thickest. The lack of flat summits to the cones suggests that these volcanoes terminated their activity within an englacial lake and did not evolve up into subaerial lava effusion. They are interpreted as glaciovolcanic pyroclastic cones (i.e. products of subglacial eruptions that arrested at a tindar stage). Due to their very low aspect ratios, relatively steep flanks and flat tops, some surmounted by a conical mound, the flat-topped mountains most closely resemble glaciovolcanic tuyas, whose eruption persisted after the volcano penetrated the surface of the surrounding Hesperian ice sheet. Finally, on the basis of their very low slopes, low heights and occurrence outside of the Dorsa Argentea Formation outcrop (in itself undiagnostic), Ghatan and Head (2002) suggested that the low domes may be subaerial volcanoes. Another possibility, that the low domes may be glaciovolcanic pillow mounds, seems unlikely: the low domes lack the steep margins that probably would have developed had they been ice confined (cf. Scanlon et al., 2014).

16.6.3 Valles Marineris

The possibility of glaciovolcanism within the prominent chasms of the Valles Marineris has long been known (e.g. Lucchitta et al., 1992, 1994), principally affecting the origin of a distinctive group known as interior layered deposits (ILDs). These ILDs are Hesperian-age, often mesa-like landforms having horizontal to dipping beds of laterally continuous materials that abut the Noachian chasmata walls, although often separated from the walls by prominent moats; some also overlie chaotic terrain. The individual layers are probably a few to tens of metres in thickness that stack up to form kilometres-thick sequences. They may reach 9 km thick in the mesas and benches of Hebes, Ophir and Candor chasmata but are somewhat thinner (1–4 km) in south Melas, Ganges, Capri, Eos and Juventae chasmata. They are the volumetrically dominant deposit type within the chasmata, with a total volume amounting to c. 60% of the volume of all deposit types present.

Interior layered deposits have principally been identified as either lake deposits, tufa (spring) mounds or of volcanic origin. Problems with a lacustrine origin include (1) a near-absence of channels on the surrounding Noachian plateau surfaces indicating flow of water into the troughs; (2) some troughs are brim-full of ILD and at higher elevations than in other chasmata, in which ILD surfaces may be 1–4 km lower than the surrounding canyon rims; (3) a lack of shorelines; and (4) adjacent mounds have very different stratigraphies. Objection 2 is important only if the troughs are interconnected. Lake waters with surfaces at elevations of 8 km (as in Ophir and Candor chasmata) would spill out of Coprates Chasma onto the surrounding plateau surfaces that have elevations of just 4–5 km; there is no evidence for such spillage. Conversely, filling a series of independent lakes would require dams to block flow between the chasmata, for which there is also no evidence. Alternatively, if the ILD are volcanic in origin, they could have been emplaced beneath ice (i.e. glaciovolcanic; Komatsu et al., 1993, 2004; Lucchitta et al., 1994; Chapman and Tanaka, 2001; Chapman, 2002, 2003; Chapman et al., 2003; Chapman and Smellie,

2007). Major climatic shifts related to periods of high obliquity can cause ice to accumulate within Valles Marineris (Section 16.4). A volcanic origin is suggested by the freestanding nature of the ILDs; the steep sides and mesa-like tops of many; the construction of the mesa caprocks by resistant horizontally layered materials (subaerial lavas?); different mesa elevations in the different chasmata (a possible glaciovolcanic criterion related to variable ice surface elevations); and a composition that may be dominated by phyllosilicates, a common alteration product of mafic volcanic rocks (e.g. water-altered palagonitised tuffs; Murchie et al., 2000; Bishop et al., 2002). Traditionally, two principal types of ILDs have been distinguished: freestanding mesa landforms in the peripheral chasmata believed to be entirely volcanic; and much thicker flat-topped to domical landforms within the central chasmata that show a distinction internally into an upper layered unit resting on a basal massive unit. In the latter only the upper layered unit was considered volcanic, with the massive basal materials possibly derived by mass wasting of the canyon walls (Lucchitta et al., 1994). However, later studies identified layering in the basal massive unit and it probably has a volcanic origin similar to the upper layered unit (Chapman and Tanaka, 2001; Chapman, 2002; Komatsu et al., 2004).

Detailed studies of the layering in ILDs in the central chasmata have shed light on their origins. Beds maintaining a constant thickness can be traced for many kilometres laterally, and in some cases tens of kilometres. In most cases bedding appears to be horizontal, but examples with beds dipping down the flanks of mesas are known (Chapman and Tanaka, 2001; Chapman, 2002). Possible cross beds up to 100 m thick (typically several tens of metres) within horizontal sequences have also been observed but are very rare. There are also several examples of deformed beds, including slumped beds and fault-related discordances (Chapman and Tanaka, 2001; Komatsu et al., 2004). The continuous thin layering lacking major erosional surfaces is consistent with a placid depositional environment (lake or englacial vault). The characteristics of the layered sequences are possibly best explained as sediment density flow deposits formed of tephra derived from explosive (hydromagmatic) eruptions in englacial lakes similar to terrestrial examples (e.g. Smellie, 2001; Schopka et al., 2006). The deformed strata may be related to slope failures that are a common and conspicuous feature of tindars (Section 9.5). Major angular discordances comprising horizontal resistant capping units sharply overlying homoclinal dipping beds are also known and may be depositional structures rather than erosional (e.g. passage zones in lava-fed deltas; Chapman and Tanaka, 2001; Komatsu et al., 2004). If some of the ILDs in the central chasmata are tuyas, they are approximately an order of magnitude larger than terrestrial tuyas (Chapman, 2003).

The freestanding mesa landforms with resistant caprocks showing horizontal layering overlying dipping beds are commonly correlated with tuyas (e.g. Komatsu et al., 1993, 2004; Lucchitta et al., 1994; Chapman and Tanaka, 2001; Chapman, 2002). Some examples have wing-like topographical projections that resemble features seen in tuyas in Russia (Komatsu et al., 2004). Although interpreted empirically as primary features caused by lava flows and associated volcaniclastic sediments expanding along meltwater tunnels beneath a capping ice sheet, the features may alternatively reflect either asymmetrical lava-fed delta progradation (i.e. subaerial progradation in a meltwater lake rather than in subglacial tunnels) or post-eruptive erosional modification of the primary volcanic landform.

16.6.4 Tropical glaciovolcanic interactions

Volcano–ice interactions have long been suspected at several of Mars' tropical volcanoes, mainly in the Tharsis and Elysium regions. A relatively poorly constrained hypothetical origin for Olympus Mons by subglacial eruptions has already been mentioned, but other interpretations of the evidence are possible (Section 16.3.6). Volcano-induced basal melting of cold-based summit ice caps was suggested to explain the presence of radiating valley networks at Ceraunius Tholus and Hecates Tholus but without coeval glaciovolcanic landforms (Fassett and Head, 2006, 2007). The interpretations of relationships between landforms at Arsia Mons and Ascraeus Mons are amongst the most sophisticated and holistic so far for any glaciovolcanic features on Mars (Kadish et al., 2008; Scanlon et al., 2014). Glaciovolcanic landforms are abundant at both volcanoes and are found within the outcrops of lobate deposits interpreted as products of cold-based glacial ice. Volcano-tectonic features on Ascraeus Mons have also been interpreted as a consequence of a combination of pressure melting and magmatic heating causing the volcanic edifice to sink in an ice-rich substratum (Murray et al., 2010). The glaciovolcanic landforms include (1) large-volume englacial lavas, including possible pillow sheets; (2) hyaloclastite mounds; (3) a flat-topped plateau with steep sides interpreted as a tuya; and (4) ice-marginal lavas.

Subglacial effusion of lava under comparatively high confining pressure is inferred to be responsible for low, geographically extensive mounds. They individually involved hundreds of cubic kilometres of erupted lava. The mounds have maximum heights of c. 100–200 m and widths typically 5–15 km or more. Their margins are variably gently sloping to steep (c. 6–12°, much steeper than on subaerial lavas observed nearby) and they are interpreted as possible pillow mounds and sheets. Some are associated with steep-sided mounds inferred to be composed of 'hyaloclastite' and morphologically similar to tindars (tephra mounds) on Earth (note: the use of hyaloclastite by Scanlon et al. (2014) is not that used in this book and probably corresponds mainly to lapilli tuff; see Sections 9.2 and 9.4.4). The tephra mounds are inferred to have formed by eruptions under comparatively low confining pressures, with volatile degassing and explosive eruption. Steep-sided sheet-like landforms with depressed centres are considered to be subglacial lavas with chilled margins that then underwent late-stage drainage (deflation) of their molten cores (Fig. 16.11). A detailed history of the formation of these subglacial lavas and geographically related non-volcanic landforms was inferred as follows (Figs. 16.12): lava was erupted under a thick glacial cover, possibly as pillows, and spread out sideways and downslope, possibly as a pillow sheet although the dimensions of the Mars examples are larger than putative terrestrial examples (cf. Smellie, 2013).

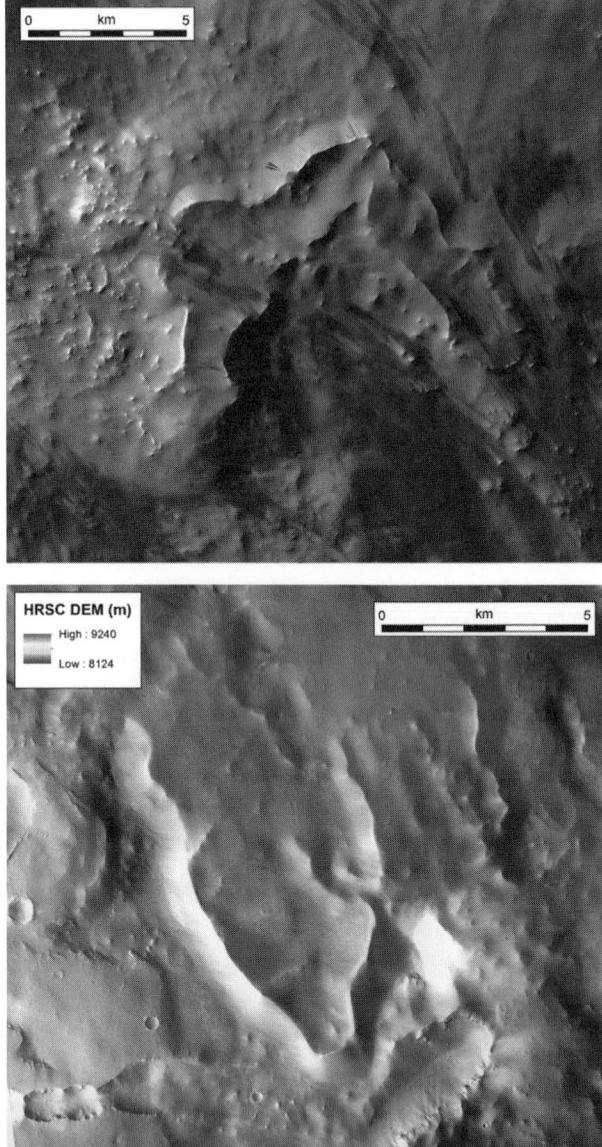

Fig. 16.11 Steep-sided lavas with sunken interiors at Arsia Mons (from Scanlon et al., 2014). The features are interpreted as subglacial lavas with steep margins due to ice confinement. The sunken interiors may be due to lava breakouts, distal drainage or magma withdrawal back into the vent.

Elevated heat over the mound or sheet caused basal melting of the overlying ice, thus decoupling the ice and causing it to surge downslope by sliding on its bed and creating a variety of erosional and depositional glacial landforms. At the margins of the ice, a jökulhlaup flowed out over the proglacial plain, carving a braided channel system. The

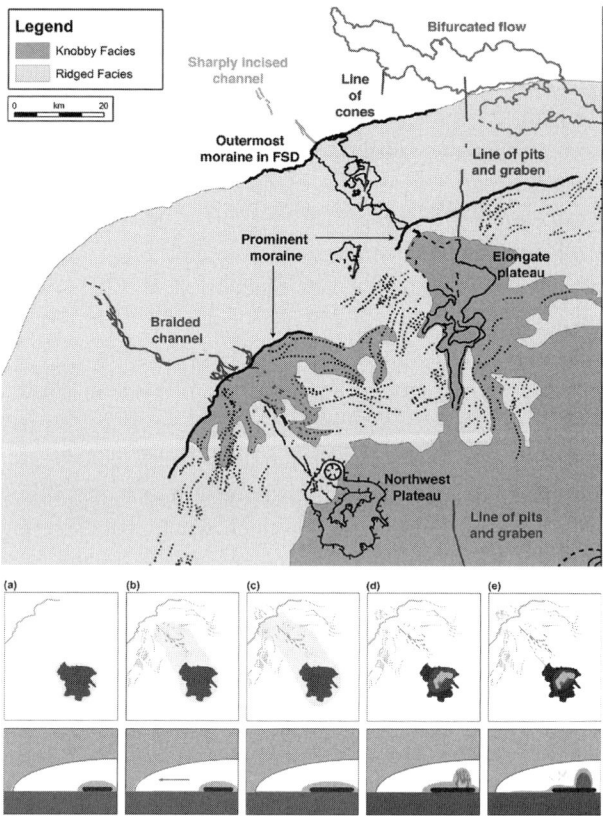

Fig. 16.12 Interpretation of the evolution of possible glaciovolcanic and associated features at Arsia Mons. (a) The Northwest Plateau (dark grey) may have been emplaced as a subglacial pillow sheet associated with a growing lens of meltwater. (b) The meltwater lens extends to the ice margin, causing the ice to stream (surge) and deposit moraines (thick black line in main map). (c) The lens drains catastrophically at the ice margin, forming jökulhlaup channels (blue lines in main map). (d) The eruption transforms from effusive to explosive, thermally eroding a tall englacial cavity above the vent. (e) Relict heat from the cooling volcanic products creates more meltwater that drains and transports debris in subglacial tunnels that is deposited as eskers (short olive green lines in main map) (from Scanlon et al., 2014). (A black and white version of this figure will appear in some formats. For the colour version, please refer to the plate section.)

sudden drainage of the melt cavity over the vent caused a reduction in pressure and the eruption to change from lava effusion to explosive activity, melting a taller cavity and constructing a tephra mound. During the waning stages of the eruption, particles reworked by meltwater from the tephra mound were deposited in meltwater tunnels carved in the basal ice where they accumulated as eskers. The eskers are sharp-crested closer to source and broader-crested further away, indicating a cogenetic relationship with the cooling volcanic edifice.

Other subglacial lavas may have flowed in meltwater tunnels and were emplaced in stages involving temporary stalling and local ponding (thickening) before continuing downslope as subglacial breakouts. Digitate lavas with stepped, layer-like margins in cliffs where they abut the fan-shaped former-ice deposits on Arsia Mons are also present and were interpreted to be successive subaerial lava flows that flowed downslope and banked and cooled against glacial ice (Fig. 16.13; cf. Harder and Russell, 2007; Section 10.2). Some of the cliffs have knobbly margins interpreted to be places where the lava overflowed onto the ice surface and subsequently collapsed into rubble when the underlying ice melted. Finally, the flat-topped plateau described by Scanlon et al. (2014) has an exceptionally smooth surface and is interpreted as a tuya. It has a volume (c. 5 km^3) and height (c. 200 m) somewhat less than terrestrial tuyas with similar areas (c. 35 km^2).

Hamilton et al. (2010, 2011) identified 167 groups of cones in western Tartarus Colles (>40 000 cones) in the Elysium–Arcadia region as rootless cones formed when ground ice beneath lava flows was volatilised. They referred to these features as volcanic rootless constructs (VRC), and suggested, based on thermodynamic modelling of the melting/vaporisation process, that only lavas c. 60 m thick would volatilise enough ice to form the explosive cones. Recognition of a glaciovolcanic origin for the cones is important as they may have sustained local hydrothermal systems for up to 1300 years, and they are a record of the climatic effects of intermediate obliquity excursions (25–32°) between 250 and 75 Ma on Mars.

16.6.5 Subsurface glaciovolcanism

The collapse of volcanic edifices and their immediate surroundings by volcano-related melting of an ice-rich substrate has already been mentioned (Sections 16.3.5 and 16.6.2). Other effects of magma intruding the cryolithosphere have been described and principally relate to the enhanced thermal fluxes surrounding magma chambers setting up hydrothermal systems. The hydrothermal cells so formed and the disposition of permeable layers in the volcanic edifice are primarily responsible for focusing enhanced heat in the summits of volcanoes (irrespective of whether the volcanoes are ice-capped; Gulick, 1998; Fassett and Head, 2006, 2007; see Section 13.6). A possible consequence that is important for making palaeoenvironmental inferences is the local transformation of the basal regime of any overlying ice from cold- to wet-based (see also Scanlon et al., 2014; Smellie et al., 2014).

Thermal effects of sill and dyke intrusion into the cryolithosphere are also important. Breaching the local permafrost seal ('cryosphere cracking': Russell and Head, 2003) is common on Mars and was probably responsible for many large-volume surface floods (see Sections 16.3.2, 16.3.3). The empirical model sketched by Squyres et al. (1987) for sill intrusion into a volatile-rich layer (Fig. 16.14) is remarkable for its broad similarities to lithofacies relationships associated with interface sills in terrestrial sheet-like sequences, i.e. subglacial sills emplaced at the junction between bedrock and an ice cover (cf. Smellie, 2008; see Section 10.6). Dykes may also feed shallow sills where

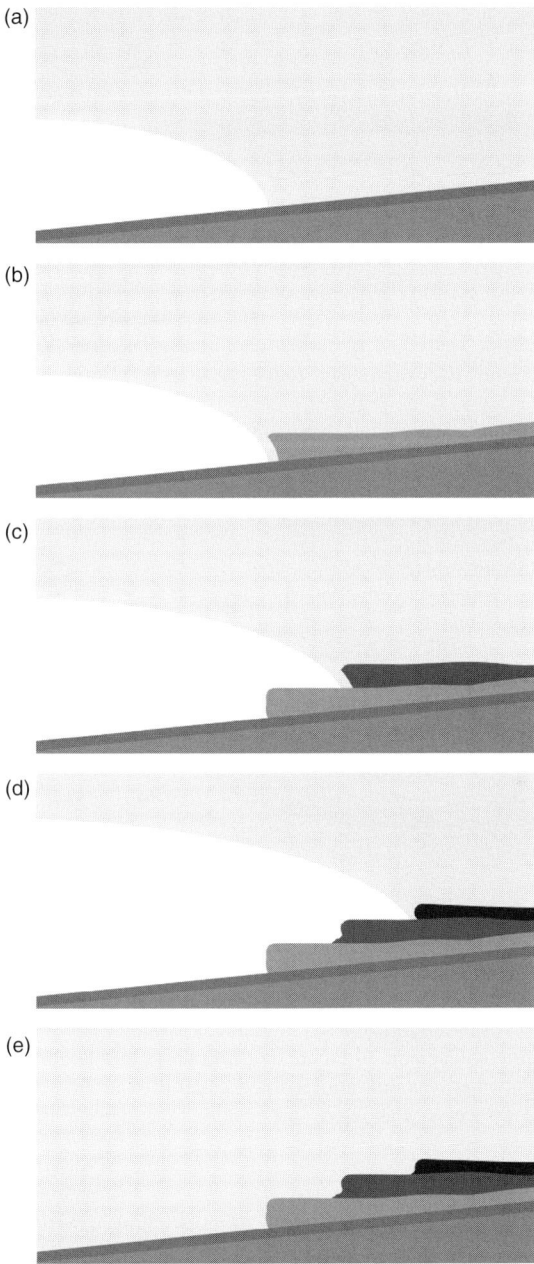

Fig. 16.13 Cartoons depicting the formation of multiple cliff lines at Arsia Mons. Successive subaerial summit lavas are envisaged flowing downslope to abut and chill against a surrounding glacier, thought to be similar to the formation of ice-confined lavas described at Hoodoo Mountain by Edwards et al. (2002; Section 11.2.1) (from Scanlon et al., 2014).

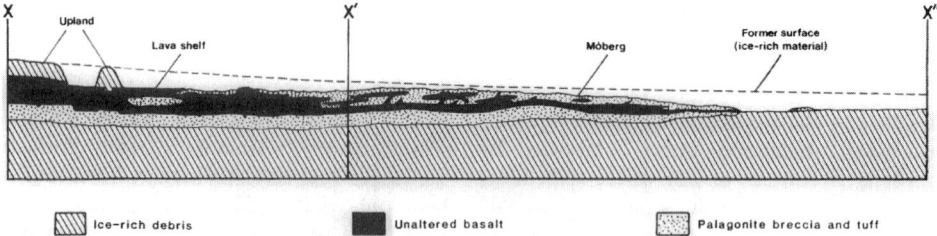

Fig. 16.14 Schematic cross section depicting a volcanic sill intruded into ice-rich material at Elysium Planitia (from Squyres et al., 1987). The ice is envisaged melting and interacting with the sill to create palagonite tuff and breccia. There is a remarkable similarity between this interpretive section and relationships shown by some terrestrial sheet-like sequences (cf. Bergh and Sigvaldason, 1991; see Section 10.6).

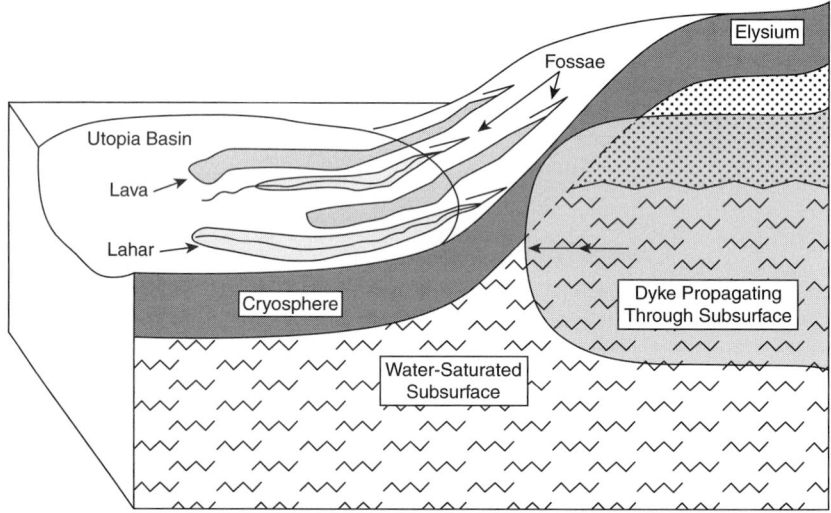

Fig. 16.15 Interpretation of hydrological and magmatic conditions associated with dyke intrusion at Elysium and consequent emplacement of lahars and lava flows into Utopia Basin (from Russell and Head, 2003). The groundwater was initially confined under pressure but was released at the surface due to dyke disruption of the cryolithosphere and creation of fossae (grabens).

they intersect a weak boundary or significant density change within the crust, leading to the dyke spreading laterally. Meltwater released by sill intrusion in the cryolithosphere may initiate violent mechanical and thermal mixing triggering fuel–coolant interactions with the sill in the upper few hundred metres of the crust. An example has been described at Hrad Vallis, where distinctive deposits 50–100 km wide flank the 8000 km-long eruption-induced depression. The deposits have lobate boundaries and show rheomorphic deformation and are inferred to consist of mud-like ejecta derived from phreatomagmatic explosions and transported and deposited from jökulhlaups (Wilson and Mouginis-Mark, 2003). The presence of a shallow sill is indicated by a superimposed bulge in the planet's topography c. 60 km wide and 100–150 m high.

During dyke emplacement, because much of the heat emitted conductively is focused at the dyke tip, melting of pore-space ice in the cryolithosphere might result in explosive phreatomagmatic activity and construction of chains of maars that may subside to form the surface grabens (Chapman, 1994). Alternatively, the grabens (known as fossae) may be caused by extensional forces and fracturing associated with the dyke as it approaches the surface (McKenzie and Nimmo, 1999; Russell and Head, 2003). Disruption of the confining cryolithosphere by dyke-induced fractures and melting may allow meltwater deluges to escape to the surface associated with mega-lahars composed of erupted magma and bedrock debris and seen on satellite images as rough-textured lobate units (Fig. 16.15).

Finally, Farrand et al. (2010) identified deposits in western Arcadia Planitia that are interpreted as glaciovolcanic cryptodomes. The domes have average diameters of 1.5 km and average heights of 160 m. Each of the deposits has a distinctive set of annular deposits comprising an outermost dark-toned aureole, a middle light-toned aureole, and an inner dark-toned apron. Spectral data for the deposits are consistent with Fe-rich basaltic compositions and the presence of tachylite. The unique dark aprons have textures thought to have been made by physical weathering of highly jointed dome rocks formed by interaction between magma and meltwater derived from ground ice.

17

Outstanding challenges and possibilities

17.1 Introduction

Publications on glaciovolcanic topics have increased exponentially during the twentieth century (Chapter 1; Russell et al., 2014) and it is likely that interest in volcano–ice interactions will continue to expand. Spectacular volcanic events such as the Eyjafjallajökull eruption in Iceland (2010) and the Villarrica and Calbuco eruptions in Chile (2015) have demonstrated that glaciovolcanic eruptions have a global reach and can have a major economic impact. Three principal avenues of research are likely to contribute substantially to improving our knowledge of glaciovolcanism: (1) Petrology and analogue modelling: greater application of modern petrological paradigms, advances in analytical/experimental/modelling techniques and theoretical understanding of volcanic processes; (2) Monitoring: improvements in real-time monitoring of glaciovolcanic eruptions and methods for mitigating hazards; and (3) Climate–volcanism linkages: improving isotopic dating, and establishing the linkages and feedbacks. For example, advances in the analytical precision and interpretation of juvenile H_2O and CO_2 contents of volcanic glasses will allow more accurate determination of syn-eruption pressures, which can then be linked to coeval ice sheet thicknesses and geometries. Understanding subglacial hydrology, particularly in areas with ice-capped active volcanoes, will allow the construction of more accurate hazard zonation maps, to the benefit of growing populations (e.g. South America). Recognition of ancient glaciovolcanic sequences on Earth will enable better testing of hypotheses related to ancient global glaciations, some of which are contentious (e.g. 'Snowball Earth'). Finally, the emerging field of large-scale, high-temperature experimentation will enable better observations and measurements of melting rates, heat transfer mechanisms and the morphologies of ice-confined deposits.

17.2 Towards a better understanding of the physics and chemistry of glaciovolcanism

17.2.1 Better integration of petrology and volcanology

Few studies have integrated both the physical and compositional evolution of glaciovolcanic sequences. The recent investigation by Pollock et al. (2014) demonstrates the importance of such combined studies. Their study of a small, pillow-dominated tindar, initially assumed to be monogenetic, used phenocryst mineralogy and major and trace element geochemistry to demonstrate that the apparently simple eruption sequence masked cryptic changes in magma compositions probably related to source region processes, and a possible polygenetic origin. Similar but less detailed studies by other workers also identified petrological changes within seemingly continuous glaciovolcanic sequences (e.g. Moore and Calk, 1991; Moore et al., 1995; Edwards et al., 2011). Such studies are rare, however. Many outstanding questions, such as realistically estimating the time required for edifice-building eruptions, will only be tackled by linking physical volcanology and petrology.

17.2.2 Improved measurements and modelling of magmatic volatiles

A much better understanding of the volatile budget of magmatic systems is important for improving our knowledge of many aspects of glaciovolcanism (e.g. Tuffen et al., 2010; Owen et al., 2012). For example, the depth at which magmatic fragmentation begins is dependent on the concentration of dissolved volatiles in the magma as it rises towards the surface. Yet few if any studies of glaciovolcanic systems have systematically measured the volatile contents of melt inclusions brought to the surface in crystals. Accurately establishing the pre-eruption volatile contents preserved in melt inclusions is critical for making interpretations of the coeval palaeo-pressures that existed at the vent due to overlying ice or water, when glaciovolcanic eruptions are initiated. Glaciovolcanic deposits should be particularly amenable to this type of study because the preservation of original compositions in melt inclusions is critically dependent on rapid cooling of the inclusion-bearing minerals (e.g. Lloyd et al., 2013), and quenching of magma is a hallmark of glaciovolcanic deposits.

At present, coeval pressures at eruption sites caused by overlying ice or water can be constrained by the abundances H_2O and CO_2 in the glassy rims of clasts or lava pillows (see Chapters 6 and 13). However, existing theoretical models for estimating the solubility of volatile species are based mainly on relatively high-pressure experiments (>0.1 GPa, which is 'high' relative to pressures during eruptions beneath 1 km or less of ice or water, equivalent to pressures of c. 10 MPa (0.01 GPa)). Thus, estimated pressures are model dependent. In addition, theory and models predict that the solubilities of all volatile species are interdependent, yet no models simultaneously consider all the coexisting major volatile species (H_2O, CO_2, SO_2, F, Cl), and few if any studies measure all of the volatiles. The strong dependence of H_2O solubility on CO_2 concentrations, coupled with difficulties measuring very low concentrations of CO_2 (i.e. <25 ppm), mean that estimated pressures calculated for any measured H_2O concentration in a glass sample can vary by a factor of two (e.g. Edwards et al., 2009b; Owen et al., 2012). This leads to significant uncertainties in reconstructed palaeo-eruption pressures, and can make discrimination of local versus regional ice ambiguous. Although techniques such as monometry or ion-probe microscopy can produce accurate measurements for volatile species present in low concentration, those techniques are not widely available. Moreover, glass is an unstable medium and there have been no studies to determine whether volatile species are effectively fixed in glass or whether they are able to migrate even at low temperatures, thus potentially yielding meaningless results.

New experiments carried out at pressures relevant to glaciovolcanic environments (e.g. 1–10 MPa; equivalent to water depths of 100–1000 m, or ice thicknesses of 110–1100 m), integrated into comprehensive models for predicting solubilities of multiple volatile species as a function of pressure, will enable the more accurate determination of volatile concentrations and significantly improve constraints on syn-eruption pressures due to overlying water and/or ice. The accurate determination of palaeo-ice sheet thickness is one of the most important, indeed unique, outputs of glaciovolcanic investigations. Fortunately, ice sheet thickness can also be constrained using lithofacies features (Section 13.5).

17.2.3 Improvements in understanding eruption initiation: effusive or explosive?

Because of the more efficient heat transfer involved during explosive glaciovolcanic eruptions (Gudmundsson, 2003), meltwater is generated and may be released more rapidly than during effusive eruptions (cf. Chapter 6). It is thus important to understand what combination of environmental and magmatic conditions will lead to explosive or effusive eruptions from the outset. Field evidence may be lacking due to poor exposure, insufficiently deep dissection of a volcanic pile or burial beneath unrelated much younger deposits. The evidence for an early effusive stage may also be erased by subsequent explosive activity. Conversely, the presence of pillow lava at low elevations on a volcanic edifice cannot be taken as unambiguous evidence for effusive initial activity since pillow lavas can drape edifices and flow several kilometres away from vent sites (e.g. Hungerford et al., 2014). Older more deeply dissected tuyas, such as examples found in British Columbia and Antarctica, are better targets for investigating eruption initiation mechanisms. The cores of glaciovolcanoes may also provide the only evidence for early-formed explosive activity since tephra deposited on a surrounding ice sheet surface will be advected away and not preserved.

17.2.4 Effects of bulk composition

Most published glaciovolcanic studies are of basaltic sequences, whereas there are dozens of unstudied or poorly studied snow- and ice-clad stratovolcanoes with intermediate compositions, often situated in remote regions (e.g. Kamchatka, Alaska, South America). Many of these are large polygenetic volcanoes with complicated eruption histories that span multiple glaciations (e.g. Crater Lake, Oregon; Bacon and Lanphere,

2006). Felsic examples are also very rare, although several occur in Iceland. Future studies of these volcanoes will greatly extend our understanding of interactions between glacierised environments and magmas with viscosities that are significantly different (higher) than basalts.

17.3 Understanding boundary conditions: how does the ice–bedrock interface influence eruptions?

The boundary conditions for glaciovolcanic eruptions can vary significantly. In most treatments, the ice–bedrock interface is generally shown to be flat (e.g. Jones, 1969b; Allen, 1980; Hickson, 2000) and most standard models assume that glaciovolcanic edifices are constructed at the ice–bedrock boundary. A few authors have explored the effects of emplacement into areas with significant topographical variations (e.g. Edwards et al., 2006; Hungerford et al., 2014) and the possibility of englacial or supraglacial lava emplacement (e.g. Edwards et al., 2002; Wilson and Head, 2002). It is well known that many (most?) glaciovolcanic deposits persist through multiple glacial-interglacial cycles. For example, glaciovolcanic edifices as old as Early Miocene (c. 29 Ma) exist in Antarctica (LeMasurier and Thomson, 1990; Wilch and McIntosh, 2000) and Pliocene tuyas and other glaciovolcanic sequences are present in British Columbia and Iceland. What has not generally been considered is what impact pre-volcanic subglacial topography, and its ability to focus glacial erosion, may have on the subsequent preservation of glaciovolcanic sequences. Workers have speculated that both glaciovolcanic and subaerial volcanic edifices may persist beneath the thickest ice sheets on Earth in the Antarctic (Behrendt et al., 2004), and that subglacial volcanism may affect the stability of ice sheets (Vogel and Tulaczyk, 2006). Conversely, products of subglacial eruptions can themselves create new topography that can subsequently channel associated ice. For example, ridge-capping andesite lava flows at Mt Rainier, which were channelled into 'valleys' that were topographic lows between ice lobes, and subsequently transformed into inverted topography after the surrounding ice melted (Lescinsky and Sisson, 1998). In such a situation, any renewed ice growth will be influenced by the new ridge-forming topography formed by the lavas. The mutual feedbacks between ice sheets, glaciovolcanic eruptions and subglacial topography have yet to be examined in any detail.

Wilson and Head (2002) postulated that a propagating dyke may significantly over-shoot the ice–bedrock interface and intrude the overlying ice. If this model is correct and happens frequently (it may have happened during the 1996 Gjálp eruption, when an en echelon fracture appeared in ice more than 500 m thick above the erupting fissure; Gudmundsson et al., 2004), it has significant implications for rates of heat transfer as well as preservation potential of the early eruption products. Rather than being preserved *in situ*, the vertical dyke will gravitationally collapse as the supporting ice walls rapidly melt back, causing fragmentation and creating a pile of massive tuff breccia (as described and depicted conceptually by Wilson et al., 2013, Fig. 13.5). Fragmental material dominated by poorly sorted tuff breccia is present in several glaciovolcanic edifices and some could have formed by collapse of initial dyke injections. Currently, few deposits with the expected characteristics derived from an englacial intrusion have been described (e.g. Edwards and Russell, 2002; Tuffen and Castro, 2009; Wilson et al., 2013).

Wilson and Head (2002) also speculated that magmatic sills could be emplaced along the ice–bedrock boundary. But, until recently, no deposits with the appropriate characteristics had been found. Smellie (2008) hypothesised that extensive glaciovolcanic sequences in the Sida–Fljotsfverfi district of southern Iceland may include sills intruded along the ice–bedrock surface (called interface sills by Smellie (2008); Section 10.6.2). Although the conceptual model for sill emplacement described by Wilson and Head (2002; also Wilson et al., 2013) is probably valid, more recent studies suggest that the Sida–Fljotsfverfi intrusions may have intruded coeval, subglacially emplaced, jökulhlaup-related lapilli tuffs rather than the ice–bedrock interface (see Sections 10.6.3 and 10.6.4). However, Smellie (2008) also speculated on the characteristics of interface sills emplaced along the ice–bedrock interface without a preceding jökulhlaup and examples may be present in the so-called pillow sheets in Iceland (see Section 10.5). Further work is needed to fully explore these hypotheses and their potential palaeoenvironmental implications (cf. Walker and Blake, 1966; Bergh and Sigvaldasson, 1991; Section 10.6.4). A related area of potentially fruitful research would be to devise analogue models that can be used to explore the dynamics of sill intrusions within unconsolidated surface sediments and ice.

17.4 Differentiating between marine and freshwater glaciovolcanic eruptions

As glaciovolcanism has occurred in coastal areas as well as inland regions (e.g. in Antarctica, Iceland and Alaska), the issue of distinguishing between shallow submarine eruptions versus eruptions beneath ice shelves or tidewater glaciers is important. Johnson and Smellie (2007) developed compositional criteria with which marine and non-marine settings may be distinguished, based on analyses of authigenic zeolite minerals formed during palagonitisation (see Sections 7.5 and 13.4.4). The study is important since it potentially provides a simple specimen-based method to define an eruptive setting using simple laboratory (microprobe) analysis rather than only using lithofacies analysis, which is sometimes complicated. However, Johnson and Smellie (2007) showed that the compositions of zeolites are not fixed but may become affected by pore fluids sourced in unrelated environments (e.g. subsequent marine transgressions) migrating along microfractures. They further showed that sample selection (i.e. type of sample and its position within a glaciovolcanic unit) is critical for reliable interpretation of the results. More work will be needed to verify and refine this technique and expand its use, perhaps in combination with studies of stable isotopes (e.g. Antibus et al., 2014). In theory, marine settings should produce not only differences in mineral compositions, but also the possibility of marine fossils, evidence of tidal/wave modification of tephra, and diagnostic passage zone features (e.g. Smellie, 2006). At present, interpreting glaciovolcanic lithofacies and associated passage zones is the most reliable method for unambiguously establishing the eruptive setting (Sections 13.4 and 13.5).

17.5 Ice melting rates and the 'space problem'

One of the most interesting and unresolved problems in glaciovolcanism is the so-called 'space problem', i.e. how magma creates space for itself when intruding ice merely by melting the ice (Smellie 2000, 2008; see Chapter 6). Because of the difference in densities between ice and water, melting ice yields water with a net decrease in volume of approximately 10%. But, although magmas have ample thermal energy to create space for themselves via melting (Allen, 1980; Höskuldsson and Sparks, 1997), observations of lava flows emplaced on top of ice (see Chapters 3 and 5) show that ice melting is not instantaneous (e.g. Edwards et al., 2013), and estimates of ice melting efficiency during subglacial eruptions suggest a ratio of only c. 5:1 (Gudmundsson, 2003). Thus, intrusion of magma at the ice–bedrock interface should raise the glacier surface, which has not yet been observed. Conversely, the frequent creation of cauldrons in the ice surface above erupting vents (e.g. the 1996 Gjálp and 2014 Bárdarbunga eruptions) indicates that, despite magma being added, mass (presumably melted ice) is being lost from the system, rapidly creating space into which the overlying ice collapses. At present, the only feasible explanation is that basal ice must not only be melted at a rate faster than is currently explicable but that meltwater must also simultaneously escape the system. Both explanations require testing, ideally during observed eruptions, which will be challenging. However, the complexity of heat and mass transfer in glaciovolcanic eruptions is fundamentally important for understanding processes occurring during the early stages of glaciovolcanic eruptions, which is critical knowledge for making accurate hazard assessments at ice-clad volcanoes.

17.6 Towards better assessment, monitoring and mitigation of glaciovolcanic hazards

Advances in remote sensing techniques will improve real-time monitoring of future glaciovolcanic eruptions, enabling testing of hypotheses regarding the dominance of specific processes (e.g. fragmentation mechanisms) and how that dominance might shift during the course of an eruption. Lessons learned from the 2010 Eyjafjallajökull eruption and new monitoring techniques tested during the 2014 Bárdarbunga eruption have helped to improve response times for assessing short- and long-range hazards such as local flooding and regional ash dispersal. Conversely, outside of Iceland where programmes such as FutureVOLC are investing significant financial resources to improve real-time eruption responses, monitoring efforts of ice-clad volcanoes are more sporadic and may even be diminishing. Efforts should be focused on assessing many of the over 150 known ice-clad volcanoes globally in order to better understand how future eruptions will affect inhabited regions, what immediate and optimal responses are required, and how eruptions may cause longer-term problems where volcanically enhanced degradation of glaciers may critically affect local supplies of freshwater. Additionally, workers in Iceland have already begun to assess how rates of glacial recession are contributing to 'excess' magma production in the mantle, the potential destabilisation of crustal magma chambers and increased rates of eruptions (see Chapter 14), and similar assessments should be made for other volcanic systems globally.

17.7 Geochronology

One of the most important issues in glaciovolcanism that urgently needs addressing is the need for improving the precision of isotopic ages of lavas in glaciovolcanic systems. Glaciovolcanic deposits are one of the most important repositories of palaeoclimate information for Earth (and probably will be for Mars). However, $^{40}Ar/^{39}Ar$ ages, for example, currently give 2-sigma errors typically 40–60 k.y. for basalts, although precision improves to a few k.y. for K-rich minerals in evolved lavas (e.g. Smellie et al., 2008; Flude et al., 2008; Martin et al., 2011). Thus, felsic lavas can yield ages that fall well within a glacial cycle (duration c. 41 k.y.), particularly cycles of Quaternary age (duration c. 100 k.y.), but most basaltic systems are not so well served. Unfortunately felsic glaciovolcanic sequences are uncommon compared to basaltic (cf. Chapters 10–12). Developing geochronological techniques that have precisions of a few k.y. (e.g. McIntosh et al., 2014), particularly for basaltic systems, will substantially improve the value and utility of glaciovolcanic studies for palaeoenvironmental purposes. For example, if ice thickness data can be linked to their positions within glacial cycles, it will become feasible to estimate how ice thicknesses vary with time, thus permitting ice-sheet volumes to be calculated with far greater accuracy than is currently possible. Such an ability would also enable meaningful comparisons of the terrestrial ice record with global climate models and the continuous marine record (e.g. Lisiecki and Raymo, 2005). New advances in exposure and burial ages are also promising for increasing precision, but they are limited by intrinsic assumptions (e.g. outcrops must have remained continuously buried or exposed to avoid ages being cumulative and reflecting more than one glacial, thus harder to interpret). Creative application of other techniques can also be useful (e.g. U/Th-He zircon ages derived from crustal xenoliths; Blondes et al., 2007) but they are not necessarily applicable to all glaciovolcanic deposits (i.e. those without crustal xenoliths). A robust solution to this analytical problem would spur the next major revolution in our understanding of long-term climate change, particularly for the Plio-Pleistocene.

17.8 Improving our understanding of planetary glaciovolcanism: Earth–Mars comparisons

Translocation of our knowledge from terrestrial glaciovolcanic environments to issues affecting Mars is an important aspect of future studies. Not only can glaciovolcanic deposits be used as palaeoclimate proxies for reconstructing past Mars climate and identifying potential regions with water for human exploitation, but volcano–ice interactions may have sustained past life on Mars (see Chapter 16). Thus, being able to confidently identify features on Mars formed during glaciovolcanic eruptions may eventually guide the selection of future landing sites for the first human missions to the Red Planet and as we continue to search for evidence of extra-terrestrial life. Moreover, field-based and theoretical studies of eruptions within cold-based glaciers are needed (see theoretical treatments by Smellie, 2009, 2013; e.g., Section 8.3). Glaciovolcanic sequences formed during eruptions within cold-based glaciers have been identified in Antarctica (Smellie et al., 2011a,b, 2014; Sections 2.2.5, 13.8.2) and are expected to be of major importance on Mars. Conversely, given the huge extent of putative past ice sheets on Mars, the apparent lack of plate tectonic processes, and the extensive subsurface cryolithosphere, we may also learn about processes that operated in Earth's distant past from Mars studies.

17.9 Towards improved integration of information from ancient glaciovolcanic deposits into planetary climate models: the past as a guide to modelling future Earth climate

One of the ultimate goals of geoscience is to understand the temporal evolution of our home planet. While we have known for more than one hundred years that great ice sheets have waxed and waned over parts of Earth throughout its history, we are still in the early stages of being able to unravel the complex interplay between astronomical forces, atmospheric chemistry and internal planetary processes that together have driven or modulated Earth's climatic variations. With the publication of this volume, we hope that we have convinced readers new to the study of glaciovolcanism of its potential to clarify our understanding of planetary palaeoenvironments and guide ever more complex modelling designed to understand the past and predict the future of Earth's climate. Finally, we hope this volume honours the monumental contributions of past workers upon whose shoulders we stand and, by highlighting some of the numerous poorly developed areas of glaciovolcanism, we also hope we will spur new and existing researchers to focus on those and other areas in considerable need of fresh and innovative approaches.

Glossary: terminology of glaciovolcanism

A specialist terminology is an important part of any science. It evolves over time in response to an ongoing requirement for words and phrases that describe unusual features and important processes that are frequently unique to a particular science. It also enables precise, efficient (i.e. rapid) and therefore more effective international communication amongst active workers. Glaciovolcanism is also characterised by several commonly used words derived from local contexts and different languages (sourced especially in British Columbia and Iceland) that many readers unfamiliar with the science will find non-intuitive and therefore hard to understand and use. To counter these effects and to reach the widest audience possible, a comprehensive glossary is presented here. It is the first published lexicon devoted solely to glaciovolcanism.

Bólstraberg An Icelandic term for pillow lava (Kjartansson, 1955). Frequently seen on geological maps in Iceland. See *pillow lava* below.

Colonnade jointing Essentially straight columnar joints with polygonal surfaces. They typically penetrate the entire thickness of a lava or are confined to a basal zone overlain by thinner curved or wavy columnar joints called *entablature*. An upper colonnade may also be present but is rarer (Long and Wood, 1986). The individual columns are typically 0.5–>1 m in diameter and crystalline, and form principally by relatively slow conductive cooling perpendicular to the cooling surface (usually bedrock below and air above).

Columnar joint A common term used to describe fractures that define columns with polygonal shapes that form especially in lava flows as a response to tensional stresses resulting from the small loss in volume as the liquid lava solidifies. The joints are generally significantly longer than their diameter in cross section, which can range from centimetres to metres, and they are thought to always propagate perpendicular to isothermal surfaces. Individual columns are generally bounded by three, four or six vertical faces. While these fractures are common in subaerial lavas and subvolcanic intrusions, they are also very common in water-cooled glaciovolcanic lithofacies. Polygonal joints is a synonym.

Cryosphere A common term used for all those areas of the Earth where the temperature is continuously below the freezing point of water, although water need not be present. If present, water occurs in solid form, i.e. as sea ice or frozen lakes, permafrost, snow, firn and glaciers/ice sheets/ice caps. Thus, cryosphere encompasses surface *and* subsurface ice.

Cryolithosphere A word used in planetary discussions, especially by Russian workers, to describe a zone of *subsurface* water in solid form (ice). Essentially synonymous with permafrost and extending to substantial (crustal) depths. Used mainly for the zone of pervasively fractured frozen crust that is extensively developed on Mars, comprising a thick blanket of impact ejecta and interbedded volcanic layers overlying heavily fractured basement rock. The Mars cryolithosphere may be up to c. 5 km thick at the equator and mid latitudes and c. 6–12 km at the poles, or even substantially thicker (up to 9 km and 10–22 km, respectively) depending on the boundary conditions used in the calculations (Clifford, 1993; Clifford et al., 2010).

Entablature jointing A type of columnar jointing formed in water-cooled lava and characterised by well-formed, wavy, curvicolumnar, often fanning, chevron or even rosetted columns with a diameter typically c. 20 cm (Long and Wood, 1986; Forbes et al., 2014a). As originally defined, the entablature is generally located directly above the basal *colonnade*. Long and Wood (1986) proposed that entablature jointing formed when water penetrated fractures on the lava surface, facilitating more rapid cooling of the flow interior. Water penetrates the master joints and disrupts the isotherms causing them to deform and resulting in the variable column orientations. The columns are much narrower and finer grained (aphanitic, sometimes with glassy margins) than the style of columnar joints known as *colonnade*.

Flow-foot breccia Subaqueously deposited volcaniclastic breccia created at the leading edge of an advancing lava-fed delta and forming crude, steep-dipping homoclinal foreset beds (Jones, 1969b). The phrase is seldom used now and should probably be allowed to quietly vanish from the modern glaciovolcanic lexicon. Also called *foreset breccia* (Jakobsson, 1978).

Glacier Generic word used to describe all forms of ice, including geographically extensive ice sheets ($>50\,000$ km^2) and ice caps ($<50\,000$ km^2) as well as ice streams, outlet glaciers and valley confined mountain glaciers (Neuendorf et al., 2011).

Glacio-isostasy Lithospheric depression or rebound due to the weight or melting of glacier ice.

Glaciovolcanic A catch-all word embracing all types of volcano interactions with ice in all its forms, including snow and firn and any meltwater derived from that ice by heating. It was first used by Kelman et al. (2002), but was formally defined by Smellie (2006) (see also Edwards et al., 2015a). It can also be spelled *glacivolcanic* but *glaciovolcanic* is better etymology. Synonymous with, but a better choice than, both *subglacial* (used in the general sense) and *intraglacial* (see below). However, all three words remain extant and reflect the user's bias or geographical location (e.g. workers in Iceland). *Subglacial* and *intraglacial* are also used with more specific meanings relating to position within an ice cover (see Jakobsson and Gudmundsson, 2008, and below).

Hackly jointing Also known as box jointing or blocky jointing. Broadly similar to *entablature* but more chaotic, it comprises arcuate intersecting fractures with a variety of orientations and cross-cutting relationships developed mainly in the outer fine to aphanitic or glassy margins of a water-cooled lava (Lescinsky and Fink, 2000). Principal joint spacing is typically c. 10 cm. Caused by rapid cooling by water penetrating down shrinkage fractures on the lava surface and causing irregular disruption to the isotherms.

Hyaloclastite Originally proposed by Rittman (1952) as a term to describe deposits of ash-sized, vitric fragments most commonly interspersed with pillow lava (*sensu stricto*). Now frequently used as an inclusive term used to describe all types of vitroclastic deposits formed both by explosive (tephra) and non-explosive (mechanically generated) hydroclastic processes. Promulgated by Fisher and Schmincke (1984; i.e. 'vitroclastic tephra produced by the interaction of water and hot magma or lava') and particularly common in USA and Iceland. Its use embracing such a wide range of vitroclastic rocks conceals significant differences important for understanding eruption mechanisms, and is not recommended (see White and Houghton, 2006).

Hyaloclastite breccia Coarse and generally fines-poor volcanic rock composed of blocky angular glassy and aphanitic, generally poorly vesicular lava fragments formed by mechanical disintegration of lava rapidly cooled by water (White and Houghton, 2006). Typically accumulates essentially *in situ* adjacent to its source lava. Commonly used to describe steep-dipping beds inclined at the angle of repose for coarse sediment (c. 30–40°; see *flow-foot breccia*, above) and formed by slope collapse, but deposits formed this way (involving significant movement of the clasts) are now called tuff breccia (White and Houghton, 2006). A principal component of lava-fed deltas.

Hyalotuff A word used by Honnorez and Kirst (1975) to distinguish tephras formed of angular vesicular glassy fragments from *hyaloclastite breccia* (see above) and mainly erupted explosively. Originally intended to be non-genetic, but in practice it is difficult to separate from genetic usage. Although the word has been used for tephras in glaciovolcanic successions (e.g. Skilling, 1994; Smellie, 2008) and it is a useful and practical field term to physically distinguish such deposits from hyaloclastite breccia, it has not been widely adopted. It should probably be discarded in favour of simpler non-genetic rock names such as tuff or lapilli tuff (e.g. White and Houghton, 2006).

Hydrovolcanic/hydromagmatic Words often used interchangeably to describe the explosive and non-explosive interaction between magma or lava and water. Although a distinction can be made between those interactions occurring deep within the Earth (*hydromagmatism*) and those at or near its surface (*hydrovolcanism*; Wohletz et al., 2013), it is often difficult to do in practice. As glaciovolcanism is mainly concerned with magma–water interactions at or near the surface, hydrovolcanism is probably more appropriate. However, interactions within the *cryosphere* or *cryolithosphere* (see above) may be considerably deeper (kilometres; e.g. on Mars) and hydromagmatism may be more appropriate in those instances. The water can be groundwater, seawater, lake water (including glacial meltwater in lakes formed during glaciovolcanic eruptions) or derived from melting permafrost. Explosive hydrovolcanic eruptions from vents flooded with water are called *Surtseyan* (see below); they typically occur during shoaling and emergence. Explosive eruptions resulting from interactions with groundwater are called phreatomagmatic or Taalian (Kokelaar, 1986). The ratio of magma to water can be very variable in the latter and results in construction of tuff rings, tuff cones or maars (Sohn, 1996). Explosive hydrovolcanic eruptions generate tephra with abundant fine ash and distinctive blocky-shaped glassy juvenile clasts with variable vesicularities together with a proportion of fluidal fragments (Heiken and Wohletz, 1985; Houghton and Wilson, 1989; Mastin et al., 2009).

Hydrosphere On Earth, the hydrosphere encompasses the combined mass of surface and subsurface water in all its forms, i.e. the entire water inventory, including ice, freshwater and saline water. On Mars, hydrosphere signifies an interconnected groundwater system in hydrostatic equilibrium that exists below a confining cryolithosphere. The hydrosphere and cryolithosphere may be in contact or else are separated by a vadose zone (water-free pore space) in which, if even a small geothermal gradient exists, evaporated groundwater rises and refreezes in the overlying cryolithosphere (Clifford, 1993; Clifford et al., 2010).

Ice-block meltout cavities A phrase introduced by Skilling (2009) to describe distinctive irregularly shaped empty cavities typically several decimetres to a few metres across showing evidence for rapid water-chilling (e.g. fine-scale hackly jointing and glassy chilled moulded surfaces). Attributed to the former presence of large detached blocks of ice engulfed in lava that subsequently melted to form the cavities seen today.

Ice cauldron Circular or elliptical depression on the surface of a glacier caused by enhanced and localised heat flux from the bedrock (Björnsson, 1975). Frequently, the first visible evidence for the start of a subglacial eruption (e.g. Gudmundsson et al., 2007). Meltwater accumulates under many cauldrons until it is either released in jökulhlaups or seeps away continuously. Cauldrons may be semi-permanent where they are linked to long-lived hydrothermal upwelling. The largest cauldrons observed so far in Iceland are 2–3 km in diameter (Björnsson, 2002).

Ice-confined A general non-specific term reflecting lava that has been laterally constrained by ice and which encompasses *ice-contact*, *ice-impounded* and *ice-constrained* (see below).

Ice-constrained Used for situations in which eruption products were laterally confined by snow or ice causing their movement to be restricted (Mee et al., 2006), for example volcanic products constructed within a glacial cavity (i.e. tindars and most tuyas).

Ice-contact Phrase used to describe situations where volcanic products have simply come into contact with snow or ice but were not confined by it (Edwards and Russell, 2002; Mee et al., 2006), such as when a subaerial lava is ponded against a downstream ice barrier. Synonymous with *ice-impounded* (see below). Numerous examples have been described by Lescinsky and Sisson (1998), Edwards and Russell (2002), and Harder and Russell (2007).

Ice-impounded Another phrase to describe volcanic products constrained but not wholly confined by ice (Kelman, 2005). Synonymous with *ice-contact* (above) and *ice-dammed* (Mathews, 1952b; Edwards and Russell, 2002). See also *ridge-forming lava*.

Ice-melt collapse pits Features recognised by Branney and Gilbert (1995) in lahar deposits from the 1991 eruption of Hudson volcano (Chile) that are indicative of the former presence of blocks of ice, indicating that the lahar formed by glaciovolcanic processes. The distinctive stratigraphy and morphology of the structures are thought to have a high preservation potential, and thus to be a useful indicator for ancient deposits.

Intraglacial A general word used to imply a subglacial (*sensu stricto*) eruption that has broken through the glacier (Mathews, 1952b; Jakobsson and Gudmundsson, 2008). Thus, subglacial eruptions may progress to intraglacial. Although strictly the word means *within ice*, it has been used because subglacial was felt to be too specific regarding vent location. Englacial has also been used with a similar meaning to intraglacial (Skilling, 1994; Werner and Schmincke, 1999; Smellie, 2000) but it also suffers by already having a well-defined glaciological sense, implying emplacement fully within ice (i.e. ice also present below the volcano, which is unlikely). *Intraglacial* has not found widespread use outside of Iceland and *glaciovolcanic* is preferred. This term is also used more generally for interbedded non-glacial and glacial (non-volcanic) deposits.

Jökulhlaup A widely used Icelandic term for a glacial lake outburst flood (Björnsson, 1975, 2002, 2009; Roberts, 2005). The floods range from those consisting almost entirely of water to hyperconcentrated water–sediment mixtures, including abundant volcanic materials mixed with water and ice related to eruptions. Most jökulhlaups are not linked with volcanic eruptions but they are generally associated most spectacularly and damagingly with the latter, sometimes involving enormous volumes of meltwater (e.g. c. 10^6 m^3 s^{-1}; Alho et al., 2005).

Kubbaberg jointing Another phrase commonly used only in Iceland for *columnar jointing of entablature type* formed in water-cooled lavas (see above; Saemundsson, 1970). Also called curvicolumnar jointing.

Lava-fed delta A distinctive volcanic landform constructed from mafic lava flowing into ponded water. In its lithofacies architecture it resembles a sedimentary Gilbert-type delta and is formed of flat-lying to gently dipping subaerial lava (structurally equivalent to sedimentary topset beds) overlying steep-dipping homoclinal foreset beds of subaqueously deposited coarse tuff breccia (Jones, 1969b). Deltas fed by pāhoehoe lava are most commonly described but 'a'ā-fed deltas are also known (Skilling, 2002; Smellie et al., 2013a).

Móberg An Icelandic word (literally 'brown rock', referring to the colour imparted to volcanic glass altered to yellow-brown *palagonite* (see below); Jakobsson, 1978) used indiscriminately to encompass mainly hyaloclastite breccia and explosively generated lapilli tuffs in *tindars* (below) but sometimes also the capping lavas in *tuyas* (below). Phrases such as *móberg ridges* and *móberg tuyas* have been used (Kjartansson, 1943, 1959) but are unnecessarily confusing and should be replaced by *tindars* and *tuyas*, respectively. Principally used in a quasi-stratigraphical context to encompass and describe the widespread subglacially erupted Pleistocene formations in Iceland, collectively called the Móberg Formation, a chronostratigraphical unit (Kjartansson, 1959;

Einarsson, 1994). The word is of historical interest, mainly stratigraphical and should no longer be used unless in a local context of Icelandic geology.

Móberg sheet A sheet-like deposit of *móberg* containing dispersed isolated pillows and pillow fragments often underlain by columnar jointed lava (Walker and Blake, 1966; Loughlin, 2002; Jakobsson and Gudmundsson, 2008). There are clear similarities to some glaciovolcanic *sheet-like sequences* (cf. Smellie, 2008; see below). Like its namesake (above), *móberg sheet* is outmoded terminology, confusing and should be avoided.

Molten fuel–coolant interaction (MFCI) A broad category of interactions between hot liquid and a substance that can cause rapid extraction of heat from the liquid. If the coolant is water, this can lead to very high rates of water-to-steam conversion and detonation of explosions. It potentially produces magma fragmentation in any water-rich environment, but it requires rapid mixing of fuel and coolant in the correct proportions. This type of interaction is the driving force for phreatomagmatic explosions (Zimanowski and Wohletz, 2000).

Palagonite/palagonitisation Palagonite is an alteration product, possibly metastable, and mainly composed of smectite (a montmorillonite clay mineral) that forms by hydration of mafic and intermediate glass (Peacock, 1926; Stroncik and Schmincke, 2002; Pauly et al., 2011). It gives the deposit a distinctive yellowish-orange to rusty-brown colour. Palagonitisation is the process by which fresh sideromelane is progressively replaced by palagonite. It consists of at least two different reaction stages: the dissolution of glass and simultaneous precipitation of amorphous gel palagonite, followed by an aging process in which the gel palagonite reacts with fluid and crystallises to smectite (Stroncik and Schmincke, 2001). The name palagonite strictly refers to the mineraloid gel stage but the term is often extended to include the smectite stage too (e.g. Johnson and Smellie, 2007). The pervasive presence of palagonite in central Iceland led Peacock (1926) to use it as a formational name (the Palagonite Formation) that is essentially synonymous with the Móberg Formation.

Passage zone A surface representing the transition between subaerial lava and underlying subaqueous tuff breccia in a lava-fed delta (Jones, 1969b; Skilling, 2002). Several different types of subaerial–subaqueous transitions have been defined in volcanic sequences (Smellie, 2006), but a passage zone (*sensu stricto*) is essentially a fossil water level coeval with the eruption that is preserved in the volcanic sequence. A passage zone showing rising and falling elevations is diagnostic of a glacial setting during eruption. Pyroclastic passage zones in subglacial tephra cones have also been postulated (Russell et al., 2014) but, unlike lava-fed delta passage zones, they are typically extremely hard to identify owing to a lack of diagnostic criteria for their recognition. The sudden appearance of fragments coated in fine ash (accretionary and armoured lapilli) within a glaciovolcanic tephra pile, which are often considered to be unlikely to survive immersion in water, may be a possible criterion. However, the survival and even sedimentary reworking of accretionary lapilli, including broken lapilli, indicate that it may be unreliable (Smellie, 1984).

Perched lava A type of *ice-impounded* lava (see above) comprising a subaerial lava that is banked against a valley wall on one side and against ice on the other (Lescinsky and

Sisson, 1998). After the ice melts away, a distinctive subvertical primary cliff face composed of surface-quenched lava is revealed. See also *ridge-forming lava*.

Phreatomagmatic A style of volcanic eruption driven by MFCI (see above). Probably an important fragmentation process during glaciovolcanic eruptions at low hydrostatic pressures.

Pillow lava A special morphology of lava where individual 'pillows' are single lava lobes, generally varying in length from 1 to 20 m, that form only in the presence of water. When viewed in cross section the lobes have elliptical shapes. Walker (1992) provided a general summary and comparison with subaerial pāhoehoe lava lobes, which generally have a lower height to width ratio than pillow lavas (because of buoyancy effects; see Section 13.4.3). The outer surface of pillow lavas is generally vitric, with thicknesses of 1–3 cm, and when viewed in cross section pillows can contain pipe vesicles or radial bands of vesicles, and inward radiating fractures that create 'pie-wedge' shapes.

Pillow mound, pillow ridge, pillow volcano Phrases used to describe a broad mound or ridge erupted subaqueously and formed of pillow lava with generally minor hyaloclastite. Both landforms can be grouped together as pillow volcanoes. Common in subglacial and submarine successions (Jones, 1969a; Schmincke and Bednarz, 1990). Also regarded by some authors as an end-member type of tindar when their overall morphology is elongate (Edwards et al., 2009b).

Pillow sheet An areally extensive (tens of kilometres) yet thin (tens of metres) sheet of pillow lava found only rarely in central Iceland (Snorrason and Vilmundardóttir, 2000) and on Mars (Scanlon et al., 2014). Pillow sheets are not well described and their status and origin has never been formally defined although they appear as mapped units on some Icelandic geology maps (Vilmundardóttir et al., 2000). They are probably synonymous with *móberg sheets*, and perhaps some sheet-like sequences (Jakobsson and Gudmundsson, 2008) or pillow mounds/volcanoes. They may also simply represent the basal parts of extensively eroded tuyas and tindars.

Pseudopillows/pseudopillow fractures/pseudopillow fracture systems All three phrases refer to distinctive cooling-related fractures consisting of arcuate master or primary fractures with an appearance and scale similar to rudimentary lava pillows (hence *pseudopillows*) and along which shorter perpendicular subsidiary or secondary polygonal fractures have initiated (Lescinsky and Fink, 2000; Forbes et al., 2012). Commoner in intermediate and felsic lavas. Formation of the master fractures may have involved water and/or steam penetrating incipient cavitation zones in ductile lava strained during flow as well as sub-solidus brittle fracturing. The multiple subsidiary fractures reflect further cooling following isotherms that were established parallel to the master fractures. Some authors have also suggested a role for expansion in the formation of pseudopillow fractures (Lodge and Lescinsky, 2009b).

Ridge-forming lava A subaerial lava whose shape and other features, particularly subvertical margins with subhorizontal cooling fractures, is strongly influenced by cooling against steep walls of ice from adjacent glaciers filling valleys on one or both sides, and which forms a ridge (inverted topography) once those glaciers have melted away (Lescinsky and Sisson, 1998). See also *ice-contact*, *ice-impounded* and *perched lava*.

Sandur An Icelandic word for an outwash plain formed from enormous quantities of sediment and icebergs deposited during repeated *jökulhlaups* (see above), many related to subglacial eruptions. The surface is reworked during normal stream- and river-flow conditions between jökulhlaup events. Extensively developed on the south coast of Iceland.

Sheet-like fractures Planar fractures orientated perpendicular to the margins of mafic or intermediate lavas that are found predominantly at locations where contact with ice is inferred (Lodge and Lescinsky, 2009b). They are typically spaced at 10–20 cm and grade into polygonal joints further into the lava interior. They may form by processes similar to but different from those that form columnar joints (Lescinsky and Fink, 2000; Lodge and Lescinsky, 2009b).

Sheeting joints/platy fractures Parallel planar fractures that are best developed towards the margins of mafic and intermediate lavas and are parallel to the lava surfaces (Lescinsky and Fink, 2000). Individual fractures may reach 5 m in length and are typically spaced at c. 1–<10 cm apart. They are orientated parallel to flow indicators (e.g. crystal orientation) and their formation is probably related to a combination of cooling and shrinkage within well-established surface-parallel isotherms at lava margins and late-stage shearing due to movement within lava interiors.

Sheet-like sequences Relatively thin but areally extensive outcrops, ribbon or sheet-like in plan view, composed of some combination of three main volcanic lithofacies: lapilli tuff, hyaloclastite and lava (Smellie, 2008). The different lithofacies form laterally widespread layers (hence sheet-like). The sequences are overwhelmingly mafic but a few more evolved examples are known with trachytic/phonolitic (Edwards and Russell, 2002) and rhyolitic compositions (Smellie et al., 2011a). Two different sheet-like sequence types have been defined (Dalsheidi and Mt Pinafore), based on differences in sequence dimensions and in the proportions and structural positions of the main lithofacies (Smellie, 2008; but see Section 10.6.3).

Stapi An Icelandic word (plural: *stapar*) introduced by Kjartansson (1943) and used to describe *tuyas*. Usage is uncommon and confined to Iceland.

Subglacial Strictly, subglacial means *beneath ice*, but the word has found widespread usage among geologists as a catch-all for all kinds of volcano–ice interactions, including where subaerial lava is simply abutted against ice (*ice-confined*, see above). An alternative, *glaciovolcanic*, is interchangeable with subglacial and is probably better terminology (see also *intraglacial*, above). Subglacial is also used extensively in the glaciological community for processes that occur at the ice–sediment/bedrock interface.

Surtseyan activity A name for explosive eruptive activity during shoaling and emergence of subaqueous vents principally resulting in construction of tuff cones (Walker, 1973b). Also applied to glaciovolcanic edifices (e.g. tephra mounds, tindars or the tindar stage of tuya construction; Smellie and Hole, 1997). The (non-glaciovolcanic) type example is Surtsey, which erupted in the 1960s off the south coast of Iceland, and the resulting edifices are known as Surtseyan. The explosivity occurs in a water-flooded vent, at high water:magma ratios, and involves a varied mixture of fresh magma, earlier-erupted lapilli and ash, and volatiles (meteoric, marine, juvenile), resulting in the ejection of relatively dense slurries in discrete jets (jetting activity) but also in fountains (continuous-uprush activity; Kokelaar, 1983).

Table mountain A phrase used in geomorphology to describe a flat-topped landform with steep sides (not necessarily volcanic); also known as a mesa. All tuya landforms are table mountains but not all table mountains are tuyas (van Bemmelen and Rutten, 1955).

Tephra mound A pyroclastic cone comprising fragmental deposits (tuff, lapilli tuff, tuff breccia) formed during explosive subaqueous eruptions. Hickson (2000) used the term 'subglacial mound' for glaciovolcanic pyroclastic cones in British Columbia. Since craters are easily removed by post-eruptive erosion (Smellie, 2013), the resulting tephra pile quickly develops a mound-like morphology. Referred to as 'conical tuyas' by Russell et al. (2014) in order to differentiate them from other types of mound-shaped tephra deposits that are not associated with glaciovolcanism. The adjective 'glaciovolcanic' or 'subglacial' needs to be included when using this term.

Tindar An Icelandic word used to describe a ridge composed principally of explosively generated tephra (tuff and lapilli tuff), frequently with a pillow lava core (Jones, 1969a, b; Smellie and Korteniemi, 2015), and occasionally with a minor cap of subaerial lava (Jones, 1969b; Jakobsson and Johnson, 2012). The adoption of *tindar* (plural (in English) *tindars*) in glaciovolcanism is widespread and is likely to continue (Jones, 1969b; Smellie, 2000; Jakobsson and Johnson, 2012) but it is etymologically incorrect since *tindar* is the Icelandic plural form of *tindur*, meaning a peak on a ridge. Some mounds or ridges entirely formed of pillow lava have also been called tindars (Jakobsson and Gudmundsson, 2008; Edwards et al., 2009b; Jakobsson and Johnson, 2012) but those examples are probably better called *pillow mounds* or *ridges* (i.e. *pillow volcanoes*) to avoid confusion (see above). The word has also been used to include single cones of subaqueously erupted tephra forming the core in a tuya (Smellie, 2000). The latter are simply tuff cones (also called *tephra mounds*; Hickson, 2000; Smellie, 2009, 2013) and *tindar* should probably be restricted to glaciovolcanic ridge-forming tephra successions erupted from multiple vents. However, the phrase 'tindar stage' is a useful shorthand notation used to describe the earlier subaqueous essentially explosive aggradational period of evolution of some glaciovolcanic centres (Smellie and Skilling, 1994; Smellie, 2000). It typically follows an initial pillow mound/ridge/volcano stage.

Tuya A steep-sided, usually monogenetic, subglacially erupted volcano with a flat upper surface (Mathews, 1947; Smellie, 2015; Stevenson and Smellie, 2015). Most tuyas have not been subjected to rigorous petrogenetic investigations, and the recognition of a 'monogenetic' origin is frequently only an assumption. Although there is a temptation to extend the term to marine-erupted shoaling to emergent mafic seamounts, which share a similar morphology ('marine tuya', e.g. Surtsey), the word should be restricted to monogenetic glaciovolcanic edifices only, consistent with its original definition and usage. The flat or gently domed shield-like top is the most distinctive characteristic of all tuyas (but see Russell et al., 2014 for a contrasting view in which many different glaciovolcanic landforms are all regarded as types of tuyas; they refer to this category as 'flat-topped tuyas'). The edifice may be circular or have rectilinear sides (van Bemmelen and Rutten, 1955). Mafic monogenetic tuyas are the most common. Their shape is due to the radially outward progradation of lava-fed deltas, with their flat tops and steep delta fronts. There are also two types of monogenetic felsic tuyas (Smellie, 2013). In *tephra-dominated* felsic tuyas, a core of explosively erupted tephra is overlain by flat-topped lavas of which the uppermost are subaerial and their steep sides reflect banking against vertical walls of ice in a piston-like cavity melted above the volcano (Tuffen et al., 2002a). By contrast, a tephra core is absent or greatly suppressed in *lava-dominated* felsic tuyas and they are tall thin pillar or blade-like edifices with a flat top (Mathews, 1951; Kelman, 2005). Less commonly, some large polygenetic volcanoes are also flat-topped and have also been called tuyas on that basis (Souther, 1991; Edwards et al., 2002; see also Russell et al., 2014). In polygenetic tuyas, which can include erupted magmas with widely varying compositions in a single centre, the edifice morphology is a compound feature constructed from the products of multiple eruptive episodes during subaerial and subglacial periods (Edwards et al., 2002; Edwards and Russell, 2002; Russell et al., 2014). The physical evolution and eruptive styles of these polygenetic edifices are strongly influenced by the proximity of large ice masses that periodically enclosed, draped or completely drowned the entire edifice. The morphology is thus is a complicated function of the history of eruptions and the evolving palaeoenvironment. The steep flanks are due in most cases to ponding of lavas against encircling ice or else marginal collapse during ice-poor periods when ice support of steep deposit fronts is removed, whereas the flat summit, commonly ice-capped, can be an expression of caldera collapse and ice infill, erosion, or essentially subhorizontal capping deposits (e.g. Hoodoo Mountain). The reasons for the tuya-like morphology are thus very different from those controlling monogenetic tuya edifices and polygenetic tuyas may not be flat-topped. Whether polygenetic edifices should be called tuyas is not clear. The usage is not widespread and should probably be discouraged.

References

Albino, F., Pinel, V. and Sigmundsson, F. 2010. Influence of surface load variations on eruption likelihood: application to two Icelandic subglacial volcanoes, Grímsvötn and Katla. *Geophysical Journal International*, 181, 1510–1524.

Alho, P., Russell, A.J., Carrivick, J.L. and Käyhkö, J. 2005. Reconstruction of the largest Holocene jökulhlaup within Jökulsa á Fjöllum, NE Iceland. *Quaternary Science Reviews*, 24, 2319–2334.

Allen, C.A. 1979. Volcano–ice interactions on Mars. *Journal of Geophysical Research*, 84, 8048–8059.

Allen, C.C. 1980. Icelandic subglacial volcanism: thermal and physical studies. *Journal of Geology*, 88, 108–117.

Alt, J.C. and Mata, P. 2000. On the role of microbes in the alteration of submarine basaltic glass: a TEM study. *Earth and Planetary Science Letters*, 181, 301–313.

Ambach, W., Blumthaler, M. and Kirchlechner, P. 1981. Application of the gravity flow theory to the percolation of melt water through firn. *Journal of Glaciology*, 27, 67–75.

Amigo, A. 2013. Estimation of tephra-fall and lahar hazards at Hudson volcano, southern Chile: insights from numerical models. In: Rose, W.I. et al. (eds) *Understanding open-vent volcanism and related hazards*. Geological Society of America Special Paper, 498, pp. 177–199.

Anderson, S.W., Stofan, E.R., Plaut, J.J. and Crown, D.A. 1998. Block size distributions on silicic lava flow surfaces: implications for emplacement conditions. *Geological Society of America Bulletin*, 110, 1258–1267.

Antibus, J.V., Panter, K.S., Wilch, T.I., Dunbar, N., McIntosh, W., Tripati, A., Bindemann, I. and Blusztajn, J. 2014. Alteration of volcaniclastic deposits at Minna Bluff: geochemical insights on mineralizing environment and climate during the Late Miocene in Antarctica. *Geochemistry, Geophysics, Geosystems*, 15, 1525–2027.

Atkins, C.B., Barrett, P.J. and Hicock, S.R. 2002. Cold glaciers erode and deposit: evidence from Allan Hills, Antarctica. *Geology*, 30, 659–662.

Aydin, A. and DeGraff, J.M. 1988. Evolution of polygonal fracture patterns in lava flows, *Science*, 239, 471–476.

Bacon, C.R. 2008. *Geologic map of Mount Mazama and Crater Lake caldera, Oregon*. United States Geological Survey, Scientific Investigations Map 2832, 45 pp. and 4 map plates.

Bacon, C. and Lanphere, M. 2006. Eruptive history and geochronology of Mount Mazama and the Crater Lake region. *Geological Society of America Bulletin*, 118, 1331–1359.

Baker, D.M.H., Head, J.W. and Marchant, D.R. 2010. Flow patterns of lobate debris aprons and lineated valley fill north of Ismeniae Fossae, Mars: evidence for extensive mid-latitude glaciations in the Late Amazonian. *Icarus*, 207, 186–209.

Baker, P.E. 1990a. South Sandwich Islands. In: LeMasurier, W.E. and Thomson, J.W. (eds) *Volcanoes of the Antarctic Plate and southern oceans*. American Geophysical Union, Antarctic Research Series, 48, pp. 361–395.

Baker, P.E. 1990b. Montagu Island. In: LeMasurier, W.E. and Thomson, J.W. (eds) *Volcanoes of the Antarctic Plate and southern oceans*. American Geophysical Union, Antarctic Research Series, 48, pp. 382–383.

Baker, P.E., McReith, I., Harvey, M.R., Roobol, M.J. and Davies, T.G. 1975. The geology of the South Shetland Islands: V. Volcanic evolution of Deception Island. *British Antarctic Survey Scientific Reports*, 78, 81 pp.

Baker, V.R. 1982. *The channels of Mars*. University of Texas Press, Austin, 198 pp.

Baker, V., Strom, R., Gulick, V., Kargel, J., Komatsu, G. and Kale, V. 1991. Ancient oceans, ice sheets and the hydrological cycle on Mars. *Nature*, 352, 589–594.

Baker, V.R., Carr, M.H., Gulick, V.C., Williams, C.R. and Marley, M.S. 1992. Channels and valley networks. In: Kieffer, H.H., Jakosky, B.M., Snyder, C.W. and Matthews, M.S. (eds) *Mars*. University of Arizona Press, Tucson, pp. 493–522.

Ball, M. and Pinkerton, H. 2006. Factors affecting the accuracy of thermal imaging cameras in volcanology. *Journal of Geophysical Research*, 111, B11203, doi:10.1029/2005JB003829.

Balme, M.R., Gallagher, C.J. and Hauber, E. 2013. Morphological evidence for geologically young thaw of ice on Mars: a review of recent studies using high-resolution imaging data. *Progress in Physical Geography*, 37, 289–324.

Banik, T.J., Wallace, P.J., Höskuldsson, Á., Miller, C.F., Bacon, C.R. and Furbish, D.J. 2014. Magma–ice–sediment interactions and the origin of lava/hyaloclastite sequences in the Sída Formation, South Iceland. *Bulletin of Volcanology*, 76, 785–803.

Barr, I.D. and Solomina, O. 2013. Pleistocene and Holocene glacier fluctuations upon the Kamchatka Peninsula. *Global and Planetary Change*, 113, 110–120.

Barrows, T.T., Hope, G.S., Prentice, M.L., Fifield, L.K. and Tims, S.G. 2011. Late Pleistocene glaciation of the Mt Giluwe volcano, Papua New Guinea. *Quaternary Science Reviews*, 30, 2676–2689.

Basile, I., Petit, J.R., Touron, S., Grousset, F.E. and Barkov, N. 2001. Volcanic layers in Antarctic (Vostok) ice cores: source identification and atmospheric implications. *Journal of Geophysical Research*, 106, 31915–31932.

Batiza, R. and White, J.D.L. 2000. Submarine lavas and hyaloclastite. In: Sigurdsson, H., Houghton, B., Rymer, H., Stix, J. and McNutt, S. (eds) *Encyclopedia of volcanoes*, 1st Edition. Academic Press, San Diego, pp. 361–381.

Beget, J, Hopkins, D.M. and Charron, S. 1996. The largest known maars on Earth, Seward Peninsula, Northwest Alaska. *Arctic*, 49, 62–69.

Behncke, B. 2004. Late Pliocene volcanic island growth and flood basalt-like lava emplacement in the Hyblean Mountains (SE Sicily). *Journal of Geophysical Research*, 109, B09201, doi:10.1029/2003JB002937.

Behrendt, J.C. 2013. The aeromagnetic method as a tool to identify Cenozoic magmatism in the West Antarctic Rift System beneath the West Antarctic Ice Sheet – A review: Thiel subglacial volcano as possible source of the ash layer in the WAISCORE. *Tectonophysics*, 585, 124–136.

Behrendt, J.C., LeMasurier, W. and Cooper, A.K. 1992. The West Antarctic rift system: a propagating rift 'captured' by a mantle plume? In: Yoshida, Y., Kaminuma, K. and

Shiraishi, K. (eds) *Recent progress in Antarctic earth science*. Terra Scientific Publishing Company (TERAPUB), Tokyo, pp. 315–322.

Behrendt, J.C., Blankenship, D.D., Finn, C.A., Bell, R.E., Sweeney, R.E., Hodge, S.R. and Brozena, J.M. 1994. Evidence for late Cenozoic flood basalts(?) in the West Antarctic rift system revealed by the CASERTZ aeromagnetic survey. *Geology*, 22, 527–530.

Behrendt, J.C., Finn, C.A., Blankenship, D.D. and Bell, R.E. 1998. Aeromagnetic evidence for a volcanic caldera(?) complex beneath the divide of the West Antarctic Ice Sheet. *Geophysical Research Letters*, 25, 4385–4388.

Behrendt, J.C., Blankenship, D.D., Morse, D.L. and Bell, R.E. 2004. Shallow-source aeromagnetic anomalies observed over the West Antarctic IceSheet compared with coincident bed topography from radar ice sounding: new evidence for glacial 'removal' of subglacially erupted late Cenozoic rift-related volcanic edifices. *Global and Planetary Change*, 42, 177–193.

Bell, C.M. 1973. The geology of Beethoven Peninsula, south-western Alexander Island. *British Antarctic Survey Bulletin*, 32, 75–83.

Belousov, A., Behncke, B. and Belousova, M. 2011. Generation of pyroclastic flows by explosive interaction of lava flows with ice/water-saturated substrate. *Journal of Volcanology and Geothermal Research*, 202, 60–72.

Belousov, A., Belousova, M., Edwards, B.R., Volynets, A. and Melnikov, D. 2015. Overview of the precusors and dynamics of the 2012–13 basaltic fissure eruption of Tolbachik volcano, Kamchatka, Russia. *Journal of Volcanology and Geothermal Research* [in press].

Benn, D.I. and Evans, D.J.A. 1998. *Glaciers and glaciation*. Arnold, London, 734 pp.

Bennett, M.R., Huddart, D. and Waller, R.I. 2006. Diamict fans in subglacial water-filled cavities: a new glacial environment. *Quaternary Science Reviews*, 25, 3050–3069.

Bennett, M.R., Huddart, D. and Gonzalez, S. 2009. Glaciovolcanic landsystem and large-scale glaciotectonic deformation along the Brekknafjöll–Jarlhettur, Iceland. *Quaternary Science Reviews*, 28, 647–676.

Bergh, S. 1985. *Structure, depositional environment and mode of emplacement of basaltic hyaloclastites and related lavas and sedimentary rocks: Plio-Pleistocene of the eastern volcanic rift zone, southern Iceland*. Nordic Volcanological Institute Report, 8502, 91 pp.

Bergh, S. and Sigvaldason, G.E. 1991. Pleistocene mass-flow deposits of basaltic hyalo-clastite on a shallow submarine shelf, South Iceland. *Bulletin of Volcanology*, 53, 597–611.

Berman, R.G. 1988. Internally-consistent thermodynamic data for minerals in the system $Na_2O-K_2O-CaO-MgO-FeO-Fe_2O_3-Al_2O_3-SiO_2-TiO_2-H_2O-CO_2$. *Journal of Petrology*, 29, 445–522.

Berman, R.G. and Brown, T.H. 1985. Heat capacity of minerals in the system $Na_2O-K_2O-CaO-MgO-FeO-Fe_2O_3-Al_2O_3-SiO_2-TiO_2-H_2O-CO_2$: representation, estimation, and high temperature extrapolation. *Contributions to Mineralogy and Petrology*, 89, 168–183.

Best, J. 1992. Sedimentology and event timing of a catastrophic volcaniclastic mass-flow, Volcan Hudson, southern Chile. *Bulletin of Volcanology*, 54, 299–318.

Bevier, M.L., Armstrong, R.L. and Souther, J.G. 1979. Miocene peralkaline volcanism in west-central British Columbia: its temporal and plate tectonic setting. *Geology*, 7, 389–392.

Bindeman, I.N. and Serebrykov, N.S. 2011. Geology, petrology and O and H isotope geochemistry of remarkably $\delta^{18}O$ depleted Paleoproterozoic rocks of the Belomorian

Belt, Karelia, Russia, attributed to global glaciation 2.4 Ga. *Earth and Planetary Science Letters*, 306, 163–174.

Bindeman, I.N., Fournelle, J.H. and Valley, J.W. 2001. Low-δ^{18}O tephra from a compositionally zoned magma body: Fisher Caldera, Unimak Island, Aleutians. *Journal of Volcanology and Geothermal Research*, 111, 35–53.

Bindeman, I.N., Ponomareva, V.V., Bailey, J.C. and Valley, J.W. 2004. Volcanic arc of Kamchatka: a province with high-δ^{18}O magma sources and large-scale ^{18}O/^{16}O depletion of the upper crust. *Geochimica et Cosmochimica Acta*, 68, 841–865.

Bindeman, I.N., Gurenko, A.A., Sigmarsson, O. and Chaussidon, M. 2008. Oxygen isotope heterogeneity and disequilibria of olivine phenocrysts in large volume Holocene basalts from Iceland: evidence for magmatic digestion and erosion of Pleistocene hyaloclastites. *Geochimica et Cosmochimica Acta*, 72, 4397–4420.

Bindeman I.N., Leonov, V.L., Izbekov, P.E., Ponomareva, V.V., Watts, K.E., Shipley, N.K., Perpelov, A.B., Bazanova, L.I., Jicha, B.R., Singer, B.S., Schmitt, A.K., Portnyagin, M.V. and Chen, C.H. 2010. Large-volume silicic volcanism in Kamchatka: Ar-Ar and U-Pb ages, isotopic, and geochemical characteristics of major pre-Holocene caldera-forming eruptions. *Journal of Volcanology and Geothermal Research*, 189, 57–80.

Bishop, J.L., Schiffman, P. and Southard, R. 2002. Geochemical and mineralogical analyses of tuffs and altered rinds of pillow basalts in Iceland and applications to Mars. In: Smellie, J.L. and Chapman, M.G. (eds) *Volcano–ice interaction on Earth and Mars*, Geological Society, London, Special Publications, 202, pp. 371–392.

Björnsson, H. 1975. Subglacial water reservoirs, jökulhlaups and volcanic eruptions. *Jökull*, 25, 1–14.

Björnsson, H. 1988. Hydrology of icecaps in volcanic regions. *Vísindafélag Íslendinga, Societas Scientiarum Islandica*, 45, 1–139.

Björnsson, H. 1992. Jökulhlaups in Iceland: prediction, characteristics and simulation. *Annals of Glaciology*, 16, 95–106.

Björnsson, H. 2002. Subglacial lakes and jökulhlaups in Iceland. *Global and Planetary Change*, 35, 255–271.

Björnsson, H. 2009. Jökulhlaups in Iceland: sources, release and drainage. In: Burr, D.M., Carling, P.A. and Baker, V.R. (eds) *Megaflooding on Earth and Mars*. Cambridge University Press, Cambridge, pp. 50–62.

Björnsson, H., Pálsson, F. and Gudmundson, M.T. 2000. Surface and bedrock topography of the Mýrdalsjökull ice cap, Iceland: the Katla caldera, eruption sites, and routes of jökulhlaups. *Jökull*, 49, 29–46.

Blake, D.H. and Löffler, E. 1971. Volcanic and glacial landforms on Mount Giluwe, Territory of Papua and New Guinea. *Geological Society of America Bulletin*, 82, 1605–1614.

Blankenship, D.D., Bell, R.E., Hodge, S.M., Brozena, J.M., Behrendt, J.C. and Finn, C.A. 1993. Active volcanism beneath the West Antarctic ice sheet and implications for icesheet stability. *Nature*, 361, 526–529.

Bleick, H.A., Coombs, M.L., Cervelli, P.F., Bull, K.F. and Wessels, R.L. 2013. Volcano–ice interactions precursory to the 2009 eruption of Redoubt Volcano, Alaska. *Journal of Volcanology and Geothermal Research*, 259, 373–388.

Blondes, M., Reiners, P.R., Edwards, B.R. and Biscontini, A. 2007. Dating young basalt eruptions using U/Th-He on xenolithic zircons. *Geology*, 35, 17–20.

Bonatti, E. and Harrison, C.G.A. 1988. Eruption styles of basalt in oceanic spreading ridges and seamounts: effect of magma temperature and viscosity. *Journal of Geophysical Research*, 93, B4, 2967–2980.

Bonnichsen, B. and Kauffman, D.F. 1987. Physical features of rhyolitic lava flows in the Snake River Plain volcanic province, southwestern Idaho. In: Fink, J.H. (ed.) *The emplacement of silicic domes and lava flows.* Geological Society of America Special Paper, 212, pp. 119–145.

Bosman, A., Casalbore, D., Romagnoli, C. and Chiocci, F.L. 2014. Formation of an 'a'ā lava delta: insights from time-lapse multibeam bathymetry and direct observations during the Stromboli 2007 eruption. *Bulletin of Volcanology*, 76, 1–12.

Bourgeois, O., Dauteuil, O. and Van Vliet-Lanoë, B. 1998. Pleistocene subglacial volcanism in Iceland: tectonic implications. *Earth and Planetary Science Letters*, 164, 165–178.

Brady, J. B. 1995. Diffusion data for silicate minerals, glasses, and liquids. In: Ahrens, T.H. (ed.) *Mineral physics and crystallography: a handbook of physical constants.* AGU Reference Shelf 2, American Geophysical Union, Washington, D.C., pp. 269–290.

Branney, M.J. and Gilbert, J.S. 1995. Ice-melt collapse pits and associated features in the 1991 lahar deposits of Volcán Hudson, Chile: criteria to distinguish eruption-induced glacier melt. *Bulletin of Volcanology*, 57, 293–302.

Branney, M.J. and Kokelaar, P. 2002. Pyroclastic density currents and the sedimentation of ignimbrites. *Geological Society, London, Memoir*, 27, 143 pp.

Brantley, S.R. 1990. *The eruption of Redoubt volcano, Alaska, December 14, 1989 – August 31, 1990.* United States Geological Survey Circular 1061, 33 pp.

Brantley, S.R. 1999. *Volcanoes of the United States.* United States Geological Survey General Publication 99_03, 40 pp.

Brantley, S.R. and Waitt, R.B. 1988. Interrelations among pyroclastic surge, pyroclastic flow, and lahars in Smith Creek valley during first minutes of 18 May 1980 eruption of Mount St. Helens, USA. *Bulletin of Volcanology*, 50, 304–326.

Brown, G.C., Everett, S.P., Rymer, H., McGarvie, D.W. and Foster, I. 1991. New light on caldera evolution: Askja, Iceland. *Geology*, 19, 352–355.

Brugman, M.M. and Meier, M.F. 1981. *Response of glaciers to the eruptions of Mount St. Helens.* United States Geological Survey Professional Paper, 1250, pp. 743–756.

Bull, K.F. and Buurman, H. 2013. An overview of the 2009 eruption of the Redoubt Volcano, Alaska. *Journal of Volcanology and Geothermal Research*, 259, 2–15.

Busby-Spera, C.J. and White, J.D.L. 1987. Variation in peperite textures associated with differing host-sediment properties. *Bulletin of Volcanology*, 49, 765–775.

Bye, A., Edwards, B.R., Hickson, C.J. 2000. Preliminary field, petrographic and geochemical analysis of volcanism at the Watts Point volcanic centre, southwestern British Columbia: a subglacial, dacite lava dome? *Geological Survey of Canada Current Research, Part A*, Paper 2000-A20, 9 pp.

Byers, F.M. 1959. Geology of Unmak and Bogoslof Islands, Aleutian Islands, Alaska. *United States Geological Survey Bulletin*, 1028-L, 267–369.

Calavache, M.L.V. 1990. Pyroclastic deposits of the November 13, 1985 eruption of Nevado del Ruiz volcano, Colombia. *Journal of Volcanology and Geothermal Research*, 41, 67–78.

Campbell, R.J. 2000. *The discovery of the South Shetland Islands 1819–1820. The journal of Midshipman C.W. Poynter.* The Hakluyt Society, London, 232 pp.

Capra, L. 2005. Abrupt climate changes as triggering mechanisms of massive volcanic collapses. *Journal of Volcanology and Geothermal Research*, 155, 329–333.

Capra, L. and Macias, J.L. 2002. The cohesive Naranjo debris flow deposit (10 km^3): a dam breakout flow derived from the Pleistocene debris-avalanche deposit of Nevado de Colima volcano (México). *Journal of Volcanology and Geothermal Research*, 117, 213–235.

Capra, L., Poblete, M.A. and Alvarado, R. 2004. The 1997 and 2001 lahars of Popocatépetl volcano (Central Mexico): textural and sedimentological constraints on their origin and hazards. *Journal of Volcanology and Geothermal Research*, 131, 351–369.

Capra, L., Roverato, M., Groppelli, G., Caballero, I., Sulpizio, R. and Norini, G. 2015. Glacier melting during lava dome growth at Nevado de Toluca volcano (Mexico): evidence of a major threat before main eruptive phases at ice-capped volcanoes. *Journal of Volcanology and Geothermal Research*, 294, 1–10.

Carr, M.H. 1996. *Water on Mars*. Oxford University Press, Oxford, 229 pp.

Carr, M.H. 2006. *The surface of Mars*. Cambridge University Press, Cambridge, 307 pp.

Carr, M.H. and Head, J.W. 2003. Oceans on Mars: an assessment of the observational evidence and possible fate. *Journal of Geophysical Research*, 108, E5, 5–42.

Carr, M.H. and Head, J.W. 2010. Geologic history of Mars. *Earth and Planetary Science Letters*, 294, 185–203.

Carrasco-Núñez, G., Vallance, J.W. and Rose, W.I. 1993. A voluminous avalanche-induced lahar from Citaltépetl volcano, Mexico: implications for hazard assessment. *Journal of Volcanology and Geothermal Research*, 59, 35–46.

Carrivick, J.L., Russell, A.J., Tweed, F.S. and Twigg, D. 2004. Palaeohydrology and sedimentary impacts of jökulhlaups from Kverkfjöll, Iceland. *Sedimentary Geology*, 172, 19–40.

Carrivick, J.L., Russell, A.J., Rushmer, E.L., Tweed, F.S., Marren, P.M., Deeming, H. and Lowe, O.J. 2009. Geomorphological evidence towards a de-glacial control on volcanism. *Earth Surface Processes and Landforms*, 34, 1164–1178.

Carrivick, J.L., Tweed, F.S., Carling, P., Alho, P., Marren, P.M., Staines, K., Russell, A.J., Rushmer, E.L. and Duller, R. 2013. Discussion of 'Field evidence and hydraulic model-ling of a large Holocene jökulhlaup at Jökulsá á Fjöllum channel, Iceland' by Douglas Howard, Sheryl Lzzadder-Beach and Timothy Beach, 2012. *Geomorphology*, 201, 512–519.

Carsewell, D.A., 1983. The volcanic rocks of the Solheimajökull area, southern Iceland. *Jökull*, 33, 61–71.

Carslaw, H.S. and Jaeger, J.C. 1959. *Conduction of heat in solids*. 2nd Edition. Oxford University Press, New York, 510 pp.

Cas, R.A.F. and Wright, J.V. 1987. *Volcanic successions: modern and ancient*. Allen and Unwin, London, 528 pp.

Cas, R.A.F., Landis, C.A. and Fordyce, R.E. 1989. A monogenetic, Surtla-type, Surtseyan volcano from the Eocene–Oligicene Waiareka–Deborah volcanics, Otago, New Zealand: a model. *Bulletin of Volcanology*, 51, 281–298.

Cashman, K.V. and Mangan, M.T. 2014. A century of studying effusive eruptions in Hawai'i. In: Poland, M.P., Takahashi, T.J. and Landowski, C.M. (eds) *Characteristics of Hawaiian volcanoes*. United States Geological Survey Professional Paper, 1801, pp. 357–394.

Castruccio, A., Clavero, J. and Rivera, A. 2010. Comparative study of lahars generated by the 1961 and 1971 eruptions of Calbuco and Villarrica volcanoes, Southern Andes of Chile. *Journal of Volcanology and Geothermal Research*, 190, 297–311.

Chapman, M.G. 1994. Evidence, age, and thickness of a frozen paleolake in Utopia Planitia, Mars. *Icarus*, 109, 393–406.

Chapman, M.G. 2002. Layered, massive and thin sediments on Mars: possible Late Noachian to Late Amazonian tephra? In: Smellie, J.L. and Chapman, M.G. (eds) *Volcano–ice interaction on Earth and Mars*. Geological Society, London, Special Publications, 202, pp. 273–293.

Chapman, M.G. 2003. Sub-ice volcanoes and ancient oceans/lakes: a Martian challenge. *Global and Planetary Change*, 35, 185–198.

Chapman, M.G. and Tanaka, K.L. 2001. Interior layered deposits on Mars: subice volcanoes? *Journal of Geophysical Research*, 106, 10087–10100.

Chapman, M.G. and Smellie, J.L. 2007. Mars interior layered deposits and terrestrial sub-ice volcanoes compared: observations and interpretations of similar geomorphic characteristics. In: Chapman, M.G. (ed.) *The geology of Mars: evidence from Earth-based analogs*. Cambridge University Press, Cambridge, pp. 178–210.

Chapman, M.G., Allen, C.C., Gudmundsson, M.T., Gulick, V.C., Jakobsson, S.P., Lucchitta, B.K., Skilling, I.P. and Waitt, R.B. 2000. Volcanism and ice interactions on Earth and Mars. In: Zimbelman, J.R. and Gregg, T.K.P. (eds) *Environmental effects on volcanic eruptions*. Kluwer Academic/Plenum Publishers, New York, pp. 39–74.

Chapman, M.G., Smellie, J.L., Gudmundsson, M.T., Gulick, V.C., Jakobsson, S.P. and Skilling, I.P. 2001. Study of volcano/ice interactions gains momentum. *Eos, Transactions, American Geophysical Union*, 82, 234–235.

Chapman, M.G., Gudmundsson, M.T., Russell, A.J. and Hare, T.M. 2003. Possible Juventae Chasma sub-ice volcanic eruptions and Maja Valles ice outburst floods, Mars: implications of MGS crater densities, geomorphology, and topography. *Journal of Geophysical Research*, 108, E10, doi:10.1029/2002JE002009.

Chappel, J. 1975. On possible relationships between upper Quaternary glaciations, geomagnetism, and vulcanism. *Earth and Planetary Science Letters*, 26, 370–376.

Charland, A., Francis, D. and Ludden, J. 1993. The relationship between the hawaiites and basalts of the Itcha Volcanic Complex, central British Columbia. *Contributions to Mineralogy and Petrology*, 121, 289–302.

Chinn, T.J., Kargel, J.S., Leonard, G.J., Haritashya, U.K. and Pleasants, M. 2014. New Zealand's glaciers. In: Kargel, J.S., Leonard, G.J., Bishop, M.P., Kääb, A. and Raup, B.H. (eds) *Global land ice measurements from space*. Springer-Verlag, Berlin, pp. 675–715.

Chirico, G.D., Favalli, M., Papale, P., Boschi, E., Pareschi, M.T. and Mamou-Mani, A. 2009. Lava flow hazard at Nyiragongo Volcano, DRC. *Bulletin of Volcanology*, 71, 375–387.

Christiansen, E.H. 1989. Lahars in the Elysium region of Mars. *Geology*, 17, 203–206.

Christensen, P.R. 2003. Formation of recent martian gullies through melting of extensive water-rich snow deposits. *Nature*, 422, 45–48.

Christiansen, R.L. and Peterson, 1981. *Chronology of the 1980 eruptive activity*. In: Lipman, P.W. and Mullineaux, D.R. (eds) *The 1980 eruptions of Mount St. Helens, Washington*. United States Geological Survey Professional Paper, 1250, pp. 17–30.

Churikova, T.G., Gordeychik, B.N., Edwards, B., Ponomareva, V.V. and Zelenin E. 2015. The Tolbachik volcanic massif: a review of the petrology, volcanology and eruption history prior to the 2012–2013 eruption. *Journal of Volcanology and Geothermal Research* [in press].

Cimarelli, C., Alatorre-Ibarguengoitia, M.A., Kueppers, U., Scheu, B. and Dingwell, D. 2013. Experimental generation of volcanic lightning. *Geology*, 38, 439–442.

Clague, J.J. and Ward, B. 2011. Pleistocene glaciation of British Columbia. In: Ehlers, J. and Gibbard, P.L. (eds) *Quaternary glaciations: extent and chronology. Part II, North America*. Developments in Quaternary Science, 2b, Elsevier, Amsterdam, pp. 563–573.

Clarke, G. and Smellie, J. (eds) 2007. Papers from the International Symposium on Earth and Planetary Ice–Volcano Interactions. *Annals of Glaciology*, 45, 199 pp.

Clark, N., Williams, M., Okamura, B., Smellie, J., Nelson, A., Knowles, T., Taylor, P., Leng, M., Zalasiewicz, J. and Haywood, M. 2010. Early Pliocene Weddell Sea seasonality determined from bryozoans. *Stratigraphy*, 7, 196–206.

Clavero, J.E., Sparks, S.J., Polanco, E. and Pringle, M.S. 2004. Evolution of Parinacoata volcano, Central Andes, Northern Chile. *Revista Geologica de Chile*, 31, 317–347.

Clifford, S.M. 1993. A model for the hydrologic and climatic behaviour of water on Mars. *Journal of Geophysical Research*, 98, 10973–11016.

Clifford, S.M. and Parker, T.J. 2001. The evolution of the Martian hydrosphere: implications for the fate of a primordial ocean and the current state of the northern plains. *Icarus*, 154, 40–79.

Clifford, S.M., Lasue, J., Heggy, E., Boisson, J., McGovern, P. and Max, M.D. 2010. Depth of the Martian cryosphere: revised estimates and implications for the existence and detection of subpermafrost groundwater. *Journal of Geophysical Research*, 115, E07001, doi:10.1029/2009JE003462.

Cockell, C.S. 2004. The uses of Martian ice. *Interdisciplinary Science Reviews*, 29, 395–407.

Colbeck, S.C. and Anderson, A.A. 1982. The permeability of a melting snow cover. *Water Resources Research*, 18, 904–908.

Cole, P.D. 1991. Migration direction of sand-wave structures in pyroclastic surge deposits: implications for depositional processes. *Geology*, 19, 1108–1111.

Comer, R.P., Solomon, S.C. and Head, J.W. 1985. Mars: thickness of the lithosphere from the tectonic response to volcanic loads. *Reviews of Geophysics*, 23, 61–92.

Conway, C.E., Townsend, D.B., Leonard, G.S., Calvert, A.T., Wilson, C.J.N. and Gamble, J.A. 2015. Lava–ice interaction on a large composite volcano: a case study from Ruapehu, New Zealand. *Bulletin of Volcanology*, 77, 21, doi:10.1007/s00445-015-0906-2.

Corr, H.F.J. and Vaughan, D.G. 2008. A recent volcanic eruption beneath the West Antarctic ice sheet. *Nature Geoscience*, 1, 122–125.

Cousins, C.R., Smellie, J.L., Jones, A.P. and Crawford, I.A. 2009. A comparative study of endolithic microborings in basaltic lavas from a transitional subglacial–marine environment. *International Journal of Astrobiology*, 8, 37–49.

Cowen, R. 2005. *History of life*. Blackwell Publishing, Oxford, 324 pp.

Craddock, C., Bastien, T.W. and Rutford, R.H. 1964. Geology of the Jones Mountains. In: Adie, R.J. (ed.) *Antarctic geology*. North-Holland Publishing Company, Amsterdam, pp. 171–187.

Craddock, R.A. and Howard, A.D. 2002. The case for rainfall on a warm, wet early Mars. *Journal of Geophysical Research*, 107, E11, 5111, doi:10.1029/2001JE001505.

Cronin, S.J., Neall, V.E., Lecointre, J.A. and Palmer, A.S. 1996. Unusual 'snow slurry' lahars from Ruapehu volcano, New Zealand, September 1995. *Geology*, 24, 1107–1110.

Crown, D.A. and Greeley, R. 1993. Volcanic geology of Hadrica Patera and the eastern Hellas region of Mars. *Journal of Geophysical Research*, 98, 3431–3451.

Cuffey, K. and Paterson, W.S.B. 2010. *The physics of glaciers*. 4th Edition. Elsevier, New York, 693 pp.

Davies, B.J., Glasser, N.F., Carrivick, J.L., Hambrey, M.J., Smellie, J.L. and Nývlt, D. 2013. Landscape evolution and ice-sheet behaviour in a semi-arid polar environment: James Ross Island, NE Antarctic Peninsula. In: Hambrey, M.J., Barker, P.F., Barrett, P.J., Bowman, V., Davies, B., Smellie, J.L. And Tranter, M. (eds) *Antarctic palaeoenvironents and Earth-surface processes*. Geological Society, London, Special Publications, 381, pp. 353–395.

DeGraff, J.M. and Aydin, A. 1987. Surface morphology of columnar joints and its significance to mechanics and direction of joint growth. *Geological Society of America Bulletin*, 99, 605–617.

DeGraff, J.M. and Aydin, A. 1993. Effect of cooling rate on growth increments and spacing of thermal contraction cracks. *Journal of Geophysical Research*, 98, 6411–6430.

DeGraff, J.M., Long, P.E. and Aydin, A. 1989. Use of joint-growth directions and rock textures to infer thermal regimes during solidification of basaltic lava flows. *Journal of Volcanology and Geothermal Research*, 38, 309–324.

Delgado Granados, H., Julio Miranda, P., lvarez, R., Cabral-Cano, E., Cárdenas Gonzalez, L., Correa Mora, F., Luna Alonso, M. and Huggel, C. 2005. Study of Ayoloco Glacier at Iztaccíhuatl volcano (Mexico): hazards related to volcanic activity–ice cover interactions. *Annals of Geomorphology*, 140, 181–193.

Delgado-Granados, H., Miranda, P.J., Núñez, G.C., Alzate, B.P., Mothes, P., Roa, H.M., Cáceres-Correa, B.E. and Ramos, J.C. 2015. Hazards at ice-clad volcanoes: phenomena, processes, and examples from Mexico, Colombia, Ecuador, and Chile. In: Haeberli, W. and Whiteman, C. (eds) *Snow and ice-related hazards, risks and disasters*. Elsevier, Amsterdam, pp. 607–646.

Dingwell, D.B. and Webb, S.L. 1990. Relaxations in silicate melts. *European Journal of Mineralogy*, 2, 427–449.

Dixon, J.E., Stolper, E.M. and Holloway, J.R. 1995. An experimental study of water and carbon dioxide solubilities in mid-ocean ridge basaltic liquids. Part I: calibration and solubilities models. *Journal of Petrology*, 36, 1607–1631.

Dixon, J.E., Filiberto, J.R., Moore, J.G. and Hickson, C.J. 2002. Volatiles in basaltic glasses from a subglacial volcano in northern British Columbia (Canada): implications for ice sheet thickness and mantle volatiles. In: Smellie, J.L. and Chapman, M.G. (eds) *Volcano–ice interaction on Earth and Mars*. Geological Society, London, Special Publications, 202, pp. 255–271.

Dorava, J.M. and Meyer, D.F. 1994. Hydrologic hazards in the lower Drift River basin associated with the 1989–90 eruptions of Redoubt Volcano, Alaska. *Journal of Volcanology and Geothermal Research*, 62, 387–407.

Dorendorf, F., Wiechert, U. and Wörner, G. 2000. Hydrated sub-arc mantle: a source for the Kluchevskoy volcano, Kamchatka/Russia. *Earth and Planetary Science Letters*, 175, 69–86.

Downie, C. 1964. Glaciations of Kilimanjaro, northeast Tanganyika. *Geological Society of America Bulletin*, 75, 1–16.

Duffey, T.S. 2008. Some recent advances in understanding the mineralogy of Earth's deep mantle. *Philosophical Transactions of the Royal Society of London*, 366, 4273–4293.

Duk-Rodkin, A., Barendregt, R.W., Froese, D.G., Weber, F., Enkin, R.J., Smith, I.R., Zazula, Grant D., Waters, P. and Klassen, R. 2004. Timing and extent of Plio-Pleistocene glaciations in North-Western Canada and East-Central Alaska. In: Ehlers, J. and Gibbard, P.L. (eds) *Quaternary glaciations: extent and chronology, Part II, North America.* Developments in Quaternary Science, 2b, Elsevier, Amsterdam, pp. 313–345.

Duk-Rodkin, A., Barendregt, R.W. and White, J. 2010. An extensive late Cenozoic terrestrial record of multiple glaciations preserved in the Tintina Trench of west-central Yukon: stratigraphy, paleomagnetism, paleosols, and pollen. *Canadian Journal of Earth Sciences*, 47, 1003–1028.

Duk-Rodkin, A. and Barendregt, R.W. 2011. Stratigraphical record of glacials/interglacials northwest Canada. In: Ehlers, J. and Gibbard, P.L. (eds) *Quaternary glaciations – extent and chronology, Part II, North America.* Developments in Quaternary Science, 2b, Elsevier, Amsterdam, pp. 661–698.

Duncan, R.A. and Helgason, J. 1998. Precise dating of the Holmatindur cooling event in eastern Iceland: evidence for mid-Miocene bipolar glaciation. *Journal of Geophysical Research*, 103, 12397–12404.

Dvigalo, V.N., Svirid, I.Yu. and Shevchenko, A.V. 2014. The first quantitative estimates of parameters for the Tolbachik fissure eruption of 2012–2013 from aerophoto-grammetric observations. *Journal of Volcanology and Seismology*, 8, 261–268.

Edwards, B.R. and Bye, A. 2003. Preliminary results of field mapping, GIS spatial analysis and major element geochemistry, Ruby Mountain volcano, Atlin volcanic district, northwestern British Columbia. *Geological Survey of Canada Current Research*, 2003-A10, 9 pp.

Edwards, B.R. and Russell, J.K. 1994. Preliminary stratigraphy for the Hoodoo Mountain volcanic center, northwestern British Columbia. *Geological Survey of Canada Current Research*, 94–1A, 69–76.

Edwards, B.R. and Russell, J.K. 1999. Northern Cordilleran volcanic province: a northern Basin and Range? *Geology*, 27, 243–246.

Edwards, B.R. and Russell, J.K. 2000. The distribution, nature and origin of Neogene-Quaternary magmatism in the Northern Cordilleran Volcanic Province, northern Canadian Cordillera. *Geological Society of America Bulletin*, 112, 1280–1295.

Edwards, B.R., Anderson, R.G., Russell, J.K., Hastings, N.L. and Guo, Y.T. 2000. *Geology of the Quaternary Hoodoo Mountain Volcanic Complex and adjacent Paleozoic and Mesozoic basement rocks; parts of Hoodoo Mountain (NTS 104B/14) and Craig River (NTS 104B/11) map areas, northwestern British Columbia.* Geological Survey of Canada, Open File Report 3721, scale 1:20 000.

Edwards, B.R. and Russell, J.K. 2002. Glacial influences on morphology and eruptive products of Hoodoo Mountain volcano, Canada. In: Smellie, J.L. and Chapman, M.G. (eds) *Volcano–ice interaction on Earth and Mars.* Geological Society, London, Special Publications, 202, pp. 179–194.

Edwards, B.R., Edwards, G. and Russell, J.K. 1995. Revised stratigraphy for the Hoodoo Mountain volcanic center, northwestern British Columbia. *Geological Survey of Canada Current Research*, 95–1A, 105–115.

Edwards, B.R., Anderson, R.G., Russell J.K., Hastings, N.L. and Guo, Y.T. 2000. *Geology of the Quaternary Hoodoo Mountain Volcanic Complex and adjacent Paleozoic and Mesozoic basement rocks; parts of Hoodoo Mountain (NTS 104B/14) and Craig River (NTS 104B/11) map areas, northwestern British Columbia.* Geological Survey of Canada, Open File Report 3721, scale 1:20 000, doi:10.4095/211646.

Edwards, B.R., Russell, J.K. and Anderson, R.G. 2002. Subglacial, phonolitic volcanism at Hoodoo Mountain volcano, northwestern Canadian Cordillera. *Bulletin of Volcanology*, 64, 254–272.

Edwards, B.R., Russell, J.K., Anderson, R.A. and Harder, M. 2003. Overview of Neogene to Recent volcanism in the Atlin volcanic district, northern Cordilleran volcanic province, northwestern British Columbia. *Geological Survey of Canada Current Research*, 2003-A8, 6 pp.

Edwards, B.R., Evenchick, C.A., McNicoll, V., Nogier, M. and Wetherell, K. 2006. Overview of the volcanology of the Bell-Irving volcanic district, northwestern Bowser

Basin, British Columbia: new examples of alpine glaciovolcanism from the northern Cordilleran volcanic province. *Geological Survey of Canada Current Research*, 2006-A3, 12 pp.

Edwards, B.R., Tuffen, H., Skilling, I.P. and Wilson, L. (eds) 2009a. Introduction to special issue on volcano–ice interactions on Earth and Mars: the state of the science. *Journal of Volcanology and Geothermal Research*, 185, 247–250.

Edwards, B.R., Skilling, I.P., Cameron, B., Haynes, C., Lloyd, A. and Hungerford, H.D. 2009b. Evolution of an englacial volcanic ridge: Pillow Ridge tindar, Mount Edziza volcanic complex, NCVP, British Columbia, Canada. *Journal of Volcanology and Geothermal Research*, 185, 251–275.

Edwards, B.R. Russell, J.K. and Simpson, K. 2011. Volcanology and petrology of Mathews Tuya, northern British Columbia, Canada: glaciovolcanic constraints on interpretations of the 0.730 Cordilleran paleoclimate. *Bulletin of Volcanology*, 73, 479–496.

Edwards, B.R., Magnússon, E., Thordarson, T., Gudmundsson, M.T., Höskuldsson, A., Oddsson, B. and Haklar, J. 2012. Interactions between lava and snow/ice during the 2010 Fimmvörðuhals eruption, south-central Iceland. *Journal of Geophysical Research*, 117, B04302, doi:10.1029/2011JB008985.

Edwards, B.R., Karson, J., Wysocki, B., Lev, E., Bindeman, I. and Kueppers, U. 2013. Insights on lava-ice/snow interactions from large-scale basaltic melt experiments. *Geology*, 41, 851–854.

Edwards, B.R., Belousov, A. and Belousova, M. 2014a. Propagation style controls lava–snow interactions, *Nature Communications*, 5, 5666. doi:10.1038/ncomms6666.

Edwards, B.R., Jicha, B., Russell, J.K. and Singer, B. 2014b. New constraints on the timing, thickness and distribution of past incarnations of the Cordilleran ice sheet from northwestern British Columbia, Canada. GSA Abstracts with Programs, Paper No. 138–6.

Edwards, B.R., Russell, J.K., Gudmundsson, M.T. 2015a. Glaciovolcanism. In: Sigurdsson, H., Houghton, B., Rymer, H., Stix, J. and McNutt, S. (eds) *The Encyclopedia of Volcanoes*, 2nd Edition. Academic Press, San Diego, pp. 377–393.

Edwards, B.R., Belousov, A., Belousova, M. and Melnikov, D. 2015b. Observations on lava, snowpack and lava–snowpack interactions during the 2012–13 Tolbachik eruption, Klyuchevskoy group of volcanoes, Kamchatka, Russia. *Journal of Volcanology and Geothermal Research* [in press].

Einarsson, T. 1949. The flowing lava: studies of its main physical and chemical characteristics. In: Einarsson, T., Kjartansson, G. and Thorarinsson, S. (eds) *The eruption of Hekla 1947–1948*. Visindafélag Islendinga, Societas Scientarium Islandica, IV, 3, 70 pp.

Einarsson, P. 1994. *Geology of Iceland: rocks and landscape*. Mál og Menning, Reykjavik, 309 pp.

Ellerman, P.J. 1992. *Depositional environments and post-depositional alteration of Cenozoic hyaloclastites in Antarctica*. PhD thesis, University of Colorado at Boulder (USA), 242 pp. [unpublished].

Endress, C. 2007. *Interpretation of glacial and glaciofluvial deposits associated with 1 Ma glaciovolcanism of the Ice Peak Formation, Mount Edziza Volcanic Complex, British Columbia, Canada*. BSc thesis, Dickinson College, Pennsylvania (USA), 165 pp. [unpublished].

Evans, D.J.A. and Benn, D.I. 2004. *A practical guide to the study of glacial sediments*. Arnold, Oxford University Press, London, 266 pp.

Evenchick, C.A., Mustard, P.S., Greig, C.J., McMechan, M.E., Ritcey, D.H., Smith, G.T. and Ferri, F. 2008. Geology, Nass River, British Columbia; Geological Survey of Canada

Open File 5705, and British Columbia Ministry of Energy, Mines and Petroleum Resources, Petroleum Geology Open File 2008–7, 1:250 000 scale.

Ewert, J.W. 2005. System for ranking relative threats of U.S. volcanoes. *Natural Hazard Reviews*, 8, 112–124.

Eyles, N., Eyles, C.H. and Miall, A.D. 1983. Lithofacies types and vertical profile models, an alternative approach to the description and environmental interpretation of glacial diamict and diamictite sequences. *Sedimentology*, 30, 393–410.

Eyles, C.H., Eyles, N. and Miall, A.D. 1985. Models of glaciomarine sedimentation and their application to the interpretation of ancient glacial sequences. *Palaeogeography, Palaeoclimatology, Palaeoecology*, 51, 15–84.

Fabel, D., Stroeven, A.P., Harbor, J., Kleman, J., Elmore, D. and Fink, D. 2002. Landscape preservation under Fennoscandian ice sheets determined from in situ produced ^{10}Be and ^{26}Al. *Earth and Planetary Science Letters*, 201, 397–406.

Fagents, S.A., Lanagan, P. and Greeley, R. 2002. Rootless cones on Mars: a consequence of lava–ground ice interaction. In: Smellie, J.L. and Chapman, M.G. (eds) *Volcano–ice interaction on Earth and Mars*. Geological Society, London, Special Publications, 202, pp. 295–318.

Fagents, S.A., Gregg, T.K.P. and Lopes, R.M.C. (eds) 2013. *Modeling volcanic processes: the physics and mathematics of volcanism*. Cambridge University Press, New York, 421 pp.

Fairchild, L.H. 1986. Quantitative analysis of lahar hazard. In: Keller, S.A.C. (ed.) *Mount St. Helens: five years later*. Eastern Washington University Press, Spokane, pp. 61–67.

Fairen, A.G. 2010. A cold and wet Mars. *Icarus*, 208, 165–175.

Fairen, A.G., Dohm, J.M., Baker, V.R., de Pablo, M.A., Ruiz, J., Ferris, J.C. and Anderson, R.C. 2003. Episodic flood inundations of the northern plains of Mars. *Icarus*, 165, 53–67.

Farrand, W.H., Lane, M.D., Edwards, B.R. and Yingst, R.A. 2010. Spectral evidence of volcanic cryptodomes on the northern plains of Mars. *Icarus*, 211, 139–156.

Fassett, C.I. and Head, J.W. 2006. Valleys on Hecates Tholus, Mars: origin by basal melting of summit snowpack. *Planetary and Space Science*, 54, 370–378.

Fassett, C.I. and Head, J.W. 2007. Valley formation on martian volcanoes in the Hesperian: evidence for melting of summit snowpack, caldera lake formation, drainage and erosion of Ceraunius Tholus. *Icarus*, 189, 118–135.

Fassett, C.I. and Head, J.W. 2008. Valley network-fed, open-basin lakes on Mars: distribution and implications for Noachian surface and subsurface hydrology. *Icarus*, 198, 37–56.

Fastook, J.L., Head, J.W., Marchant, D.R. and Forget, F. 2008. Tropical mountain glaciers on Mars: altitude-dependence of ice accumulation conditions, formation times, glacier dynamics, and implications for planetary spin-axis/orbital history. *Icarus*, 198, 305–317.

Fedotov, S.A. and Masurenkov, Yu.P. 1991. *Active volcanoes of Kamchatka*, Volume 1. Nauka, Moscow, 302 pp.

Fink, J.H. and Manley, C.R. 1987. Origin of pumiceous and glassy textures in rhyolite flows and domes. In: Fink, J.H. (ed.) *The emplacement of silicic domes and lava flows*. Geological Society of America Special Paper, 212, pp. 119–145.

Fisher, R.V. 1961. Proposed classification of volcaniclastic sediments and rocks. *Geological Society of America Bulletin*, 72, 1409–1414.

Fisher, R.V. and Schmincke, H.-U. 1984. *Pyroclastic rocks*. Springer-Verlag, Berlin, 472 pp.

Fiske, R.S., Hopson, C.A. and Waters, A.C. 1963. *Geology of Mount Rainier National Park, Washington.* United States Geological Survey Professional Paper, 444, 93 pp.

Flovenz, Ó.G. and Saemundsson, K. 1993. Heat flow and geothermal processes in Iceland. *Tectonophysics*, 225, 123–138.

Flude, S., Burgess, R. and McGarvie, D.W. 2008. Silicic volcanism at Ljósufjöll, Iceland: insights into evolution and eruptive history from Ar–Ar dating. *Journal of Volcanology and Geothermal Research*, 169, 154–175.

Flude, S., McGarvie, D.W., Burgess, R. and Tindle, A.G. 2010. Rhyolites at Kerlingarfjöll, Iceland: the evolution and lifespan of silicic central volcanoes. *Bulletin of Volcanology*, 72, 523–538.

Forbes, A.E.S., Blake, S., McGarvie, D.W. and Tuffen, H. 2012. Pseudopillow fracture systems in lavas: insights into cooling mechanisms and environments from lava flow fractures. *Journal of Volcanology and Geothermal Research*, 245–246, 68–80.

Forbes, A.E.S., Blake, S. and Tuffen, H. 2014a. Entablature: fracture types and mechanisms. *Bulletin of Volcanology*, 76, 1–13.

Forbes, A.E.S., Blake, S., Tuffen, H. and Wilson, A. 2014b. Fractures in a trachyandesitic lava at Öræfajökull, Iceland, used to infer subglacial emplacement in 1727–8 eruption. *Journal of Volcanology and Geothermal Research*, 288, 8–18.

Fountain, A.G. 1989. The storage of water in, and hydraulic characteristics of, the firn of South Cascade Glacier, Washington State, U.S.A. *Annals of Glaciology*, 36, 69–75.

Fountain, A.G. and Walder, J.S. 1998. Water flow through temperate glaciers. *Reviews of Geophysics*, 36, 299–328.

Fountain, A.G., Jacobel, R.W., Schlichting, R. and Jansson, P. 2005. Fractures as the main pathways of water flow in temperate glaciers. *Nature*, 433, 618–621.

Fournelle, J.H., Marsh, B.D. and Myers, J.D. 1994. Age, character, and significance of Aleutian arc volcanism. In: Plafker, G. and Berg, H.C. (eds) *The geology of Alaska.* Geological Society of America, Boulder, Colorado, pp. 723–757.

Francis, D. and Ludden, J. 1990. The mantle source for olivine nephelinite, basanite and alkali olivine basalt at Fort Selkirk, Yukon. *Journal of Petrology*, 31, 371–400.

Fraser, C.I., Terauds, A., Smellie, J., Convey, P. and Chown, S.L. 2014. Geothermal activity helps life survive glacial cycles. *Proceedings of the National Aademy of Sciences*, doi/10.1073/pas.1321437111.

Fretzdorff, S., Worthington, T.J., Haase, K.M. and Hékinian, R. 2004. Magmatism in the Bransfield Basin: rifting of the South Shetland arc? *Journal of Geophysical Research*, 109, B12208, doi:10.1029/2004JB003046.

Fridleifsson, I.B., Furnes, H. and Atkins, F.B. 1982. Subglacial volcanics: on the control of magma chemistry on pillow dimensions. *Journal of Volcanology and Geothermal Research*, 13, 103–117.

Froese, D.G., Barendregt, R.W., Enkins, R.J. and Baker, J. 2000. Paleomagnetic evidence for multiple Late Pliocene–Early Pleistocene glaciations in the Klondike area, Yukon Territory. *Canadian Journal of Earth Sciences*, 37, 863–877.

Fuller, E.R. and Head, J.W. 2002. Amazonis Planitia: the role of geologically recent volcanism and sedimentation in the formation of the smoothest plains on Mars. *Journal of Geophysical Research*, 107, E10, 5081, doi:10.1029/2002JE001842.

Fuller, R.E. 1931. Aqueous chilling of basaltic lava on the Columbia River Plateau. *American Journal of Science*, 21, 281–300.

Fuller, R.E. 1934. Structural features in the Columbia River lavas of Central Washington. *Journal of Geology*, 42, 311–328.

Furman, T., Meyer, P.S. and Frey, F. 1992. Evolution of Icelandic central volcanoes: evidence from the Austurhorn intrusion, southeastern Iceland. *Bulletin of Volcanology*, 55, 45–62.

Furman, T., Bryce, J.G., Karson, J. and Iotti, A. 2004. East African Rift System (EARS) plume structure: insights from Quaternary mafic lavas of Turkana, Kenya. *Journal of Petrology*, 45, 1069–1088.

Furnes, H. 1978. Element mobility during palagonitization of a subglacial hyaloclastite in Iceland. *Chemical Geology*, 22, 249–264.

Furnes, H. 1984. Chemical changes during progressive subaerial palagonitization of a subglacial olivine tholeiite hyaloclastite: a microprobe study. *Chemical Geology*, 43, 271–264.

Furnes, H. and Fridleifsson, I.B. 1974. Tidal effects on the formation of pillow lava/hyaloclastite deltas. *Geology*, 2, 381–384.

Furnes, H., Fridliefsson, I.B. and Atkins, F.B. 1980. Subglacial volcanics: on the formation of acid hyaloclastites. *Journal of Volcanology and Geothermal Research*, 8, 95–110.

Furnes, H., Thorseth, I.H., Torsvik, T., Muehlenbachs, K., Staudigel, H. and Tumyr, P. 2002. Identifying bio-interaction with basaltic glass in oceanic crust and implications for estimating the depth of the oceanic biosphere: a review. In: Smellie, J.L. and Chapman, M.G. (eds) *Volcano–ice interaction on Earth and Mars*. Geological Society, London, Special Publications, 202, pp. 407–421.

Gabrielse, H. 1969. *Geology of the Jennings River map area, British Columbia (104-O)*. Geological Survey of Canada Paper 68–55, 37 pp.

Gabrielse, H. 1998. Geology of the Cry Lake and Dease Lake map areas, northern British Columbia (104-J). *Geological Survey of Canada Bulletin*, 504, 144 pp.

Ganci, G., Vicari, A., Cappello, A. and del Negro, C. 2012. An emergent strategy for volcanic hazard assessment: from thermal satellite monitoring to lava flow modeling. *Remote Sensing of the Environment*, 119, 197–207.

Gardeweg, M., Sparks, R. and Mathews, S. 1998. Evolution of Lascar volcano, northern Chile. *Journal of the Geological Society, London*, 155, 89–104.

Gardner, C.A., Neal, C.A., Waitt, R.B. and Janda, R.J. 1994. Proximal pyroclastic deposits from the 1989–90 eruption of Redoubt volcano, Alaska: stratigraphy, distribution, and physical characteristics. In: Miller, T.P. and Chouet, B.A. (eds) *The 1989–1990 eruption of Redoubt Volcano, Alaska. Journal of Volcanology and Geothermal Research*, 62, 213–250.

Gaździcki, A., Tatur, A., Hara, U. and del Valle, R.A. 2004. The Weddell Sea Formation: post-late Pliocene terrestrial glacial deposits on Seymour Island, Antarctic peninsula. *Polish Polar Research*, 25, 189–204.

Gee, M.A.M., Taylor, R.N., Thirwall, M.F. and Murton, B.J. 1998. Glacioisostacy controls on chemical and isotopic characteristics of tholeiites from the Reykjanes Peninsula, SW Iceland. *Earth and Planetary Science Letters*, 164, 1–5.

Geirsdóttir, Á. 1991. Diamictites of late Pliocene age in western Iceland. *Jökull*, 40, 3–25.

Geirsdóttir, Á. and Eiríksson, J. 1994. Growth of an intermittent ice sheet in Iceland during the late Pliocene and early Pleistocene. *Quaternary Research*, 42, 115–130.

Geirsdóttir, Á., Miller, G.H. and Andrews, J.T. 2007. Glaciation, erosion, and landscape evolution of Iceland. *Journal of Geodynamics*, 43, 170–186.

Gellatly, A.F., Gordon, J.E., Whalley, W.B. and Hansom, J.G. 1988. Thermal regime and geomorphology of plateau ice caps in northern Norway: observations and implications. *Geology*, 16, 983–986.

Ghatan, G.J. and Head, J.W. 2002. Candidate subglacial volcanoes in the south polar region of Mars: morphology, morphometry, and eruption conditions. *Journal of Geophysical Research*, 107, E7, 10.1029/2001JE001519.

Ghatan, G.J., Head, J.W. and Pratt, S. 2003. Cavi Angusti, Mars: characterization and assessment of possible formation mechanisms. *Journal of Geophysical Research*, 108, E5, doi:10.1029/2002JE001972.

Ghiorso, M.S. and Sack, R.O. 1995. Chemical mass transfer in magmatic processes. IV. A revised and internally consistent thermodynamic model for the interpolation and extrapolation of liquid–solid equilibria in magmatic systems at elevated temperatures and pressures. *Contributions to Mineralogy and Petrology*, 119, 197–212

Ghiorso, M.S., Hirschmann, M.M., Reiners, P.W. and Kress, V.C. 2002. The pMELTS: a revision of MELTS aimed at improving calculation of phase relations and major element partitioning involved in partial melting of the mantle at pressures up to 3 GPa. *Geochemistry, Geophysics, Geosystems* 3, 5, doi:10.1029/2001GC000217.

Gifkins, C., McPhie, J. and Allen, R. 2002. Pumiceous rhyolitic peperite in ancient submarine volcanic successions. *Journal of Volcanology and Geothermal Research*, 114, 181–203.

Gilbert, J.S., Stasiuk, M.V., Lane, S., Adam, C.R., Murphy, M.D., Sparks, R.S.J. and Naranjo, J.A. 1996. Non-explosive, constructional evolution of the ice-filled caldera at Volcán Sollipulli, Chile. *Bulletin of Volcanology*, 58, 67–83.

Giordano, D., Romano, C., Papale, P. and Dingwell, D.B. 2004. The viscosity of trachytes, and comparison with basalts, phonolites, and rhyolites. *Chemical Geology*, 213, 49–61.

Giordano, D., Nichols, A.R.L. and Dingwell, D.B. 2005. Glass transition temperatures of natural hydrous melts: a relationship with shear viscosity and implications for the welding process. *Journal of Volcanology and Geothermal Research*, 142, 105–188.

Giordano, D., Mangicapra, A., Potuzak, M., Russell, J.K., Romano, C., Dingwell, D.B. and Di Muro, A. 2006. An expanded nonArrhenian model for silicate melt viscosity: a treatment for metaluminous, peraluminous and peralkaline melts. *Chemical Geology*, 229, 42–56.

Giordano, D., Russell, J.K. and Dingwell, D. 2008. Viscosity of magmatic liquids: a model. *Earth and Planetary Science Letters*, 271, 123–134.

Giordano, D., Ardia, P., Romano, C., Dingwell, D.B., Di Muro, A., Schmidt, M.W., Mangiacapra, A. and Hess, H.-U. 2009. The rheological evolution of alkaline Vesuvius magmas and comparison with alkaline series from the Phlegrean Fields, Etna, Stromboli and Teide. *Geochimica et Cosmochimica Acta*, 73, 6613–6630.

Girina, O. 2013. Chronology of Bezymianny Volcano activity, 1956–2010. *Journal of Volcanology and Geothermal Research*, 263, 22–41.

Glazner, A.F., Manley, C.R., Marron, J.S. and Rojstaczer, S. 1999. Fire or ice: anticorrelation of volcanism and glaciation in California over the past 800,000 years. *Geophysical Research Letters*, 26, 1759–1762.

Glen, J.W. 1952. Experiments on the deformation of ice. *Journal of Glaciology*, 2, 111–114.

Glen, J.W. 1954. The stability of ice-dammed lakes and other water-filled holes in glaciers. *Journal of Glaciology*, 2, 316–318.

Goehring, L. and Morris, S.W. 2008. Scaling of columnar joints in basalt. *Journal of Geophysical Research*, 113, B10203, doi:10.1029/2007JB005018.

Golledge, N.R., Levy, R.H., McKay, R.M., Fogwill, C.J., White, D.A., Graham, A.G.C., Smith, J.A., Hillenbrand, C.-D., Licht, K.J., Denton, G.H., Ackert, R.P., Maas, S.M. and

Hall., B.L. 2013. Glaciology and geological signature of the Last Glacial Maximum Antarctic ice sheet. *Quaternary Science Reviews*, 78, 225–247.

Gonnermann, H.M. and Manga, M. 2013. Dynamics of magma ascent in the volcanic conduit. In: Fagents, S.A., Gregg, T.K.P. and Lopes, R.M.C. (eds) *Modeling volcanic processes: the physics and mathematics of volcanism.* Cambridge University Press, Cambridge, pp. 55–84.

González-Ferrán, O. 1995. *Volcanes de Chile.* Instituto Geográfico Militar, Santiago, 640 pp.

Goodfellow, B.W. 2007. Relict non-glacial surfaces in formerly glaciated landscapes. *Earth-Science Reviews*, 80, 47–73.

Gorbach, N., Portnyagin, M. and Tembrel, I. 2013. Volcanic structure and composition of Old Shiveluch volcano, Kamchatka. *Journal of Volcanology and Geothermal Research*, 263, 193–208.

Gordeev, E.I., Murav'ev Ya. D., Samoilenko S.B., Volynets A.O., Melnikov D.V. and Dvigalo V.N. 2013. The Tolbachik fissure eruption of 2012–2013: preliminary results. *Doklady Earth Sciences*, 452, 1046–1050.

Gore, D. 1992. Ice-damming and fluvial erosion in the Vestfold Hills, East Antarctica. *Antarctic Science*, 4, 227–234.

Gore, D.B. and Pickard, J. 1998. Proglacial hydrology and drainage, southeastern Vestfold Hills, East Antarctica. *Proceedings of the Linnean Society of New South Wales*, 119, 181–196.

Gottsmann, J. and Dingwell, D.B. 2001. Cooling dynamics of spatter-fed phonolite obsidian flows on Tenerife, Canary Islands. *Journal of Volcanology and Geothermal Research*, 105, 323–342.

Gourgard, A. and Thouret, J.-C. 1990. Magma mixing and petrogenesis of the 13 November 1985 eruptive products at Nevado del Ruiz (Colombia). *Journal of Volcanology and Geothermal Research*, 41, 79–96.

Graettinger, A.H., Ellis, M.K., Skilling, I.P., Reath, K., Ramsey, M.S., Lee, R.J., Hughes, C.G. and McGarvie, D.W. 2013a. Remote sensing and geologic mapping of glaciovolcanic deposits in the region surrounding Askja (Dyngjufjöll) volcano, Iceland. *International Journal of Remote Sensing*, 34, 7178–7198.

Graettinger, A.H., Skilling, I.P., McGarvie, D.W. and Cameron, B. 2013b. Environmental reconstruction of basaltic glaciovolcanic deposits at Askja Volcano, Iceland, using lithofacies and geomorphology. IAVCEI 2013 General Assembly, Kagoshima, Japan (Abstract 4W_3K-P5, p. 1077).

Graettinger, A.H., Skilling, I., McGarvie, D. and Höskuldsson, A. 2013c. Subaqueous basaltic magmatic explosions trigger phreatomagmatism: a case study from Askja, Iceland. *Journal of Volcanology and Geothermal Research*, 264, 17–35.

Green, N. 1990. Late Cenozoic Volcanism in the Mount Garibaldi and Garibaldi Lake volcanic fields, Garibaldi volcanic belt, southwestern British Columbia. *Geoscience Canada*, 17, 171–175.

Gregg, T.K.P. and Fink, J.H. 1995. Quantification of submarine lava-flow morphology through analog experiments. *Geology*, 23, 73–76.

Gregg, T.K.P. and Fink, J.H. 2000. A laboratory investigation into the effects of slope on lava flow morphology. *Journal of Volcanology and Geothermal Research*, 96, 145–159.

Griffiths, R.W. and Fink, J.H. 1992. Solidification and morphology of submarine lavas: a dependence on extrusion rate. *Journal of Geophysical Research*, 97, 19729–19737.

Grönvold, K. 1972. Structural and petrological studies in the Kerlingarfjöll region, central Iceland. DPhil thesis, Oxford University (UK), 237 pp. [unpublished].

Grönvold, K., Larsen, G., Einarsson, P., Thorarinsson, S. and Saemundsson, K., 1983. The 1980–81 Hekla eruption. *Bulletin of Volcanology*, 46, 349–363.

Grosse, P., van Wyk de Vries, B., Petrinovic, I.A., Euillades, P.A. and Alvarado, G.E. 2009. Morphometry and evolution of arc volcanoes. *Geology*. 37, 651–654.

Grosse, P., van Wyk de Vries, B., Euillades, P.A., Kervyn, M. and Petrinovic, I.A. 2012. Systematic morphometric characterization of volcanic edifices using digital elevation models. *Geomorphology*, 136, 114–131.

Grosse, P., van Wyk de Vries, B., Euillades, P.A. and Euillades, L.D. 2014. A global database of composite volcano morphometry. *Bulletin of Volcanology*, 76, 784.

Grossenbacher, K.A. and McDuffie, S.M. 1995. Conductive cooling of lava: columnar joint diameter and stria width as functions of cooling rate and thermal gradient. *Journal of Volcanology and Geothermal Research*, 69, 95–103.

Grove, E.W. 1974. Deglaciation: a possible triggering mechanism for recent volcanism. In: González-Ferrán, O. (ed.) *Proceedings of the Symposium on Andean and Antarctic Volcanology Problems, Santiago, Chile*, IAVCEI, Rome, pp. 88–97.

Gualda G.A.R., Ghiorso M.S., Lemons R.V. and Carley T.L. 2012. Rhyolite-MELTS: a modified calibration of MELTS optimized for silica-rich, fluid-bearing magmatic systems. *Journal of Petrology*, 53, 875–890.

Gudmundsson, A. 1986. Mechanical aspects of post-glacial volcanism and tectonics of the Reykjanes Peninsula, Southwest Iceland. *Journal of Geophysical Research*, 91, 12 711–12 721.

Gudmundsson, M.T. 2003. Melting of ice by magma-ice-water interactions during sub-glacial eruptions as an indicator of heat transfer in subaqueous eruptions. In: White, J.D.L., Smellie, J.L. and Clague, D.A. (eds) *Subaqueous explosive volcanism*. American Geophysical Union, Geophysical Monograph Series, 140, pp. 61–72.

Gudmundsson, M.T. 2005. Subglacial volcanic activity in Iceland. In: Caseldine, C.J., Russell, A., Hardardóttir, J. and Knudsen, Ó. (eds) *Iceland: modern processes, past environments*. Elsevier, Amsterdam, pp. 127–151.

Gudmundsson, M.T. and Bjornsson, H. 1991. Eruptions in Grímsvötn, Vatnajökull, Iceland, 1934–1991. *Jökull*, 41, 21–45.

Gudmundsson, A., Oskarsson, N., Gronvold, K., Saemundsson, K., Sigurdsson, O., Stefansson, R., Gislason, S.R., Einarsson, P., Brandsdóttir, B., Larsen, G., Johannesson, H. and Thordarson, T. 1992. The 1991 eruption of Hekla. *Bulletin of Volcanology*, 54, 238–246.

Gudmundsson, M.T., Sigmundsson, F. and Björnsson, H. 1997. Ice–volcano interaction in the 1996 Gjálp eruption, Vatnajökull, Iceland. *Nature*, 389, 954–957.

Gudmundsson, M.T., Pálsson, F., Björnsson, H. and Högnadóttir, Th. 2002. The hyaloclastite ridge formed in the subglacial 1996 eruption in Gjálp, Vatnajökull, Iceland: present day shape and future preservation. In: Smellie, J.L. and Chapman, M.G. (eds) *Volcano–ice interaction on Earth and Mars*. Geological Society, London, Special Publications, 202, pp. 319–335.

Gudmundsson, M.T., Sigmundsson, F., Björnson, H. and Högnadóttir, Th. 2004. The 1996 eruption at Gjálp, Vatnajökull icecap, Iceland: efficiency of heat transfer, ice deformation and subglacial water pressure. *Bulletin of Volcanology*, 66, 46–65.

Gudmundsson, M.T., Högndóttir, Th., Kristinsson, A.B. and Gudbjörnsson, S. 2007. Geothermal activity in the subglacial Katla caldera, Iceland, 1999–2005, studied with radar altimetry. In: Clarke, G. and Smellie, J. (eds) *Papers from the International Symposium on Earth and Planetary Ice–Volcano Interactions held in Reykjavik, Iceland, on 19–23 June, 2006. Annals of Glaciology*, 45, 66–72.

Gudmundsson, M.T., Larsen, G., Höskuldsson, Á. and Gylfason, Á.G. 2008. Volcanic hazards in Iceland. *Jökull*, 58, 251–268.

Gudmundsson, M.T., Pedersen, R., Vogfjörd, K., Thorbjarnardóttir, B., Jakobsdóttir, S. and Roberts, M.J. 2010. Eruptions of Eyjafjallajökull Volcano, Iceland. *Eos, Transactions, American Geophysical Union*, 91, 190–191.

Gudmundsson, M.T., Thordarson, T., Höskuldsson, Á., Larsen G., Björnsson, H., Prata, A.J., Oddsson, B., Magnússon, E., Högnadóttir, T., Pedersen, G.N., Hayward, C.L., Stevenson, J.A. and Jónsdóttir, I. 2012a. Ash generation and distribution from the April-May 2010 eruption of Eyjafjallajökull, Iceland. *Nature Scientific Reports*, 2, 572; doi:10.1038/srep00572.

Gudmundsson, M.T., Palsson, F, Thordarsson, T., Hoskuldsson, A., Larsen, G., Hognadottir, T., Oddson, B., Oladottir, B.A., Godnason, J. 2012b. Water/mass fractions in phreatomagmatic eruption plumes: constraints from the Grimsvotn 2011 eruption. Abstract V11B-4718 presented at the 2012 Fall Meeting, American Geophysical Union, San Francisco, CA, 15–19 December.

Guillou, H., Vliet-Lanoë, B.V., Gudmundsson, A. and Nomade, S. 2010. New unspiked K-Ar ages of Quaternary sub-glacial and sub-aerial volcanic activity in Iceland. *Quaternary Geochronology*, 5, 10–19.

Gulick, V.C. 1993. *Magmatic intrusions and hydrothermal systems: Implications for the formation of Martian fluvial valleys.* PhD thesis, University of Arizona (USA), 146 pp. [unpublished].

Gulick, V.C. 1998. Magmatic intrusions and a hydrothermal origin for fluvial valleys on Mars. *Journal of Geophysical Research*, 103, 19365–19387.

Hall, K.J. 1982. Rapid deglaciation as an initiator of volcanic activity: an hypothesis. *Earth Surface Processes and Landforms*, 7, 45–51.

Hall, K., Meiklejohn, I. and Bumby, A. 2011. Marion Island volcanism and glaciation. *Antarctic Science*, 23, 155–163.

Hall, M. and Mothes, P. 2008. The rhyolitic–andesitic eruptive history of Cotopaxi volcano, Ecuador. *Bulletin of Volcanology*, 70, 675–702.

Hambrey, M.J. 1994. *Glacial environments.* UBC Press, London, 299 pp.

Hambrey, M.J. and Glasser, N.F. 2003. Glacial sediments: processes, environments and facies. In: Middleton, G.V. (ed.) *Encyclopedia of sediments and sedimentary rocks.* Kluwer, Dordrecht, pp. 316–331.

Hambrey, M.J. and Glasser, N.F. 2012. Discriminating glacier thermal and dynamic regimes in the sedimentary record. *Sedimentary Geology*, 251, 1–33.

Hambrey, M.J. and Smellie, J.L. 2006. Distribution, lithofacies and environmental context of Neogene glacial sequences on James Ross and Vega islands, Antarctic Peninsula. In: Francis, J.E., Pirrie, D. and Crame, J.SA. (eds) *Cretaceous–Tertiary high-latitude palaeoenvironments, James Ross Basin, Antarctica.* Geological Society, London, Special Publications, 258, 187–200.

Hambrey, M.J., Smellie, J.L., Nelson, A.E. and Johnson, J.S. 2008. Late Cenozoic glacier–volcano interaction on James Ross Island and adjacent areas, Antarctic Peninsula region. *Bulletin of the American Geological Society*, 120, 709–731.

Hamilton, C.W., Fagents, S.A. and Wilson, L. 2010. Explosive lava–water interactions in Elysium Planitia, Mars: geologic and thermodynamic constraints on the formation of the Tartarus Colles cone groups. *Journal of Geophysical Research*, 115, doi:10.1029/2009JE003546.

Hamilton, C.W., Fagents, S.A. and Thordarson, T. 2011. Lava–ground ice interactions in Elysium Planitia, Mars: geomorphological and geospatial analysis of the Tartarus Colles cone groups. *Journal of Geophysical Research*, 116, doi:10.1029/2010JE003657.

Hamilton, T.D. 1994. Late Cenozoic glaciation of Alaska. In: Plafker, G. and Berg, H.C. (eds) *The Geology of Alaska.* Geological Society of America, Boulder, Colorado, pp. 813–844.

Hamilton, T.S. 1981. *Late Cenozoic alkaline volcanics of the Level Mountain Range, north-western British Columbia*. PhD thesis, University of Alberta, Edmonton (Canada), 490 pp. [unpublished].

Hamilton, W. 1972. *The Hallett Volcanic Province, Antarctica*. United States Geological Survey Professional Paper, 456-C, 62 pp.

Hammond, P.E. 1987. Lone Butte and Crazy Hills: subglacial volcanic complexes, Cascade Range, Washington. In: Hill, M.L. (ed.) *Geological Society of America Centennial Field Guide: Volume 1, Cordilleran Section*, pp. 339–344.

Hardarson, B.S. and Fitton, J.G. 1991. Increased mantle melting beneath Snaefellsjokull Volcano during late Pleistocene deglaciation. *Nature*, 353, 62–64.

Hardarson, B.S., Fitton, J.G. and Hjartarson, Á. 2008. Tertiary volcanism in Iceland. *Jökull*, 58, 161–178.

Harder, M. and Russell, J.K. 2007. Basanite glaciovolcanism at Llangorse Mountain, northern British Columbia, Canada. *Bulletin of Volcanology*, 69, 329–340.

Harris, A. 2013a. *Thermal remote sensing of active volcanoes: a user's manual.* Cambridge University Press, Cambridge, 736 pp.

Harris, A.J.L. 2013b. Lava flows. In Fagents, S.A., Gregg, T.K.P., Lopes, R.M.C. (eds) *Modeling volcanic processes: the physics and mathematics of volcanism*, Cambridge University Press, Cambridge, pp. 85–106.

Hauksdóttir, S., Enegren, E.G. and Russell, J.K. 1994. Recent basaltic volcanism in the Iskut-Unuk rivers area, northwestern British Columbia. *Geological Survey of Canada Current Research*, 1994-A, 57–68.

Haywood, A.M., Smellie, J.L., Ashworth, A.C., Cantrill, D.J., Florindo, F., Hambrey, M.J., Hill, D., Hillenbrand, C.-D., Hunter, S.J., Larter, R.D., Lear, C.H., Passchier, S. and van de Wal, R. 2009. Middle Miocene to Pliocene history of Antarctica and the Southern Ocean. In: Siegert, M.J. and Florindo, F. (eds) *Developments in Earth & Environmental Sciences, 8, Antarctic Climate Evolution*. Elsevier, Oxford, pp. 401–463.

Head, J.W. and Marchant, D.R. 2003. Cold-based mountain glaciers on Mars: Western Arsia Mons. *Geology*, 31, 641–644.

Head, J.W. and Pratt, S. 2001. Extensive Hesperian-aged south polar ice sheet on Mars: evidence for massive melting and retreat, and lateral flow and ponding of meltwater. *Journal of Geophysical Research*, 106, 12275–12299.

Head, J.W. and Wilson, L. 2002. Mars: a review and synthesis of general environments and geological settings of magma–H_2O interactions. In: Smellie, J.L. and Chapman, M.G. (eds) *Volcano–ice interaction on Earth and Mars*. Geological Society, London, Special Publications, 202, pp. 27–57.

Head, J.W. and Wilson, L. 2003. Deep submarine pyroclastic eruptions: theory and predicted landform and deposits. *Journal of Volcanology and Geothermal Research*, 121, 155–193.

Head, J.W. and Wilson, L. 2007. Heat transfer in volcano–ice interactions on Mars: synthesis of environments and implications for processes and landforms. In: Clarke, G. and Smellie, J. (eds) *Papers from the International Symposium on Earth and Planetary*

Ice–Volcano Interactions held in Reykjavik, Iceland, on 19–23 June, 2006. Annals of Glaciology, 45, 1–13.

Head, J.W., Hiesinger, H., Ivanov, M.A., Kreslavsky, M.A., Pratt, S. and Thomson, B.J. 1999. Possible ancient oceans on Mars: evidence from Mars Orbiter Laser Altimeter data. *Science*, 286, 2134–2137.

Head, J.W., Mustard, J.F., Kreslavsky, M.A., Milliken, R.E. and Marchant, D.R. 2003. Recent ice ages on Mars. *Nature*, 426, doi:10.1038/nature02114.

Head, J.W., Neukum, J.F., Jaumann, R., Hiesinger, H., Hauber, E., Carr, M. et al. 2005. Tropical to mid-latitude snow and ice accumulation, flow and glaciations of Mars. *Nature*, 434, 346–351.

Heiken, G. and Wohletz, K. 1985. *Volcanic ash*. University of California Press, Berkeley, 246 pp.

Heiken, G. and Wohletz, K. 1987. Tephra deposits associated with silicic domes and lava flows. In: Fink, J. (ed.) *The emplacement of silicic domes and lava flows*. Geological Society of America Special Paper, 212, pp. 55–76.

Helgason, J. 1999. Formation of Olympus Mons and the aureole-escarpment problem on Mars. *Geology*, 27, 231–234.

Helgason, J. 2007. *Skaftafell. Bedrock geology. 1:25 000-scale geological map*. Ekra Geological Consulting.

Helgason, J. and Duncan, R.A. 2001. Glacial-interglacial history of the Skaftafell region, southeast Iceland, 0–5 Ma. *Geology*, 29, 179–182.

Herron, M.H. and Langway, C.C. 1980. Firn densification: an empirical model. *Journal of Glaciology*, 25, 373–385.

Hetenyi, G., Taisne, B., Garel, F., Médard, E., Bosshard, S. and Mattsson, H.B. 2012. Scales of columnar jointing in igneous rocks: field measurements and controlling factors. *Bulletin of Volcanology*, 74, 457–482.

Hickson, C.J. 1987. *Quaternary volcanism in the Wells Gray–Clearwater area, east central British Columbia*. PhD thesis, University of British Columbia (Canada), 357 pp. [unpublished]

Hickson, C.J. 2000. Physical controls and resulting morphological forms of Quaternary ice-contact volcanoes in western Canada. *Geomorphology*, 32, 239–261.

Hickson, C.J. and Vigouroux, N. 2014. Volcanism and glacial interaction in the Wells Gray-Clearwater volcanic field, east-central British Columbia, *Geological Society of America Field Guides*, 38, 169–191.

Hickson, C.J., Moore, J.G., Calk, L. and Metcalfe, P. 1995. Intraglacial volcanism in the Wells Gray–Clearwater volcanic field, east-central British Columbia, Canada. *Canadian Journal of Earth Science*, 32, 838–851.

Hildreth, W. 2007. *Quaternary magmatism in the Cascades: geologic perspectives*. United States Geological Survey Professional Paper, 1744, 125 pp.

Hindmarsh, R.C.A., van der Wateren, F.M. and Verbers, A.L.L.M. 1998. Sublimation of ice through sediment in Beacon Valley, Antarctica. *Geografiska Annaler*, 80A, 209–219.

Hoare, J.M. and Coonrad, W.L. 1978. A tuya in Togiak Valley, southwest Alaska. *United States Geological Survey Journal of Research*, 6, 193–201.

Hodges, C.A. and Moore, H.J. 1979. The subglacial birth of Olympus Mons and its aureoles. *Journal of Geophysical Research*, 84, 8061–8074.

Hoffman, P.F., Kaufman, A.J., Halverson, G.P. and Schrag, D.P. 1998. A Neoproterozoic snowball Earth. *Science*, 281, 1342–1346.

Hofmeister, A.M., Whittington, A.G., Goldsand, J. and Criss, R.G. 2014. Effects of chemical composition and temperature on transport properties of silica-rich glasses and melts. *American Mineralogist*, 99, 564–577.

Hole, M.J. 1988. Post-subduction alkaline volcanism along the Antarctic Peninsula. *Journal of the Geological Society, London*, 145, 985–988.

Hole, M.J. 1990. Geochemical evolution of Pliocene-Recent post-subduction alkali basalts from Seal Nunataks, Antarctic Peninsula. *Journal of Volcanology and Geothermal Research*, 40, 149–167.

Hole, M.J., Storey, B.C. and LeMasurier, W.E. 1994. Tectonic setting and geochemistry of Miocene alkalic basalts from the Jones Mountains, West Antarctica. *Antarctic Science*, 6, 85–92.

Hole, M.J., Saunders, A.D., Rogers, G. and Sykes, M.A. 1995. The relationship between alkaline magmatism, lithospheric extension and slab window formation along continental destructive plate margins. In: Smellie, J.L. (ed.) *Volcanism associated with extension at consuming plate margins*. Geological Society, London, Special Publications, 81, pp. 265–285.

de Hon, R.A. and Pani, E.A. 1993. Duration and rates of discharge: Maja Valles, Mars. *Journal of Geophysical Research*, 98, 9129–9138.

Honnorez, J. and Kirst, P. 1975. Submarine basaltic volcanism: morphometric parameters for discriminating hyaloclastites from hyalotuffs. *Bulletin of Volcanology*, 39, 1–25.

Hooke, R.LeB. 1989. Englacial and subglacial hydrology: a qualitative review. *Arctic and Alpine Research*, 21, 221–233.

Hooten, J.A. and Ort, M.H. 2002. Peperite as a record of early-stage phreatomagmatic fragmentation processes: an example from the Hopi Buttes volcanic field, Navajo Nation, Arizona, USA. *Journal of Volcanology and Geothermal Research*, 114, 95–106.

Höskuldsson, Á. 1992. *A Late Pleistocene subglacial caldera formation at Cerro Las Cumbres, eastern Mexico*, PhD thesis, Université Blaise Pascal, Clermont-Ferrand (France), 178–196 pp. [unpublished].

Höskuldsson, A. and Sparks, R.S.J. 1997. Thermodynamics and fluid dynamics of effusive subglacial eruptions. *Bulletin of Volcanology*, 59, 219–230.

Höskuldsson, A., Sparks, R.S.J and Carroll, M.R. 2006. Constraints on the dynamics of subglacial basalt eruptions from geological and geochemical observations at Kverkfjöll, NE Iceland. *Bulletin of Volcanology*, 68, 689–701.

Höskuldsson, Á., Óskarsson, N., Pedersen, R., Gronvold, K., Vogfjord, K. and Olafsdóttir, R. 2007. The millennium eruption of Hekla in February 2000. *Bulletin of Volcanology*, 70, 169–182.

Houghton, F. and Wilson, C.J.N. 1989. A vesicularity index for pyroclastic deposits. *Bulletin of Volcanology*, 51, 451–462.

Howard, D., Luzzadder-Beach, S. and Beach, T. 2012. Field evidence and hydraulic modeling of a large Holocene jökulhlaup at Jökulsá á Fjöllum channel, Iceland. *Geomorphology*, 147–148, 73–85.

Hubberten, H.-W., Morche, W., Westall, F., Fütterer, D.K. and Keller, J. 1991. Geochemical investigations of volcanic ash layers from southern Atlantic Legs 113 and 114. *Proceedings of the Ocean Drilling Project, Scientific Results*, 114, 733–749.

Hughes, T., Parkinson, C., and Brecher, H. (1974). Ice dynamics study of a glacial surge induced by the August 1970 eruption on Deception Island, Antarctica. In: González-Ferrán, O. (ed.) *Proceedings of the Symposium on Andean and Antarctic Volcanology Problems, Santiago, Chile*, IAVCEI, Rome, pp. 112–133.

Hungerford, J.D.G., Edwards, B.R., Skilling, I.P. and Cameron, B. 2014. Evolution of a subglacial basaltic lava flow field: Tennena Cone volcanic center, Mount Edziza Volcanic Complex, British Columbia, Canada. *Journal of Volcanology and Geothermal Research*, 272, 39–58.

Hull, D. 1999. *Fractography: observing, measuring and interpreting fracture surface topography.* Cambridge University Press, New York, 366 pp.

Hunns, S.R. and McPhie, J. 1999. Pumiceous peperite in a submarine volcanic succession at Mount Chalmers, Queensland, Australia. *Journal of Volcanology and Geothermal Research*, 88, 239–254.

Huppert, H.E. and Sparks, R.S.J. 1988. Melting the roof of a chamber containing a hot, turbulently convecting fluid. *Journal of Fluid Mechanics*, 188, 107–131.

Huybers, P. and Langmuir, C. 2009. Feedback between deglaciation, volcanism and atmospheric CO_2. *Earth and Planetary Science Letters*, 286, 479–491.

Hynek, B.M. and Phillips, R.J. 2003. New data reveal mature, integrated drainage systems on Mars indicative of past precipitation. *Geology*, 31, 575–760.

Illangasekare, T.H., Walter, R.J., Meier, M.F. and Pfeffer, W.T. 1990. Modeling of meltwater infiltration in subfreezing snow. *Water Resources Research*, 26, 1001–1012.

Incorpera, F.P., DeWitt, D.P., Bergman, T.L. and Lavine, A.S. 2010. *Fundamentals of heat and mass transfer*, 9th Edition, John Wiley and Sons, New York, 729 pp.

Isherwood, R.J., Jozwiak, L.M., Jansen, J.C. and Andrews-Hanna, J.C. 2013. The volcanic history of Olympus Mons from paleo-topography and flexural modelling. *Earth and Planetary Science Letters*, 363, 88–96.

Ito, E., White, W.M. and Göpel, C. 1987. The O, Sr, Nd and Pb isotope geochemistry of MORB. *Chemical Geology*, 62, 157–176.

Ivanov, M.A. and Head, J.W. 2003. Syrtis Major and Isidis Basin contact: Morphological and topographic characteristics of Syrtis Major lava flows and materials of the Vastitas Borealis Formation. *Journal of Geophysical Research*, 108, E6, 5063, doi:10.1029/2002JE001994.

Jaeger, J.C. 1968. Cooling and solidification of igneous rocks. In: Hess, H.H. and Poldervaart, A. (eds) *Basalts: the Poldervaart treatise on rocks of basaltic composition.* Interscience Publishers Inc., New York, pp. 503–536.

Jackson, L.E. 1989. Pleistocene subglacial volcanism near Fort Selkirk, Yukon Territory. In: *Geological Survey of Canada Paper 89–1E*, pp. 251–256.

Jackson, L.E. Jr., Barendregt, R.W., Baker, J. and Irving, E. 1996. Early Pleistocene volcanism and glaciation in central Yukon: a new chronology from field studies and paleomagnetism. *Canadian Journal of Earth Science*, 33, 904–916.

Jackson, L.E. Jr., Nelson, F.E., Huscroft, C.A., Villeneuve, M., Barendregt, R.W., Storer, J.E. and Ward, B.C. 2012. Pliocene and Pleistocene volcanic interaction with Cordilleran ice sheets, damming of the Yukon River and vertebrate paleontology, Fort Selkirk volcanic group, west-central Yukon, Canada. *Quaternary International*, 260, 3–20.

Jakobsson, S.P. 1978. Environmental factors controlling the palagonitization of the Surtsey tephra, Iceland. *Bulletin of the Geological Society of Denmark*, Special Issue 27, 91–105.

Jakobsson, S.P. and Gudmundsson, M.T. 2008. Subglacial and intraglacial volcanic formations in Iceland. *Jökull*, 58, 179–196.

Jakobsson, S.P. and Moore, J.G. 1986. Hydrothermal minerals and alteration rates at Surtsey volcano, Iceland. *Geological Society of America Bulletin*, 97, 648–659.

Jakobsson, S.P., Jonsson, J. and Shido, F. 1978. Petrology of the western Reykjanes Peninsula, Iceland. *Bulletin of Volcanology*, 56, 516–528.

Jakobsson, S.P., Jonasson, K. and Sigurdsson, I.A. 2008. The three igneous rock series of Iceland. *Jökull*, 58, 117–138.

Jakobsson, S.P. and Johnson, G.L. 2012. Intraglacial volcanism in the Western Volcanic Zone, Iceland. *Bulletin of Volcanology*, 74, 1141–1160.

Jakosky, B.M. and Carr, M.H. 1985. Possible precipitation of ice at low latitudes of Mars during periods of high obliquity. *Nature*, 315, 559–561.

Janda, R.J., Scott, KM., Nolan, K.M. and Martinson, H.A. 1981. Lahar movement, effects, and deposits. In: Lipman, P.W. and Mullineaux, D.R. (eds) *The 1980 eruptions of Mount St. Helens, Washington*. United States Geological Survey Professional Paper, 1250, pp.461–478.

Jansen, E. and Skoholm, J. 1991. Reconstruction of glaciation over the past 6 Myr from iceborne deposits in the Norwegian Sea. *Nature*, 349, 600–603.

Jarosch, A., Gudmundsson, M.T., Högnadóttir, T. and Axelsson, G. 2008. Progressive cooling of the hyaloclastite ridge at Gjálp, Iceland, 1996–2005. *Journal of Volcanology and Geothermal Research*, 170, 218–229.

Jaupart, C. 2000. Magma ascent at shallow levels. In: Sigurdsson, H., Houghton, B., Rymer, H., Stix, J. and McNutt, S. (eds) *Encyclopedia of volcanoes*, 1st Edition. Academic Press, San Diego, pp. 237–248.

Jaupart, C. and Mareschal, J.-C. 2010. *Heat generation and transport in the Earth*. Cambridge University Press, Cambridge, 477 pp.

Jellinek, A.M., Manga, M. and Saar, M.O. 2004. Did melting of glaciers cause volcanic eruptions in eastern California? Probing the mechanics of dike formation. *Journal of Geophysical Research*, 109, doi:10.1029/2004JB002978.

Jercinovic, M.J. 1988. *Alteration of basaltic glasses from British Columbia, Iceland, and the deep sea*. PhD thesis, University of Albuquerque, New Mexico (USA), 475 pp. [unpublished].

Jercinovic, M.J., Keil, K., Smith, M.R. and Schmitt, R.A. 1990. Alteration of basaltic glasses from north-central British Columbia, Canada. *Geochimica et Cosmochimica Acta*, 54, 2679–2696.

Johnson, J.S. and Smellie, J.L. 2007. Zeolite compositions as proxies for eruptive paleoenvironment. *Geochemistry, Geophysics, Geosystems*, 8, Q03009, doi:10.1029/2006GC001450.

Johnson, J.S., Smellie, J.L., Nelson, A.E. and Stuart, F.M. 2009. History of the Antarctic Peninsula Ice Sheet since the early Pliocene: evidence from cosmogenic dating of Pliocene lavas on James Ross Island, Antarctica. *Global and Planetary Change*, 69, 205–213.

Jones, J.G. 1966. Intraglacial volcanoes of south-west Iceland and their significance in the interpretation of the form of marine basaltic volcanoes. *Nature*, 212, 586–588.

Jones, J.G. 1969a. Pillow lavas as depth indicators. *American Journal of Science*, 267, 181–195.

Jones, J.G. 1969b. Intraglacial volcanoes of the Laugarvatn region, south-west Iceland, I. *Quarterly Journal of the Geological Society, London*, 124, 197–211.

Jones, J.G. 1970. Intraglacial volcanoes of the Laugarvatn region, southwest Iceland, II. *Journal of Geology*, 78, 127–140.

Jones, J.G. and Nelson, P.H.H. 1970. The flow of basalt lava from air into water: its structural expression and stratigraphic significance. *Geological Magazine*, 107, 13–19.

Jonkers, H.A. 1998. The Cockburn Island Formation; Late Pliocene interglacial sedimentation in the James Ross Island Basin, northern Antarctic Peninsula. *Newsletters on Stratigraphy*, 36, 63–76.

Jonkers, H.A., Lirio, J.M., del Valle, R.A. and Kelley, S.P. 2002. Age and environment of Miocene–Pliocene glaciomaine deposits, James Ross Island, Antarctica. *Geological Magazine*, 139, 577–594.

Jónsson, J., 1982. Notes on the Katla volcanological debris flows. *Jökull*, 32, 61–68.

Jordan, T.A., Ferraccioli, F., Jones, P.C., Smellie, J.L., Ghidella, M. and Corr, H. 2009. Airborne gravity reveals interior of Antarctic volcano. *Physics of the Earth and Planetary Interiors*, 175, 127–136.

Jørgensen, K.A. 1980. The Thorsmörk ignimbrite: an unusual comenditic pyroclastic flow in southern Iceland. *Journal of Volcanology and Geothermal Research*, 8, 7–22.

Jude-Eton, T.C., Thordarson, Th., Gudmundsson, M.T. and Oddsson, B. 2012. Dynamics, stratigraphy and proximal dispersal of supraglacial tephra during the ice-confined 2004 eruption at Grímsvötn volcano, Iceland. *Bulletin of Volcanology*, 74, 1057–1082.

Julio Miranda, P., Gonzalez-Huesca, A.E., Delgado Granados, H. and Kääb, A. 2005. Glacier melting and lahar formation during January 22, 2001 eruption, Popocatépetl volcano (Mexico). *Annals of Geomorphology*, 140, 93–102.

Jull, M. and McKenzie, D. 1996. The effect of deglaciation on mantle melting beneath Iceland. *Journal of Geophysical Research*, 101, 21815–21828.

Kadish, S.J. and Head, J.W. 2011. Impacts into non-polar ice-rich paleodeposits on Mars: excess ejecta craters, perched craters and pedestal craters as clues to Amazonian climate history. *Icarus*, 215, 34–46.

Kadish, S.J., Head, J.W., Parsons, R.L. and Marchant, D.R. 2008. The Ascraeus Mons fan-shaped deposit: volcano–ice interactions and the climatic implications of cold-based tropical mountain glaciations. *Icarus*, 197, 84–109.

Kargel, J.S., Baker, V.R., Beget, J.E., Lockwood, J.F., Pewe, T.L., Shaw, J.S. and Strom, R.G. 1995. Evidence for ancient continental glaciations in the Martian northern plains. *Journal of Geophysical Research*, 100, 5351–5368.

Kargel, J., Leonard, G., Wheate, R. and Edwards, B. 2014. ASTER and SEM change assessment of changing glaciers near Hoodoo Mountain, British Columbia, Canada. In: Kargel, J.S., Leonard, G.J, Bishop, M.P., Kääb, A. and Raup, B.H. (eds) *Global land ice measurements from space*. Springer-Verlag, Berlin, pp. 353–373.

Katsui, Y., Kawachi, S., Kondo, Y., Ikeda, Y., Nakagawa, M., Gotoh, Y., Yamigishi, H., Yamazaki, T. and Sumita, M. 1990. The 1988–89 explosive eruption of Tokachi-dake, Central Hokkaido, its sequence and mode. *Bulletin of the Volcanological Society of Japan*, 35, 111–129.

Kaufman, D.S. and Manley, W.F. 2004. Pleistocene maximum and Late Wisconsinan glacier extents across Alaska, U.S.A. In: Ehlers, J. and Gibbard, P.L. (eds) *Quaternary glaciations: extent and chronology. Part II, North America*. Developments in Quaternary Science, 2b, Elsevier, Amsterdam, pp. 9–27.

Kelemen, P.B., Hirth, G., Shimizu, N., Spiegelman, M. and Dick, H.J.B. 1997. A review of melt migration processes in the asthenospheric mantle beneath oceanic spreading centers. *Philosophical Transactions of the Royal Society of London*, A355, 283–318.

Keller, R.A., Fisk, M.R., Smellie, J.L., Strelin, J.A. and Lawver, L.A. 2002. Geochemistry of back-arc basin volcanism in Bransfield Strait, Antarctica: subducted contributions and along-axis variations. *Journal of Geophysical Research*, 107, B8, doi:10.1029/2001JB000444.

Kelman, M.C. 2005. *Glaciovolcanism at the Mount Cayley volcanic field, Garibaldi volcanic belt, southwestern British Columbia*. PhD thesis, The University of British Columbia (Canada), 258 pp. [unpublished].

Kelman, M.C., Russell, J.K. and Hickson, C.J. 2002. Effusive intermediate glaciovolcanism in the Garibaldi Volcanic Belt, southwestern British Columbia, Canada. In: Smellie, J.L. and Chapman, M.G. (eds) *Volcano–ice interaction on Earth and Mars*. Geological Society, London, Special Publications, 202, pp. 195–211.

Kerr, F.A. 1948. *Lower Stikine and Western Iskut River Areas, British Columbia*. Canada Department of Mines and Resources Geological Survey Memoir, 246, 94 pp.

Kilgour, G., Manville, V., Della Pasqua, F., Graettinger, A., Hodgson, K.A. and Jolly, G.E. 2010. The 25 September eruption of Mount Ruapehu, New Zealand: directed ballistics, surtseyan jets, and ice-slurry lahars. *Journal of Volcanology and Geothermal Research*, 191, 1–14.

Kjartansson, G. 1943. Geology of Árnessýsla. (In Icelandic). In: *Árnesingasaga I*, Árnesingafélagið, Reykjavík, pp. 1–250.

Kjartansson, G. 1951. Water flood and mudflows. In: Einarsson, T., Kjartansson, G. and Thorarinsson, S. (eds) *The eruption of Hekla 1947–1948*. Visindafélag Islendinga, Societas Scientarium Islandica, II, 4, 51 pp.

Kjartansson, G. 1955. Bölstraberg pillow lava in Iceland. *Natturfræðingurinn*, 25, 227–240.

Kjartansson, G. 1959. The Moberg Formation; II. In: Thorarinsson, S. (ed.) On the geology and geomorphology of Iceland. *Geografiska Annaler*, 41, 139–143.

Klemen, J. and Glasser, N.F. 2007. The subglacial thermal organisation (STO) of ice sheets. *Quaternary Science Reviews*, 26, 585–597.

Klingelhöfer, F., Hort, M., Kümpel, H.-J. and Schminke, H.-U. 1999. Constraints on the formation of submarine lava flows from numerical model calculations. *Journal of Volcanology and Geothermal Research*, 92, 215–229.

Knauth, L.P. and Burt, D.M. 2002. Eutectic brines on Mars: origin and possible relation to young seepage features. *Icarus*, 158, 267–271.

Kokelaar, B.P. 1983. The mechanism of Surtseyan volcanism. *Journal of the Geological Society, London*, 140, 939–944.

Kokelaar, P. 1986. Magma–water interactions in subaqueous and emergent basaltic volcanism. *Bulletin of Volcanology*, 48, 275–289.

Kokelaar, B.P. and Busby, C. 1992. Subaqueous explosive eruption and welding of pyroclastic deposits. *Science*, 257, 196–200.

Komatsu, G., Geissler, P.E., Strom, R.G. and Singer, R.B. 1993. Stratigraphy and erosional landforms of layered deposits in Valles Marineris, Mars. *Journal of Geophysical Research*, 98, 11105–11121.

Komatsu, G., Ori, G.G., Ciarcelluti, P. and Litasov, Y.D. 2004. Interior layered deposits of Valles Marineris, Mars: analogous subice volcanism related to Baikal rifting, southern Siberia. *Planetary and Space Science*, 52, 167–187.

Komatsu, G., Arzhannikov, S.G., Arzhannikova, A.V. and Ershov, K. 2007a. Geomorphology of subglacial volcanoes in the Azas Plateau, the Tuva Republic, Russia. *Geomorphology*, 88, 312–328.

Komatsu, G., Arzhannikov, S.G., Arzhannikova, A.V. and Ori, G.G 2007b. Origin of glacial–fluvial landforms in the Azas Plateau volcanic field, the Tuva Republic, Russia: role of ice–magma interaction. *Geomorphology*, 88, 352–366.

Kratzmann, D.J., Carey, S., Scasso, R. and Naranjo, J-A. 2009. Compositional variations and magma mixing in the 1991 eruptions of Hudson volcano, Chile. *Bulletin of Volcanology*, 71, 419–439.

Kratzmann, D.J., Carey, S.N., Fero, J., Scasso, R.A. and Naranjo, J-A. 2010. Simulations of tephra dispersal from the 1991 explosive eruptions of Hudson volcano, Chile. *Journal of Volcanology and Geothermal Research*, 190, 337–352.

Kreslavsky, M.A. and Head, J.W. 2002. Fate of outflow channel effluents in the northern lowlands of Mars: the Vastitas Borealis Formation as a sublimation residue from frozen ponded bodies of water. *Journal of Geophysical Research*, 107, E12, 5121, doi:10.1029/2001JE001831.

Kuehn, C., Guest, B., Russell, J.K. and Benowitz, J.A. 2015. The Satah Mountain and Baldface Mountain volcanic fields: Pleistocene hotspot volcanism in the Anahim volcanic belt, west-central British Columbia. *Bulletin of Volcanology*, 77:19, doi:10.1007/s00445-015-0907-1.

Kueppers, U. Cimarelli, C., Hess, K.-W., Tadduecci, J., Wadsworth, F.B. and Dingwell, D. 2014. The thermal stability of Eyjafjallajokull ash versus turbine ingestion test sands. *Journal of Applied Volcanology*, doi:10.1186/2191-5040-3-4.

Lachlan-Cope, T., Smellie, J.L. and Ladkin, R. 2001. Discovery of a recurrent lava lake on Saunders Island (South Sandwich Islands) using AVHRR imagery. *Journal of Volcanology and Geothermal Research*, 112, 105–116.

LaMoreaux, K.A. 2008. *Recognizing ice contact trachyte-phonolite lavas at Mount Edziza volcanic complex, British Columbia, Canada*. MSc thesis, University of Pittsburgh, Pennsylvania (USA), 162 pp. [unpublished].

Lanagan, P.D., McEwen, A.S., Keszthelyi, L.P. and Thordarson, T. 2001. Rootless cones on Mars indicating the presence of shallow equatorial ground ice in recent times. *Geophysical Research Letters*, 28, 2365–2367.

Lange, R.L. and Carmichael, I.S.E. 1990. Thermodynamic properties of silicate liquids with emphasis on density, thermal expansion, and compressibility. In: Nicholls, J. and Russell, J.K. (eds) *Modern Methods of Igneous Petrology*. Mineralogical Society of America Reviews in Mineralogy, 24, pp. 25–64.

Lara, L. 2004. *Geología del Volcán Lanín, Región de la Araucanía.* Servicio Nacional de Geología y Minería, Santiago. Carta Geológica de Chile, Serie Geología Básica, No. 88, 14 pp., 1 map, scale 1:50.000.

Lara, L.E., Naranjo, J.A. and Moreno, H. 2004. Lanín volcano (39.5°S), Southern Andes: geology and morphostructural evolution. *Revista Geológica de Chile*, 31, 2, 241–257.

Larsen, W. 1940. Petrology of interglacial volcanics from the Andes of northern Patagonia. *Bulletin of the Geological Institute of Upsala*, 28, 405 pp.

Larsen, G. 2000. Holocene eruptions within the Katla volcanic system, south Iceland: characteristics and environmental impact. *Jökull*, 49, 1–28.

Larsen, G. 2002. A brief overview of eruptions from ice-covered and ice-capped volcanic systems in Iceland during the past 11 centuries: frequency, periodicity and implications. In: Smellie, J.L. and Chapman, M.G. (eds) *Volcano–ice interaction on Earth and Mars.* Geological Society, London, Special Publications, 202, pp. 81–90.

Larsen, J.F., Neal, C., Schaefer, J., Beget, J. and Nye, C. 2007. Late Pleistocene Holocene caldera-forming eruptions of Okmok caldera, Aleutian Islands, Alaska. In: Eichelberger, J., Gordeev, E., Izbekov, P., Kasahara, M. and Lees, J. (eds) *Volcanism and subduction: the Kamchatka region*, American Geophysical Union Monograph, 172, pp. 343–364.

Laskar, J., Correia, A.C.M., Gastineau, M., Joutel, F., Levrard, B. and Robutel, P. 2004. Long term evolution and chaotic diffusion of the insolation quantities of Mars. *Icarus*, 170, 343–364.

Leat, P.T., Smellie, J.L., Millar, I.L. and Larter, R.D. 2003. Magmatism in the South Sandwich arc. In: Larter, R.D. and Leat, P.T. (eds) *Intra-oceanic subduction systems: tectonic and agamic processes*. Geological Society, London, Special Publications, 219, pp. 285–313.

Leat, P.T., Fretwell, P.T., Tate, A.J., Larter, R.D., Martin, T.J., Smellie, J.L., Jokat, W. and Bohrmann, G. 2014. *Bathymetry and geological setting of the South Sandwich Islands volcanic arc (various scales)*. BAS GEOMAP 2 series, Sheet 6, British Antarctic Survey, Cambridge, UK.

LeMaitre, R.W., Streckeisen, A., Zanettin, B., LeBas, M.J., Bonin, B., Bateman, P., Bellieni, G., Dudek, A., Efremova, S., Keller, J., Lameyre, J., Sabine, P.A., Schmid, R., Sørensen, H. and Wooley, A.R. 2002. *Igeous rocks: a classification and glossary of terms. Recommendations of the International Union of Geological Sciences Subcommission on the Systematics of Igneous Rocks*. Cambridge University Press, Cambridge, 236 pp.

LeMasurier, W.E. 1972a. Volcanic record of Cenozoic glacial history of Marie Byrd Land. In: Adie, R.J. (ed.) *Antarctic geology and geophysics*. Universitetsforlaget, Oslo, 251–260.

LeMasurier, W.E. 1972b. Volcanic record of Antarctic glacial history: implications with regard to Cenozoic sea levels. In: Price, R.J. and Sugden, D.E. (eds) *Polar geomorphology*. Institute of British Geographers, Special Publications, 4, 59–74.

LeMasurier, W.E. 1990. Marie Byrd Land: summary. In: LeMasurier, W.E. and Thomson, J.W. (eds) *Volcanoes of the Antarctic Plate and southern oceans*. American Geophysical Union, Antarctic Research Series, 48, pp. 147–163.

LeMasurier, W.E. 2002. Architecture and evolution of hydrovolcanic deltas in Marie Byrd Land, Antarctica. In: Smellie, J.L. and Chapman, M.G. (eds) *Volcano–ice interaction on Earth and Mars*. Geological Society, London, Special Publications, 202, pp. 115–148.

LeMasurier, W.E. and Rex, D.C. 1982. Volcanic record of Cenozoic glacial history in Marie Byrd Land and western Ellsworth Land: revised chronology and evaluation of tectonic factors. In: Craddock, C. (ed.) *Antarctic geoscience*. University of Wisconsin Press, Madison, Wisconsin, pp. 725–734.

LeMasurier, W.E. and Thomson, J.W. (eds) 1990. *Volcanoes of the Antarctic Plate and southern oceans*. American Geophysical Union, Antarctic Research Series, 48, 487 pp.

LeMasurier, W.E., Harwood, D.M. and Rex, D.C. 1994. Geology of Mount Murphy Volcano: an 8-m.y. history of interaction between a rift volcano and the west Antarctic ice sheet. *Geological Society of America Bulletin*, 106, 265–280.

LeMasurier, W.E., Futa, K., Hole, M.J. and Kawachi, Y. 2003. Polybaric evolution of phonolite, trachyte, and rhyolite volcanoes in eastern Marie Byrd Land, Antarctica: controls on peralkalinity and silica saturation. *International Geological Review*, 45, 1055–1099.

Lescinsky, D.T. 1999. *Interactions between glacial ice and lava flows*. PhD thesis, Arizona State University (USA), 209 pp. [unpublished].

Lescinsky, D.T. and Sisson, T.W. 1998. Ridge-forming ice-bounded lava flows at Mount Rainier, Washington. *Geology*, 26, 351–354.

Lescinsky, D.T. and Fink, J.H. 2000. Lava and ice interaction at stratovolcanoes: use of characteristic features to determine past glacial extents and future volcanic hazards. *Journal of Geophysical Research*, 105, 23711–23726.

Lewis, A.R., Marchant, D.R., Ashworth, A.C., Hemming, S.R. and Machlus, M.L. 2007. Major middle Miocene global climate change: evidence from East Antarctica and the Transantarctic Mountains. *Geological Society of America Bulletin*, 119, 1449–1461.

Licciardi, J.M., Kurz, M.D. and Curtice, J.M. 2007. Glacial and volcanic history of Icelandic table mountains from cosmogenic ^{3}He exposure ages. *Quaternary Science Reviews*, 26, 1529–1546.

Lipman, P.W. and Banks, N.B. 1987. Aa flow dynamics. Mauna Loa 1984. In: Decker, R.W., Wright, T.L. and Stauffer, P.H. (eds) *Volcanism in Hawaii*. United States Geological Survey Professional Paper, 1350, pp. 1527–1567.

Lisiecki, L.E. and Raymo, M.E. 2005. A Pliocene-Pleistocene stack of 57 globally distributed benthic δ^{18}O records. *Paleoceanography*, 20, doi:10.1029/2004PA001071.

Liu, E.J., Cashman, K.V., Rust, A.C. and Gislason, S.R. 2015. The role of bubbles in generating fine ash during hydromagmatic eruptions. *Geology*, 43, 239–242.

Llewellin, E.W. and Manga, M. 2005. Bubble suspension rheology and implications for conduit flow. In: Sahagian, D. (ed.) *Volcanic eruption mechanisms; insights from inter-comparison of models of conduit processes. Journal of Volcanology and Geothermal Research*, 143, 205–217.

Lloyd, A. 2007. *Plagioclase growth rates in pillow basalts: a study in thermal modelling and crystal size distribution*. BSc thesis, Dickinson College, Pennsylvania (USA), 54 pp. [unpublished].

Lloyd, A.S., Plank, T., Ruprecht, P., Hauri, E.H. and Rose, W. 2013. Volatile loss from melt inclusions in pyroclasts of differing sizes. *Contributions to Mineralogy and Petrology*, 165, 129–153.

Lloyd-Davies, M.T., Atkins, C.B., van der Meer, J.J.M., Barrett, P.J. and Hicock, S.R. 2009. Evidence for cold-based glacial activity in the Allan Hills, Antarctica. *Quaternary Science Reviews*, 28, 3124–3137.

Lodge, R.W.D and Lescinsky, D.T. 2009a. Fracture patterns at lava–ice contacts on Kokostick Butte, OR, and Mazama Ridge, Mount Rainier, WA: implications for flow emplacement and cooling histories. *Journal of Volcanology and Geothermal Research*, 185, 298–310.

Lodge, R.W.D. and Lescinsky, D.T. 2009b. Anisotropic stress accumulation in cooling lava flows and resulting fracture patterns: insights from starch–water desiccation experiment. *Journal of Volcanology and Geothermal Research*, 185, 323–336.

Long, P.E. and Wood, B.J. 1986. Structures, textures and cooling histories of Columbia River basalt flows. *Geological Society of America Bulletin*, 97, 1144–1155.

Loock, S., van Wyk de Vries, B. and Hénot, J.-M. 2010. Clinker formation in basaltic and trachybasaltic lava flows. *Bulletin of Volcanology*, 72, 859–870.

López-Martínez, J. and Serrano, E. 2002. Geomorphology. In: Smellie, J.L., López-Martínez, J., et al. 2002. *Geology and geomorphology of Deception Island. BAS GEOMAP Series, Sheets 6-A and 6-B, 1:25 000, supplementary text*. British Antarctic Survey, Cambridge, UK, pp. 31–39.

Lore, J., Gao, H. and Aydin, A. 2000. Viscoelastic thermal stress in cooling basalt flows. *Journal of Geophysical Research*, 105, 23695–23709.

Lough, A.C., Wiens, D.A., Barcheck, C.G., Anandakrishnan, S., Aster, R.C., Blankenship, D.D., Huerta, A.D., Nyblade, A., Young, D.A. and Wilson, T.J. 2013. Seismic detection of an active subglacial magmatic complex in Marie Byrd Land, Antarctica. *Nature Geoscience*, doi:10.1038/NGO1992.

Loughlin, S.C., 2002. Facies analysis of proximal subglacial and proglacial volcaniclastic successions at the Eyjafjallajökull central volcano, southern Iceland. In: Smellie, J.L., Chapman, M.G. (eds) *Volcano–ice interaction on Earth and Mars.* Geological Society, London, Special Publications, 202, pp. 149–178.

Lowe, D.R. 1982. Sediment gravity flows: II. Depositional models with special reference to the deposits of high-density turbidity currents. *Journal of Sedimentary Petrology*, 52, 279–297.

Lowell, P. 1906. *Mars and its canals.* The Macmillan Company, New York, 385 pp.

Lube, G., Cronin, S. and Procter, J.N. 2009. Explaining the extreme mobility of volcano ice-slurry flows, Ruapehu volcano, New Zealand. *Geology*, 37, 15–18.

Lucchitta, B.K. 1982. Ice sculpture in the Martian outflow channels. *Journal of Geophysical Research*, 87, 9951–9973.

Lucchitta, B.K. 2001. Antarctic ice streams and outflow channels on Mars. *Geophysical Research Letters*, 28, 403–406.

Lucchitta, B.K., McEwen, A.S., Clow, G.D., Geissler, P.E., Singer, R.B., Schulz, R.A. and Squyres, S.W. 1992. Valles Marineris. In: Keifer, H.H., Jakosky, B.M., Snyder, C.W. and Matthews, M.S. (eds) *Mars.* University of Arizona Press, Tucson, pp. 453–492.

Lucchitta, B.K., Isbell, N.K. and Howington-Kraus, A. 1994. Topography of Valles Marineris: implications for erosional and structural history. *Journal of Geophysical Research*, 99, 3783–3798.

MacDonald, G.A. 1963. Physical properties of erupting Hawaiian magmas. *Geological Society of America Bulletin*, 74, 1071–1078.

Mackenzie, D.E. and Johnson, R.W. 1984. *Pleistocene volcanoes of the western Papua New Guinea highlands: morphology, geology, petrography and modal chemical analyses.* Australian Bureau of Mineral Resources, Geology and Geophysics Report, 246, 271 pp.

Maclennan, J., Jull, M., McKenzie, D., Slater, L. and Gronvold, K. 2002. The link between volcanism and deglaciation in Iceland. *Geochemistry, Geophysics, Geosystems*, 3, 1062, doi:10.1029/2001GC000282.

Magnússon, E., Gudmundsson, M.T., Roberts, M.J., Sigurdsson, G., Höskuldsson, F. and Oddsson, B. 2012. Ice–volcano interactions during the 2010 Eyjafjallajökull eruption, as revealed by airborne imaging radar. *Journal of Geophysical Research*, 117, B07405, doi:10.1029/2012JB009250.

Mahood, G.A., Ring, J.H., Manganelli, S. and McWilliams, M.O. 2010. New $^{40}Ar/^{39}Ar$ ages reveal contemporaneous mafic and silicic eruptions during the past 160,000 years at Mammoth Mountain and Long Valley Caldera, California. *Geological Society of America Bulletin*, 122, 396–407.

Maizels, J.K. 1993. Lithofacies variations within sandur deposits: the role of runoff regime, flow dynamics and sediment supply charateristics. *Sedimentary Geology*, 85, 299–325.

Major, J.J. and Newhall, C.G. 1989. Snow and ice perturbation during historical volcanic eruptions and the formation of lahars and floods: a global review. *Bulletin of Volcanology*, 52, 1–27.

Major, J.J., Pierson, T.C. and Scott, K.M. 2005. Debris flows at Mount St. Helens, Washington, USA. In: Jakob, M. and Hungr, O. (eds) *Debris-flow Hazards and Related Phenomena.* Springer, Berlin, pp. 685–731.

Manglik, A. and Singh, R.N. 1995. Postintrusive thermal evolution of continental crust: a moving boundary approach. *Journal of Geophysical Research*, 100, 18031–18043.

Manville, V. and Cronin, S. 2007. Breakout lahar from New Zealand's Crater Lake. *Eos, Transactions, American Geophysical Union*, 88, 441–2.

Manville, V. and Wilson, C.J.N. 2004. Vertical density currents: a review of their potential role in deposition of deep-sea ash layers. *Journal of the Geological Society, London*, 161, 947–958.

Marchant, D.R., Lewis, A., Phillips, W.M., Souchez, R., Denton, G.H., Sugden, D.E. and Landis, G.P. 2002. Formation of patterned ground and sublimation till over Miocene glacier ice in Beacon Valley, southern Victoria Land, Antarctica. *Geological Society of America Bulletin*, 114, 718–730.

Marenssi, S.A., Casadio, S. and Santillana, S.N. 2010. Record of Late Miocene glacial deposits on Isla Marambio (Seymour Island), Antarctic Peninsula. *Antarctic Science*, 22, 193–198.

Mark, B.G. and Osmaston, H.A. 2008. Quaternary glaciation in Africa: key chronologies and climatic implications. *Journal of Quaternary Science*, 23, 589–608.

Marsh, B.D. 1981. On the crystallinity, probability of occurrence, and rheology of lava and magma. *Contributions to Mineralogy and Petrology*, 78, 85–98.

Martin, A., Paquette, J.L., Bosse, V., Ruffet, G., Tiepolo, M. and Sigmarsson, O. 2011. Geodynamics of rift–plume interaction in Iceland as constrained by new ^{40}Ar/^{39}Ar and in situ U-Pb zircon ages. *Earth and Planetary Science Letters*, 311, 28–38.

Martínez-Alonso, A., Mellon, M.T., Banks, M.E., Keszthelyi, L.P., McEwen, A.S. and the HiRISE Team. 2011. Evidence of volcanic and glacial activity in Chryse and Acidalia Planitiae, Mars. *Icarus*, 212, 597–621.

Mastin, L.G., Christiansen, R.L., Thornber, C., Lowenstern, J. and Beeson, M. 2004. What makes hydromagmatic eruptions violent? Some insights from the Keanakāko'I Ash, Kīlauea Volcano, Hawai'i. *Journal of Volcanology and Geothermal Research*, 137, 15–31.

Mastin, L.G., Spieler, O. and Downey, W.S. 2009. An experimental study of hydromagmatic fragmentation through energetic, non-explosive magma–water mixing. *Journal of Volcanology and Geothermal Research*, 180, 161–170.

Matchan, E.L. and Phillips, D. 2014. High precision multi-collector ^{40}Ar/^{39}Ar dating of young basalts: Mount Rouse volcano (SE Australia) revisited. *Quaternary Geochronology*, 22, 57–64.

Mathews, W.H. 1947. 'Tuyas': flat-topped volcanoes in northern British Columbia. *American Journal of Science*, 245, 560–570.

Mathews, W.H., 1951. The Table, a flat-topped volcano in southern BC. *American Journal of Science*, 249, 830–841.

Mathews, W.H. 1952a. Ice-dammed lavas from Clinker Mountain, southwestern British Columbia. *American Journal of Science*, 250, 553–565.

Mathews, W. H. 1952b. Mount Garibaldi, a supraglacial Pleistocene volcano in southwestern British Columbia. *American Journal of Science*, 250, 81–103.

Mathews, W.H. 1958. Geology of the Mount Garibaldi map-area, southwestern British Columbia, Canada: Part II. Geomorphology and Quaternary volcanic rocks. *Geological Society of America Bulletin*, 69, 179–198.

Mathews, W.H. 1987. Garibaldi area, southwestern British Columbia; volcanoes versus glacier ice. In: Hill, M.L. (ed.) *Geological Society of America Centennial Field Guide: Volume 1, Cordilleran Section*, pp. 403–406.

Matthews, R.K. 1969. Tectonic implications of glacio-eustatic sea level fluctuations. *Earth and Planetary Science Letters*, 5, 459–462.

Mattox, T.N. and Mangan, M.T. 1997. Littoral hydrovolcanic explosions: a case study of lava–seawater interaction at Kilauea Volcano. *Journal of Volcanology and Geothermal Research*, 75, 1–17.

Mattsson, H.B., Caricchi, L., Almqvist, B.S.G., Caddick, M.J., Bosshard, S.A., Hetényi, G. and Hirt, A.M. 2011. Melt migration in basalt columns driven by crystallization-induced pressure gradients. *Nature Communications*, 2, 299, doi:10.1038/ncomms1298.

McGarvie, D.W. 1984. Torfajökull: a volcano dominated by magma mixing. *Geology*, 12, 685–688.

McGarvie, D. 2009. Rhyolitic volcano–ice interactions in Iceland. *Journal of Volcanology and Geothermal Research*, 185, 367–389.

McGarvie, D.W., Macdonald, R., Pinkerton, H. and Smith, R.L. 1990. Petrogenetic evolution in the Torfajökull volcanic complex, Iceland. 2. The role of magma mixing. *Journal of Petrology*, 31, 461–481.

McGarvie, D.W., Burgess, R., Tindle, A.G., Tuffen, H. and Stevenson, J.A. 2006. Pleistocene rhyolitic volcanism at Torfajökull, Iceland: eruption ages, glaciovolcanism, and geochemical evolution. *Jökull*, 56, 57–75.

McGarvie, D.W., Stevenson, J.A., Burgess, T., Tuffen, H. and Tindle, A.G. 2007. Volcano–ice interactions at Prestahnúkur, Iceland: rhyolite eruption during the last interglacial–glacial transition. In: Clarke, G and Smellie, J. (eds) *Papers from the International Symposium on Earth and Planetary Ice–Volcano Interactions held in Reykjavik, Iceland, on 19–23 June, 2006. Annals of Glaciology*, 45, 38–47.

McGarvie, D.W., Skilling, I.P., Graettinger, A.H., Guillou, H. and Höskuldsson, A. 2013. Rapid growth of a basaltic volcano beneath an ice sheet: Askja, Iceland. IAVCEI 2013 General Assembly, Kagoshima, Japan (Abstract 3A2_3K-O12, p. 1070).

McGovern, P.J. and Solomon, S.C. 1993. State of stress, faulting and eruption characteristics of large volcanoes on Mars. *Journal of Geophysical Research*, 98, 23533–23579.

McGuire, W. J., Howarth, R.J., Firth, C.R., Solow, A.R., Pullen, A.D., Saunders, S.J., Stewart, I.S. and Vita-Finzi, C. 1997. Correlation between rate of sea-level change and frequency of explosive volcanism in the Mediterranean. *Nature*, 389, 473–476.

McIntosh, W.C., Dunbar, N., Iverson, N. and Heizler, M. 2014. *New generation mass-spectrometers offer improved $^{40}Ar/^{39}Ar$ dating of tephra*. Tephra 2014 conference, Portland, Oregon, USA, 4–7 August 2014 [abstract].

McKenzie, D. and Bickle, M.J. 1988. The volume and composition of melt generated by extension of the lithosphere. *Journal of Petrology*, 29, 625–679.

McKenzie, D. and Nimmo, F. 1999. The generation of martian floods by the melting of ground ice above dykes. *Nature*, 397, 231–233.

McLoughlin, N., Furnes, H., Banerjee, N.R., Muehlenbachs, K. and Staudigel, H. 2009. Ichnotaxonomy of microbial trace fossils in volcanic glass. *Journal of the Geological Society, London*, 166, 159–169.

McPhie, J., Doyle, M. and Allen, R. 1993. *Volcanic textures: a guide to the interpretation of textures in volcanic rocks*. CODES Key Centre, University of Tasmania, Hobart, 196 pp.

Mee, K., Tuffen, H. and Gilbert, J.S. 2006. Snow-contact volcanic facies and their use in determining past eruptive environments at Nevados de Chillán volcano, Chile. *Bulletin of Volcanology*, 68, 363–376.

Mee, K., Gilbert, J.S., McGarvie, D.W., Naranjo, J.A. and Pringle, M.S. 2009. Palaeoenvironment reconstruction, volcanic evolution and geochronology of the Cerro

Blanco subcomplex, Nevados de Chillán volcanic complex, central Chile. *Bulletin of Volcanology*, 71, 933–952.

Mège, D. and Masson, P. 1996. A plume tectonics model for the Tharsis province, Mars. *Planetary and Space Science*, 44, 1499–1546.

Mège, D. and Bourgeois, O. 2011. Equatorial glaciations on Mars revealed by gravitational collapse of Valles Marineris wallslopes. *Earth and Planetary Science Letters*, 310, 182–191.

Melson, W.G., Allan, J.F., Jerez, D.R., Nelen, J., Calvache, M.L., Williams, S.N., Fournelle, J. and Perfit, M. 1990. Water contents, temperatures and diversity of the magmas of the catastrophic eruption of Nevado del Ruiz, Colombia, November 13, 1985. *Journal of Volcanology and Geothermal Research*, 41, 97–126.

Menzies, J. (ed.) 1995. *Modern glacial environments: processes, dynamics and sediments*. Butterworth Heinemann, Oxford, 621 pp.

Menzies, J. (ed.) 1996. *Past glacial environments: sediments, forms and techniques*. Butterworth Heinemann, Oxford, 598 pp.

Metcalfe, P. 1987. *Petrogenesis of Quaternary alkaline lavas in Wells Gray Provincial Park, B.C., and constraints on the petrology of the Subcordilleran mantle*. PhD thesis, University of Alberta, Edmonton (Canada), 790 pp. [unpublished].

Miller, C.D. 1980. Potential hazards from future eruptions of Mount Shasta volcano, northern California. *United States Geological Survey Bulletin*, 1503, 43 pp. and map.

Miller, T.P. and Richter, D.H. 1994. Quaternary volcanism in the Alaska Peninsula and Wrangell Mountains, Alaska. In: Plafker, G. and Berg, H.C. (eds) *The geology of Alaska*. Geological Society of America, Boulder, Colorado, pp. 759–779.

Miller, K.G., Sugarman, P.J., Browning, J.V., Kominz, M.A., Olsson, R.K., Feigenson, M.D. and Hernandez, J.C. 2004. Upper Cretaceous sequences and sea-level history, New Jersey Coastal Plain. *Geological Society of America Bulletin*, 116, 368–393.

Mitchell, N.C., Beier, C., Rosin, P.L., Quartau, R. and Tempera, F. 2008. Lava penetrating water: submarine lava flows around the coasts of Pico Island, Azores. *Geochemistry, Geophysics, Geosystems*, doi:10.1029/2007/GC001725.

Mojica, J. Columenares, F., Villarroel, C., Macia, C. and Moreno, M. 1986. Characteristicas del flujo de lodo ocurrido el 12 de Noviembre de 1985 en el valle de Armero (Tolima, Columbia): historia y comentarios de los flujos de 1595 y 1845. *Geología Columbia*, 14, 107–140.

Moore, D.P. 1976. *The Rubble Creek landslide, Garibaldi, British Columbia*. MSc thesis, University of British Columbia (Canada), 97 pp. [unpublished].

Moore, J.G. 1985. Structure and eruptive mechanisms at Surtsey Volcano, Iceland. *Geological Magazine*, 122, 649–661.

Moore, H.J. 1987. Preliminary estimates of the rheological properties of 1984 Mauna Loa lava. In: Decker, R.W., Wright, T.L. and Stauffer, P.H. (eds) *Volcanism in Hawaii. United States Geological Survey Professional Paper*, 1350, pp. 1569–1588.

Moore, J.G. and Calk, L. 1991. Degassing and differentiation in subglacial volcanoes, Iceland. *Journal of Volcanology and Geothermal Research*, 46, 157–180.

Moore, J.G., Phillips, R.L., Grigg, R.W., Peterson, D.W. and Swanson, D.A. 1973. Flow of lava into the sea, 1969–1971, Kilauea volcano, Hawaii. *Geological Society of America Bulletin*, 84, 537–546.

Moore, J.G., Hickson, C.J. and Calk, L. 1995. Tholeiitic-alkalic transition at subglacial volcanoes, Tuya region, British Columbia. *Journal of Geophysical Research*, 100, 24577–24592.

Moreno, H. 1994. The May 17–19 Llaima volcano eruption, Southern Andes (38° 42′71″S 71°41′W). *Revista Geologica de Chile*, 21, 167–171.

Moreno, H. and Fuentealba, G. 1994. The May 17–19 1994 Llaima volcano eruption, southern Andes (38°42′S-71°44′W). *Revista Geologica de Chile*, 21, 167–171.

Mortimer, N., Dunlap, W.J., Isaac, M.J., Sutherland, R.P. and Faure, K. 2008. Basal Adare volcanics, Robertson Bay, North Victoria Land, Antarctica: Late Miocene intraplate basalts of subaqueous origin. In: Cooper, A.K. and Raymond, C.R. (eds) *Antarctica: A Keystone in a Changing World*. Online Proceedings of the 10th ISAES, United States Geological Survey Open-File Report 2007–1047, Short Research Paper 045, 7 pp.

Mothes, P.A., Hall, M.L., Andrade, D., Sameniego, P., Pierson, T.C., Ruiz, A.G. and Yepes, H. 2004. Character, stratigraphy and magnitude of historical lahars of Cotopaxi volcano (Ecuador). *Acta Volcanologica*, 16, 85–108.

Motoki, A., Orihashi, Y., Naranjo, J.A., Hirata, D., Skvarca, P. and Anma, R. 2006. Geologic reconnaissance at Lautaro volcano, Chilean Patagonia. *Revista Geologica de Chile*, 33, 177–187.

Mulder, T. and Alexander, J. 2001. The physical character of subaqueous sedimentary density flows and their deposits. *Sedimentology*, 48, 269–299.

Mullineaux, D.R. and Crandall, D.R. 1981. The eruption history of Mt. St. Helens. In: Lipman, P.W. and Mullineaux, D.R. (eds) *The 1980 eruptions of Mount St. Helens, Washington*. United States Geological Survey Professional Paper, 1250, 3–15.

Murchie, S., Kirkland, L., Erad, S., Mustard, J. and Robinson, M. 2000. Near-infrared spectral variations of Martian surface materials from ISM imaging spectrometer data. *Icarus*, 147, 444–471.

Murray, J.B., Muller, J.-P., Neukum, G. and nine authors plus the HRSC Co-Investigator Team. 2005. Evidence from the Mars Express High Resolution Stereo Camera for a frozen sea close to Mars' equator. *Nature*, 434, 352–355.

Murray, J.B., van Wyk de Vries, B., Marquez, A., Williams, D.A., Byrne, P., Muller, J.-P. and Kim, J.-R. 2010. Late-stage water eruptions from Ascraeus Mons volcano, Mars: implications for its structure and history. *Earth and Planetary Science Letters*, 294, 479–491.

Murtagh, R.M. and White, J.D.L. 2013. Pyroclast characteristics of a subaqueous to emergent Surtseyan eruption, Black Point volcano, California. *Journal of Volcanology and Geothermal Research*, 267, 75–91.

Naish, T., Powell, R., Levy, R. and 53 authors. 2009. Obliquity-paced Pliocene West Antarctic ice sheet oscillations. *Nature*, 458, doi:10.1038/nature07867.

Nakada, M. and Yokose, H. 1992. Ice age as a trigger of active Quaternary volcanism and tectonism. *Tectonophysics*, 212, 321–329.

Naranjo, J.A. and Moreno, H. 1991. Actividad explosive postglacial en el volcan Llaima Andes del Sur (38°45′S). *Revista Geologica de Chile*, 18, 69–80.

Naranjo, J.A., Sigurdsson, H., Carey, S.N. and Fritz, W. 1986. Eruption of the Nevado del Ruiz volcano, Colombia, on 13 November 1985: tephra fall and lahars. *Science*, 233, 961–963.

Naranjo, J.A., Moreno, H. and Banks, N.G. 1993. La erupción del volcán Hudson en 1991 (467S), Región XI, Aisén, Chile. *Servicio Nacional Geologica y Minería Boletin*, 44, 1–50.

Naranjo, J.A. and Stern, C.R. 1998. Holocene explosive activity at Hudson volcano, southern Andes. *Bulletin of Volcanology*, 59, 291–306.

Naranjo, J.A. and Moreno, H. 2004. Laharic debris-flows from Villarrica Volcano. *Boletín Servicio Nacional de Geología y Minería*, 61, 28–38.

Nardini, I., Armienti, P., Rocchi, S. and Burgess, R. 2003. ^{40}Ar-^{39}Ar chronology and petrology of the Miocene rift-related volcanism of Daniell Peninsula (northern Victoria Land, Antarctica). *Terra Antartica*, 10, 39–62.

NOAA (National Oceanic and Atmospheric Administration) 2010. *Mount Redoubt volcanic eruptions March-April 2009. Service Review*, United States Department of Commerce, 51 pp.

Nawrocki, J., Pańczyk, M. and Williams, I.S. 2011. Isotopic ages of selected magmatic rocks from King George Island (West Antarctica) controlled by magnetostratigraphy. *Geological Quarterly*, 55, 301–322.

Nelson, A.E., Smellie, J.L., Hambrey, M.J., Williams, M., Vautravers, M., Salzmann, U., McArthur, J.M. and Regelous, M. 2009. Neogene glacigenic debris flows on James Ross Island, northern Antarctic Peninsula, and their implications for regional climate history. *Quaternary Science Reviews*, 28, 3138–3160.

Nelson, P.H.H. 1975. The James Ross Island Volcanic Group of north-east Graham Land. *British Antarctic Survey Scientific Reports*, 54, 62 pp.

Neuendorf, K.E., Mehl, J.P. and Jackson, J.A. 2011. *Glossary of geology*. American Geosciences Institute, Alexandra, Virginia, 779 pp.

Neuffer, D.P., Schultz, R.A. and Watters, R.J. 2006. Mechanisms of slope failure on Pyramid Mountain, a subglacial volcano in Wells Gray Provincial Park, British Columbia. *Canadian Journal of Earth Science*, 43, 147–155.

Newman, S. and Lowenstern, J.B. 2002. VolatileCalc: a silicate melt-H_2O-CO_2 solution model written in Visual Basic for Excel. *Computers and Geosciences*, 597–604.

Nichols, A.R.L., Carroll, M.R. and Höskuldsson, A. 2002. Is the Iceland hot spot also wet? Evidence from the water contents of undegassed submarine and subglacial pillow basalts. *Earth and Planetary Science Letters*, 202, 77–87.

Nielsen, N. 1937. A volcano under an ice cap, Vatnajökull, Iceland, 1934–36. *Geographical Journal*, 90, 6–23.

Ninkovich, D., Heezen, B.C., Conolly, J.R. and Burckle, L.H. 1964. South Sandwich tephra in deep-sea sediments. *Deep Sea Research*, 11, 605–619.

Noe-Nygaard, A. 1940. Sub-glacial volcanic activity in ancient and recent times (Studies in the palagonite-system of Iceland no. 1). *Folia Geographica Danica*, 1, 67 pp.

Nowell, D., Jones, C. and Pyle, D. 2006. Episodic Quaternary volcanism in France and Germany. *Journal of Quaternary Science*, 21, 645–675.

Nye, J.F. 1951. The flow of glaciers and ice sheets as a problem in plasticity. *Proceedings of the Royal Society, Series A*, 207, 554–572.

Nye, J.F. 1952a. The mechanics of glacier flow. *Journal of Glaciology*, 2, 82–93.

Nye, J.F. 1952b. A method of calculating the thickness of the ice-sheets. *Nature*, 169, 529.

Nye, J.F. 1953. The flow law of ice from measurements in glacier tunnels, laboratory experiments, and the Jungfraufirn borehole experiment. *Proceedings of the Royal Society, Series A*, 219, 447–489.

Nye, J.F. and Frank, F.C. 1973. Hydrology of the intergranular veins in a temperate glacier. In: *Cambridge Symposium, 1969, Hydrology of Glaciers*. International Association of Hydrological Sciences, Publication 95, 157–161.

Nyland, R.E., Panter, K.S., Rocchi, S., Vincenzo, G., Di Del Carlo, P., Tiepolo, M., Field, B. and Gorsevski, P. 2013. Volcanic activity and its link to glaciation cycles: single-grain age and geochemistry of Early to Middle Miocene volcanic glass from ANDRILL AND-2A core, Antarctica. *Journal of Volcanology and Geothermal Research*, 250, 106–128.

Nývlt, D., Košler, J., Mlčoch, B., Mixa, P., Lisá, L., Bubík, M. and Hendriks, B.W.H. 2011. The Medel Formation: evidence for Late Miocene climatic cyclicity at the northern tip of the Antarctic Peninsula. *Palaeogeography, Palaeoclimatology, Palaeoecology*, 299, 363–384.

Oddsson, B., Gudmundsson, M.T., Hognadottir, T, Magnusson, E. and Hoskuldsson. A. 2012. Lava–ice interaction during the advance of a trachyandsitic lava flow down the Gígjökull outlet glacier in the April–May 2010 Eyjafjallajökull eruption, Iceland. Abstract NH11B-1131 presented at the 2010 Fall Meeting, American Geophysical Union, San Francisco, CA, 13–17 December.

Oehler, J.-F., van Wyk de Vries, B. and Labazuy, P. 2005. Landslides and spreading of oceanic hot-spot and arc shield volcanoes on Low Strength Layers (LSLs): an analogue modelling approach. *Journal of Volcanology and Geothermal Research*, 144, 169–189.

Ono, Y., Aoki, T., Hasegawa, H. and Dali, L. 2005. Mountain glaciation in Japan and Taiwan at the global Last Glacial Maximum. *Quaternary International*, 138–139, 79–92.

Orihashi, Y., Naranjo, J.A., Motoki, A., Sumino, H., Hirata, D., Anma, R. and Nagao, K. 2004. The Quaternary volcanic activities of Hudson and Lautaro volcanoes, Chilean Patagonia: new constraints from K-Ar ages. *Revista Geológica de Chile*, 31, 207–224.

Orth, K and McPhie, J. 2003. Textures formed during emplacement and cooling of a Palaeoproterozoic, small-volume rhyolitic sill. *Journal of Volcanology and Geothermal Research*, 128, 341–362.

Óskarsson, B.V. and Riishuus, M.S. 2013. The mode of emplacement of Neogene flood basalts in eastern Iceland: facies architecture and structure of the Hólmar and Grjótá olivine basalt groups. *Journal of Volcanology and Geothermal Research*, 267, 92–118.

Owen, J., Tuffen, H. and McGarvie, D.W. 2012. Using dissolved H_2O in rhyolitic glasses to estimate palaeo-ice thickness during a subglacial eruption at Bláhnúkur (Torfajökull, Iceland). *Bulletin of Volcanology*, 74, 1355–1378.

Ozerov, A.Y., Karpov, G.A., Droznin, V.A., Dvigalo, V.N., Demyanchuk, Yu.V., Ivanov, V.V., Belousov, A.B., Firstove, P.P., Gavrilov, V.A., Yashchuk, V.V. and Okrugina, A.M. 1997. The September 7 – October 2, 1994 eruption of Klyuchevskoi volcano, Kamchatka. *Volcanology and Seismology*, 18, 501–516.

Pagli, C. and Sigmundsson, F. 2008. Will present day glacier retreat increase volcanic activity? Stress induced by recent glacier retreat and its effect on magmatism at the Vatnajokull ice cap, Iceland. *Geophysical Research Letters*, 35, 9, L09304, doi:10.1029/2008GL033510.

Panter, K.S., McIntosh, W.C. and Smellie, J.L. 1994. Volcanic history of Mount Sidley, a major alkaline volcano in Marie Byrd Land, Antarctica. *Bulletin of Volcanology*, 56, 361–376.

Panter, K., Kyle, P. and Smellie, J. 1997. Petrogenesis of a phonolite-trachyte succession at Mount Sidley, Marie Byrd Land, Antarctica. *Journal of Petrology*, 38, 1225–1253.

Papale, P., Moretti, R. and Barbato, D. 2006. The compositional dependence of the saturation surface of H_2O+CO_2 fluids in silicate melts. *Chemical Geology*, 229, 78–95.

Parfitt, L. and Wilson, L. 2011. *Fundamentals of physical volcanology*. Wiley-Blackwell, Oxford, 252 pp.

Parker, T.J., Gorsline, D.S., Saunders, R.S., Pieri, D. and Schneeberger, D.M. 1993. Coastal morphology of the Martian northern plains. *Journal of Geophysical Research*, 98, 11061–11078.

Paterson, W.S.B. 1994. *The physics of glaciers*. Pergamon, Oxford, 480 pp.

Patrick, M.R. and Smellie, J.L. 2013. A spaceborne inventory of volcanic activity in Antarctica and southern oceans, 2000–2010. *Antarctic Science*, 25, 475–500.

Patrick, M.R., Smellie, J.L., Harris, A.J.L., Wright, R., Dean, K., Garbal, I.L. and Pilger, E. 2005. First recorded eruption of Mount Belinda volcano (Montagu Island), South Sandwich Islands. *Bulletin of Volcanology*, 67, 415–422.

Patterne, M. and Guichard, F. 1993. Triggering of volcanic pulses in the Campanian area, south Italy, by periodic deep magma infux. *Journal of Geophysical Research*, 98, 1861–1873.

Pattyn, F. 2010. Antarctic subglacial conditions inferred from a hybrid ice sheet/ice stream model. *Earth and Planetary Science Letters*, 295, 451–461.

Pauly, B.D., Schiffman, P., Zierenberg, R.A. and Clague, D.A. 2011. Environmental and chemical controls on palagonitization. *Geochemistry, Geophysics, Geosystems*, 12, Q12017, doi:10.1029/2011GC003639.

Peacock, M.A. 1926. The volcano-glacial palagonite formation of Iceland. *Geological Magazine*, 63, 385–399.

Pearce, J.A., Baker, P.E., Harvey, P.K. and Luff, I.W. 1995. Geochemical evidence for subduction fluxes, mantle melting and fractional crystallization beneath the South Sandwich Island arc. *Journal of Petrology*, 36, 1073–1109.

Pedersen, G.B.M. and Grosse, P. 2014. Morphometry of subaerial shield volcanoes and glaciovolcanoes from Reykjanes Peninsula, Iceland: effects of eruption environment. *Journal of Volcanology and Geothermal Research*, 282, 115–133.

Philpotts, A.R. and Ague, J.J. 2009. *Principles of igneous and metamorphic petrology*, 2nd Edition. Cambridge University Press, New York, 667 pp.

Pierson, T.C. 1985. Initiation and flow behavior of the 1980 Pine Creek and Muddy River lahars, Mount St. Helens, Washington. *Geological Society of America Bulletin*, 96, 1056–1069.

Pierson, T.C. and Waitt, R.B. 1999. Dome-collapse rockslide and multiple sediment-water flows generated by a small explosive eruption on February 2–3, 1983. In: Pierson, T.C. (ed.) *Hydrologic consequences of hot-rock/snowpack interactions at Mount St. Helens volcano, Washington, 1982–84*. United States Geological Survey Professional Paper, 1586, 53–68.

Pierson, T.C., Janda, R.J., Thouret, J.-C. and Borrero, C.A. 1990. Perturbation and melting of snow and ice by the 13 November 1985 eruption of Nevado del Ruiz, Colombia, and consequent mobilization, flow and deposition of lahars. *Journal of Volcanology and Geothermal Research*, 41, 17–66.

Pinkerton, H. and Stevenson, R.J. 1992. Methods of determining the rheological properties of magmas at subliquidus temperatures. *Journal of Volcanology and Geothermal Research*, 53, 47–66.

Pirrie, D., Crame, J.A., Riding, J.B., Butcher, A.R. and Taylor, P.D. 1997. Miocene glaciomarine sedimentation in the northern Antarctic Peninsula region: the stratigraphy and sedimentology of the Hobbs Glacier Formation, James Ross Island. *Geological Magazine*, 136, 745–762.

Pistolesi, M., Cioni, R., Rosi, M., Cashman, K.V., Rossotti, A. and Aguilera, E. 2013. Evidence for lahar-triggering mechanisms in complex stratigraphic sequences: the post-twelfth century eruptive activity of Cotopaxi volcano, Ecuador. *Bulletin of Volcanology*, 75, 698–716.

Pjetursson, H. 1900. The glacial palagonite-formation of Iceland. *Scottish Geographical Magazine*, 16, 265–293.

Pollock, M., Edwards, B.R., Hauksdóttir, S., Alcorn, R. and Bowman, L. 2014. Geochemical and lithostratigraphic constraints on the formation of pillow-dominated tindars from Undirhlíðar quarry, Reykjanes Peninsula, southwest Iceland. *Lithos*, 200–201, 317–333.

Ponomoreva, V., Churikova, T., Melekestsev, I., Braitseva, O., Pevzner, M. and Sulerzhistsky, L. 2007. Late Pleistocene-Holocene volcanism of the Kamchatka Peninsula, northwest Pacific region. In: Eichelberger, J., Gordeev, E., Izbekov, P., Kasahara, M. and Lees, J. (eds) *Volcanism and subduction: the Kamchatka region*, American Geophysical Union Monograph, 172, pp. 165–198.

Porter, S.C. 1979a. Quaternary stratigraphy and chronology of Mauna Kea, Hawaii: a 380,000-yr record of mid-Pacific volcanism and ice-cap glaciation: summary. *Geological Society of America Bulletin, Part I*, 90, 609–611.

Porter, S.C. 1979b. Quaternary stratigraphy and chronology of Mauna Kea, Hawaii: a 380,000-yr record of mid-Pacific volcanism and ice-cap glaciation. *Geological Society of America Bulletin, Part II*, 90, 908–1093.

Porter, S.C. 1979c. *Geologic map of Mauna Kea volcano, Hawaii*. The Geological Society of America Map and Chart Series, MC-30.

Porter, S.C. 1987. Pleistocene subglacial eruptions on Mauna Kea. In: Decker, R.W., Wright, T.L. and Stauffer, P.H. (eds) *Volcanism in Hawaii*. United States Geological Survey Professional Paper, 1350, 587–598.

Porter, S.C., Stuiver, M. and Yang, I.C. 1977. Chronology of Hawaiian glaciations. *Science*, 195, 61–63.

Postma, G. 1986. Classification of sediment gravity-flow deposits based on flow conditions during sedimentation. *Geology*, 14, 291–294.

Prestvik, T. 1979. *Geology of the Öraefi district, southeastern Iceland*. Nordic Volcanological Institute Report, N. 79 01, 28 pp.

Prestvik, T. 1985. *Petrology of Quaternary volcanic rocks from Öraefi, southeast Iceland*. Rapporter fra Geologisk Institutt, Universitetet i Trondheim Norges Tekniske Høgskole (Reports of the Department of Geology, University of Trondheim (Norway)), 21, 81 pp.

Putirka, K. 2008. Thermometers and barometers for volcanic systems. In: Putirka, K. and Tepley, F. (eds) *Minerals, inclusions and volcanic processes*. Reviews in Mineralogy and Geochemistry, Mineralogical Society of America/Geochemical Society, 69, pp. 61–120.

Ramalho, R.S., Quartau, R., Trenhaile, A.S., Mitchell, N.C., Woodroffe, C.D. and Ávila, S.P. 2013. Coastal erosion on volcanic oceanic islands: a complex interplay between volcanism, erosion, sedimentation, sea-level change and biogenic production. *Earth-Science Reviews*, 127, 140–170.

Rampino, M.R., Self, S. and Fairbridge, R.W. 1979. Can rapid climate change cause volcanic eruptions? *Science*, 206, 826–829.

Raymo, M.E., Ganley, K., Carter, S., Oppo, D.W. and McManus, J. 1998. Millennial-scale climate instability during the early Pleistocene epoch. *Nature*, 392, 699–702.

Raymond, C.F. and Nolan, M. 2002. Drainage of a glacial lake through an ice spillway. In: Nakawo, M., Raymond, C.F. and Fountain, A. (eds) *Debris-covered glaciers*, International Association of Hydrological Sciences, Publication 264, 199–207.

Reinardy, B.T.I., Pudsey, C.J., Hillenbrand, C.-D., Murray, T. and Evans, J. 2009. Contrasting sources for glacial and interglacial shelf sediments used to interpret changing iceflow directions in the Larsen Basin, Northern Antarctic Peninsula. *Marine Geology*, 266, 156–171.

Rittman, A. 1952. Nomenclature of volcanic rocks. *Bulletin of Volcanology*, 12, 75–102.

Roberts, M.J. 2005. Jökulhlaups: a reassessment of floodwater flow through glaciers. *Reviews of Geophysics*, 43, RG1002, doi:10.1029/2003RG000147.

Rocchi, S., LeMasurier, W.E. and Di Vincenzo, G. 2006. Oligocene to Holocene erosion and glacial history in Marie Byrd Land, west Antarctica, inferred from exhumation of the Dorrel Rock intrusive complex and volcano morphologies. *Geological Society of America Bulletin*, 118, 991–1005.

Romano, R. and Sturiale, C. 1982. The historical eruptions of Mt. Etna. *Memorie della Societa Geologica Italiana*, 23, 75–97.

Romine, W.L., Whittington, A.G., Nabelek, P.I. and Hofmeister, A.M. 2012. Thermal diffusivity of rhyolitic glasses and melts: effects of temperature, crystals and dissolved water. *Bulletin of Volcanology*, 74, 2273–2287.

Rossi, M.J. 1996. Morphology and mechanism of eruption of postglacial shields in Iceland. *Bulletin of Volcanology*, 57, 530–540.

Roth, R. and Joos, F. 2012. Model limits on the role of volcanic carbon emissions in regulating glacial-interglacial CO_2 variations. *Earth and Planetary Science Letters*, 329–330, 141–149.

Rowland, S.K. and Walker, G.P.L. 1990. Pahoehoe and aa in Hawaii: volumetric flow rate controls the lava structure. *Bulletin of Volcanology*, 52, 615–628.

Rowley, P.D., Laudon, T.S., La Prade, K.E. and LeMasurier, W.E. 1990. Hudson Mountains. In: LeMasurier, W.E. and Thomson, J.W. (eds) *Volcanoes of the Antarctic plate and southern oceans*. American Geophysical Union, Antarctic Research Series, 48, pp. 289–293.

Ruddiman, W.F., and Wright, H.E. 1987. *North America and adjacent oceans during the last deglaciation.* Geological Society of America, Boulder, Colorado, The Geology of North America, v. K-3, pp. 1–12.

Russell, J.K. and Hauksdóttir, S. 2001. Estimates of crustal assimilation in Quaternary lavas from the northern Cordillera, British Columbia. *Canadian Mineralogist*, 39, 275–297.

Russell, J.K., Stasiuk, M.V., Page, T., Nicholls, J., Rust, A., Cross, G., Schmok, J., Edwards, B.R., Hickson, C.J. and Maxwell, M. 1998. Radar studies of the Hoodoo icecap, Iskut River region, British Columbia. *Geological Survey of Canada Current Research*, 98-1A, pp. 55–63.

Russell, J.K., Edwards, B.R. and Porritt, L.A. 2013. Pyroclastic passage zones in glaciovolcanic sequences. *Nature Communications*, 4, doi:10.1038/ncomms2829.

Russell, J.K., Edwards, B.R., Porritt, P. and Ryane, C. 2014. Tuyas: a descriptive genetic classification. *Quaternary Science Reviews*, 87, 70–81.

Russell, P.S. and Head, J.W. 2003. Elysium-Utopia flows as mega-lahars: a model of dike intrusion, cryosphere cracking, and water-sediment release. *Journal of Geophysical Research*, 108, E6 5064, doi:10.1029/2002JE001995.

Rutford, R.H. and McIntosh, W.C. 2008. Jones Mountains, Antarctica: evidence for Tertiary glaciation revisited. In: Cooper, A.K. and Raymond, C.R. (eds) *Antarctica: a keystone in a changing world. Online Proceedings of the 10th ISAES*, United States Geological Survey Open-File Report 2007-1047, Extended abstract 203, 5 pp.

Ryane, C., Edwards, B.R. and Russell, J.K. 2011. The volcanic stratigraphy of Kima'Kho Mountain: a Pleistocene tuya, northwestern British Columbia, *Geological Survey of Canada Current Research*, 2011–104, 12 pp.

Ryane, C., Edwards, B.R. and Russell, J.K. 2013. The stratigraphy of Kima'Kho Mountain, Tuya-Kawdy volcanic field, northwestern British Columbia: a Pleistocene tuya. *Geological Survey of Canada Current Research*, 2011-A3, 12 pp.

Saemundsson, K. 1970. Interglacial lava flows in the lowlands of southern Iceland and the problem of two-tiered columnar jointing. *Jökull*, 20, 62–77.

Saemundsson, K. 1979. Outline geology of Iceland. *Jökull*, 29, 7–28.

Saemundsson, K. and Noll, H. 1974. K/Ar ages of rocks from Húsafell, Western Iceland, and the development of the Húsafell central volcano. *Jökull* 24, 40–59.

Salzmann, U., Riding, J.B., Nelson, A.E. and Smellie, J.L. 2011. How likely was a green Antarctic Peninsula during warm Pliocene interglacials? A critical reassessment based on new playnofloras from James Ross Island. *Palaeogeography, Palaeoclimatology, Palaeoecology*, 309, 73–82.

Scanlon, K.E., Head, J.W., Wilson, L. and Marchant, D.R. 2014. Volcano-ice interactions in the Arsia Mons tropical mountain glacier deposits. *Icarus*, 237, 315–339.

Scarpa, R., and Tilling, R.I. (1996). *Monitoring and mitigation of volcano hazards*, Springer-Verlag, Berlin, 843 pp.

Scasso, R.A. and Carey, S. 2005. Morphology and formation of glassy volcanic ash from the August 12–15, 1991 eruption of Hudson volcano, Chile. *Latin American Journal of Sedimentology and Basin Analysis*, 12, 3–21.

Schaefer, J.R., Scott, W.E., Evans, W.C., Jorgenson, J., McGimsey, R.G. and Wang, B. 2008. The 2005 catastrophic acid crater lake drainage, lahar, and acidic aerosol formation at Mount Chiginagak volcano, Alaska, USA: field observations and preliminary water and vegetation chemistry results. *Geochemistry, Geophysics, Geosystems*, 9, 29 pp., Q07018, doi:10.1029/2007GC001900.

Schäfer, J.M., Baur, H., Denton, G.H., Ivy-Ochs, S., Marchant, D.R., Schlüchter, C. and Wieler, R. 2000. The oldest ice on Earth in Beacon Valley, Antarctica: new evidence from surface exposure dating. *Earth and Planetary Science Letters*, 179, 91–99.

Shea, T., Houghton, B.F., Gurioli, L., Cashman, K.V., Hammer, J.E. and Hobden, B.J. 2009. Textural studies of vesicles in volcanic rocks: an integrated methodology. *Journal of Volcanology and Geothermal Research*, 190, 271–289.

Shean, D.E., Head, J.W. and Marchant, D.R. 2005. Origin and evolution of a cold-based tropical mountain glacier on Mars: the Pavonis Mons fan-shaped deposit. *Journal of Geophysical Research*, 110, E05001, doi:10.1029/2004JE002360.

Shean, D.E., Head, J.W., Fastook, J.L. and Marchant, D.R. 2007. Recent glaciation at high elevations on Arsia Mons, Mars: implications for the formation and evolution of large tropical mountain glaciers. *Journal of Geophysical Research*, 112, E03004, doi:10.1029/2006JE002761.

Scheinder, D., Delgado Granados, H., Huggel, C. and Kääb, A. 2008. Assessing lahars from ice-capped volcanoes using ASTER satellite data, the SRTM DTM and two different flow models: case study on Iztaccíhuatl (Central Mexico). *Natural Hazards and Earth System Sciences*, 8, 559–571.

Schiffman, P., Southard, R.J., Eberl, D.D. and Bishop, J.L. 2002. Distinguishing palagonitized from pedogenically-altered basaltic Hawaiian tephra: mineralogical and geochemical criteria. In: Smellie, J.L. and Chapman, M.G. (eds) *Volcano–ice interaction on Earth and Mars*. Geological Society, London, Special Publications, 202, pp. 393–406.

Schilling, S.P., Carrara, P.E., Thompson, R.A. and Iwatsubo, E.Y. 2004. Posteruption glacier development within the crater of Mount St. Helens, Washington, USA. *Quaternary Research*, 61, 325–329.

Schmidt, P., Lund, B., Hieronymus, C., Maclennan, J., Arnadóttir, T., and Pagli, C. 2013. Effects of present-day deglaciation in Iceland on mantle melt production rates. *Journal of Geophysical Research*, 118, 3366–3379.

Schmincke, H.-U. and Bednarz, U. 1990. Pillow, sheet-flow and breccia volcanoes and volcano-tectonic cycles in the Extrusive Series of the northeastern Troodos ophiolite (Cyprus). In: Malpas, J., Moores, E.M., Panayiotou, A. and Xenophontos, C. (eds) *Ophiolites: oceanic crustal analogues*. Geological Survey Department, Nicosia, Cyprus, pp. 185–206.

Schon, S.C. and Head, J.W. 2012. Decameter-scale pedestal craters in the tropics of Mars: evidence for the recent presence of very young regional ice deposits in Tharsis. *Earth and Planetary Science Letters*, 317–318, 68–75.

Schopka, H.H., Gudmundsson, M.T. and Tuffen, H 2006. The formation of Helgafell, southwest Iceland, a monogenetic subglacial hyaloclastite ridge: sedimentology, hydrology and volcano–ice interaction. *Bulletin of Volcanology*, 152, 359–377.

Scott, K.M. 1988. Lahars and lahar-runout flows in the Toutle-Cowlitz River system, Mount St. Helens, Washington: origins, behavior, and sedimentology. *United States Geological Survey Professional Paper*, 1447-A, 76 pp.

Scott, K.M. 1989. Magnitude and frequency of lahars and lahar-runout flows in the Toutle-Cowlitz River system. *United States Geological Survey Professional Paper*, 1447-B, 33 pp.

Scott, K.M. and McGimsey, R.G. 1994. Character, mass distribution and origin of tephra-fall deposits of the 1989–1990 eruption of Redoubt Volcano, south-central Alaska. In: Miller, T.P. and Chouet, B.A. (eds) *The 1989–1990 eruption of Redoubt Volcano, Alaska*. *Journal of Volcanology and Geothermal Research*, 62, 251–272.

Self, S., Keszthelyi, L. and Thordarson, T. 1998. The importance of pahoehoe. *Annual Reviews of the Earth and Planetary Sciences*, 26, 81–110.

Shaw, H.R. 1972. Viscosities of magmatic silicate liquids: an empirical method of prediction. *American Journal of Science*, 272, 870–893.

Shreve, R.L. 1972. Movement of water in glaciers. *Journal of Glaciology*, 62, 205–214.

Siebert, L., Simkin, T. and Kimberly, P. 2010. *Volcanoes of the world*. University of California Press, Oakland, 568 pp.

Siegert, M.J. and Dowdeswell, J.A. 1996. Spatial variations in heat at the base of the Antarctic ice sheet from analysis of the thermal regime above sub-glacial lakes. *Journal of Glaciology*, 42, 501–509.

Sigmarsson, O., Vlastelic, I., Andreasen, R., Bindeman, I., Devidal, J-L., Moune, S., Keiding, J.K., Larsen, G., Höskuldsson, A. and Thordarson, T. 2011. Remobilization of silicic intrusion by mafic magmas during the 2010 Eyjafjallajökull eruption. *Solid Earth*, 2, 271–281.

Sigmundsson, F., Pinel, V., Lund, B., Albino, F., Pagli, C., Geirsson, H. and Sturkell, E. 2010a. Climate effects on volcanism: influence on magmatic systems of loading and unloading from ice mass variations, with examples from Iceland. *Philosophical Transactions of the Royal Society*, A368, 2519–34.

Sigmundsson, F., Hreinsdóttir, S., Hooper, A., Árnadóttir, T., Pedersen, R., Roberts, M.J., Óskarsson, N., Auriac, A., Decriem, J., Einarsson, P., Geirsson, H., Hensch, M., Ófeigsson, B.G., Sturkell, E., Sveinbjörnsson, H. and Feigl, K.L. 2010b. Intrusion triggering of the 2010 Eyjafjallajökull explosive eruption. *Nature*, 468, 426–430.

Sigurdsson, H., Carey, S., Palais, J.M. and Devine, J. 1990. Pre-eruption compositional gradients and mixing of andesite and dacite magma erupted from Nevado del Ruiz Volcano, Colombia in 1985. *Journal of Volcanology and Geothermal Research*, 41, 127–151.

Sigurdsson, H., Houghton, B.F., McNutt, S.R., Rymer, H. and Stix, J. 2000. *Encyclopedia of Volcanoes*, 1st Edition. Academic Press, San Diego, 1417 pp.

Sigvaldason, G.E., Annerta, K. and Nilsson, M. 1992. Effect of glacier loading/unloading on volcanism: postglacial volcanic production rate of the Dyngjufjoll area, central Iceland. *Bulletin of Volcanology*, 54, 385–392.

Simpson, K. 1996. *The geology, geochemistry and geomorphology of Mathews tuya: a subglacial volcano in northwestern British Columbia*. BSc thesis, University of British Columbia (Canada), 97 pp. [unpublished].

Simpson, K. and McPhie, J. 2001. Fluidal-clast breccia generated by submarine fire fountaining, Trooper Creek Formation, Queensland, Australia. *Journal of Volcanology and Geothermal Research*, 109, 339–355.

Sims, K.W.W., Maclennan, J., Blichart-Toft, J., Mervine, E.M., Blusztajn, J. and Gronvold, K. 2013. Short length scale mantle heterogeneity beneath Iceland probed by glacial modulation of melting. *Earth and Planetary Science Letters*, 379, 146–157.

Singer, B.S., Jicha, B.R., Harper, M.A., Naranjo, J.A., Lara, L.E. and Moreno-Roa, H. 2008. Eruptive history, geochronology, and magmatic evolution of the Puyehue-Cordón Calle volcanic complex, Chile. *Geological Society of America Bulletin*, 120, 599–618.

Sinton, J., Gronvold, K., and Saemundsson, K. 2005. Postglacial eruptive history of the Western Volcanic Zone, Iceland. *Geochemistry, Geophysics, Geosystems*, 6, Q12009, doi:10.1029/2005GC001021.

Skilling, I.P. 1994. Evolution of an englacial volcano: Brown Bluff, Antarctica. *Bulletin of Volcanology*, 56, 573–591.

Skilling, I.P. 2002. Basaltic pahoehoe lava-fed deltas: large-scale characteristics, clast generation, emplacement processes and environmental discrimination. In: Smellie, J.L. and Chapman, M.G. (eds) *Volcano–ice interaction on Earth and Mars*. Geological Society, London, Special Publications, 202, pp. 91–113.

Skilling, I.P. 2009. Subglacial to emergent basaltic volcanism at Hlödufell, southwest Iceland: a history of ice-confinement. *Journal of Volcanology and Geothermal Research*, 186, 276–289.

Skilling, I.P., White, J.D.L. and McPhie, J. 2002. Peperite: a review of magma–sediment mingling. *Journal of Volcanology and Geothermal Research*, 114, 1–17.

Skilling, I.P., McGarvie, D.W., Graettinger, A. and Höskuldsson, A. 2013. Multiple fissure-fed construction of a glaciovolcanic complex at the Askja volcano, Iceland. IAVCEI 2013 General Assembly, Kagoshima, Japan (Abstract 4W_3K-P4, p. 1076).

Slater, L., Jull, M., McKenzie, D. and Gronvold, K. 1998. Deglaciation effects on mantle melting under Iceland: results from the northern volcanic zone. *Earth and Planetary Science Letters*, 164, 151–164.

Smellie, J.L. 1984. Accretionary lapilli and highly vesiculated pumice in the Ballantrae ophiolite complex: ash-fall products of subaerial eruptions. *British Geological Survey Reports*, 16, 3640.

Smellie, J.L. 1999. Lithostratigraphy of Miocene–Recent, alkaline volcanic fields in the Antarctic Peninsula and eastern Ellsworth Land. *Antarctic Science*, 11, 362–378.

Smellie, J.L. 2000. Subglacial eruptions. In: Sigurdsson, H., Houghton, B.F., McNutt, S.R., Rymer, H. and Stix, J. (eds) *Encyclopedia of volcanoes*, 1st Edition. Academic Press, San Diego, pp. 403–418.

Smellie, J.L. 2001. Lithofacies architecture and construction of volcanoes erupted in englacial lakes: Icefall Nunatak, Mount Murphy, eastern Marie Byrd Land, Antarctica. In: White, J.D.L. and Riggs, N. (eds) *Volcaniclastic sedimentation in lacustrine settings*. International Association of Sedimentologists Special Publications, 30, pp. 9–34.

Smellie, J.L. 2002. The 1969 subglacial eruption on Deception Island (Antarctica): events and processes during an eruption beneath a thin glacier and implications for volcanic hazards. In: Smellie, J.L. and Chapman, M.G. (eds) *Volcano–ice interaction on Earth and Mars*. Geological Society, London, Special Publications, 202, pp. 59–79.

Smellie, J.L. 2006. The relative importance of supraglacial versus subglacial meltwater escape in basaltic subglacial tuya eruptions: an important unresolved conundrum. *Earth-Science Reviews*, 74, 241–268.

Smellie, J.L. 2007. Quaternary vulcanism: subglacial landforms. In: Elias, S.A. (ed.) *Encyclopedia of Quaternary sciences*. Elsevier, Amsterdam, pp. 784–798.

Smellie, J.L. 2008. Basaltic subglacial sheet-like sequences: evidence for two types with different implications for the inferred thickness of associated ice. *Earth-Science Reviews*, 88, 60–88.

Smellie, J.L. 2009. Terrestrial sub-ice volcanism: landform morphology, sequence characteristics and environmental influences, and implications for candidate Mars examples. In: Chapman, M.G. and Leszthely, L. (eds.) *Preservation of random mega-scale events on Mars and Earth: influence on geologic history*. Geological Society of America Special Papers, 453, pp. 55–76.

Smellie J.L. 2013. Quaternary vulcanism: subglacial landforms. In: Elias, S.A. (ed.) *The encyclopedia of Quaternary science*, 2nd edition, Vol. 1, Elsevier, Amsterdam, pp. 780–802.

Smellie, J.L. 2015. Lava-fed delta. In: Hargitai, H., Kereszturi, Á. et al. (eds) *Encyclopedia of planetary landforms*. Springer, New York.

Smellie, J.L. and Chapman, M.G. (eds) 2002. Volcano–ice interaction on Earth and Mars. *Geological Society*, London, Special Publications, 202, 431 pp.

Smellie, J.L. and Hole, M.J. 1997. Products and processes in Pliocene – Recent, subaqueous to emergent volcanism in the Antarctic Peninsula: examples of englacial Surtseyan volcano reconstruction. *Bulletin of Volcanology*, 58, 628–646.

Smellie, J.L. and Korteniemi, J. 2015. Tindar. In: Hargitai, H., Kereszturi, A. et al. (eds) *Encyclopedia of planetary landforms*. Springer, New York.

Smellie J.L. and Skilling, I.P. 1994. Products of subglacial eruptions under different ice thicknesses: two examples from Antarctica. *Sedimentary Geology*, 91, 115–129.

Smellie, J.L., Pankhurst, R.J., Hole, M.J. and Thomson, J.W. 1988. Age, distribution and eruptive conditions of late Cenozoic alkaline volcanism in the Antarctic Peninsula and eastern Ellsworth Land: review. *British Antarctic Survey Bulletin*, 80, 21–49.

Smellie, J.L., Hole, M.J. and Nell, P.A.R. 1993. Late Miocene valley-confined subglacial volcanism in northern Alexander Island, Antarctic Peninsula. *Bulletin of Volcanology*, 55, 273–288.

Smellie, J.L., López-Martínez, J., et al. 2002. *Geology and geomorphology of Deception Island*. BAS GEOMAP Series, Sheets 6-A and 6-B, 1:25 000, supplementary text. British Antarctic Survey, Cambridge, UK, 77 pp.

Smellie, J.L., McArthur, J.M., McIntosh, W.C. and Esser, R. 2006a. Late Neogene interglacial events in the James Ross Island region, northern Antarctic Peninsula, dated by Ar/Ar and Sr-isotope stratigraphy. *Palaeogeography, Palaeoclimatology, Palaeoecology*, 242, 169–187.

Smellie, J.L., McIntosh, W.C., Esser, R. and Fretwell, P. 2006b. The Cape Purvis volcano, Dundee Island (northern Antarctic Peninsula): late Pleistocene age, eruptive processes and implications for a glacial palaeoenvironment. *Antarctic Science*, 18, 399–408.

Smellie, J.L., McIntosh, W.C. and Esser, R. 2006c. Eruptive environment of volcanism on Brabant Island: evidence for thin wet-based ice in northern Antarctic Peninsula during the late Quaternary. *Palaeogeography, Palaeoclimatology, Palaeoecology*, 231, 233–252.

Smellie, J.L., Johnson, J.S., McIntosh, W.C., Esser, R., Gudmundsson, M.T., Hambrey, M.J. and van Wyk de Vries, B. 2008. Six million years of glacial history recorded in volcanic lithofacies of the James Ross Island Volcanic Group, Antarctic Peninsula. *Palaeogeography, Palaeoclimatology, Palaeoecology*, 260, 122–148.

Smellie, J.L., Haywood, A.M., Hillenbrand, C.-D., Lunt, D.J. and Valdes, P.J. 2009. Nature of the Antarctic Peninsula Ice Sheet during the Pliocene: geological evidence and modelling results compared. *Earth-Science Reviews*, 94, 79–94.

Smellie J.L., Rocchi, S. and Armienti, P. 2011a. Late Miocene volcanic sequences in northern Victoria Land, Antarctica: products of glaciovolcanic eruptions under different thermal regimes. *Bulletin of Volcanology*, 73, 1–25.

Smellie J.L., Rocchi, S., Gemelli, M., Di Vincenzo, G. and Armienti, P. 2011b. Late Miocene East Antarctic ice sheet characteristics deduced from terrestrial glaciovolcanic sequences in northern Victoria Land, Antarctica. *Palaeogeography, Palaeoclimatology, Palaeoecology*, 307, 129–149.

Smellie, J.L., Wilch, T.I. and Rocchi, S. 2013a. 'A'ā lava-fed deltas: a new reference tool in paleoenvironmental studies. *Geology*, 41, 403–406.

Smellie, J.L., Johnson, J.S. and Nelson, A.E. 2013b. *Geological map of James Ross Island. 1. James Ross Island Volcanic Group (1:125 000 scale)*. BAS GEOMAP 2 Series, Sheet 5, British Antarctic Survey, Cambridge, UK.

Smellie, J.L., Rocchi, S., Wilch, T.I., Gemelli, M., Di Vincenzo, G., McIntosh, W., Dunbar, N., Panter, K. and Fargo, A. 2014. Glaciovolcanic evidence for a polythermal Neogene East Antarctic Ice Sheet. *Geology*, 42, 39–41.

Snorrason, S.P. and Vilmundardóttir, E.G. 2000. Pillow lava sheets: origins and flow patterns. In: Gulick, V.C. and Gudmundsson, M.T. (eds) *Volcano/ice interaction on Earth and Mars.* Conference abstracts, Reykjavik, Iceland, August 13–18, p. 45.

Sohn, Y.K. 1996. Hydrovolcanic processes forming basaltic tuff rings and cones on Cheju Island, Korea. *Bulletin of Volcanology*, 108, 1199–1211.

Song, S.R., Jones, K.W., Lindquist, W.B., Dowd, B.A. and Sahagian, D.L. 2001. Synchrotron X-ray computed microtomography: studies on vesiculated basaltic rock. *Bulletin of Volcanology*, 63, 252–263.

Souness, C. and Hubbard, B. 2012. Mid-latitude glaciations on Mars. *Progress in Physical Geography*, 36, 238–261.

Souness, C., Hubbard, B., Milliken, R.E. and Quincey, D. 2012. An inventory and population-scale analysis of martian glacier-like forms. *Icarus*, 217, 243–255.

Souther, J.G. 1991. Hoodoo Mountan. In: Wood, C.A. and Kienle, J. (eds) *Volcanoes of North America, United States and Canada.* Cambridge University Press, Cambridge, pp. 127–128.

Souther, J.G. 1992. *The late Cenozoic Mount Edziza volcanic complex.* Geological Survey of Canada Memoir, 420, 320 pp.

Souther, J.G. and Hickson, C.J. 1984. Crystal fractionation of basalt comendite series of the Mount Edziza volcanic complex. *Journal of Volcanology and Geothermal Research*, 21, 79–106.

Souther, J.G. and Yorath, C.J. 1991. Neogene assemblages. In: Gabrielse, H. and Yorath, C.J. (eds) *Geology of the Cordilleran Orogen in Canada*. Geological Survey of Canada, 4, pp. 373–401.

Souther, J.G., Armstrong, R.L. and Harakal, J. 1984. Chronology of the peralkaline, late Cenozoic Mount Edziza volcanic complex, northern British Columbia, Canada. *Geological Society of America Bulletin*, 95, 337–349.

Sparks, R. 1978. The dynamics of bubble formation and growth in magmas: a review and analysis. *Journal of Volcanology and Geothermal Research*, 3, 1–37.

Spera, F. 2000. Physical properties of magma. In: Sigurdsson, H., Houghton, B.F., McNutt, S.R., Rymer, H. and Stix, J. (eds) *Encyclopedia of volcanoes*, 1st Edition. Academic Press, San Diego, pp. 171–190.

Spooner, I.S., Osborn, G.D., Barendregt, R.W. and Irving, E. 1995. A record of an early Pleistocene glaciation on the Mount Edziza Plateau, northwest British Columbia. *Canadian Journal of Earth Science*, 32, 2046–2056.

Spörli, K.B. and Rowland, J.V. 2006. 'Column on column' structures as indicators of lava/ice interaction, Ruapehu andesite volcano, New Zealand. *Journal of Volcanology and Geothermal Research*, 157, 294–310.

Squyres, S.W., Wilhelms, D.E. and Moosman, A.C. 1987. Large-scale volcano–ground ice interactions on Mars. *Icarus*, 385–408.

Staudigel, H. and Hart, S.R. 1983. Alteration of basaltic glass: mechanisms and significance for the oceanic crust–seawater budget. *Geochimica et Cosmochimica Acta*, 47, 337–350.

Staudigel, H. and Schmincke, H.-U. 1984. The Pliocene seamount series of La Palma/Canary Islands. *Journal of Geophysical Research*, 89, 11195–11215.

Stearns, H.T. 1945. Glaciation of Mauna Kea, Hawaii. *Geological Society of America Bulletin*, 56, 267–274.

Stearns, H.T. 1966. *Geology of the State of Hawaii*. Pacific Books, Palo Alto, 266 pp.

Stefánsdóttir, M.B. and Gíslason, S.R. 2005. The erosion and suspended matter/seawater interaction during and after the 1996 outburst flood friom the Vatnajökull Glacier, Iceland. *Earth and Planetary Science Letters*, 237, 433–452.

Stern, C.R. 2004. Active Andean volcanism: its geologic and tectonic setting. *Revista Geologíca de Chile*, 31, 161–206.

Stevenson, J.A., McGarvie, D.W., Smellie, J.L. and Gilbert, J.S. 2006. Subglacial and ice-contact volcanism at Vatnafjall, Öraefajökull, Iceland. *Bulletin of Volcanology*, 68, 737–752.

Stevenson, J.A., Smellie, J.L., McGarvie, D.W. and Gilbert, J.S. 2009. Subglacial intermediate volcanism at Kerlingarfjöll, Iceland: magma-water interactions beneath thick ice. *Journal of Volcanology and Geothermal Research*, 185, 337–351.

Stevenson, J.A., Gilbert, J.S., McGarvie, D.W. and Smellie, J.L. 2011. Explosive rhyolite tuya formation: classic examples from Kerlingarfjöll, Iceland. *Quaternary Science Reviews*, 30, 192–209.

Stevenson, J.A., Mitchell, N.C., Mochrie, F., Cassidy, M. and Pinkerton, H. 2012. Lava penetrating water: the different behaviours of pāhoehoe and ʻaʻā at the Nesjahraun, Thingvellir, Iceland. *Bulletin of Volcanology*, 74, 33–46.

Stevenson, J.A. and Smellie, J.L. 2015. Tuya. In: Hargitai, H., Kereszturi, A. et al. (eds) *Encyclopedia of Planetary Landforms*, Springer, New York.

Stewart, M.L., Russell, J.K. and Hickson, C.J. 2003. Discrimination of hot versus cold avalanche deposits: implications for hazard assessment at Mount Meager, B.C. *Natural Hazards and Earth System Science*, 3, 713–724.

Stroncik, N.A. and Schmincke, H.-U. 2001. Evolution of palagonite: crystallization chemical changes, and element budget. *Geochemistry, Geophysics, Geosystems*. 2, 1017, doi:10.1029/2000GC000102.

Stroncik, N.A. and Schmincke, H.-U. 2002. Palagonite: a review. *International Journal of Earth Sciences*, 91, 680–697.

Stump, E., Sheridan, M.F., Borg, S.G. and Sutter, J.F. 1980. Early Miocene subglacial basalts, the East Antarctic ice sheet and uplift of the Transantarctic Mountains. *Science*, 207, 757–759.

Sturkell, E., Einarsson, P., Sigmundsson, P., Hreinsdóttir, F. and Geirsson, S. 2003a. Deformation of Grímsvötn volcano, Iceland: 1998 eruption and subsequent inflation. *Geophysical Research Letters*, 30, 1182, doi:10.129/2002GL016460.

Sturkell, E., Sigmundsson, F. and Einarsson, P. 2003b. Recent unrest and magma movements at Eyjafjallajökull and Katla volcanoes, Iceland. *Journal of Geophysical Research*, 108, 2369, doi:10.1029/2001JB000917.

Sturkell, E., Einarsson, P., Sigmundsson, F., Geirsson, H., Ólafsson, H., Pedersen, R., De Zeeuwan van Dalfsen, E., Linde, A.T., Sacks, I.S. and Stefánsson, R. 2006. Volcano geodesy and magma dynamics in Iceland. *Journal of Volcanology and Geothermal Research*, 150, 14–34.

Sturkell, E., Einarsson, O., Sigmundsson, F., Hooper, A., Ófeigsson, B.G., Geirsson, H. and Ólafsson, H. 2010. Katla and Eyjafjallajökull volcanoes. *Developments in Quaternary Sciences*, 13, 5–21.

Sugden, D.E. 1996. The East Antarctic Ice Sheet: unstable ice or unstable ideas? *Transactions of the Institute of British Geographers*, 21, 443–454.

Sugden, D.E., Marchant, D.R. and Denton, G.H. 1993. The case for a stable East Antarctic Ice Sheet. *Geografiska Annaler*, 75A, 151–351.

Sugden, D.E., Marchant, D.R., Potter, N., Souchez, R.A., Denton, G.H., Swisher, C.C. and Tison, J.L. 1995. Preservation of Miocene glacier ice in East Antarctica. *Nature*, 376, 412–414.

Sykes, M.A. 1988. New K-Ar age determinations on the James Ross Island Volcanic Group, north-east Graham Land, Antarctica. *British Antarctic Survey Bulletin*, 80, 51–56.

Taddeucci, J., Scarlato, P., Montanaro, C., Cimarelli, C., Del Bello, E., Freda, C., Andronico, D., Gudmundsson, M.T. and Dingwell, D.B. 2011. Aggregation-dominated ash settling from the Eyjafjallajökull volcanic cloud illuminated by field and laboratory high-speed imaging. *Geology*, 39, 891–894.

Tanaka, K.L. 1985. Ice-lubricated gravity spreading of the Olympus Mons aureole deposits. *Icarus*, 62, 191–206.

Thorarinsson, S. 1953. Some new aspects of the Grimsvötn problem. *Journal of Glaciology*, 2, 267–274.

Thorarinsson, S. 1958. The Öraefajökull eruption of 1362. *Acta Naturalia Islandica*, 2, 100 pp.

Thorarinsson, S. 1967. *The eruptions of Hekla in historical times: a tephrochronological study*. Societas Scientiarium Islandica, Reykjavik, 170 pp.

Thorarinsson, S., Einarsson, T., Sigvaldason, G. and Elisson, G. 1964. The submarine eruption off the Vestmann Islands 1963–1964. *A preliminary report. Bulletin Volcanologique*, 27, 435–445.

Thordarson, T. and Höskuldsson, A. 2002. *Iceland*. Terra Publishing, Liverpool/Dunedin Academic Press, Edinburgh, 200 pp.

Thordarson, Th. and Höskuldsson, Á. 2008. Postglacial volcanism in Iceland. *Jökull*, 58, 197–228.

Thordarson, T. and Larsen, G. 2007. Volcanism in Iceland in historical time: volcano types, eruption styles and eruptive history. *Journal of Geodynamics*, 43, 118–152.

Thordarson, T. and Self, S. 1998. The Roza Member, Columbia River Basalt Group: a gigantic pahoehoe lava flow field formed by endogenous processes? *Journal of Geophysical Research*, 103, B11, 27411–27445.

Thordarson, T. and Sigmarsson, O. 2009. Effusive activity in the 1963–1967 Surtsey eruption, Iceland: flow emplacement and growth of small lava shields. In: Thordarson, T., Self, S., Larsen, G., Rowland, S. and Hoskuldsson, A. (eds) *Studies in volcanology: the legacy of George Walker.* Special Publications of IAVCEI, 2, Geological Society, London, pp. 53–84.

Thorkelson, D.J., Madsen, J.K. and Sluggett, C.L. 2011. Mantle flow through Northern Cordilleran slab window revealed by volcanic geochemistry. *Geology*, 39, 267–270.

Thorseth, I.H., Furnes, H. and Tumyr, O. 1991. A textural and chemical study of Icelandic palagonite of varied composition and its bearing on the mechanism of the glass-palagonite transformation. *Geochimica et Cosmochimica Acta*, 55, 731–749.

Thorseth, I.H., Furnes, H. and Heidal, M. 1992. The importance of microbiological activity in the alteration of natural basaltic glass. *Geochimica et Cosmochimica Acta*, 56, 845–850.

Thouret, J.-C. 1990. Effects of the November 13, 1985 eruption on the snow pack and ice cap of Nevado del Ruiz volcano, Colombia. *Journal of Volcanology and Geothermal Research*, 41, 177–201.

Tómasson, H. 1996. The jökulhlaup from Katla in 1918. *Annals of Glaciology*, 22, 249–254.

Touma, J. and Wisdom, J. 1993. The chaotic obliquity of Mars. *Science*, 259, 1294–1297.

Trabant, D.C., Waitt, R.B. and Major, J.J. 1994. Disruption of Drift glacier and origin of floods during the 1989–90 eruptions of Redoubt Volcano, Alaska. *Journal of Volcanology and Geothermal Research*, 62, 369–386.

Tribble, G.W. 1991. Underwater observations of active lava flows from Kilauea volcano, Hawaii. *Geology*, 19, 633–636.

Tripati, A.K., Backman, J., Elderfield, H. and Ferretti, P. 2005. Eocene bipolar glaciation associated with global carbon cycle changes. *Nature*, 436, 341–346.

Tucker, D.S. and Scott, K.M. 2009. Structures and facies associated with the flow of subaerial basaltic lava into a deep freshwater lake: the Sulphur Creek lava flow, North Cascades, Washington. *Journal of Volcanology and Geothermal Research*, 185, 311–322.

Tuffen, H. 2007. Models of ice melting and edifice growth at the onset of subglacial basaltic eruptions. *Journal of Geophysical Research*, 112, B03203, doi:10.1029/2006JB004523.

Tuffen, H. and Castro, J.M. 2009. The emplacement of an obsidian dyke through thin ice: Hrafntinnuhryggur, Krafla Iceland. *Journal of Volcanology and Geothermal Research*, 185, 352–366.

Tuffen, H., Gilbert, J. and McGarvie, D. 2001. Products of an effusive subglacial rhyolite eruption: Bláhnúkur, Torfajökull, Iceland. *Bulletin of Volcanology*, 63, 179–190.

Tuffen, H., McGarvie, D.W., Gilbert, J.S. and Pinkerton, H. 2002a. Physical volcanology of a subglacial-to-emergent rhyolitic tuya at Raudufossafjöll, Torfajökull, Iceland. In:

Smellie, J.L. and Chapman, M.G. (eds) *Volcano–ice interactions on Earth and Mars.* Geological Society, London, Special Publications, 202, pp. 213–236.

Tuffen, H., Pinkerton, H., McGarvie, D.W. and Gilbert, J.S. 2002b. Melting at the glacier base during a small-volume subglacial rhyolite eruption: evidence from Bláhnúkur, Iceland. *Sedimentary Geology*, 149, 183–198.

Tuffen, H., McGarvie, D.W. and Gilbert, J.S. 2007. Will subglacial rhyolite eruptions be explosive or intrusive? Some insights from analytical models. In: Clarke, G. and Smellie, J. (eds) *Papers from the International Symposium on Earth and Planetary Ice–Volcano Interactions held in Reykjavik, Iceland, on 19–23 June, 2006. Annals of Glaciology*, 45, pp. 87–94.

Tuffen, H., McGarvie, D.W., Pinkerton, H., Gilbert, J.S. and Brooker, R.A. 2008. An explosive–intrusive subglacial rhyolite eruption at Dalakvísl, Torfajökull, Iceland. *Bulletin of Volcanology*, 70, 841–860.

Tuffen, H., Owens, J. and Denton, J. 2010. Magma degassing during subglacial eruptions and its use to reconstruct palaeo-ice thickness. *Earth Science Reviews*, 99, 1–18.

Tuffen, H., James, M.R., Castro, J.M. and Schipper, I. 2013. Exceptional mobility of an advancing rhyolitic obsidian flow at Cordón Caulle volcano in Chile. *Nature Communications*, 4, 2709, doi:10.1038/ncomms3709.

Turnbull, M., Russell, J.K., Edwards, B.R. and Porritt, L. 2014. Diverse passage zones at Kima'Kho volcano: records of englacial lake dynamics during a prolonged subglacial volcanic eruption. GSA Abstracts with Programs, Paper No. 138–12.

Umeda, K. and Ban, M. 2012. Quaternary volcanism along the volcanic front in northeast Japan. In: Stoppa, F. (ed.) *Updates in volcanology: a comprehensive approach to volcanological problems.* InTech. Available from: http://www.intechopen.com/books/ updates-involcanology-a-comprehensive-approach-to-volcanological-problems/quatern ary-volcanism-along-the-volcanicfront-in-northeast-japan.

Umino, S., Nonaka, M. and Kauahikaua, J. 2006. Emplacement of subaerial pahoehoe lava sheet flows into water: 1990 Kūpaianaha flow of Kilauea volcano at Kaimū Bay, Hawai'i. *Bulletin of Volcanology*, 69, 125–139.

van Bemmelen, R.W. and Rutten, M.G. 1955. *Tablemountains of Northern Iceland.* E.J. Brill, Leiden, 217 pp.

Vázquez-Selem, L. and Klaus Heine, K. 2004, Late Quaternary glaciation of Mexico. In: Ehlers, J. and Gibbard, P.L. (eds) *Quaternary glaciations: extent and chronology. Part III, North America.* Developments in Quaternary Science, 2b, Elsevier, Amsterdam, pp. 233–242.

Vilmundardóttir, E.G., Snorrason, S.P. and Larsen, G. 2000. *Geological map of the sub-glacial volcanic area southwest of Vatnajökull icecap, Iceland, 1:50.000.* Orkustofnun and Landsvirkjun, Reykjavik.

Vinogradov, V.N. and Muravyev, Ya.D. 1985. Lava–ice interaction during the 1983 eruption at Klyuchevskoy volcano. *Volcanology and Seismology*, 1, 29–46.

Vogel, S.W. and Tulaczyk, S. 2006. Ice-dynamical constraints on the existence and impact of subglacial volcanism on West Antarctic ice sheet stability. *Geophysical Research Letters*, 33, L23502, doi:10.1029/2006GL027345.

Vogfjörd, K.S., Jakobsdóttir, S.S., Gudmundsson, G.B., Roberts, M.J., Agustsson, K., Arason, T., Geirsson, H., Karlsdóttir, S., Hjaltadóttir, S., Olafsdóttir, U., Thorbjarnardóttir, B., Skaftadóttir, T., Sturkell, E., Jonasdóttir, E. B., Hafsteinsson, G., Sveinbjornsson, H., Stefansson, R. and Jonsson, T.V. 2005. Forecasting and monitoring a subglacial eruption in Iceland. *Eos, Transactions, American Geophysical Union*, 86, 245–252.

Waitt, R.B. 1989. Swift snowmelt and floods (lahars) caused by great pyroclastic surge at Mount St. Helens, Washington, 18 May 1980. *Bulletin of Volcanology*, 52, 138–157.

Waitt, R.B., 1995, Hybrid wet flows formed by hot pyroclasts interacting with snow during Crater Peak (Mt. Spurr) eruptions, summer 1992. In Keith, T.E.C. (ed.) *The 1992 eruptions of Crater Peak at Mount Spurr volcano, Alaska*. United States Geological Survey Bulletin, 2139, pp. 107–118.

Waitt, R.B., Pierson, T.C., MacLeod, N.S., Janda, R.J., Voight, B. and Holcomb, R.T. 1983. Eruption-triggered avalanche, flood, and lahar at Mount St. Helens: effects of winter snowpack. *Science*, 221, 1394–1397.

Waitt, R.B., Gardner, C.A., Pierson, T.C., Major, J.J. and Neal, C.A. 1994. Unusual ice diamicts emplaced during 15 December 1989 eruption of Redoubt Volcano, Alaska. In: Miller, T.P. and Chouet, B.A. (eds) *The 1989–1990 eruption of Redoubt Volcano, Alaska. Journal of Volcanology and Geothermal Research*, 62, 409–428.

Walder, J.S. 1986. Hydraulics of subglacial cavities. *Journal of Glaciology*, 32, 439–445.

Walder, J.S., LaHusen, R.G., Vallance, J.W. and Schilling, S.P. 2007. Emplacement of a silicic lava dome through a crater glacier: Mount St. Helens, 2004–2006. In: Clarke, G. and Smellie, J. (eds) *Papers from the International Symposium on Earth and Planetary Ice–Volcano Interactions. Annals of Glaciology*, 45, pp. 14–20.

Walder, J.S., Schilling, S.P., Vallance, J.W., LaHusen, R.G. 2008. Effects of lava-dome growth on the Crater Glacier of Mount St. Helens, Washington. In: Sherrod, D.R., Scott, W.E. and Stauffer, P.H. (eds) *A volcano rekindled: the renewed eruption of Mount St. Helens, 2004–2006*. United States Geological Survey Professional Paper, 1750, pp. 257–276.

Walker, A.J. 2011. *Rhyolite volcanism at Öraefajökull volcano, S.E. Iceland: a window on Quaternary climate change*. PhD thesis, University of Manchester (UK), 325 pp. [unpublished].

Walker, G.P.L. 1965. Some aspects of Quaternary volcanism in Iceland. *Transactions of the Leicester Literary and Philosophical Society*, 59, 25–40.

Walker, G.P.L. 1970. Compound and simple lava flows and flood basalts. *Bulletin of Volcanology*, 35, 579–590.

Walker, G.P.L. 1973a. Lengths of lava flows. *Philosophical Transactions of the Royal Society, London*, A274, 107–116.

Walker, G.P.L. 1973b. Explosive volcanic eruptions: a new classification scheme. *Geologische Rundschau*, 62, 431–446.

Walker, G.P.L. 1992. Morphometric study of pillow-size spectrum among pillow lavas. *Bulletin of Volcanology*, 54, 459–474.

Walker, G.P.L. and Blake, D.H. 1966. The formation of a palagonite breccia mass beneath a valley glacier in Iceland. *Journal of the Geological Society, London*, 122, 45–61.

Walker, R.G. 1975. Generalised facies models for resedimented conglomerates of turbidite association. *Geological Society of America Bulletin*, 86, 737–748.

Walker, R.G. 1978. Deep-water sandstone facies and ancient submarine fans: models for exploration for stratigraphic traps. *American Association of Petroleum Geologists*, 62, 932–966.

Wallmann, P.C., Mahood, G.A. and Pollard, D.D. 1988. Mechanical models for correlation of ring-fracture eruptions at Pantelleria, Strait of Sicily, with glacial sea level drawdown. *Bulletin of Volcanology*, 50, 327–339.

Waters, T.R., McGovern, P.J. and Rossman, P.I. 2007. Hemispheres apart: the crustal dichotomy of Mars. *Annual Review of Earth and Planetary Sciences*, 35, 621–652.

Watson, K.D. and Mathews, W.H. 1944. The Tuya-Teslin area, northern British Columbia. *British Columbia Department of Mines Bulletin*, 19, 52 pp.

Waythomas, C.F., Pierson, T.C., Major, J.J. and Scott, W.E. 2013. Voluminous ice-rich and water-rich lahars generated during the 2009 eruption of Redoubt Volcano, Alaska. *Journal of Volcanology and Geothermal Research*, 239, 389–413.

Welch, B.C., Dwyer, K., Helgen, M., Waythomas, C.F. and Jacobel, R.W. 2007. Geophysical survey of the intra-caldera icefield of Mt. Veniaminof, Alaska. In: Clarke, G. and Smellie, J. (eds) *Papers from the International Symposium on Earth and Planetary Ice–Volcano Interactions. Annals of Glaciology*, 45, pp. 58–65.

Weller, D., Miranda, C.G., Moreno, P.I., Villa-Martínez, R., Stern, C.R. 2014. The large late-glacial Ho eruption of the Hudson volcano, southern Chile. *Bulletin of Volcanology*, 76, 831.

Werner, R., Schmincke, H.-U. and Sigvaldason, G. 1996. A new model for the evolution of table mountains: volcanological and petrological evidence from Herdubreid and Herdubreidartögl volcanoes (Iceland). *Bulletin of Volcanology*, 85, 390–397.

Werner, R. and Schmincke, H.-U. 1999. Englacial versus lacustrine origin of volcanic table mountains: evidence from Iceland. *Bulletin of Volcanology*, 60, 335–354.

White, J.D.L. 2000. Subaqueous eruption-fed density currents and their deposits. *Precambrian Research*, 101, 87–109.

White, J.D.L. and Houghton, B. 2000. Surtseyan and related phreatomagmatic eruptions. In: Sigurdsson, H., Houghton, B., Rymer, H., Stix, J. and McNutt, S. (eds) *Encyclopedia of Volcanoes*, 1st Edition. Academic Press, San Diego, pp. 495–512.

White, J.D.L. and Houghton, B.F. 2006. Primary volcaniclastic rocks. *Geology*, 34, 677–680.

White, J.D.L., Smellie, J.L. and Clague, D.A. (eds) 2003. *Explosive subaqueous volcanism*. American Geophysical Union, Geophysical Monograph, 140, 379 pp.

White, S.E. 2010. Glaciers of Mexico. In: Williams, R.S. and Ferrigno, J.G. (eds) *Satellite image atlas of glaciers of the world*. United States Geological Survey Professional Paper, 1386-J, pp. 383–405.

White, W.M. 2013. *Geochemistry*. Wiley-Blackwell, Oxford, 672 pp.

Wilch, T.I. and McIntosh, W.C. 2000. Eocene and Oligocene volcanism at Mount Petras, Marie Byrd Land: implications for middle Cenozoic ice sheet reconstructions in West Antarctica. *Antarctic Science*, 12, 477–491.

Wilch, T.I. and McIntosh, W.C. 2002. Lithofacies analysis amd ^{40}Ar/^{39}Ar geochronology of ice–volcano interactions at Mt. Murphy and the Crary Mountains, Marie Byrd Land, Antarctica. In: Smellie, J.L. and Chapman, M.G. (eds) *Volcano–ice interaction on Earth and Mars*. Geological Society, London, Special Publications, 202, pp. 237–253.

Wilch, T.I. and McIntosh, W.C. 2008. Miocene–Pliocene ice–volcano interactions at monogenetic volcanoes near Hobbs Coast, Marie Byrd Land, Antarctica. In: Cooper, A.K. and Raymond, C.R. (eds) *Antarctica: a keystone in a changing world*. Online Proceedings of the 10th ISAES, United States Geological Survey Open-File Report 2007-1047, Short Research Paper 074, 7 pp.

Wilding, M.C., Smellie, J.L., Morgan, S., Lesher, C.E. and Wilson, L. 2004. Cooling process recorded in subglacially erupted rhyolite glasses: rapid quenching, thermal buffering, and the formation of meltwater. *Journal of Geophysical Research*, 109, B08201, doi:10.1029/2003JB002721.

Williams, M., Smellie, J.L., Johnson, J.S. and Blake, D.B. 2006. Late Miocene Asterozoans (Echinodermata) from the James Ross Island Volcanic Group. *Antarctic Science*, 18, 117–122.

Williams, M., Nelson, A.E., Smellie, J.L., Leng, M.J., Johnson, A.L.A., Jarram, D.R., Haywood, A.M., Peck, V.L., Zalasiewicz, J., Bennett, C. and Schöne, B.R. 2010. Sea ice extent and seasonality for the Early Pliocene northern Weddell Sea. *Palaeogeography, Palaeoclimatology, Palaeoecology*, 292, 306–318.

Wilson, G.S., 1995, The Neogene East Antarctic Ice Sheet: a dynamic or stable feature? *Quaternary Science Reviews*, 14, 101–123.

Wilson, L. and Head, J.W. 1983. A comparison of volcanic eruption processes on Earth, Moon, Mars, Io and Venus. *Nature*, 302, 663–669.

Wilson, L. and Head, J.W. 1994. Mars: review and analysis of volcanic eruption theory and relationships to observed landforms. *Reviews of Geophysics*, 32, 221–263.

Wilson, L. and Head, J.W. 2002. Heat transfer and melting in subglacial basaltic eruptions: implications for volcanic deposit morphology and meltwater volumes. In: Smellie, J.L. and Chapman, M.G. (eds) *Volcano–ice interaction on Earth and Mars.* Geological Society, London, Special Publications, 202, pp. 5–26.

Wilson, L. and Head, J.W. 2007a. Explosive volcanic eruptions on Mars: tephra and accretionary lapilli formation, dispersal and recognition in the geologic record. *Journal of Volcanology and Geothermal Research*, 163, 83–97.

Wilson, L., Head, J.W. 2007b. Heat transfer in volcano-ice interactions: synthesis and applications to processes and landforms on Earth. *Annals of Glaciology*, 45, 83–86.

Wilson, L. and Mouginis-Mark, P.J. 2003. Phreatomagmatic explosive origin of Hrad Vallis, Mars. *Journal of Geophysical Research*, 108, doi:10.1029/2002JE001927.

Wilson, L., Smellie, J.L. and Head, J.W. 2013. Volcano–ice interactions. In: Fagents, S.A., Gregg, T.K.P. and Rosaly, M.C. (eds) *Modeling volcanic processes: the physics and mathematics of volcanism*. Cambridge University Press, Cambridge, pp. 275–299.

Winchester, S. 2003. *Krakatoa: the day the world exploded*. Harper-Collins, New York, 464 pp.

Wohletz, K.H. 2003. Water/magma interaction: physical considerations for the deep sea environment. In: White, J.D.L., Smellie, J.L. and Clague, D.A. (eds) *Explosive subaqueous volcanism*. American Geophysical Union, Geophysical Monograph, 140, pp. 25–49.

Wohletz, K.H. and Sheridan, M.F. 1983. Hydrovolcanic explosions II. Evolution of tuff rings and cones. *American Journal of Science*, 283, 385–413.

Wohletz, K., Zimanowski, B. and Büttner, R. 2013. Magma–water interactions. In: Fagents, S.A., Gregg, T.K.P. and Lopes, R.M.C. (eds) *Modeling volcanic processes: the physics and mathematics of volcanism*. Cambridge University Press, Cambridge, pp. 230–257.

Wolf, T. 1878. *Memoria sobre el Cotopaxi y su ultima erupcion*. Guayaquil, Ecuador, 48 pp.

Wolfe, E.W., Wise, W.S. and Dalrymple, G.B. 1997. *The geology and petrology of Mauna Kea Volcano, Hawaii: a study of post-shield volcanism*. United States Geological Survey Professional Paper, 1557, 129 pp.

Wood, C.A. and Kienle, J. 1993. *Volcanoes of North America: United States and Canada*. Cambridge University Press, Cambridge, 364 pp.

Woodcock, D.C., Gilbert, J.S. and Lane, S.J. 2012. Particle–water heat transfer during explosive volcanic eruptions. *Journal of Geophysical Research*, 117, doi:10.1029/2012JB009240.

Wörner, G. and Viereck, L. 1987. Subglacial to emergent volcanism at Shield Nunatak, Mt. Melbourne Volcanic Field, Antarctica. *Polarforschung*, 57, 27–41.

Wright-Grassham, A.C. 1987. *Volcanic geology, mineralogy, and petrogenesis of the Discovery Volcanic Sub-Province, southern Victoria Land, Antarctica*. PhD thesis, New Mexico Institute of Mining and Technology, Socorro (USA), 460 pp. [unpublished].

Wright, R., Flynn, L.P., Garbeil, H., Harris, A.J.L. and Pilger, E. 2002. Automated volcanic eruption detection using MODIS. *Remote Sensing of Environment*, 82, 135–155.

Wright, R., Flynn, L.P., Garbeil, H., Harris, A.J.L. and Pilger, E. 2004. MODVOLC: near-real-time thermal monitoring of global volcanism. *Journal of Volcanology and Geothermal Research*, 135, 29–49.

Wynn-Williams, D.D., Cabrol, N.A., Grin, E.A., Haberle, R.M. and Stoker, C.R. 2001. Brines in seepage channels as eluants for subsurface relict biomolecules on Mars? *Astrobiology*, 1, 165–184.

Xia, Z. and Woo, M.-K. 1992. Theoretical analysis of snow-dam decay. *Journal of Glaciology*, 38, 191–199.

Yamagishi, H. and Dimroth, E. 1985. A comparison of Miocene and Archaean rhyolite hyaloclastites: evidence for a hot and fluid rhyolite lava. *Journal of Volcanology and Geothermal Research*, 23, 337–355.

Yarmolyuk, V.V., Lebedev, V.I., Arkelyants, M.M., Prudnikov, S.G., Sugorakova, A.M. and Kovalenko, V.I. 1999. Neovolcanism in Eastern Tuva: chronology of volcanic events based on K–Ar dating. *Doklady Akademii Nauk*, 368, 907–911 (in Russian).

Yarmolyuk, V.V., Lebedev, V.I., Sugorakova, A.M., Bragin, V.Y., Litasov, Y.D., Prudnikov, S.T., Arakelyants, M.M., Lebedev, V.A., Ivanov, V.G. and Kozlovskii, A.M. 2001. The Eastern Tuva region of recent volcanism in Central Asia: periods, products and types of volcanic activity. *Volcanology and Seismology*, 3, 3–32 (in Russian).

Yount, M.E., Miller, T.P., Emanuel, R.P. and Wilson, F.H. 1985. Eruption in the ice-filled caldera of Mount Veniaminof. *United States Geological Survey Circular*, 945, 58–60.

Zielinski, G.A., Mayewski, L.D., Meeker, L.D., Whitlow, M.S. and Twickler, A. 1996. A 110,000-year record of explosive volcanism from the GISP2 (Greenland) ice core. *Quaternary Research*, 45, 109–188.

Zimanowski, B. and Büttner, R. 2003. Phreatomagmatic explosions in subaqueous volcanism. In: White, J.D.L., Smellie, J.L. and Clague, D.A. (eds) *Explosive subaqueous volcanism*. American Geophysical Union, Geophysical Monograph, 140, pp. 51–60.

Zimanowski, B. and Wohletz, K. 2000. Physics of phreatomagmatism, I. *Terra Nostra*, 6, 515–523.

Zimanowski, B., Büttner, R., Lorenz, V. and Häfele, H.-G. 1997. Fragmentation of basaltic melt in the course of explosive volcanism. *Journal of Geophysical Research*, 102, 803–814.

Index